Graduate Texts in Mathematics   173

Graduate Texts in Mathematics

Series Editors:

Sheldon Axler
San Francisco State University, San Francisco, CA, USA

Kenneth Ribet
University of California, Berkeley, CA, USA

Graduate Texts in Mathematics bridge the gap between passive study and creative understanding, offering graduate-level introductions to advanced topics in mathematics. The volumes are carefully written as teaching aids and highlight characteristic features of the theory. Although these books are frequently used as textbooks in graduate courses, they are also suitable for individual study.

More information about this series at http://www.springer.com/series/136

Reinhard Diestel

Graph Theory

Fifth Edition

 Springer

Reinhard Diestel
Mathematisches Seminar der
Universität Hamburg
Hamburg, Germany

ISSN 0072-5285   ISSN 2197-5612   (electronic)
Graduate Texts in Mathematics
ISBN 978-3-662-53621-6   (hardcover)
ISBN 978-3-662-57560-4   (softcover)
DOI 10.1007/978-3-662-53622-3

Library of Congress Control Number: 2017936668

Mathematics Subject Classification (2010): 05-01, 05Cxx

Printed on acid-free paper

This Springer imprint is published by Springer Nature
The registered company is Springer-Verlag GmbH Germany
The registered company address is: Heidelberger Platz 3, 14197 Berlin, Germany

To Dagmar

Preface

Almost two decades have passed since the appearance of those graph theory texts that still set the agenda for most introductory courses taught today. The canon created by those books has helped to identify some main fields of study and research, and will doubtless continue to influence the development of the discipline for some time to come.

Yet much has happened in those 20 years, in graph theory no less than elsewhere: deep new theorems have been found, seemingly disparate methods and results have become interrelated, entire new branches have arisen. To name just a few such developments, one may think of how the new notion of list colouring has bridged the gulf between invariants such as average degree and chromatic number, how probabilistic methods and the regularity lemma have pervaded extremal graph theory and Ramsey theory, or how the entirely new field of graph minors and tree-decompositions has brought standard methods of surface topology to bear on long-standing algorithmic graph problems.

Clearly, then, the time has come for a reappraisal: *what are, today, the essential areas, methods and results that should form the centre of an introductory graph theory course aiming to equip its audience for the most likely developments ahead?*

I have tried in this book to offer material for such a course. In view of the increasing complexity and maturity of the subject, I have broken with the tradition of attempting to cover both theory and applications: this book offers an introduction to the theory of graphs as part of (pure) mathematics; it contains neither explicit algorithms nor 'real world' applications. My hope is that the potential for depth gained by this restriction in scope will serve students of computer science as much as their peers in mathematics: assuming that they prefer algorithms but will benefit from an encounter with pure mathematics of *some* kind, it seems an ideal opportunity to look for this close to where their heart lies!

In the selection and presentation of material, I have tried to accommodate two conflicting goals. On the one hand, I believe that an introductory text should be lean and concentrate on the essential, so as to offer guidance to those new to the field. As a graduate text, moreover, it should get to the heart of the matter quickly: after all, the idea is to convey at least an impression of the depth and methods of the subject. On the other hand, it has been my particular concern to write with sufficient detail to make the text enjoyable and easy to read: guiding questions and ideas will be discussed explicitly, and all proofs presented will be rigorous and complete.

A typical chapter, therefore, begins with a brief discussion of what are the guiding questions in the area it covers, continues with a succinct account of its classic results (often with simplified proofs), and then presents one or two deeper theorems that bring out the full flavour of that area. The proofs of these latter results are typically preceded by (or interspersed with) an informal account of their main ideas, but are then presented formally at the same level of detail as their simpler counterparts. I soon noticed that, as a consequence, some of those proofs came out rather longer in print than seemed fair to their often beautifully simple conception. I would hope, however, that even for the professional reader the relatively detailed account of those proofs will at least help to minimize reading time...

If desired, this text can be used for a lecture course with little or no further preparation. The simplest way to do this would be to follow the order of presentation, chapter by chapter: apart from two clearly marked exceptions, any results used in the proof of others precede them in the text.

Alternatively, a lecturer may wish to divide the material into an easy basic course for one semester, and a more challenging follow-up course for another. To help with the preparation of courses deviating from the order of presentation, I have listed in the margin next to each proof the reference numbers of those results that are used in that proof. These references are given in round brackets: for example, a reference (4.1.2) in the margin next to the proof of Theorem 4.3.2 indicates that Lemma 4.1.2 will be used in this proof. Correspondingly, in the margin next to Lemma 4.1.2 there is a reference [4.3.2] (in square brackets) informing the reader that this lemma will be used in the proof of Theorem 4.3.2. Note that this system applies between different sections only (of the same or of different chapters): the sections themselves are written as units and best read in their order of presentation.

The mathematical prerequisites for this book, as for most graph theory texts, are minimal: a first grounding in linear algebra is assumed for Chapter 1.9 and once in Chapter 5.5, some basic topological concepts about the Euclidean plane and 3-space are used in Chapter 4, and a previous first encounter with elementary probability will help with Chapter 11. (Even here, all that is assumed formally is the knowledge of basic definitions: the few probabilistic tools used are developed in the text.) There are two areas of graph theory which I find both fascinating and important, especially from the perspective of pure mathematics adopted here, but which are not covered in this book: these are algebraic graph theory and infinite graphs.

At the end of each chapter, there is a section with exercises and another with bibliographical and historical notes. Many of the exercises were chosen to complement the main narrative of the text: they illustrate new concepts, show how a new invariant relates to earlier ones, or indicate ways in which a result stated in the text is best possible. Particularly easy exercises are identified by the superscript $^-$, the more challenging ones carry a $^+$. The notes are intended to guide the reader on to further reading, in particular to any monographs or survey articles on the theme of that chapter. They also offer some historical and other remarks on the material presented in the text.

Ends of proofs are marked by the symbol □. Where this symbol is found directly below a formal assertion, it means that the proof should be clear after what has been said—a claim waiting to be verified! There are also some deeper theorems which are stated, without proof, as background information: these can be identified by the absence of both proof and □.

Almost every book contains errors, and this one will hardly be an exception. I shall try to post on the Web any corrections that become necessary. The relevant site may change in time, but will always be accessible via the following two addresses:

http://www.springer-ny.com/supplements/diestel/
http://www.springer.de/catalog/html-files/deutsch/math/3540609180.html

Please let me know about any errors you find.

Little in a textbook is truly original: even the style of writing and of presentation will invariably be influenced by examples. The book that no doubt influenced me most is the classic GTM graph theory text by Bollobás: it was in the course recorded by this text that I learnt my first graph theory as a student. Anyone who knows this book well will feel its influence here, despite all differences in contents and presentation.

I should like to thank all who gave so generously of their time, knowledge and advice in connection with this book. I have benefited particularly from the help of N. Alon, G. Brightwell, R. Gillett, R. Halin, M. Hintz, A. Huck, I. Leader, T. Łuczak, W. Mader, V. Rödl, A.D. Scott, P.D. Seymour, G. Simonyi, M. Škoviera, R. Thomas, C. Thomassen and P. Valtr. I am particularly grateful also to Tommy R. Jensen, who taught me much about colouring and all I know about k-flows, and who invested immense amounts of diligence and energy in his proofreading of the preliminary German version of this book.

March 1997   *RD*

About the second edition

Naturally, I am delighted at having to write this addendum so soon after this book came out in the summer of 1997. It is particularly gratifying to hear that people are gradually adopting it not only for their personal use but more and more also as a course text; this, after all, was my aim when I wrote it, and my excuse for agonizing more over presentation than I might otherwise have done.

There are two major changes. The last chapter on graph minors now gives a complete proof of one of the major results of the Robertson-Seymour theory, their theorem that excluding a graph as a minor bounds the tree-width if and only if that graph is planar. This short proof did not exist when I wrote the first edition, which is why I then included a short proof of the next best thing, the analogous result for path-width. That theorem has now been dropped from Chapter 12. Another addition in this chapter is that the tree-width duality theorem, Theorem 12.4.3, now comes with a (short) proof too.

The second major change is the addition of a complete set of hints for the exercises. These are largely Tommy Jensen's work, and I am grateful for the time he donated to this project. The aim of these hints is to help those who use the book to study graph theory on their own, but *not* to spoil the fun. The exercises, including hints, continue to be intended for classroom use.

Apart from these two changes, there are a few additions. The most noticable of these are the formal introduction of depth-first search trees in Section 1.5 (which has led to some simplifications in later proofs) and an ingenious new proof of Menger's theorem due to Böhme, Göring and Harant (which has not otherwise been published).

Finally, there is a host of small simplifications and clarifications of arguments that I noticed as I taught from the book, or which were pointed out to me by others. To all these I offer my special thanks.

The Web site for the book has followed me to

http://www.math.uni-hamburg.de/home/diestel/books/graph.theory/

I expect this address to be stable for some time.

Once more, my thanks go to all who contributed to this second edition by commenting on the first—and I look forward to further comments!

December 1999   *RD*

About the third edition

There is no denying that this book has grown. Is it still as 'lean and concentrating on the essential' as I said it should be when I wrote the preface to the first edition, now almost eight years ago?

I believe that it is, perhaps now more than ever. So why the increase in volume? Part of the answer is that I have continued to pursue the original dual aim of offering two different things between one pair of covers:

- a reliable first introduction to graph theory that can be used either for personal study or as a course text;

- a graduate text that also offers some depth on the most important topics.

For each of these aims, some material has been added. Some of this covers new topics, which can be included or skipped as desired. An example at the introductory level is the new section on packing and covering with the Erdős-Pósa theorem, or the inclusion of the stable marriage theorem in the matching chapter. An example at the graduate level is the Robertson-Seymour structure theorem for graphs without a given minor: a result that takes a few lines to state, but one which is increasingly relied on in the literature, so that an easily accessible reference seems desirable. Another addition, also in the chapter on graph minors, is a new proof of the 'Kuratowski theorem for higher surfaces'—a proof which illustrates the interplay between graph minor theory and surface topology better than was previously possible. The proof is complemented by an appendix on surfaces, which supplies the required background and also sheds some more light on the proof of the graph minor theorem.

Changes that affect previously existing material are rare, except for countless local improvements intended to consolidate and polish rather than change. I am aware that, as this book is increasingly adopted as a course text, there is a certain desire for stability. Many of these local improvements are the result of generous feedback I got from colleagues using the book in this way, and I am very grateful for their help and advice.

There are also some local additions. Most of these developed from my own notes, pencilled in the margin as I prepared to teach from the book. They typically complement an important but technical proof, when I felt that its essential ideas might get overlooked in the formal write-up. For example, the proof of the Erdős-Stone theorem now has an informal post-mortem that looks at how exactly the regularity lemma comes to be applied in it. Unlike the formal proof, the discussion starts out from the main idea, and finally arrives at how the parameters to be declared at the start of the formal proof must be specified. Similarly, there is now a discussion pointing to some ideas in the proof of the perfect graph theorem. However, in all these cases the formal proofs have been left essentially untouched.

The only substantial change to existing material is that the old Theorem 8.1.1 (that cr^2n edges force a TK^r) seems to have lost its nice (and long) proof. Previously, this proof had served as a welcome opportunity to explain some methods in sparse extremal graph theory. These methods have migrated to the connectivity chapter, where they now live under the roof of the new proof by Thomas and Wollan that $8kn$ edges make a $2k$-connected graph k-linked. So they are still there, leaner than ever before, and just presenting themselves under a new guise. As a consequence of this change, the two earlier chapters on dense and sparse extremal graph theory could be reunited, to form a new chapter appropriately named as *Extremal Graph Theory*.

Finally, there is an entirely new chapter, on infinite graphs. When graph theory first emerged as a mathematical discipline, finite and infinite graphs were usually treated on a par. This has changed in recent years, which I see as a regrettable loss: infinite graphs continue to provide a natural and frequently used bridge to other fields of mathematics, and they hold some special fascination of their own. One aspect of this is that proofs often have to be more constructive and algorithmic in nature than their finite counterparts. The infinite version of Menger's theorem in Section 8.4 is a typical example: it offers algorithmic insights into connectivity problems in networks that are invisible to the slick inductive proofs of the finite theorem given in Chapter 3.3.

Once more, my thanks go to all the readers and colleagues whose comments helped to improve the book. I am particularly grateful to Imre Leader for his judicious comments on the whole of the infinite chapter; to my graph theory seminar, in particular to Lilian Matthiesen and Philipp Sprüssel, for giving the chapter a test run and solving all its exercises (of which eighty survived their scrutiny); to Agelos Georgakopoulos for much proofreading elsewhere; to Melanie Win Myint for recompiling the index and extending it substantially; and to Tim Stelldinger for nursing the whale on page 404 until it was strong enough to carry its baby dinosaur.

May 2005   *RD*

About the fourth edition

In this fourth edition there are few substantial additions of new material, but many improvements.

As with previous new editions, there are countless small and subtle changes to further elucidate a particular argument or concept. When prompted by reader feedback, for which I am always grateful, I still try to recast details that have been found harder than they should be. These can be very basic; a nice example, this time, is the definition of a minor in Chapter 1.

At a more substantial level, there are several new and simpler proofs of classical results, in one case reducing the already shortened earlier proof to half its length (and twice its beauty). These newly added proofs include the marriage theorem, the tree packing theorem, Tutte's cycle space and wheel theorem, Fleischner's theorem on Hamilton cycles, and the threshold theorem for the edge probability guaranteeing a specified type of subgraph. There are also one or two genuinely new theorems. One of these is an ingenious local degree condition for the existence of a Hamilton cycle, due to Asratian and Khachatrian, that implies a number of classical hamiltonicity theorems.

In some sections I have reorganized the material slightly, or rewritten the narrative. Typically, these are sections that had grown over the previous three editions, and this was beginning to affect their balance of material and momentum. As the book remains committed to offering not just a collection of theorems and proofs, but tries whenever possible to indicate a somewhat larger picture in which these have their place, maintaining its original freshness and flow remains a challenge that I enjoy trying to meet.

Finally, the book has its own dedicated website now, at

http://diestel-graph-theory.com/

Potentially, this offers opportunities for more features surrounding the book than the traditional free online edition and a dwindling collection of misprints. If you have any ideas and would like to see them implemented, do let me know.

May 2010   *RD*

About the fifth edition

This fifth edition of the book is again a major overhaul, in the spirit of its first and third edition.

I have rewritten Chapter 12 on graph minors to take account of recent developments. In addition to many smaller updates it offers a new proof of the tree-width duality theorem, due to Mazoit, which has not otherwise been published. More fundamentally, I have added a section on tangles. Originally devised by Robertson and Seymour as a technical device for their proof of the graph minor theorem, tangles have turned out to be much more fundamental than this: they define a new paradigm for identifying highly connected parts in a graph. Unlike earlier attempts at defining such substructures—in terms of, say, highly connected subgraphs, minors, or topological minors—tangles do not attempt to pin down this substructure in terms of vertices, edges, or connecting paths, but seek to capture it indirectly by orienting all the low-order separations of the graph towards it. In short, we no longer ask *what* exactly the highly connected region is, but only *where* it is. For many applications, this is exactly what matters. Moreover, this more abstract notion of high local connectivity can easily be transported to contexts outside graph theory. This, in turn, makes graph minor theory applicable beyond graph theory itself in a new way, via tangles. I have written the new section on tangles from this modern perspective.

Chapter 2 has a newly written section on tree packing and covering. I rewrote it from scratch to take advantage of a beautiful new unified theorem containing both aspects at once: the *packing-covering theorem* of Bowler and Carmesin. While their original result was proved for matroids, its graph version has a very short and self-contained proof. This proof is given in Chapter 2.4, and again is not found in print elsewhere.

Chapter 8, on infinite graphs, now treats the topological aspects of locally finite graphs more thoroughly. It puts the Freudenthal compactification of a graph G into perspective by describing it, in addition, as an inverse limit of the finite contraction minors of G. Readers with a background in group theory will find this familiar.

As always, there are countless small improvements to the narrative, proofs, and exercises. My thanks go to all those who suggested these.

Finally, I have made two adjustments to help ensure that the exercises remain usable in class at a time of instant internet access. The Hints appendix still exists, but has been relegated to the professional electronic edition so that lecturers can decide which hints to give and which not. Similarly, exercises asking for a proof of a named theorem no longer mention this name, so that the proof cannot simply be searched for. However if you know the name and wish to find the exercise, the index still has a name entry that will take you to the right page.

July 2016   *RD*

Contents

---

* Sections marked by an asterisk are recommended for a first course.
  Of sections marked $^{(*)}$, the beginning is recommended for a first course.

1

The Basics

This chapter gives a gentle yet concise introduction to most of the terminology used later in the book. Fortunately, much of standard graph theoretic terminology is so intuitive that it is easy to remember; the few terms better understood in their proper setting will be introduced later, when their time has come.

Section 1.1 offers a brief but self-contained summary of the most basic definitions in graph theory, those centred round the notion of a graph. Most readers will have met these definitions before, or will have them explained to them as they begin to read this book. For this reason, Section 1.1 does not dwell on these definitions more than clarity requires: its main purpose is to collect the most basic terms in one place, for easy reference later. For deviations for multigraphs see Section 1.10.

From Section 1.2 onwards, all new definitions will be brought to life almost immediately by a number of simple yet fundamental propositions. Often, these will relate the newly defined terms to one another: the question of how the value of one invariant influences that of another underlies much of graph theory, and it will be good to become familiar with this line of thinking early.

By $\mathbb{N}$ we denote the set of natural numbers, including zero. The set $\mathbb{Z}/n\mathbb{Z}$ of integers modulo n is denoted by $\mathbb{Z}_n$; its elements are written $\qquad \mathbb{Z}_n$ as $\bar{i} := i + n\mathbb{Z}$. When we regard $\mathbb{Z}_2 = \{\bar{0}, \bar{1}\}$ as a field, we also denote it as $\mathbb{F}_2 = \{0, 1\}$. For a real number x we denote by $\lfloor x \rfloor$ the greatest integer $\leqslant x$, and by $\lceil x \rceil$ the least integer $\geqslant x$. Logarithms written as $\qquad \lfloor x \rfloor, \lceil x \rceil$ 'log' are taken at base 2; the natural logarithm will be denoted by 'ln'. $\qquad$ log, ln The expressions $x := y$ and $y =: x$ mean that x is being defined as y.

A set $\mathcal{A} = \{A_1, \ldots, A_k\}$ of disjoint subsets of a set A is a *partition* $\qquad$ *partition* of A if the union $\bigcup \mathcal{A}$ of all the sets $A_i \in \mathcal{A}$ is A and $A_i \neq \emptyset$ for every i. $\qquad \bigcup \mathcal{A}$ Another partition $\{A'_1, \ldots, A'_\ell\}$ of A *refines* the partition $\mathcal{A}$ if each A'_i is contained in some A_j. By $[A]^k$ we denote the set of all k-element subsets $\qquad [A]^k$ of A. Sets with k elements will be called *k-sets*; subsets with k elements are *k-subsets*. $\qquad$ *k-set*

© Reinhard Diestel 2017
R. Diestel, *Graph Theory*, Graduate Texts in Mathematics 173,
DOI 10.1007/978-3-662-53622-3_1

1.1 Graphs

graph
A *graph* is a pair $G = (V, E)$ of sets such that $E \subseteq [V]^2$; thus, the elements of E are 2-element subsets of V. To avoid notational ambiguities, we shall always assume tacitly that $V \cap E = \emptyset$. The elements of V are the

vertex
edge
vertices (or *nodes*, or *points*) of the graph G, the elements of E are its *edges* (or *lines*). The usual way to picture a graph is by drawing a dot for each vertex and joining two of these dots by a line if the corresponding two vertices form an edge. Just how these dots and lines are drawn is considered irrelevant: all that matters is the information of which pairs of vertices form an edge and which do not.

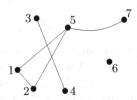

Fig. 1.1.1. The graph on $V = \{1, \ldots, 7\}$ with edge set
$$E = \{\{1, 2\}, \{1, 5\}, \{2, 5\}, \{3, 4\}, \{5, 7\}\}$$

on
A graph with vertex set V is said to be a graph *on* V. The vertex

$V(G), E(G)$
set of a graph G is referred to as $V(G)$, its edge set as $E(G)$. These conventions are independent of any actual names of these two sets: the vertex set W of a graph $H = (W, F)$ is still referred to as $V(H)$, not as $W(H)$. We shall not always distinguish strictly between a graph and its vertex or edge set. For example, we may speak of a vertex $v \in G$ (rather than $v \in V(G)$), an edge $e \in G$, and so on.

order
The number of vertices of a graph G is its *order*, written as $|G|$; its

$|G|, \|G\|$
number of edges is denoted by $\|G\|$. Graphs are *finite*, *infinite*, *countable* and so on according to their order. Except in Chapter 8, our graphs will be finite unless otherwise stated.

$\emptyset$
trivial
graph
For the *empty graph* $(\emptyset, \emptyset)$ we simply write $\emptyset$. A graph of order 0 or 1 is called *trivial*. Sometimes, e.g. to start an induction, trivial graphs can be useful; at other times they form silly counterexamples and become a nuisance. To avoid cluttering the text with non-triviality conditions, we shall mostly treat the trivial graphs, and particularly the empty graph $\emptyset$, with generous disregard.

incident
A vertex v is *incident* with an edge e if $v \in e$; then e is an edge *at* v.

ends
The two vertices incident with an edge are its *endvertices* or *ends*, and an edge *joins* its ends. An edge $\{x, y\}$ is usually written as xy (or yx). If $x \in X$ and $y \in Y$, then xy is an X–Y *edge*. The set of all X–Y edges

$E(X, Y)$
in a set E is denoted by $E(X, Y)$; instead of $E(\{x\}, Y)$ and $E(X, \{y\})$ we simply write $E(x, Y)$ and $E(X, y)$. The set of all the edges in E at a

$E(v)$
vertex v is denoted by $E(v)$.

Two vertices x, y of G are *adjacent*, or *neighbours*, if $\{x, y\}$ is an edge   *adjacent*
of G. Two edges $e \neq f$ are *adjacent* if they have an end in common. If all   *neighbour*
the vertices of G are pairwise adjacent, then G is *complete*. A complete   *complete*
graph on n vertices is a K^n; a K^3 is called a *triangle*. K^n

Pairwise non-adjacent vertices or edges are called *independent*.
More formally, a set of vertices or of edges is *independent* if no two of its   *inde-*
elements are adjacent. Independent sets of vertices are also called *stable*.   *pendent*

Let $G = (V, E)$ and $G' = (V', E')$ be two graphs. A map $\varphi \colon V \to V'$
is a *homomorphism* from G to G' if it preserves the adjacency of vertices,   *homo-*
that is, if $\{\varphi(x), \varphi(y)\} \in E'$ whenever $\{x, y\} \in E$. Then, in particular,   *morphism*
for every vertex x' in the image of φ its inverse image $\varphi^{-1}(x')$ is an
independent set of vertices in G. If φ is bijective and its inverse φ^{-1} is
also a homomorphism (so that $xy \in E \Leftrightarrow \varphi(x)\varphi(y) \in E'$ for all $x, y \in V$),
we call φ an *isomorphism*, say that G and G' are *isomorphic*, and write   *isomorphic*
$G \simeq G'$. An isomorphism from G to itself is an *automorphism* of G. $\simeq$

We do not normally distinguish between isomorphic graphs. Thus,
we usually write $G = G'$ rather than $G \simeq G'$, speak of *the* complete $=$
graph on 17 vertices, and so on. If we wish to emphasize that we are
only interested in the isomorphism type of a given graph, we informally
refer to it as an *abstract graph*.

A class of graphs that is closed under isomorphism is called a *graph
property*. For example, 'containing a triangle' is a graph property: if   *property*
G contains three pairwise adjacent vertices then so does every graph
isomorphic to G. A map taking graphs as arguments is called a *graph
invariant* if it assigns equal values to isomorphic graphs. The number   *invariant*
of vertices and the number of edges of a graph are two simple graph
invariants; the greatest number of pairwise adjacent vertices is another.

We set $G \cup G' := (V \cup V', E \cup E')$ and $G \cap G' := (V \cap V', E \cap E')$. $G \cup G'$
If $G \cap G' = \emptyset$, then G and G' are *disjoint*. If $V' \subseteq V$ and $E' \subseteq E$, then $G \cap G'$

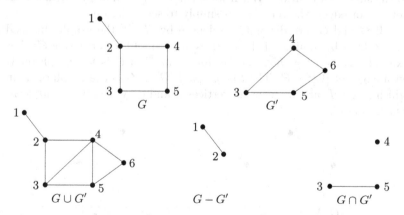

Fig. 1.1.2. Union, difference and intersection; the vertices 2,3,4
induce (or span) a triangle in $G \cup G'$ but not in G

subgraph
$G' \subseteq G$

G' is a *subgraph* of G (and G a *supergraph* of G'), written as $G' \subseteq G$. Less formally, we say that G *contains* G'. If $G' \subseteq G$ and $G' \neq G$, then G' is a *proper subgraph* of G.

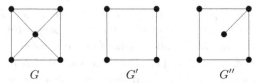

Fig. 1.1.3. A graph G with subgraphs G' and G'':
G' is an induced subgraph of G, but G'' is not

induced
subgraph

G[U]

spanning

If $G' \subseteq G$ and G' contains all the edges $xy \in E$ with $x, y \in V'$, then G' is an *induced subgraph* of G; we say that V' *induces* or *spans* G' in G, and write $G' =: G[V']$. Thus if $U \subseteq V$ is any set of vertices, then $G[U]$ denotes the graph on U whose edges are precisely the edges of G with both ends in U. If H is a subgraph of G, not necessarily induced, we abbreviate $G[V(H)]$ to $G[H]$. Finally, $G' \subseteq G$ is a *spanning* subgraph of G if V' spans all of G, i.e. if $V' = V$.

–

+

edge-
maximal

minimal
maximal

If U is any set of vertices (usually of G), we write $G - U$ for $G[V \smallsetminus U]$. In other words, $G - U$ is obtained from G by *deleting* all the vertices in $U \cap V$ and their incident edges. If $U = \{v\}$ is a singleton, we write $G - v$ rather than $G - \{v\}$. Instead of $G - V(G')$ we simply write $G - G'$. For a subset F of $[V]^2$ we write $G - F := (V, E \smallsetminus F)$ and $G + F := (V, E \cup F)$; as above, $G - \{e\}$ and $G + \{e\}$ are abbreviated to $G - e$ and $G + e$. We call G *edge-maximal* with a given graph property if G itself has the property but no graph (V, F) with $F \supsetneq E$ does.

More generally, when we call a graph *minimal* or *maximal* with some property but have not specified any particular ordering, we are referring to the subgraph relation. When we speak of minimal or maximal sets of vertices or edges, the reference is simply to set inclusion.

*G * G'*

comple-
ment $\overline{G}$

line graph
L(G)

If G and G' are disjoint, we denote by $G * G'$ the graph obtained from $G \cup G'$ by joining all the vertices of G to all the vertices of G'. For example, $K^2 * K^3 = K^5$. The *complement* $\overline{G}$ of G is the graph on V with edge set $[V]^2 \smallsetminus E$. The *line graph* $L(G)$ of G is the graph on E in which $x, y \in E$ are adjacent as vertices if and only if they are adjacent as edges in G.

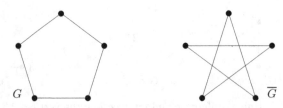

Fig. 1.1.4. A graph isomorphic to its complement

1.2 The degree of a vertex

Let $G = (V, E)$ be a (non-empty) graph. The set of neighbours of a
vertex v in G is denoted by $N_G(v)$, or briefly by $N(v)$.[1] More generally $N(v)$
for $U \subseteq V$, the neighbours in $V \setminus U$ of vertices in U are called *neighbours
of U*; their set is denoted by $N(U)$.

The *degree* (or *valency*) $d_G(v) = d(v)$ of a vertex v is the number   *degree* $d(v)$
$|E(v)|$ of edges at v; by our definition of a graph,[2] this is equal to the
number of neighbours of v. A vertex of degree 0 is *isolated*. The number   *isolated*
$\delta(G) := \min \{ d(v) \mid v \in V \}$ is the *minimum degree* of G, the number $\delta(G)$
$\Delta(G) := \max \{ d(v) \mid v \in V \}$ its *maximum degree*. If all the vertices $\Delta(G)$
of G have the same degree k, then G is *k-regular*, or simply *regular*. A   *regular*
3-regular graph is called *cubic*.   *cubic*

The number

$$d(G) := \frac{1}{|V|} \sum_{v \in V} d(v)$$
          $d(G)$

is the *average degree* of G. Clearly,   *average
degree*

$$\delta(G) \leqslant d(G) \leqslant \Delta(G) \,.$$

The average degree quantifies globally what is measured locally by the
vertex degrees: the number of edges of G per vertex. Sometimes it will
be convenient to express this ratio directly, as $\varepsilon(G) := |E|/|V|$. $\varepsilon(G)$

The quantities d and ε are, of course, intimately related. Indeed,
if we sum up all the vertex degrees in G, we count every edge exactly
twice: once from each of its ends. Thus

$$|E| = \tfrac{1}{2} \sum_{v \in V} d(v) = \tfrac{1}{2} d(G) \cdot |V| \,,$$

and therefore

$$\varepsilon(G) = \tfrac{1}{2} d(G) \,.$$

Proposition 1.2.1. *The number of vertices of odd degree in a graph is*   [10.3.1]
always even.

Proof. As $|E| = \tfrac{1}{2} \sum_{v \in V} d(v)$ is an integer, $\sum_{v \in V} d(v)$ is even.   □

---

[1] Here, as elsewhere, we drop the index referring to the underlying graph if the
reference is clear.

[2] but not for multigraphs; see Section 1.10

If a graph has large minimum degree, i.e. everywhere, locally, many edges per vertex, it also has many edges per vertex globally: $\varepsilon(G) = \frac{1}{2}d(G) \geqslant \frac{1}{2}\delta(G)$. Conversely, of course, its average degree may be large even when its minimum degree is small. However, the vertices of large degree cannot be scattered completely among vertices of small degree: as the next proposition shows, every graph G has a subgraph whose average degree is no less than the average degree of G, and whose minimum degree is more than half its average degree:

[1.4.3]
[7.2.2]
Proposition 1.2.2. *Every graph G with at least one edge has a subgraph H with $\delta(H) > \varepsilon(H) \geqslant \varepsilon(G)$.*

Proof. To construct H from G, let us try to delete vertices of small degree one by one, until only vertices of large degree remain. Up to which degree $d(v)$ can we afford to delete a vertex v, without lowering ε? Clearly, up to $d(v) = \varepsilon$: then the number of vertices decreases by 1 and the number of edges by at most ε, so the overall ratio ε of edges to vertices will not decrease.

Formally, we construct a sequence $G = G_0 \supseteq G_1 \supseteq \ldots$ of induced subgraphs of G as follows. If G_i has a vertex v_i of degree $d(v_i) \leqslant \varepsilon(G_i)$, we let $G_{i+1} := G_i - v_i$; if not, we terminate our sequence and set $H := G_i$. By the choices of v_i we have $\varepsilon(G_{i+1}) \geqslant \varepsilon(G_i)$ for all i, and hence $\varepsilon(H) \geqslant \varepsilon(G)$.

What else can we say about the graph H? Since $\varepsilon(K^1) = 0 < \varepsilon(G)$, none of the graphs in our sequence is trivial, so in particular $H \neq \emptyset$. The fact that H has no vertex suitable for deletion thus implies $\delta(H) > \varepsilon(H)$, as claimed. $\qquad\square$

1.3 Paths and cycles

path

A *path* is a non-empty graph $P = (V, E)$ of the form

$$V = \{x_0, x_1, \ldots, x_k\} \qquad E = \{x_0x_1, x_1x_2, \ldots, x_{k-1}x_k\},$$

where the x_i are all distinct. The vertices x_0 and x_k are *linked* by P and are called its *endvertices* or *ends*; the vertices $x_1, \ldots, x_{k-1}$ are the *inner* vertices of P. The number of edges of a path is its *length*, and the path of length k is denoted by P^k. Note that k is allowed to be zero; thus, $P^0 = K^1$.

length
P^k

We often refer to a path by the natural sequence of its vertices,[3]

---

[3] More precisely, by one of the two natural sequences: $x_0 \ldots x_k$ and $x_k \ldots x_0$ denote the same path. Still, it often helps to fix one of these two orderings of $V(P)$ notationally: we may then speak of things like the 'first' vertex on P with a certain property, etc.

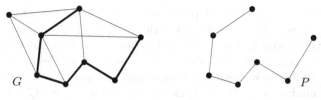

Fig. 1.3.1. A path $P = P^6$ in G

writing, say, $P = x_0 x_1 \ldots x_k$ and calling P a path *from x_0 to x_k* (as well as *between x_0 and x_k*).

For $0 \leqslant i \leqslant j \leqslant k$ we write $xPy,\ \mathring{P}$

$$Px_i := x_0 \ldots x_i$$
$$x_i P := x_i \ldots x_k$$
$$x_i P x_j := x_i \ldots x_j$$

and

$$\mathring{P} := x_1 \ldots x_{k-1}$$
$$P\mathring{x}_i := x_0 \ldots x_{i-1}$$
$$\mathring{x}_i P := x_{i+1} \ldots x_k$$
$$\mathring{x}_i P \mathring{x}_j := x_{i+1} \ldots x_{j-1}$$

for the appropriate subpaths of P. We use similar intuitive notation for the concatenation of paths; for example, if the union $Px \cup xQy \cup yR$ of three paths is again a path, we may simply denote it by $PxQyR$. $PxQyR$

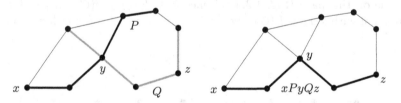

Fig. 1.3.2. Paths P, Q and $xPyQz$

Given sets A, B of vertices, we call $P = x_0 \ldots x_k$ an *A–B path* if A–B path
$V(P) \cap A = \{x_0\}$ and $V(P) \cap B = \{x_k\}$. As before, we write a–B path
rather than $\{a\}$–B path, etc. Two or more paths are *independent* if   inde-
none of them contains an inner vertex of another. Two a–b paths, for   pendent
instance, are independent if and only if a and b are their only common
vertices.

Given a graph H, we call P an *H-path* if P is non-trivial and meets H-path
H exactly in its ends. In particular, the edge of any H-path of length 1
is never an edge of H.

cycle

If $P = x_0 \ldots x_{k-1}$ is a path and $k \geqslant 3$, then the graph $C :=$ $P + x_{k-1}x_0$ is called a *cycle*. As with paths, we often denote a cycle by its (cyclic) sequence of vertices; the above cycle C might be written

length

C^k

as $x_0 \ldots x_{k-1}x_0$. The *length* of a cycle is its number of edges (or vertices); the cycle of length k is called a *k-cycle* and denoted by C^k.

girth $g(G)$

The minimum length of a cycle (contained) in a graph G is the *girth*

circum-ference

$g(G)$ of G; the maximum length of a cycle in G is its *circumference*. (If G does not contain a cycle, we set the former to ∞, the latter to zero.)

chord

An edge which joins two vertices of a cycle but is not itself an edge of the cycle is a *chord* of that cycle. Thus, an *induced cycle* in G, a cycle in

induced cycle

G forming an induced subgraph, is one that has no chords (Fig. 1.3.3).

Fig. 1.3.3. A cycle C^8 with chord xy, and induced cycles C^6, C^4

If a graph has large minimum degree, it contains long paths and cycles (see also Exercise 9):

[1.4.3]
[7.2.2]

Proposition 1.3.1. *Every graph G contains a path of length $\delta(G)$ and a cycle of length at least $\delta(G)+1$ (provided that $\delta(G) \geqslant 2$).*

Proof. Let $x_0 \ldots x_k$ be a longest path in G. Then all the neighbours of x_k lie on this path (Fig. 1.3.4). Hence $k \geqslant d(x_k) \geqslant \delta(G)$. If $i < k$ is minimal with $x_i x_k \in E(G)$, then $x_i \ldots x_k x_i$ is a cycle of length at least $\delta(G)+1$. $\square$

Fig. 1.3.4. A longest path $x_0 \ldots x_k$, and the neighbours of x_k

Minimum degree and girth, on the other hand, are not related (unless we fix the number of vertices): as we shall see in Chapter 11, there are graphs combining arbitrarily large minimum degree with arbitrarily large girth.

distance $d(x,y)$

The *distance* $d_G(x,y)$ in G of two vertices x, y is the length of a shortest x–y path in G; if no such path exists, we set $d(x,y) := \infty$. The greatest distance between any two vertices in G is the *diameter* of G,

diameter diam(G)

denoted by diam(G). Diameter and girth are, of course, related:

Proposition 1.3.2. *Every graph G containing a cycle satisfies $g(G) \leqslant 2\operatorname{diam}(G)+1$.*

Proof. Let C be a shortest cycle in G. If $g(G) \geqslant 2\operatorname{diam}(G)+2$, then C has two vertices whose distance in C is at least $\operatorname{diam}(G)+1$. In G, these vertices have a lesser distance; any shortest path P between them is therefore not a subgraph of C. Thus, P contains a C-path xPy. Together with the shorter of the two x–y paths in C, this path xPy forms a shorter cycle than C, a contradiction. $\square$

A vertex is *central* in G if its greatest distance from any other vertex is as small as possible. This distance is the *radius* of G, denoted by $\operatorname{rad}(G)$. Thus, formally, $\operatorname{rad}(G) = \min_{x \in V(G)} \max_{y \in V(G)} d_G(x,y)$. As one easily checks (exercise), we have

$$\operatorname{rad}(G) \leqslant \operatorname{diam}(G) \leqslant 2\operatorname{rad}(G).$$

Diameter and radius are not related to minimum, average or maximum degree if we say nothing about the order of the graph. However, graphs of large diameter and minimum degree must be large (larger than forced by each of the two parameters alone; see Exercise 10), and graphs of small diameter and maximum degree must be small:

Proposition 1.3.3. *A graph G of radius at most k and maximum degree at most $d \geqslant 3$ has fewer than $\frac{d}{d-2}(d-1)^k$ vertices.*

Proof. Let z be a central vertex in G, and let D_i denote the set of vertices of G at distance i from z. Then $V(G) = \bigcup_{i=0}^{k} D_i$. Clearly $|D_0| = 1$ and $|D_1| \leqslant d$. For $i \geqslant 1$ we have $|D_{i+1}| \leqslant (d-1)|D_i|$, because every vertex in D_{i+1} is a neighbour of a vertex in D_i (why?), and each vertex in D_i has at most $d-1$ neighbours in D_{i+1} (since it has another neighbour in D_{i-1}). Thus $|D_{i+1}| \leqslant d(d-1)^i$ for all $i < k$ by induction, giving

$$|G| \leqslant 1 + d\sum_{i=0}^{k-1}(d-1)^i = 1 + \frac{d}{d-2}\bigl((d-1)^k - 1\bigr) < \frac{d}{d-2}(d-1)^k.$$

$\square$

Similarly, we can bound the order of G from below by assuming that both its minimum degree and girth are large. For $d \in \mathbb{R}$ and $g \in \mathbb{N}$ let

$$n_0(d,g) := \begin{cases} 1 + d\displaystyle\sum_{i=0}^{r-1}(d-1)^i & \text{if } g =: 2r+1 \text{ is odd;} \\[2ex] 2\displaystyle\sum_{i=0}^{r-1}(d-1)^i & \text{if } g =: 2r \text{ is even.} \end{cases}$$

central

radius
$\operatorname{rad}(G)$

[9.4.1]
[9.4.2]

It is not difficult to prove that a graph of minimum degree δ and girth g has at least $n_0(\delta, g)$ vertices (Exercise 7). Interestingly, one can obtain the same bound for its average degree:

Theorem 1.3.4. (Alon, Hoory & Linial 2002)
Let G be a graph. If $d(G) \geqslant d \geqslant 2$ and $g(G) \geqslant g \in \mathbb{N}$ then $|G| \geqslant n_0(d, g)$.

One aspect of Theorem 1.3.4 is that it guarantees the existence of a short cycle compared with $|G|$. Using just the easy minimum degree version of Exercise 7, we get the following rather general bound:

[2.3.1] **Corollary 1.3.5.** *If $\delta(G) \geqslant 3$ then $g(G) < 2 \log |G|$.*

Proof. If $g := g(G)$ is even then

$$n_0(3, g) = 2\,\frac{2^{g/2} - 1}{2 - 1} = 2^{g/2} + (2^{g/2} - 2) > 2^{g/2},$$

while if g is odd then

$$n_0(3, g) = 1 + 3\,\frac{2^{(g-1)/2} - 1}{2 - 1} = \frac{3}{\sqrt{2}}\,2^{g/2} - 2 > 2^{g/2}.$$

As $|G| \geqslant n_0(3, g)$, the result follows. $\qquad\qquad\qquad\square$

walk   A *walk* (of *length k*) in a graph G is a non-empty alternating sequence $v_0 e_0 v_1 e_1 \ldots e_{k-1} v_k$ of vertices and edges in G such that $e_i = \{v_i, v_{i+1}\}$ for all $i < k$. If $v_0 = v_k$, the walk is *closed*. If the vertices in a walk are all distinct, it defines an obvious path in G. In general, every walk between two vertices contains[4] a path between these vertices (proof?).

1.4 Connectivity

connected   A graph G is called *connected* if it is non-empty and any two of its vertices are linked by a path in G. If $U \subseteq V(G)$ and $G[U]$ is connected, we also call U itself connected (in G). Instead of 'not connected' we usually say 'disconnected'.

[1.5.2] **Proposition 1.4.1.** *The vertices of a connected graph G can always be enumerated, say as $v_1, \ldots, v_n$, so that $G_i := G[v_1, \ldots, v_i]$ is connected for every i.*

---

[4] We shall often use terms defined for graphs also for walks, as long as their meaning is obvious.

Proof. Pick any vertex as v_1, and assume inductively that $v_1, \ldots, v_i$ have been chosen for some $i < |G|$. Now pick a vertex $v \in G - G_i$. As G is connected, it contains a v–v_1 path P. Choose as v_{i+1} the last vertex of P in $G - G_i$; then v_{i+1} has a neighbour in G_i. The connectedness of every G_i follows by induction on i. □

Let $G = (V, E)$ be a graph. A maximal connected subgraph of G is a *component* of G. Clearly, the components are induced subgraphs, and   *component* their vertex sets partition V. Since connected graphs are non-empty, the empty graph has no components.

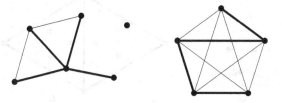

Fig. 1.4.1. A graph with three components, and a minimal spanning connected subgraph in each component

If $A, B \subseteq V$ and $X \subseteq V \cup E$ are such that every A–B path in G contains a vertex or an edge from X, we say that X *separates* the sets A *separate* and B in G. Note that this implies $A \cap B \subseteq X$. We say that X *separates* two vertices a, b if it separates the sets $\{a\}, \{b\}$ but $a, b \notin X$, and that X *separates* G if X separates some two vertices in G. A separating set of vertices is a *separator*. Separating sets of edges have no generic name,   *separator* but some such sets do; see Section 1.9 for the definition of *cuts* and *bonds*. A vertex which separates two other vertices of the same component is a   *cutvertex* *cutvertex*, and an edge separating its ends is a *bridge*. Thus, the bridges   *bridge* in a graph are precisely those edges that do not lie on any cycle.

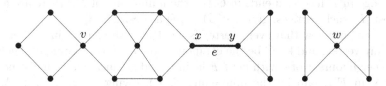

Fig. 1.4.2. A graph with cutvertices v, x, y, w and bridge $e = xy$

The unordered pair $\{A, B\}$ is a *separation* of G if $A \cup B = V$ and G *separation* has no edge between $A \smallsetminus B$ and $B \smallsetminus A$. Clearly, the latter is equivalent to saying that $A \cap B$ separates A from B. If both $A \smallsetminus B$ and $B \smallsetminus A$ are non-empty, the separation is *proper*. The number $|A \cap B|$ is the *order* of the separation $\{A, B\}$; the sets A, B are its *sides*.

G is called *k-connected* (for $k \in \mathbb{N}$) if $|G| > k$ and $G - X$ is connected   *k-connected* for every set $X \subseteq V$ with $|X| < k$. In other words, no two vertices of G

are separated by fewer than k other vertices. Every (non-empty) graph is 0-connected, and the 1-connected graphs are precisely the non-trivial connected graphs. The greatest integer k such that G is k-connected is the *connectivity* $\kappa(G)$ of G. Thus, $\kappa(G) = 0$ if and only if G is disconnected or a K^1, and $\kappa(K^n) = n-1$ for all $n \geqslant 1$.

connectivity
$\kappa(G)$

If $|G| > 1$ and $G - F$ is connected for every set $F \subseteq E$ of fewer than ℓ edges, then G is called *ℓ-edge-connected*. The greatest integer ℓ such that G is ℓ-edge-connected is the *edge-connectivity* $\lambda(G)$ of G. In particular, we have $\lambda(G) = 0$ if G is disconnected.

ℓ-edge-
connected
edge-
connectivity
$\lambda(G)$

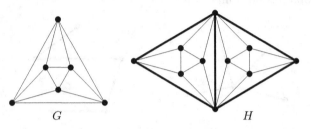

Fig. 1.4.3. The *octahedron* G (left) with $\kappa(G) = \lambda(G) = 4$,
and a graph H with $\kappa(H) = 2$ but $\lambda(H) = 4$

[3.2.1]

Proposition 1.4.2. *If G is non-trivial then $\kappa(G) \leqslant \lambda(G) \leqslant \delta(G)$.*

Proof. The second inequality follows from the fact that all the edges incident with a fixed vertex separate G. To prove the first, let F be a set of $\lambda(G)$ edges such that $G - F$ is disconnected. Such a set exists by definition of λ; note that F is a minimal separating set of edges in G. We show that $\kappa(G) \leqslant |F|$.

Suppose first that G has a vertex v that is not incident with an edge in F. Let C be the component of $G - F$ containing v. Then the vertices of C that are incident with an edge in F separate v from $G - C$. Since no edge in F has both ends in C (by the minimality of F), there are at most $|F|$ such vertices, giving $\kappa(G) \leqslant |F|$ as desired.

Suppose now that every vertex is incident with an edge in F. Let v be any vertex, and let C be the component of $G - F$ containing v. Then the neighbours w of v with $vw \notin F$ lie in C and are incident with distinct edges in F (again by the minimality of F), giving $d_G(v) \leqslant |F|$. As $N_G(v)$ separates v from any other vertices in G, this yields $\kappa(G) \leqslant |F|$— unless there are no other vertices, i.e. unless $\{v\} \cup N(v) = V$. But v was an arbitrary vertex. So we may assume that G is complete, giving $\kappa(G) = \lambda(G) = |G| - 1$. $\qquad\square$

By Proposition 1.4.2, high connectivity requires a large minimum degree. Conversely, large minimum degree does not ensure high connectivity, not even high edge-connectivity (examples?). It does, however, imply the existence of a highly connected subgraph:

[7.2.3]
[11.2.3]

Theorem 1.4.3. (Mader 1972)
Let $0 \neq k \in \mathbb{N}$. Every graph G with $d(G) \geqslant 4k$ has a $(k+1)$-connected subgraph H such that $\varepsilon(H) > \varepsilon(G) - k$.

(1.2.2)
(1.3.1)
γ

Proof. Put $\gamma := \varepsilon(G)$ ($\geqslant 2k$), and consider the subgraphs $G' \subseteq G$ such that

$$|G'| \geqslant 2k \quad \text{and} \quad \|G'\| > \gamma \left(|G'| - k\right). \qquad (*)$$

Such graphs G' exist since G is one; let H be one of smallest order.

H

No graph G' as in $(*)$ can have order exactly $2k$, since this would imply that $\|G'\| > \gamma k \geqslant 2k^2 > \binom{|G'|}{2}$. The minimality of H therefore implies that $\delta(H) > \gamma$: otherwise we could delete a vertex of degree at most γ and obtain a graph $G' \subsetneq H$ still satisfying $(*)$. In particular, we have $|H| \geqslant \gamma$. Dividing the inequality of $\|H\| > \gamma|H| - \gamma k$ from $(*)$ by $|H|$ therefore yields $\varepsilon(H) > \gamma - k$, as desired.

It remains to show that H is $(k+1)$-connected. If not, then H has a proper separation $\{U_1, U_2\}$ of order at most k; put $H[U_i] =: H_i$. Since any vertex $v \in U_1 \smallsetminus U_2$ has all its $d(v) \geqslant \delta(H) > \gamma$ neighbours from H in H_1, we have $|H_1| \geqslant \gamma \geqslant 2k$. Similarly, $|H_2| \geqslant 2k$. As by the minimality of H neither H_1 nor H_2 satisfies $(*)$, we further have

H_1, H_2

$$\|H_i\| \leqslant \gamma \left(|H_i| - k\right)$$

for $i = 1, 2$. But then

$$
\begin{aligned}
\|H\| &\leqslant \|H_1\| + \|H_2\| \\
&\leqslant \gamma \left(|H_1| + |H_2| - 2k\right) \\
&\leqslant \gamma \left(|H| - k\right) \qquad (\text{as } |H_1 \cap H_2| \leqslant k),
\end{aligned}
$$

which contradicts $(*)$ for H. $\qquad\qquad\qquad\qquad\qquad\qquad \square$

1.5 Trees and forests

An *acyclic* graph, one not containing any cycles, is called a *forest*. A connected forest is called a *tree*. (Thus, a forest is a graph whose components are trees.) The vertices of degree 1 in a tree are its *leaves*,[5] the others are its *inner vertices*. Every non-trivial tree has a leaf—consider, for example, the ends of a longest path. This little fact often comes in handy, especially in induction proofs about trees: if we remove a leaf from a tree, what remains is still a tree.

forest
tree
leaf

---

[5] . . . except that the *root* of a tree (see below) is never called a leaf, even if it has degree 1.

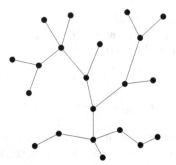

Fig. 1.5.1. A tree

[1.6.1]
[1.9.5]
[4.2.9]

Theorem 1.5.1. *The following assertions are equivalent for a graph T:*

   (i) *T is a tree;*

  (ii) *Any two vertices of T are linked by a unique path in T;*

 (iii) *T is minimally connected, i.e. T is connected but $T - e$ is disconnected for every edge $e \in T$;*

 (iv) *T is maximally acyclic, i.e. T contains no cycle but $T + xy$ does, for any two non-adjacent vertices $x, y \in T$.* $\qquad\square$

      The proof of Theorem 1.5.1 is straightforward, and a good exercise for anyone not yet familiar with all the notions it relates. Extending our notation for paths from Section 1.3, we write xTy for the unique path in a tree T between two vertices x, y (see (ii) above).

xTy

      A common application of Theorem 1.5.1 is that every connected graph contains a spanning tree: take a minimal connected spanning subgraph and use (iii), or take a maximal acyclic subgraph and apply (iv). Figure 1.4.1 shows a spanning tree in each of the three components of the graph depicted. When T is a spanning tree of G, the edges in $E(G) \setminus E(T)$ are the *chords* of T in G.

chord

Corollary 1.5.2. *The vertices of a tree can always be enumerated, say as $v_1, \ldots, v_n$, so that every v_i with $i \geqslant 2$ has a unique neighbour in $\{v_1, \ldots, v_{i-1}\}$.*

(1.4.1)

Proof. Use the enumeration from Proposition 1.4.1. $\qquad\square$

[1.9.5]
[2.4.4]
[4.2.9]

Corollary 1.5.3. *A connected graph with n vertices is a tree if and only if it has $n - 1$ edges.*

Proof. Induction on i shows that the subgraph spanned by the first i vertices in Corollary 1.5.2 has $i - 1$ edges; for $i = n$ this proves the forward implication. Conversely, let G be any connected graph with n vertices and $n - 1$ edges. Let G' be a spanning tree in G. Since G' has $n - 1$ edges by the first implication, it follows that $G = G'$. $\qquad\square$

Corollary 1.5.4. *If T is a tree and G is any graph with $\delta(G) \geqslant |T| - 1$, then $T \subseteq G$, i.e. G has a subgraph isomorphic to T.*

[9.2.1]
[9.2.3]

Proof. Find a copy of T in G inductively along its vertex enumeration from Corollary 1.5.2. □

Sometimes it is convenient to consider one vertex of a tree as special; such a vertex is then called the *root* of this tree. A tree T with a fixed root r is a *rooted tree*. Writing $x \leqslant y$ for $x \in rTy$ then defines a partial ordering on $V(T)$, the *tree-order* associated with T and r. We shall think of this ordering as expressing 'height': if $x < y$ we say that x lies *below y* in T, we call

root

tree-order

up/above
down/below

$$\lceil y \rceil := \{ x \mid x \leqslant y \} \quad \text{and} \quad \lfloor x \rfloor := \{ y \mid y \geqslant x \}$$

$\lceil t \rceil, \lfloor t \rfloor$

the *down-closure* of y and the *up-closure* of x, and so on. A set $X \subseteq V(T)$ that equals its up-closure, i.e. which satisfies $X = \lfloor X \rfloor := \bigcup_{x \in X} \lfloor x \rfloor$, is *closed upwards*, or an *up-set* in T. Similarly, there are *down-closed* sets, or *down-sets* etc..

down-closure
up-closure

Note that the root of T is the least element in its tree-order, the leaves are its maximal elements, the ends of any edge of T are comparable, and the down-closure of every vertex is a *chain*, a set of pairwise comparable elements. (Proofs?) The vertices at distance k from the root have *height k* and form the kth *level* of T.

chain

height, level

A rooted tree T contained in a graph G is called *normal* in G if the ends of every T-path in G are comparable in the tree-order of T. If T spans G, this amounts to requiring that two vertices of T must be comparable whenever they are adjacent in G; see Figure 1.5.2.

normal tree

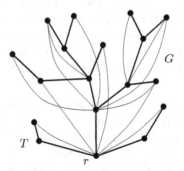

Fig. 1.5.2. A normal spanning tree with root r

A normal tree T in G can be a powerful tool for examining the structure of G, because G reflects the separation properties of T:

[8.2.3]
[8.6.8]

Lemma 1.5.5. *Let T be a normal tree in G.*

(i) *Any two vertices $x, y \in T$ are separated in G by the set $\lceil x \rceil \cap \lceil y \rceil$.*

(ii) *If $S \subseteq V(T) = V(G)$ and S is down-closed, then the components of $G - S$ are spanned by the sets $\lfloor x \rfloor$ with x minimal in $T - S$.*

Proof. (i) Let P be any x–y path in G; we show that P meets $\lceil x \rceil \cap \lceil y \rceil$. Let $t_1, \ldots, t_n$ be a minimal sequence of vertices in $P \cap T$ such that $t_1 = x$ and $t_n = y$ and t_i and t_{i+1} are comparable in the tree-order of T for all i. (Such a sequence exists: the set of *all* vertices in $P \cap T$, in their natural order as they occur on P, has this property because T is normal and every segment $t_i P t_{i+1}$ is either an edge of T or a T-path.) In our minimal sequence we cannot have $t_{i-1} < t_i > t_{i+1}$ for any i, since t_{i-1} and t_{i+1} would then be comparable, and deleting t_i would yield a smaller such sequence. Thus, our sequence has the form

$$x = t_1 > \ldots > t_k < \ldots < t_n = y$$

for some $k \in \{1, \ldots, n\}$. As $t_k \in \lceil x \rceil \cap \lceil y \rceil \cap V(P)$, our proof is complete.

(ii) Consider a component C of $G - S$, and let x be a minimal element of its vertex set. Then $V(C)$ has no other minimal element x': as x and x' would be incomparable, any x–x' path in C would by (i) contain a vertex below both, contradicting their minimality in $V(C)$. Hence as every vertex of C lies above some minimal element of $V(C)$, it lies above x. Conversely, every vertex $y \in \lfloor x \rfloor$ lies in C, for since S is down-closed, the ascending path xTy lies in $T - S$. Thus, $V(C) = \lfloor x \rfloor$.

Let us show that x is minimal not only in $V(C)$ but also in $T - S$. The vertices below x form a chain $\lceil t \rceil$ in T. As t is a neighbour of x, the maximality of C as a component of $G - S$ implies that $t \in S$, giving $\lceil t \rceil \subseteq S$ since S is down-closed. This completes the proof that every component of $G - S$ is spanned by a set $\lfloor x \rfloor$ with x minimal in $T - S$.

Conversely, if x is any minimal element of $T - S$, it is clearly also minimal in the component C of $G - S$ to which it belongs. Then $V(C) = \lfloor x \rfloor$ as before, i.e., $\lfloor x \rfloor$ spans this component. $\square$

Normal spanning trees are also called *depth-first search trees*, because of the way they arise in computer searches on graphs (Exercise 26). This fact is often used to prove their existence, which can also be shown by a very short and clever induction (Exercise 25). The following constructive proof, however, illuminates better how normal trees capture the structure of their host graphs.

[6.5.3]
[8.2.4]

Proposition 1.5.6. *Every connected graph contains a normal spanning tree, with any specified vertex as its root.*

Proof. Let G be a connected graph and $r \in G$ any specified vertex. Let T be a maximal normal tree with root r in G; we show that $V(T) = V(G)$.

Suppose not, and let C be a component of $G - T$. As T is normal, $N(C)$ is a chain in T. Let x be its greatest element, and let $y \in C$ be adjacent to x. Let T' be the tree obtained from T by joining y to x; the tree-order of T' then extends that of T. We shall derive a contradiction by showing that T' is also normal in G.

Let P be a T'-path in G. If the ends of P both lie in T, then they are comparable in the tree-order of T (and hence in that of T'), because then P is also a T-path and T is normal in G by assumption. If not, then y is one end of P, so P lies in C except for its other end z, which lies in $N(C)$. Then $z \leqslant x$, by the choice of x. For our proof that y and z are comparable it thus suffices to show that $x < y$, i.e. that $x \in rT'y$. This, however, is clear since y is a leaf of T' with neighbour x. $\qquad\square$

1.6 Bipartite graphs

Let $r \geqslant 2$ be an integer. A graph $G = (V, E)$ is called *r-partite* if V admits a partition into r classes such that every edge has its ends in different classes: vertices in the same partition class must not be adjacent. Instead of '2-partite' one usually says *bipartite*.

r-partite

bipartite

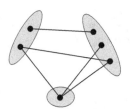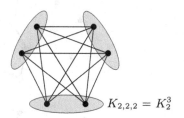

$K_{2,2,2} = K_2^3$

Fig. 1.6.1. Two 3-partite graphs

An *r*-partite graph in which *every* two vertices from different partition classes are adjacent is called *complete*; the complete *r*-partite graphs for all r together are the *complete multipartite* graphs. The complete *r*-partite graph $\overline{K^{n_1}} * \ldots * \overline{K^{n_r}}$ is denoted by $K_{n_1,\ldots,n_r}$; if $n_1 = \ldots = n_r =: s$, we abbreviate this to K_s^r. Thus, K_s^r is the complete *r*-partite graph in which every partition class contains exactly s vertices.[6] (Figure 1.6.1 shows the example of the octahedron K_2^3; compare its drawing with that in Figure 1.4.3.) Graphs of the form $K_{1,n}$ are

complete r-partite

$K_{n_1,\ldots,n_r}$

K_s^r

---

[6] Note that we obtain a K_s^r if we replace each vertex of a K^r by an independent *s*-set; our notation of K_s^r is intended to hint at this connection.

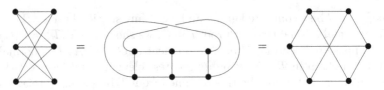

Fig. 1.6.2. Three drawings of the bipartite graph $K_{3,3} = K_3^2$

star

centre

odd cycle

called *stars*; the vertex in the singleton partition class of this $K_{1,n}$ is the star's *centre*.

Clearly, a bipartite graph cannot contain an *odd cycle*, a cycle of odd length. In fact, the bipartite graphs are characterized by this property:

[1.9.4]
[5.3.1]
[6.4.2]

Proposition 1.6.1. *A graph is bipartite if and only if it contains no odd cycle.*

(1.5.1)

Proof. Let $G = (V, E)$ be a graph without odd cycles; we show that G is bipartite. Clearly a graph is bipartite if all its components are bipartite or trivial, so we may assume that G is connected. Let T be a spanning tree in G, pick a root $r \in T$, and denote the associated tree-order on V by $\leqslant_T$. For each $v \in V$, the unique path rTv has odd or even length. This defines a bipartition of V; we show that G is bipartite with this partition.

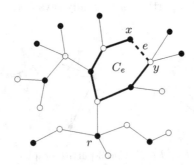

Fig. 1.6.3. The cycle C_e in $T+e$

Let $e = xy$ be an edge of G. If $e \in T$, with $x <_T y$ say, then $rTy = rTxy$ and so x and y lie in different partition classes. If $e \notin T$ then $C_e := xTy + e$ is a cycle (Fig. 1.6.3), and by the case treated already the vertices along xTy alternate between the two classes. Since C_e is even by assumption, x and y again lie in different classes.   □

1.7 Contraction and minors

In Section 1.1 we saw two fundamental containment relations between graphs: the 'subgraph' relation, and the 'induced subgraph' relation. In this section we meet two more: the 'minor' relation, and the 'topological minor' relation. Let X be a fixed graph.

A *subdivision* of X is, informally, any graph obtained from X by 'subdividing' some or all of its edges by drawing new vertices on those edges. In other words, we replace some edges of X with new paths between their ends, so that none of these paths has an inner vertex in $V(X)$ or on another new path. When G is a subdivision of X, we also say that G *is a TX*.[7] The original vertices of X are the *branch vertices of the TX*; its new vertices are called *subdividing vertices*. Note that subdividing vertices have degree 2, while branch vertices retain their degree from X.

subdivision
TX of X

branch vertices

If a graph Y contains a TX as a subgraph, then X is a *topological minor* of Y (Fig. 1.7.1).

topological minor

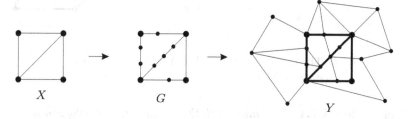

Fig. 1.7.1. The graph G is a TX, a *subdivision* of X.
As $G \subseteq Y$, this makes X a *topological minor* of Y.

Similarly, replacing the vertices x of X with disjoint connected graphs G_x, and the edges xy of X with non-empty sets of G_x–G_y edges, yields a graph that we shall call an *IX*.[8] More formally, a graph G *is an IX* if its vertex set admits a partition $\{\, V_x \mid x \in V(X)\,\}$ into connected subsets V_x such that distinct vertices $x, y \in X$ are adjacent in X if and only if G contains a V_x–V_y edge. The sets V_x are the *branch sets of the IX*. Conversely, we say that X arises from G by *contracting* the subgraphs G_x and call it a *contraction minor* of Y.

IX

branch sets

contraction minor, $\preccurlyeq$
model

If a graph Y contains an IX as a subgraph, then X is a *minor* of Y, the IX is a *model* of X in Y, and we write $X \preccurlyeq Y$ (Fig. 1.7.2).

---

[7] The 'T' stands for 'topological'. Although, formally, TX denotes a whole class of graphs, the class of all subdivisions of X, it is customary to use the expression as indicated to refer to an arbitrary member of that class.

[8] The 'I' stands for 'inflated'. As before, while IX is formally a class of graphs, those admitting a vertex partition $\{\, V_x \mid x \in V(X)\,\}$ as described below, we use the expression as indicated to refer to an arbitrary member of that class.

Thus, X is a minor of Y if and only if there is a map φ from a subset of $V(Y)$ onto $V(X)$ such that for every vertex $x \in X$ its inverse image $\varphi^{-1}(x)$ is connected in Y and for every edge $xx' \in X$ there is an edge in Y between the branch sets $\varphi^{-1}(x)$ and $\varphi^{-1}(x')$ of its ends. If the domain of φ is all of $V(Y)$, and $xx' \in X$ whenever $x \neq x'$ and Y has an edge between $\varphi^{-1}(x)$ and $\varphi^{-1}(x')$ (so that Y is an IX), we call φ a *contraction*   *contraction* of Y onto X.

Since branch sets can be singletons, every subgraph of a graph is also its minor. In infinite graphs, branch sets are allowed to be infinite. For example, the graph shown in Figure 8.1.1 is an IX with X an infinite star.

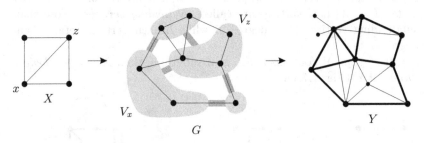

Fig. 1.7.2. The graph G is a model of X in Y, which makes X a *minor* of Y.

[12.6.1] **Proposition 1.7.1.** *The minor relation $\preccurlyeq$ and the topological-minor relation are partial orderings on the class of finite graphs, i.e. they are reflexive, antisymmetric and transitive.* $\square$

G/P If G is an IX, then $P = \{\, V_x \mid x \in X \,\}$ is a partition of $V(G)$, and we write $X =: G/P$ for this contraction minor of G. If $U = V_x$ is the only
G/U non-singleton branch set, we write $X =: G/U$, write v_U for the vertex
v_U $x \in X$ to which U contracts, and think of the rest of X as an induced subgraph of G. The 'smallest' non-trivial case of this is that U contains exactly two vertices forming an edge e, so that $U = e$. We then say that
contracting $X = G/e$ arises from G by *contracting the edge* e; see Figure 1.7.3.
an edge

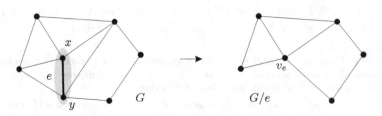

Fig. 1.7.3. Contracting the edge $e = xy$

Since the minor relation is transitive, every sequence of single vertex
or edge deletions or contractions yields a minor. Conversely, every minor
of a given finite graph can be obtained in this way:

Corollary 1.7.2. *Let X and Y be finite graphs. X is a minor of Y if
and only if there are graphs $G_0, \ldots, G_n$ such that $G_0 = Y$ and $G_n = X$
and each G_{i+1} arises from G_i by deleting an edge, contracting an edge,
or deleting a vertex.*

Proof. Induction on $|Y| + \|Y\|$. $\square$

Finally, we have the following relationship between minors and topological minors:

Proposition 1.7.3.
  (i) *Every TX is also an IX (Fig. 1.7.4); thus, every topological minor
       of a graph is also its (ordinary) minor.*
 (ii) *If $\Delta(X) \leqslant 3$, then every IX contains a TX; thus, every minor
       with maximum degree at most 3 of a graph is also its topological
       minor.* $\square$

[4.4.2]
[7.3.1]
[12.7.3]

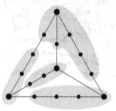

Fig. 1.7.4. A subdivision of K^4 viewed as an IK^4

Now that we have met all the standard relations between graphs,
we can also define what it means to embed one graph in another. Basi-
cally, an *embedding* of G in H is an injective map $\varphi \colon V(G) \to V(H)$ that   *embedding*
preserves the kind of structure we are interested in. Thus, φ embeds G
in H 'as a subgraph' if it preserves the adjacency of vertices, and 'as an
induced subgraph' if it preserves both adjacency and non-adjacency. If
φ is defined on $E(G)$ as well as on $V(G)$ and maps the edges xy of G to
independent paths in H between $\varphi(x)$ and $\varphi(y)$, it embeds G in H 'as
a topological minor'. Similarly, an embedding φ of G in H 'as a minor'
would be a map from $V(G)$ to disjoint connected vertex sets in H (rather
than to single vertices) so that H has an edge between the sets $\varphi(x)$ and
$\varphi(y)$ whenever xy is an edge of G. Further variants are possible; depending on the context, one may wish to define embeddings 'as a spanning
subgraph', 'as an induced minor' and so on, in the obvious way.

1.8 Euler tours

Any mathematician who happens to find himself in the East Prussian city of Königsberg (and in the 18th century) will lose no time to follow the great Leonhard Euler's example and inquire about a round trip through the old city that traverses each of the bridges shown in Figure 1.8.1 exactly once.

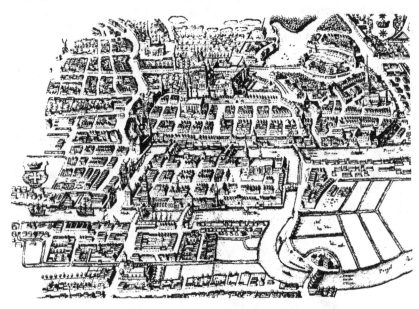

Fig. 1.8.1.   The bridges of Königsberg (anno 1736)

Eulerian   Thus inspired,[9] let us call a closed walk in a graph an *Euler tour* if it traverses every edge of the graph exactly once. A graph is *Eulerian* if it admits an Euler tour.

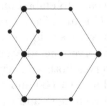

Fig. 1.8.2. A graph formalizing the bridge problem

[2.1.5]
[10.3.1]
Theorem 1.8.1. (Euler 1736)
A connected graph is Eulerian if and only if every vertex has even degree.

---

[9]   Anyone to whom such inspiration seems far-fetched, even after contemplating Figure 1.8.2, may seek consolation in the *multigraph* of Figure 1.10.1.

Proof. The degree condition is clearly necessary: a vertex appearing k times in an Euler tour (or $k+1$ times, if it is the starting and finishing vertex and as such counted twice) must have degree $2k$.

Conversely, we show by induction on $\|G\|$ that every connected graph G with all degrees even has an Euler tour. The induction starts trivially with $\|G\| = 0$. Now let $\|G\| \geqslant 1$. Since all degrees are even, we can find in G a non-trivial closed walk that contains no edge more than once. (How exactly?) Let W be such a walk of maximal length, and write F for the set of its edges. If $F = E(G)$, then W is an Euler tour. Suppose, therefore, that $G' := G - F$ has an edge.

For every vertex $v \in G$, an even number of the edges of G at v lies in F, so the degrees of G' are again all even. Since G is connected, G' has an edge e incident with a vertex on W. By the induction hypothesis, the component C of G' containing e has an Euler tour. Concatenating this with W (suitably re-indexed), we obtain a closed walk in G that contradicts the maximal length of W. $\qquad\square$

1.9 Some linear algebra

[8.7]

Let $G = (V, E)$ be a graph with n vertices and m edges, say $V = \{v_1, \ldots, v_n\}$ and $E = \{e_1, \ldots, e_m\}$. The *vertex space* $\mathcal{V}(G)$ of G is the vector space over the 2-element field $\mathbb{F}_2 = \{0, 1\}$ of all functions $V \to \mathbb{F}_2$. Every element of $\mathcal{V}(G)$ corresponds naturally to a subset of V, the set of those vertices to which it assigns a 1, and every subset of V is uniquely represented in $\mathcal{V}(G)$ by its indicator function. We may thus think of $\mathcal{V}(G)$ as the power set of V made into a vector space: the sum $U + U'$ of two vertex sets $U, U' \subseteq V$ is their symmetric difference (why?), and $U = -U$ for all $U \subseteq V$. The zero in $\mathcal{V}(G)$, viewed in this way, is the empty (vertex) set $\emptyset$. Since $\{\{v_1\}, \ldots, \{v_n\}\}$ is a basis of $\mathcal{V}(G)$, its *standard basis*, we have $\dim \mathcal{V}(G) = n$.

vertex space $\mathcal{V}(G)$

$G = (V, E)$

$+$

In the same way as above, the functions $E \to \mathbb{F}_2$ form the *edge space* $\mathcal{E}(G)$ of G: its elements correspond to the subsets of E, vector addition amounts to symmetric difference, $\emptyset \subseteq E$ is the zero, and $F = -F$ for all $F \subseteq E$. As before, $\{\{e_1\}, \ldots, \{e_m\}\}$ is the *standard basis* of $\mathcal{E}(G)$, and $\dim \mathcal{E}(G) = m$. Given two elements F, F' of the edge space, viewed as functions $E \to \mathbb{F}_2$, we write

edge space $\mathcal{E}(G)$

standard basis

$$\langle F, F' \rangle := \sum_{e \in E} F(e) F'(e) \in \mathbb{F}_2 \, .$$

$\langle F, F' \rangle$

This is zero if and only if F and F' have an even number of edges in common; in particular, we can have $\langle F, F \rangle = 0$ with $F \neq \emptyset$. Given a subspace $\mathcal{F}$ of $\mathcal{E}(G)$, we write

$$\mathcal{F}^\perp := \big\{ D \in \mathcal{E}(G) \mid \langle F, D \rangle = 0 \text{ for all } F \in \mathcal{F} \big\}.$$

$\mathcal{F}^\perp$

This is again a subspace of $\mathcal{E}(G)$ (the space of all vectors solving a certain set of linear equations—which?), and one can show that

$$\dim \mathcal{F} + \dim \mathcal{F}^\perp = m.$$

cycle space
$\mathcal{C}(G)$

The *cycle space* $\mathcal{C} = \mathcal{C}(G)$ is the subspace of $\mathcal{E}(G)$ spanned by all the cycles in G—more precisely, by their edge sets.[10] The dimension of $\mathcal{C}(G)$ is sometimes called the *cyclomatic number* of G.

The elements of $\mathcal{C}$ are easily recognized by the degrees of the subgraphs they form. Moreover, to generate the cycle space from cycles we only need disjoint unions rather than arbitrary symmetric differences:

[4.5.1]
[8.7.3]

Proposition 1.9.1. *The following assertions are equivalent for edge sets* $D \subseteq E$:

(i) $D \in \mathcal{C}(G)$;

(ii) D *is a (possibly empty) disjoint union of edge sets of cycles in* G;

(iii) *All vertex degrees of the graph* (V, D) *are even.*

Proof. Since cycles have even degrees and taking symmetric differences preserves this, (i)→(iii) follows by induction on the number of cycles used to generate D. The implication (iii)→(ii) follows by induction on $|D|$: if $D \neq \emptyset$ then (V, D) contains a cycle C, whose edges we delete for the induction step. The implication (ii)→(i) is immediate from the definition of $\mathcal{C}(G)$. $\square$

cut

A set F of edges is a *cut* in G if there exists a partition[11] $\{V_1, V_2\}$ of V such that $F = E(V_1, V_2)$. The edges in F are said to *cross* this partition. The sets V_1, V_2 are the *sides* of the cut. Recall that for $V_1 = \{v\}$ this cut is denoted by $E(v)$. A minimal non-empty cut in G is a *bond*.

bond

[4.6.3]

Proposition 1.9.2. *Together with* $\emptyset$, *the cuts in* G *form a subspace* $\mathcal{B} = \mathcal{B}(G)$ *of* $\mathcal{E}(G)$. *This space is generated by cuts of the form* $E(v)$.

Proof. Let $\mathcal{B}$ denote the subspace of $\mathcal{E}(G)$ generated by the cuts of the form $E(v)$. Every cut of G, with vertex partition $\{V_1, V_2\}$ say, equals $\sum_{v \in V_1} E(v)$ and hence lies in $\mathcal{B}$. Conversely, every set $\sum_{u \in U} E(u) \in \mathcal{B}$ is either empty, e.g. if $U \in \{\emptyset, V\}$, or it is the cut $E(U, V \setminus U)$. $\square$

---

[10] For simplicity, we shall not always distinguish between the edge sets $F \in \mathcal{E}(G)$ and the subgraphs (V, F) they induce in G. When we wish to be more precise, such as in Chapter 8.6, we shall use the word '*circuit*' for the edge set of a cycle.

[11] Recall that partition classes in this book are non-empty. The empty set of edges, therefore, is a cut only if the graph is disconnected.

The space $\mathcal{B}$ from Proposition 1.9.2 is the *cut space*, or *bond space*, of G. It is not difficult to find among the cuts $E(v)$ an explicit basis for $\mathcal{B}$, and thus to determine its dimension (Exercise 40). Note that the bonds are for $\mathcal{B}$ what cycles are for $\mathcal{C}$: the minimal non-empty elements.

cut space $\mathcal{B}(G)$

The 'non-empty' condition in the definition of a bond bites only if G is disconnected. If G is connected, its bonds are just its minimal cuts, and these are easy to recognize: a cut in a connected graph is minimal if and only if both sides of the corresponding vertex partition induce connected subgraphs (Exercise 36). If G is disconnected, its bonds are the minimal cuts of its components.

In analogy to Proposition 1.9.1, bonds and disjoint unions suffice to generate the cut space:

Lemma 1.9.3. *Every cut is a disjoint union of bonds.*

[4.6.2]
[6.5.2]

Proof. We apply induction on the size of the cut F considered. For $F = \emptyset$ the assertion is trivial (with the empty union). If $F \neq \emptyset$ is not itself a bond, it properly contains some other non-empty cut F'. By Proposition 1.9.2, also $F \smallsetminus F' = F + F'$ is a smaller non-empty cut. By the induction hypothesis, both F' and $F \smallsetminus F'$ are disjoint unions of bonds, and hence so is F. $\qquad\square$

Exercise 39 indicates how to construct the bonds for Lemma 1.9.3 explicitly. In Chapter 3.1 we shall prove some more details about the possible positions of the cycles and bonds of a graph within its overall structure (Lemmas 3.1.2 and 3.1.3).

Theorem 1.9.4. *The cycle space $\mathcal{C}$ and the cut space $\mathcal{B}$ of any graph satisfy*

[4.6]

$$\mathcal{C} = \mathcal{B}^\perp \quad \text{and} \quad \mathcal{B} = \mathcal{C}^\perp .$$

Proof. Consider a graph $G = (V, E)$. Clearly, any cycle in G has an even number of edges in each cut. This implies $\mathcal{C} \subseteq \mathcal{B}^\perp$ and $\mathcal{B} \subseteq \mathcal{C}^\perp$.

(1.6.1)
(1.10)

To prove $\mathcal{B}^\perp \subseteq \mathcal{C}$, recall from Proposition 1.9.1 that for every edge set $F \notin \mathcal{C}$ there exists a vertex v incident with an odd number of edges in F. Then $\langle E(v), F \rangle = 1$, so $E(v) \in \mathcal{B}$ implies $F \notin \mathcal{B}^\perp$. This completes the proof of $\mathcal{C} = \mathcal{B}^\perp$.

To prove $\mathcal{C}^\perp \subseteq \mathcal{B}$, let $F \in \mathcal{C}^\perp$ be given. Consider the multigraph[12] H obtained from G by contracting the edges in $E \smallsetminus F$. Any cycle in H has all its edges in F. Since we can extend it to a cycle in G by edges from $E \smallsetminus F$, the number of these edges must be even. Hence H is bipartite, by Proposition 1.6.1. Its bipartition induces a bipartition (V_1, V_2) of V such that $E(V_1, V_2) = F$, showing $F \in \mathcal{B}$ as desired. $\qquad\square$

---

[12] See Section 1.10: such contractions might create loops in F, but bipartite multigraphs have no loops. The proof of Proposition 1.6.1 works for multigraphs too.

fundamental cycle/cut

(1.5.1)

Consider a connected graph $G = (V, E)$ with a spanning tree $T \subseteq G$. For every chord $e \in E \smallsetminus E(T)$ there is a unique cycle C_e in $T + e$, the *fundamental cycle* of e with respect to T. Similarly, for every edge $f \in T$ the forest $T - f$ has exactly two components (Theorem 1.5.1 (iii)). The set $D_f \subseteq E$ of edges of G between these components is a bond in G, the *fundamental cut* of f with respect to T.

Notice that $f \in C_e$ if and only if $e \in D_f$, for all edges $e \notin T$ and $f \in T$. This is an indication of some deeper duality, which the following theorem explores further.

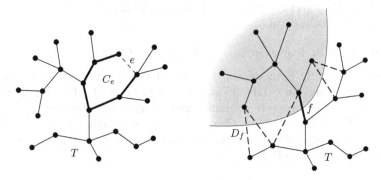

Fig. 1.9.1. The fundamental cycle C_e, and the fundamental cut D_f

[4.5.1]

Theorem 1.9.5. *Let G be a connected graph with n vertices and m edges, and let $T \subseteq G$ a spanning tree.*

 (i) *The fundamental cuts and cycles of G with respect to T form bases of $\mathcal{B}(G)$ and $\mathcal{C}(G)$, respectively.*

 (ii) *Hence, $\dim \mathcal{B}(G) = n - 1$ and $\dim \mathcal{C}(G) = m - n + 1$.*

(1.5.3)

Proof. (i) Note that an edge $f \in T$ lies in D_f but in no other fundamental cut, while an edge $e \notin T$ lies in C_e but in no other fundamental cycle. Hence the fundamental cuts and cycles form linearly independent sets in $\mathcal{B} = \mathcal{B}(G)$ and $\mathcal{C} = \mathcal{C}(G)$, respectively.

Let us show that the fundamental cycles generate every cycle C. By our initial observation, $D := C + \sum_{e \in C \smallsetminus T} C_e$ is an element of $\mathcal{C}$ that contains no edge outside T. But by Proposition 1.9.1, the only element of $\mathcal{C}$ contained in T is $\emptyset$. So $D = \emptyset$, giving $C = \sum_{e \in C \smallsetminus T} C_e$.

Similarly, every cut D is a sum of fundamental cuts. Indeed, the element $D + \sum_{f \in D \cap T} D_f$ of $\mathcal{B}$ contains no edge of T. As $\emptyset$ is the only element of $\mathcal{B}$ missing T, this implies $D = \sum_{f \in D \cap T} D_f$.

(ii) By (i), the fundamental cuts and cycles form bases of $\mathcal{B}$ and $\mathcal{C}$. As there are $n - 1$ fundamental cuts (Corollary 1.5.3), there are $m - n + 1$ fundamental cycles. $\square$

incidence matrix

The *incidence matrix* $B = (b_{ij})_{n \times m}$ of a graph $G = (V, E)$ with $V = \{v_1, \ldots, v_n\}$ and $E = \{e_1, \ldots, e_m\}$ is defined over $\mathbb{F}_2$ by

$$b_{ij} := \begin{cases} 1 & \text{if } v_i \in e_j \\ 0 & \text{otherwise.} \end{cases}$$

As usual, let B^t denote the transpose of B. Then B and B^t define linear maps $B: \mathcal{E}(G) \to \mathcal{V}(G)$ and $B^t: \mathcal{V}(G) \to \mathcal{E}(G)$ with respect to the standard bases. As is easy to check, B maps an edge set $F \subseteq E$ to the set of vertices incident with an odd number of edges in F, while B^t maps a set $U \subseteq V$ to set of edges with exactly one end in U. In particular:

Proposition 1.9.6.

   (i) *The kernel of B is $\mathcal{C}(G)$.*

   (ii) *The image of B^t is $\mathcal{B}(G)$.*   □

More on this in the exercises and notes at the end of this chapter.

adjacency matrix

The *adjacency matrix* $A = (a_{ij})_{n \times n}$ of G is defined by

$$a_{ij} := \begin{cases} 1 & \text{if } v_i v_j \in E \\ 0 & \text{otherwise.} \end{cases}$$

Viewed as a linear map $\mathcal{V} \to \mathcal{V}$, the adjacency matrix maps a given set $U \subseteq V$ to the set of vertices with an odd number of neighbours in U.

Let D denote the real diagonal matrix $(d_{ij})_{n \times n}$ with $d_{ii} = d(v_i)$ and $d_{ij} = 0$ otherwise. Our last proposition establishes a connection between A and B (now viewed as real matrices), which can be verified simply from the definition of matrix multiplication:

Proposition 1.9.7. $BB^t = A + D$.   □

It is also instructive to check that $A + D$, with entries taken mod 2, defines the same map $\mathcal{V} \to \mathcal{V}$ as the composition of the maps of B and B^t (Exercise 48).

1.10 Other notions of graphs

For completeness, we now mention a few other notions of graphs which feature less frequently or not at all in this book.

hypergraph

A *hypergraph* is a pair (V, E) of disjoint sets, where the elements of E are non-empty subsets (of any cardinality) of V. Thus, graphs are special hypergraphs.

directed graph

A *directed graph* (or *digraph*) is a pair (V, E) of disjoint sets (of *vertices* and *edges*) together with two maps $\text{init}: E \to V$ and $\text{ter}: E \to V$

<table>
<tr><td>

init(e)
ter(e)

loop

orientation

oriented
graph

multigraph

</td><td>

assigning to every edge e an *initial vertex* init(e) and a *terminal vertex* ter(e). The edge e is said to be *directed from* init(e) *to* ter(e). Note that a directed graph may have several edges between the same two vertices x, y. Such edges are called *multiple edges*; if they have the same direction (say from x to y), they are *parallel*. If init(e) = ter(e), the edge e is called a *loop*.

A directed graph D is an *orientation* of an (undirected) graph G if $V(D) = V(G)$ and $E(D) = E(G)$, and if {init(e), ter(e)} = {x, y} for every edge $e = xy$. Intuitively, such an *oriented graph* arises from an undirected graph simply by directing every edge from one of its ends to the other. Put differently, oriented graphs are directed graphs without loops or multiple edges.

A *multigraph* is a pair (V, E) of disjoint sets (of *vertices* and *edges*) together with a map $E \to V \cup [V]^2$ assigning to every edge either one or two vertices, its *ends*. Thus, multigraphs too can have loops and multiple edges: we may think of a multigraph as a directed graph whose edge directions have been 'forgotten'. To express that x and y are the ends of an edge e we still write $e = xy$, though this no longer determines e uniquely.

</td></tr>
</table>

A graph is thus essentially the same as a multigraph without loops or multiple edges. Somewhat surprisingly, proving a graph theorem more generally for multigraphs may, on occasion, simplify the proof. Moreover, there are areas in graph theory (such as plane duality; see Chapters 4.6 and 6.5) where multigraphs arise more naturally than graphs, and where any restriction to the latter would seem artificial and be technically complicated. We shall therefore consider multigraphs in these cases, but without much technical ado: terminology introduced earlier for graphs will be used correspondingly.

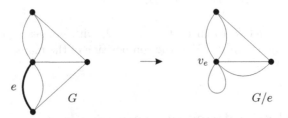

Fig. 1.10.1. Contracting the edge e in the multigraph corresponding to Fig. 1.8.1

A few differences, however, should be pointed out. A multigraph may have cycles of length 1 or 2: loops, and pairs of multiple edges (or *double edges*). A loop at a vertex makes it its own neighbour, and contributes 2 to its degree; in Figure 1.10.1, we thus have $d(v_e) = 6$. The ends of loops and parallel edges in a multigraph G are considered as

separating that edge from the rest of G. The vertex v of a loop e, therefore, is a cutvertex unless $(\{v\}, \{e\})$ is a component of G, and $(\{v\}, \{e\})$
is a 'block' in the sense of Chapter 3.1. Thus, a multigraph with a loop
is never 2-connected, and any 3-connected multigraph is in fact a graph.

The notion of edge contraction is simpler in multigraphs than in
graphs. If we contract an edge $e = xy$ in a multigraph $G = (V, E)$ to a
new vertex v_e, there is no longer a need to delete any edges other than
e itself: edges parallel to e become loops at v_e, while edges xv and yv
become parallel edges between v_e and v (Fig. 1.10.1). Thus, formally,
$E(G/e) = E \smallsetminus \{e\}$, and only the incidence map $e' \mapsto \{\text{init}(e'), \text{ter}(e')\}$ of
G has to be adjusted to the new vertex set in G/e. Contracting a loop
thus has the same effect as deleting it.

The notion of a minor adapts accordingly. The contraction minor
G/P defined by a partition P of $V(G)$ into connected sets has precisely
those edges of G that join distinct partition classes. If there are several
such edges between the same two classes, they become parallel edges
of G/P. However, we do not normally give G/P any loops resulting from
edges of G whose ends lie in the same partition class U. This would require us to say which of the edges of $G[U]$ are contracted (assuming they
induce a connected spanning subgraph of $G[U]$), or at least how many
are, which seems futile if we do not care about loops in G/P anyway.

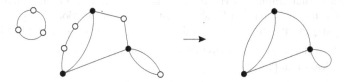

Fig. 1.10.2. Suppressing the white vertices

If v is a vertex of degree 2 in a multigraph G, then by *suppressing v*   *suppressing*
we mean deleting v and adding an edge between its two neighbours.[13]   *a vertex*
(If its two incident edges are identical, i.e. form a loop at v, we add no
edge and obtain just $G - v$. If they go to the same vertex $w \neq v$, the
added edge will be a loop at w. See Figure 1.10.2.) Since the degrees
of all vertices other than v remain unchanged when v is suppressed,
suppressing several vertices of G always yields a well-defined multigraph
that is independent of the order in which those vertices are suppressed.

Finally, it should be pointed out that authors who usually work with
multigraphs tend to call them 'graphs'; in their terminology, our graphs
would be called 'simple graphs'.

---

[13] This is just a clumsy combinatorial paraphrase of the topological notion of
amalgamating the two edges at v into one edge, of which v becomes an inner point.

Exercises

1.⁻ What is the number of edges in a K^n?

2. Let $d \in \mathbb{N}$ and $V := \{0,1\}^d$; thus, V is the set of all 0–1 sequences of length d. The graph on V in which two such sequences form an edge if and only if they differ in exactly one position is called the d-*dimensional cube*. Determine the average degree, number of edges, diameter, girth and circumference of this graph.

    (Hint for the circumference: induction on d.)

3. Let G be a graph containing a cycle C, and assume that G contains a path of length at least k between two vertices of C. Show that G contains a cycle of length at least $\sqrt{k}$.

4.⁻ Is the bound in Proposition 1.3.2 best possible?

5. Let v_0 be a vertex in a graph G, and $D_0 := \{v_0\}$. For $n = 1, 2, \ldots$ inductively define $D_n := N_G(D_0 \cup \ldots \cup D_{n-1})$. Show that $D_n = \{ v \mid d(v_0, v) = n \}$ and $D_{n+1} \subseteq N(D_n) \subseteq D_{n-1} \cup D_{n+1}$ for all $n \in \mathbb{N}$.

6. Show that $\mathrm{rad}(G) \leqslant \mathrm{diam}(G) \leqslant 2\,\mathrm{rad}(G)$ for every graph G.

7. Prove the weakening of Theorem 1.3.4 obtained by replacing average with minimum degree. Deduce that $|G| \geqslant n_0(d/2, g)$ for every graph G as given in the theorem.

8. Show that graphs of girth at least 5 and order n have a minimum degree of $o(n)$. In other words, show that there is a function $f : \mathbb{N} \to \mathbb{N}$ such that $f(n)/n \to 0$ as $n \to \infty$ and $\delta(G) \leqslant f(n)$ for all such graphs G.

9.⁺ Show that every connected graph G contains a path or cycle of length at least $\min\{2\delta(G), |G|\}$.

10. Show that a connected graph of diameter k and minimum degree d has at least about $kd/3$ vertices but need not have substantially more.

11.⁻ Show that the components of a graph partition its vertex set. (In other words, show that every vertex belongs to exactly one component.)

12.⁻ Show that every 2-connected graph contains a cycle.

13. Determine $\kappa(G)$ and $\lambda(G)$ for $G = P^m, C^n, K^n, K_{m,n}$ and the d-dimensional cube (Exercise 2); $d, m, n \geqslant 3$.

14.⁻ Is there a function $f : \mathbb{N} \to \mathbb{N}$ such that, for all $k \in \mathbb{N}$, every graph of minimum degree at least $f(k)$ is k-connected?

15.⁺ Let α, β be two graph invariants with positive integer values. Formalize the two statements below, and show that each implies the other:

   (i) β is bounded above by a function of α;

   (ii) α can be forced up by making β large enough.

   Show that the statement

   (iii) α is bounded below by a function of β

   is not equivalent to (i) and (ii). Which small change will make it so?

16.[+] Show for every $k \in \mathbb{N}$ that every graph of minimum degree $2k$ has a $(k+1)$-edge-connected subgraph.

17. Consider the proof of Theorem 1.4.3. Would it not seem more natural to assume in the second statement of $(*)$ that $\varepsilon(G') > \gamma - k$, as required for H in the statement of the theorem?

   (i) Look how this alteration would change the proof: which parts would carry over, which could be adapted, and which would fail?

   (ii) Explain how the use of an assumption of the form $m \geqslant c_k n - b_k$ rather than $m \geqslant c_k n$ helps to obtain a contradiction in the final inequality of the proof.

18.[+] (Ex. 16–17 continued) Find the smallest integer $b = b(k)$ such that every graph of order n with more than $kn + b$ edges has a $(k+1)$-edge-connected subgraph, for every $k \in \mathbb{N}$.

19. Prove Theorem 1.5.1.

20.[−] Show that every tree T has at least $\Delta(T)$ leaves.

21. Show that a tree without a vertex of degree 2 has more leaves than other vertices. Can you find a very short proof that does not use induction?

22. Let F, F' be forests on the same set of vertices, with $\|F\| < \|F'\|$. Show that F' has an edge e such that $F + e$ is again a forest.

23. Show that the tree-order associated with a rooted tree T is indeed a partial order on $V(T)$, and verify the claims made about this partial order in the text.

24. Show that a graph is 2-edge-connected if and only if it has a *strongly connected* orientation, one in which every vertex can be reached from every other vertex by a directed path.

25.[+] Find a short inductive proof for the existence of normal spanning trees in finite connected graphs.

26.[+] Let G be a connected graph, and let $r \in G$ be a vertex. Starting from r, move along the edges of G, going whenever possible to a vertex not visited so far. If there is no such vertex, go back along the edge by which the current vertex was first reached (unless the current vertex is r; then stop). Show that the edges traversed form a normal spanning tree in G with root r.

   (This procedure has earned those trees the name of *depth-first search* trees.)

27. Let $\mathcal{T}$ be a set of subtrees of a tree T, and $k \in \mathbb{N}$.

   (i) Show that if the trees in $\mathcal{T}$ have pairwise non-empty intersection then their overall intersection $\bigcap \mathcal{T}$ is non-empty.

   (ii) Show that either $\mathcal{T}$ contains k disjoint trees or there is a set of at most $k - 1$ vertices of T meeting every tree in $\mathcal{T}$.

28. Show that every automorphism of a tree fixes a vertex or an edge.

29.⁻ Do the partition classes of a regular bipartite graph always have the same size?

30.⁻ Show that a graph is bipartite if and only if every *induced* cycle has even length.

31. Prove or disprove that a graph is bipartite if and only if no two adjacent vertices have the same distance from any other vertex.

32.⁺ Find a function $f: \mathbb{N} \to \mathbb{N}$ such that, for all $k \in \mathbb{N}$, every graph of average degree at least $f(k)$ has a bipartite subgraph of minimum degree at least k.

33. Show that the minor relation $\preccurlyeq$ defines a partial ordering on any set of pairwise non-isomorphic finite graphs. Is the same true for infinite graphs?

34.⁻ If we had been careless, we might have defined a walk as an alternating sequence of vertices and edges, $v_0 e_0 v_1 e_1 \ldots e_{k-1} v_k$ say, such that every edge e_i is incident with both v_i and v_{i+1}. Show that, based on this definition, Theorem 1.8.1 would fail.

35. Prove or disprove that every connected graph contains a walk that traverses each of its edges exactly once in each direction.

36. Show that a cut in a connected graph G is a bond if and only if both parts of the corresponding bipartition of $V(G)$ are connected in G.

37. Show that the cycle space of a graph is spanned by

   (i) its induced cycles;

   (ii) its geodesic cycles.

   (A cycle $C \subseteq G$ is *geodesic* in G if, for every two vertices of C, their distances in G equals their distance in C.)

38.⁻ Show directly, without using generating bonds $E(v)$, that the cuts of a graph together with the empty set form a subspace of its edge space. How does the vertex partition of a sum of two given cuts arise from their vertex partitions?

39. Let F be a cut in G, with vertex partition $\{V_1, V_2\}$. For $i = 1, 2$ let $C_1^i, \ldots, C_{k(i)}^i$ denote the components of $G[V_i]$. Use the C_j^i to define bonds whose disjoint union is F.

40. Given a graph G, find among all cuts of the form $E(v)$ a basis for the cut space of G.

41. Prove that the cycles and the cuts in a graph together generate its entire edge space, or find a counterexample.

42.⁻ Show the following duality between the fundamental cycles C_e and the fundamental cuts D_f in a graph with respect to some fixed spanning tree: $e \in D_f \Leftrightarrow f \in C_e$.

43. Show that in a connected graph the minimal edge sets containing an edge from every spanning tree are precisely its bonds.

44. Let F be a set of edges in a graph G.

  (i) Show that F extends to an element of $\mathcal{B}(G)$ if and only if it
      contains no odd cycle.

  (ii)$^+$ Show that F extends to an element of $\mathcal{C}(G)$ if and only if it
      contains no odd cut.

45.$^+$ In a graph G let a, b be two vertices that are separated by a cut F of
     k edges but cannot be separated by fewer edges. Show that F is not a
     sum of cuts of fewer than k edges.

     (Hint: Show that the cuts not separating a from b form a subspace
     of $\mathcal{B}(G)$. To prove this, design an 'invariant' of cuts in G, depending on
     a and b, that is constant on sums but distinguishes cuts that separate
     a from b from those that do not.)

46.$^+$ Prove Gallai's theorem that the edge set of any graph G can be written
     as a disjoint union $E(G) = C \cup D$ with $C \in \mathcal{C}(G)$ and $D \in \mathcal{B}(G)$.

47. Show that a set of vertices lies in the image of the incidence matrix of
     a connected graph if and only if it has even cardinality.

48. (i) Generalize Proposition 1.9.6 by describing the images under B
     and B^t of given sets $F \subseteq E$ and $U \subseteq V$, as indicated in the text.

     (ii) Reprove Proposition 1.9.7 for matrices with values in $\mathbb{F}_2$ by showing
     that BB^t and $A + D$ define the same the maps $\mathcal{V} \to \mathcal{V}$.

49. Let $A = (a_{ij})_{n \times n}$ be the adjacency matrix of the graph G. Show that
     the matrix $A^k = (a'_{ij})_{n \times n}$ displays, for all $i, j \leqslant n$, the number a'_{ij} of
     walks of length k from v_i to v_j in G.

Notes

The terminology used in this book is mostly standard. Alternatives do exist,
and some of these are stated when a concept is first defined.

Our formal definition of a graph $G = (V, E)$ with $E \subseteq [V]^2$ is intended to
convey two messages: that the edges are undirected (since $\{u, v\} = \{v, u\}$ for
sets), and that there are neither loops (since $\{v, v\} \notin [V]^2$ because $|\{v, v\}| = 1$)
nor multiple edges (since two sets are equal as soon as they have the same
elements). This formal definition—like any other—occasionally clashes with
other standard terminology.[14] But avoiding all such possible clashes would
make the terminology so unwieldy that it would defeat the purpose of clarity.

There is one small point where our notation deviates slightly from standard usage. Complete graphs, paths, cycles etc. of given order are usually
denoted by K_n, P_k, C_ℓ and so on, but we use superscripts instead of sub-
scripts. This has the advantage of leaving the variables K, P, C etc. free for
ad-hoc use: we may now enumerate components as $C_1, C_2, \ldots$, speak of paths
$P_1, \ldots, P_k$, and so on—without any danger of confusion.

---

[14] For example, when $e = \{u, v\}$ is an edge of G, then $G - e$ and $G - \{u, v\}$ mean
two different things: in $G - e$ we deleted the edge e but kept the vertices u and v,
whereas in $G - \{u, v\}$ we deleted the vertices u, v and all their incident edges.

Theorem[15] 1.3.4 was proved by N. Alon, S. Hoory and N. Linial, The Moore bound for irregular graphs, *Graphs Comb.* **18** (2002), 53–57. The proof uses an ingenious argument counting random walks along the edges of the graph considered.

The main assertion of Theorem 1.4.3, that an average degree of at least $4k$ forces a k-connected subgraph, is from W. Mader, Existenz n-fach zusammenhängender Teilgraphen in Graphen genügend großer Kantendichte, *Abh. Math. Sem. Univ. Hamburg* **37** (1972) 86–97.

The intuition behind the notion of contraction is topological. When G is an IX and we view graphs topologically, we can reobtain X from G by contracting in each of the connected graphs $G[V_x]$ a spanning tree T_x to a single vertex x, and deleting any loops or multiple edges that arise in the contraction.

For the history of the Königsberg bridge problem, and Euler's actual part in its solution, see N.L. Biggs, E.K. Lloyd & R.J. Wilson, *Graph Theory 1736–1936*, Oxford University Press 1976.

Of the large subject of algebraic methods in graph theory, Section 1.9 does not convey an adequate impression. A good introduction is N.L. Biggs, *Algebraic Graph Theory* (2nd edn.), Cambridge University Press 1993. A more comprehensive account is given by C.D. Godsil & G.F. Royle, *Algebraic Graph Theory*, Springer GTM 207, 2001. Surveys on the use of algebraic methods can also be found in the *Handbook of Combinatorics* (R.L. Graham, M. Grötschel & L. Lovász, eds.), North-Holland 1995. See also Chung's book cited below.

In algebraic graph theory one usually takes as the elements of the vertex and edge space the functions mapping the vertices, respectively the oriented edges, to the reals. Then there are 2^m standard bases of $\mathcal{E}$ and 2^m incidence matrices, one for every choice of edge orientations. (No more, since we require that such functions ψ satisfy $\psi(e, u, v) = -\psi(e, v, u)$ for every pair of inverse orientations of the same edge e.) For every fixed choice of orientations, the corresponding incidence matrix represents with respect to the corresponding basis of $\mathcal{E}$ the *boundary map* $\partial : \mathcal{E} \to \mathcal{V}$ that assigns to every (basis element for the) oriented edge (e, u, v) the map $V \to \mathbb{R}$ assigning 1 to v and -1 to u and 0 to every other vertex (and which extends linearly to all of $\mathcal{E}$). Similarly, the transpose of the incidence matrix represents the *coboundary map* $\delta : \mathrm{Hom}(\mathcal{V}, \mathbb{R}) \to \mathrm{Hom}(\mathcal{E}, \mathbb{R})$ mapping φ to $\varphi \circ \partial$; thus, δ is dual to ∂ in the linear algebra sense. The product of the incidence matrix and its transpose is now $BB^t = D - A$, the *Laplacian* of G. Note that, unlike B, the Laplacian is independent of our choice of basis for $\mathcal{E}$, i.e., of our initial choice of orientations that defined our basis. It plays a fundamental role in algebraic graph theory and its connections to other areas of mathematics; see F.R.K. Chung, *Spectral Graph Theory*, AMS 1997 for much more.

---

[15] In the interest of readability, the end-of-chapter notes in this book give references only for Theorems, and only in cases where these references cannot be found in a monograph or survey cited for that chapter.

2

Matching
Covering
and Packing

Suppose we are given a graph and are asked to find in it as many independent edges as possible. How should we go about this? Will we be able to pair up all its vertices in this way? If not, how can we be sure that this is indeed impossible? Somewhat surprisingly, this basic problem does not only lie at the heart of numerous applications, it also gives rise to some rather interesting graph theory.

A set M of independent edges in a graph $G = (V, E)$ is called a *matching*. M is a matching *of* $U \subseteq V$ if every vertex in U is incident with an edge in M. The vertices in U are then called *matched* (by M); vertices not incident with any edge of M are *unmatched*.

matching
matched

A k-regular spanning subgraph is called a *k-factor*. Thus, a subgraph $H \subseteq G$ is a 1-factor of G if and only if $E(H)$ is a matching of V. The problem of how to characterize the graphs that have a 1-factor, i.e. a matching of their entire vertex set, will be our main theme in the first two sections of this chapter.

factor

A generalization of the matching problem is to find in a given graph G as many disjoint subgraphs as possible that are each isomorphic to an element of a given class $\mathcal{H}$ of graphs. This is known as the *packing* problem. It is related to the *covering* problem, which asks how few vertices of G suffice to meet all its subgraphs isomorphic to a graph in $\mathcal{H}$. Clearly, we need at least as many vertices for such a cover as the maximum number k of $\mathcal{H}$-graphs that we can pack disjointly into G. If there is no cover by just k vertices, perhaps there is always a cover by at most $f(k)$ vertices, where $f(k)$ may depend on $\mathcal{H}$ but not on G? In

packing
covering

© Reinhard Diestel 2017
R. Diestel, *Graph Theory*, Graduate Texts in Mathematics 173,
DOI 10.1007/978-3-662-53622-3_2

Section 2.3 we shall prove that when $\mathcal{H}$ is the class of cycles, then there is such a function f.

In Section 2.4 we consider packing and covering in terms of edges: we ask how many edge-disjoint spanning trees we can find in a given graph, and how few trees in it will cover all its edges. In Section 2.5 we prove a path cover theorem for directed graphs, which implies the well-known duality theorem of Dilworth for partial orders.

2.1 Matching in bipartite graphs

$G = (V, E)$
A, B
a, b etc.

For this whole section, we let $G = (V, E)$ be a fixed bipartite graph with bipartition $\{A, B\}$. Vertices denoted as a, a' etc. will be assumed to lie in A, vertices denoted as b etc. will lie in B.

How can we find a matching in G with as many edges as possible? Let us start by considering an arbitrary matching M in G. A path in G which starts in A at an unmatched vertex and then contains, alternately, *alternating path* edges from $E \smallsetminus M$ and from M, is an *alternating path* with respect to M. Note that the path is allowed to be trivial, i.e. to consist of its starting vertex only. An alternating path P that ends in an unmatched vertex *augmenting path* of B is called an *augmenting path* (Fig. 2.1.1), because we can use it to turn M into a larger matching: the symmetric difference of M with $E(P)$ is again a matching (consider the edges at a given vertex), and the set of matched vertices is increased by two, the ends of P.

Fig. 2.1.1. Augmenting the matching M by the alternating path P

Alternating paths play an important role in the practical search for large matchings. In fact, if we start with any matching and keep applying augmenting paths until no further such improvement is possible, the matching obtained will always be an optimal one, a matching with the largest possible number of edges (Exercise 1). The algorithmic problem of finding such matchings thus reduces to that of finding augmenting paths—which is an interesting and accessible algorithmic problem.

Our first theorem characterizes the maximal cardinality of a matching in G by a kind of duality condition. Let us call a set $U \subseteq V$ a *(vertex)* *cover* *cover* of E if every edge of G is incident with a vertex in U.

Theorem 2.1.1. (König 1931)
The maximum cardinality of a matching in G is equal to the minimum cardinality of a vertex cover of its edges.

Proof. Let M be a matching in G of maximum cardinality. From every edge in M let us choose one of its ends: its end in B if some alternating path ends in that vertex, and its end in A otherwise (Fig. 2.1.2). We shall prove that the set U of these $|M|$ vertices covers E; since any vertex cover of E must cover M, there can be none with fewer than $|M|$ vertices, and so the theorem will follow.

M

U

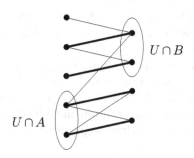

Fig. 2.1.2. The vertex cover U

Note that if an alternating path P ends in a vertex $b \in B$, then $b \in U$: as M is a largest matching, P is not an augmenting path, so b is matched to some $a \in A$ and was put in U when we considered the edge $ab \in M$ while constructing U.

To show that U covers E, let an edge $ab \in E$ be given. If $a \in U$ we are done, so assume that $a \notin U$. To prove $b \in U$, it suffices to show that some alternating path ends in b. If a is unmatched, then ab is such a path. If not, we have $ab' \in M$ for some $b' \in B$. Since $a \notin U$, there exists an alternating path P ending in b'. Depending on whether or not $b \in P$, either Pb or $Pb'ab$ is an alternating path ending in b.   □

Let us return to our main problem, the search for some necessary and sufficient conditions for the existence of a 1-factor. In our present case of a bipartite graph, we may as well ask more generally when G contains a matching of A; this will define a 1-factor of G if $|A| = |B|$, a condition that has to hold anyhow if G is to have a 1-factor.

A condition clearly necessary for the existence of a matching of A is that every subset of A has enough neighbours in B, i.e. that

$$|N(S)| \geqslant |S| \qquad \text{for all } S \subseteq A.$$

marriage condition

The following *marriage theorem* says that this obvious necessary condition is in fact sufficient:

Theorem 2.1.2. (Hall 1935)
G contains a matching of A if and only if $|N(S)| \geqslant |S|$ for all $S \subseteq A$.

We give three proofs, of rather different character.[1] In each proof we
assume that G satisfies the marriage condition and find a matching of A.

M

a

First proof. We show that for every matching M of G that leaves a
vertex $a \in A$ unmatched there is an augmenting path with respect to M.

Let A' be the set of vertices in A that can be reached from a by a
non-trivial alternating path, and $B' \subseteq B$ the set of all penultimate ver-
tices of such paths. The last edges of these paths lie in M, so $|A'| = |B'|$.
Hence by the marriage condition, there is an edge from a vertex v in
$S = A' \cup \{a\}$ to a vertex b in $B \smallsetminus B'$.

As $v \in A' \cup \{a\}$, there is an alternating path P from a to v. Then
Pvb is an alternating path from a to b; notice that $b \notin P$, since the
vertices of P in B lie in B'. If b was matched, by $a'b \in M$ say, then
$Pvba'$ would be an alternating path putting b in B'. But $b \notin B'$, so b is
unmatched, and Pvb is the desired augmenting path. $\square$

Second proof. We apply induction on $|A|$. For $|A| = 1$ the assertion
is true. Now let $|A| \geqslant 2$, and assume that the marriage condition is
sufficient for the existence of a matching of A when $|A|$ is smaller.

If $|N(S)| \geqslant |S| + 1$ for every non-empty set $S \subsetneq A$, we pick an edge
$ab \in G$ and consider the graph $G' := G - \{a, b\}$. Then every non-empty
set $S \subseteq A \smallsetminus \{a\}$ satisfies

$$|N_{G'}(S)| \geqslant |N_G(S)| - 1 \geqslant |S|,$$

so by the induction hypothesis G' contains a matching of $A \smallsetminus \{a\}$. Together with the edge ab, this yields a matching of A in G.

Suppose now that A has a non-empty proper subset A' with $|B'| =
|A'|$ for $B' := N(A')$. By the induction hypothesis, $G' := G[A' \cup B']$
contains a matching of A'. But $G - G'$ satisfies the marriage condition
too: for any set $S \subseteq A \smallsetminus A'$ with $|N_{G-G'}(S)| < |S|$ we would have
$|N_G(S \cup A')| < |S \cup A'|$, contrary to our assumption. Again by induction, $G - G'$ contains a matching of $A \smallsetminus A'$. Putting the two matchings
together, we obtain a matching of A in G. $\square$

H

For our last proof, let H be an edge-minimal subgraph of G that
satisfies the marriage condition and contains A. Note that $d_H(a) \geqslant 1$
for every $a \in A$, by the marriage condition with $S = \{a\}$.

---

[1] The theorem can also be derived easily from König's theorem; see Exercise 5.

Third proof. We show that $d_H(a) = 1$ for every $a \in A$. The edges of H then form a matching of A, since by the marriage condition no two such edges can share a vertex in B.

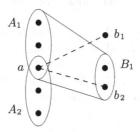

Fig. 2.1.3. B_1 contains b_2 but not b_1

Suppose a has distinct neighbours b_1, b_2 in H. By definition of H, the graphs $H - ab_1$ and $H - ab_2$ violate the marriage condition. So for $i = 1, 2$ there is a set $A_i \subseteq A$ containing a such that $|A_i| > |B_i|$ for $B_i := N_{H-ab_i}(A_i)$ (Fig. 2.1.3). Since $b_1 \in B_2$ and $b_2 \in B_1$,

$$
\begin{aligned}
|N_H(A_1 \cap A_2 \smallsetminus \{a\})| &\leqslant |B_1 \cap B_2| \\
&= |B_1| + |B_2| - |B_1 \cup B_2| \\
&= |B_1| + |B_2| - |N_H(A_1 \cup A_2)| \\
&\leqslant |A_1| - 1 + |A_2| - 1 - |A_1 \cup A_2| \\
&= |A_1 \cap A_2| - 2 \\
&= |A_1 \cap A_2 \smallsetminus \{a\}| - 1 .
\end{aligned}
$$

Hence H violates the marriage condition, contrary to assumption. $\qquad\square$

This last proof has a pretty 'dual', which begins by showing that $d_H(b) \leqslant 1$ for every $b \in B$. See Exercise 6 and its hint for details.

Corollary 2.1.3. *Every k-regular ($k \geqslant 1$) bipartite graph has a 1-factor.*

Proof. If G is k-regular, then clearly $|A| = |B|$; it thus suffices to show by Theorem 2.1.2 that G contains a matching of A. Now every set $S \subseteq A$ is joined to $N(S)$ by a total of $k|S|$ edges, and these are among the $k|N(S)|$ edges of G incident with $N(S)$. Therefore $k|S| \leqslant k|N(S)|$, so G does indeed satisfy the marriage condition. $\qquad\square$

In some real-life applications, matchings are not chosen on the basis of global criteria for the entire graph but evolve as the result of independent decisions made locally by the participating vertices. A typical situation is that vertices are not indifferent to which of their incident edges are picked to match them, but prefer some to others. Then if M

is a matching and $e = ab$ is an edge not in M such that both a and b prefer e to their current matching edge (if they are matched), then a and b may agree to change M locally by including e and discarding their earlier matching edges. The matching M, although perhaps of maximum size, would thus be *unstable*.

preferences   More formally, call a family $(\leqslant_v)_{v \in V}$ of linear orderings $\leqslant_v$ on $E(v)$ a *set of preferences* for G. Then call a matching M in G *stable* if for
stable   every edge $e \in E \smallsetminus M$ there exists an edge $f \in M$ such that e and f have
matching   a common vertex v with $e <_v f$. The following result is sometimes called the *stable marriage theorem*; see Exercises 16 and 17 for a discussion of alternative proofs.

[5.4.4]   **Theorem 2.1.4.** (Gale & Shapley 1962)
For every set of preferences, G has a stable matching.

Proof. Call a matching M in G *better* than a matching $M' \neq M$ if M makes the vertices in B happier than M' does, that is, if every vertex b in an edge $f' \in M'$ is incident also with some $f \in M$ such that $f' \leqslant_b f$. We shall construct a sequence of better and better matchings. Since these can increase the happiness of a fixed vertex b at most $d(b)$ times, this process will terminate.

Given a matching M, call a vertex $a \in A$ *acceptable to* $b \in B$ if $e = ab \in E \smallsetminus M$ and any edge $f \in M$ at b satisfies $f <_b e$. Call $a \in A$ *happy with M* if a is unmatched or its matching edge $f \in M$ satisfies $f >_a e$ for all edges $e = ab$ such that a is acceptable to b.

Starting with the empty matching, let us construct a sequence of matchings that keep all the vertices in A happy. Given such a matching M, consider a vertex $a \in A$ that is unmatched but acceptable to some $b \in B$. (If no such a exists, terminate the sequence.) Add to M the $\leqslant_a$-maximal edge ab such that a is acceptable to b, and discard from M any other edge at b.

Clearly, each matching in our sequence is better than the previous and keeps the vertices in A happy (which they initially are, when $M = \emptyset$). So the sequence continues until it terminates with a matching M such that no unmatched vertex in A is acceptable to any of its neighbours in B. As every matched vertex in A is happy with M, this matching is stable.   □

Despite its seemingly narrow formulation, the marriage theorem counts among the most frequently applied graph theorems, both outside graph theory and within. Often, however, recasting a problem in the setting of bipartite matching requires some clever adaptation. As a simple example, we now use the marriage theorem to derive one of the earliest results of graph theory, a result whose original proof is not all that simple, and certainly not short:

Corollary 2.1.5. (Petersen 1891)
Every regular graph of positive even degree has a 2-factor.

Proof. Let G be any $2k$-regular graph ($k \geqslant 1$), without loss of generality (1.8.1)
connected. By Theorem 1.8.1, G contains an Euler tour $v_0 e_0 \ldots e_{\ell-1} v_\ell$,
with $v_\ell = v_0$. We replace every vertex v by a pair (v^-, v^+), and every
edge $e_i = v_i v_{i+1}$ by the edge $v_i^+ v_{i+1}^-$ (Fig. 2.1.4). The resulting bipartite
graph G' is k-regular, so by Corollary 2.1.3 it has a 1-factor. Collapsing
every vertex pair (v^-, v^+) back into a single vertex v, we turn this 1-
factor of G' into a 2-factor of G. □

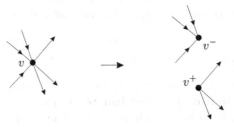

Fig. 2.1.4. Splitting vertices in the proof of Corollary 2.1.5

2.2 Matching in general graphs

Given a graph G, let us denote by $\mathcal{C}_G$ the set of its components, and by $\mathcal{C}_G$
$q(G)$ the number of its *odd components*, those of odd order. If G has a $q(G)$
1-factor, then clearly

$$q(G - S) \leqslant |S| \qquad \text{for all } S \subseteq V(G),$$

Tutte's condition

since every odd component of $G - S$ will send a factor edge to S.

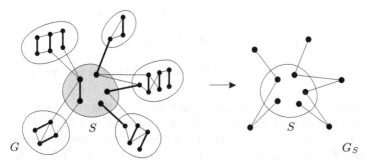

Fig. 2.2.1. Tutte's condition $q(G - S) \leqslant |S|$ for $q = 3$, and the
contracted graph G_S from Theorem 2.2.3.

Again, this obvious necessary condition for the existence of a 1-factor
is also sufficient:

Theorem 2.2.1. (Tutte 1947)
A graph G has a 1-factor if and only if $q(G - S) \leqslant |S|$ for all $S \subseteq V(G)$.

V, E
bad set

Proof. Let $G = (V, E)$ be a graph without a 1-factor. Our task is to find a *bad set* $S \subseteq V$, one that violates Tutte's condition.

We may assume that G is edge-maximal without a 1-factor. Indeed, if G' is obtained from G by adding edges and $S \subseteq V$ is bad for G', then S is also bad for G: any odd component of $G' - S$ is the union of components of $G - S$, and one of these must again be odd.

What does G look like? Clearly, if G contains a bad set S then, by its edge-maximality and the trivial forward implication of the theorem,

all the components of $G - S$ are complete and every vertex $\qquad$ (∗)
$s \in S$ is adjacent to all the vertices of $G - s$.

But also conversely, if a set $S \subseteq V$ satisfies (∗) then either S or the empty set must be bad: if S is not bad we can join the odd components of $G - S$ disjointly to S and pair up all the remaining vertices—unless $|G|$ is odd, in which case $\emptyset$ is bad.

So it suffices to prove that G has a set S of vertices satisfying (∗).

S

Let S be the set of vertices that are adjacent to every other vertex. If this set S does not satisfy (∗), then some component of $G - S$ has non-

a, b, c

adjacent vertices a, a'. Let a, b, c be the first three vertices on a shortest a–a' path in this component; then $ab, bc \in E$ but $ac \notin E$. Since $b \notin S$,

d
M_1, M_2

there is a vertex $d \in V$ such that $bd \notin E$. By the maximality of G, there is a matching M_1 of V in $G + ac$, and a matching M_2 of V in $G + bd$.

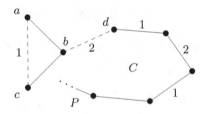

Fig. 2.2.2. Deriving a contradiction if S does not satisfy (∗)

v

Let $P = d \ldots v$ be a maximal path in G starting at d with an edge from M_1 and containing alternately edges from M_1 and M_2 (Fig. 2.2.2). If the last edge of P lies in M_1, then $v = b$, since otherwise we could continue P. Let us then set $C := P + bd$. If the last edge of P lies in M_2, then by the maximality of P the M_1-edge at v must be ac, so $v \in \{a, c\}$; then let C be the cycle $dPvbd$. In each case, C is an even cycle with every other edge in M_2, and whose only edge not in E is bd. Replacing in M_2 its edges on C with the edges of $C - M_2$, we obtain a matching of V contained in E, a contradiction. $\qquad\square$

Corollary 2.2.2. (Petersen 1891)
Every bridgeless cubic graph has a 1-factor.

Proof. We show that any bridgeless cubic graph G satisfies Tutte's condition. Let $S \subseteq V(G)$ be given, and consider an odd component C of $G - S$. Since G is cubic, the degrees (in G) of the vertices in C sum to an odd number, but only an even part of this sum arises from edges of C. So G has an odd number of S–C edges, and therefore has at least 3 such edges (since G has no bridge). The total number of edges between S and $G - S$ thus is at least $3q(G - S)$. But it is also at most $3|S|$, because G is cubic. Hence $q(G - S) \leqslant |S|$, as required.   □

   In order to shed a little more light on the techniques used in matching theory, we now give a second proof of Tutte's theorem. In fact, we shall prove a slightly stronger result, a result that places a structure interesting from the matching point of view on an *arbitrary* graph. If the graph happens to satisfy the condition of Tutte's theorem, this structure will at once yield a 1-factor.

   A non-empty graph $G = (V, E)$ is called *factor-critical* if G has no 1-factor but for every vertex $v \in G$ the graph $G - v$ has a 1-factor. We call a vertex set $S \subseteq V$ *matchable* to $\mathcal{C}_{G-S}$ if the (bipartite[2]) graph G_S, which arises from G by contracting the components $C \in \mathcal{C}_{G-S}$ to single vertices and deleting all the edges inside S, contains a matching of S. (Formally, G_S is the graph with vertex set $S \cup \mathcal{C}_{G-S}$ and edge set $\{ sC \mid \exists c \in C : sc \in E \}$; see Fig. 2.2.1.)

factor-critical

matchable

G_S

Theorem 2.2.3. *Every graph $G = (V, E)$ contains a vertex set S with the following two properties:*

   (i) *S is matchable to $\mathcal{C}_{G-S}$;*

   (ii) *Every component of $G - S$ is factor-critical.*

Given any such set S, the graph G contains a 1-factor if and only if $|S| = |\mathcal{C}_{G-S}|$.

For any given G, the assertion of Tutte's theorem follows easily from this result. Indeed, by (i) and (ii) we have $|S| \leqslant |\mathcal{C}_{G-S}| = q(G - S)$ (since factor-critical graphs have odd order); thus Tutte's condition of $q(G - S) \leqslant |S|$ implies $|S| = |\mathcal{C}_{G-S}|$, and the existence of a 1-factor follows from the last statement of Theorem 2.2.3.

Proof of Theorem 2.2.3. Note first that the last assertion of the theorem follows at once from the assertions (i) and (ii): if G has a 1-factor, we have $q(G - S) \leqslant |S|$ and hence $|S| = |\mathcal{C}_{G-S}|$ as above;

(2.1.2)

---

[2] except for the—permitted—case that S or $\mathcal{C}_{G-S}$ is empty

conversely if $|S| = |\mathcal{C}_{G-S}|$, then the existence of a 1-factor follows straight from (i) and (ii).

We now prove the existence of a set S satisfying (i) and (ii), by induction on $|G|$. For $|G| = 0$ we may take $S = \emptyset$. Now let G be given with $|G| > 0$, and assume the assertion holds for graphs with fewer vertices.

Consider the sets $T \subseteq V$ for which Tutte's condition fails worst, i.e. for which

$$d(T) := d_G(T) := q(G - T) - |T|$$

is maximum, and let S be a largest such set T. Note that $d(S) \geqslant d(\emptyset) \geqslant 0$.

We first show that every component $C \in \mathcal{C}_{G-S} =: \mathcal{C}$ is odd. If $|C|$ is even, pick a vertex $c \in C$, and consider $T := S \cup \{c\}$. As $C - c$ has odd order it has at least one odd component, which is also a component of $G - T$. Therefore

$$q(G - T) \geqslant q(G - S) + 1 \quad \text{while} \quad |T| = |S| + 1 \,,$$

so $d(T) \geqslant d(S)$ contradicting the choice of S.

Next we prove the assertion (ii), that every $C \in \mathcal{C}$ is factor-critical. Suppose there exist $C \in \mathcal{C}$ and $c \in C$ such that $C' := C - c$ has no 1-factor. By the induction hypothesis (and the fact that, as shown earlier, for fixed G our theorem implies Tutte's theorem) there exists a set $S' \subseteq V(C')$ with

$$q(C' - S') > |S'| \,.$$

Since $|C|$ is odd and hence $|C'|$ is even, the numbers $q(C' - S')$ and $|S'|$ are either both even or both odd, so they cannot differ by exactly 1. We may therefore sharpen the above inequality to

$$q(C' - S') \geqslant |S'| + 2 \,,$$

giving $d_{C'}(S') \geqslant 2$. Then for $T := S \cup \{c\} \cup S'$ we have

$$d(T) \geqslant d(S) - 1 - 1 + d_{C'}(S') \geqslant d(S) \,,$$

where the first '-1' comes from the loss of C as an odd component and the second comes from including c in the set T. As before, this contradicts the choice of S.

It remains to show that S is matchable to $\mathcal{C}_{G-S}$. If not, then by the marriage theorem there exists a set $S' \subseteq S$ that sends edges to fewer than $|S'|$ components in $\mathcal{C}$. Since the other components in $\mathcal{C}$ are also components of $G - (S \setminus S')$, the set $T = S \setminus S'$ satisfies $d(T) > d(S)$, contrary to the choice of S. $\square$

Let us consider once more the set S from Theorem 2.2.3, together with any matching M in G. As before, we write $C := C_{G-S}$. Let us denote by k_S the number of edges in M with at least one end in S, and by k_C the number of edges in M with both ends in $G - S$. Since each $C \in \mathcal{C}$ is odd, at least one of its vertices is not incident with an edge of the second type. Therefore every matching M satisfies

S
$\mathcal{C}$

k_S, k_C

$$k_S \leqslant |S| \quad \text{and} \quad k_C \leqslant \tfrac{1}{2}\Big(|V| - |S| - |\mathcal{C}|\Big). \tag{1}$$

Moreover, G contains a matching M_0 with equality in both cases: first choose $|S|$ edges between S and $\bigcup \mathcal{C}$ according to (i), and then use (ii) to find a suitable set of $\tfrac{1}{2}(|C| - 1)$ edges in every component $C \in \mathcal{C}$. This matching M_0 thus has exactly

M_0

$$|M_0| = |S| + \tfrac{1}{2}\Big(|V| - |S| - |\mathcal{C}|\Big) \tag{2}$$

edges.

Now (1) and (2) together imply that *every* matching M of maximum cardinality satisfies both parts of (1) with equality: by $|M| \geqslant |M_0|$ and (2), M has at least $|S| + \tfrac{1}{2}(|V| - |S| - |\mathcal{C}|)$ edges, which implies by (1) that neither of the inequalities in (1) can be strict. But equality in (1), in turn, implies that M has the structure described above: by $k_S = |S|$, every vertex $s \in S$ is the end of an edge $st \in M$ with $t \in G - S$, and by $k_C = \tfrac{1}{2}(|V| - |S| - |\mathcal{C}|)$ exactly $\tfrac{1}{2}(|C| - 1)$ edges of M lie in C, for every $C \in \mathcal{C}$. Finally, since these latter edges miss only one vertex in each C, the ends t of the edges st above lie in different components C for different s.

The seemingly technical Theorem 2.2.3 thus hides a wealth of structural information: it contains the essence of a detailed description of all maximum-cardinality matchings in all graphs. A reference to the full statement of this structural result, known as the *Gallai-Edmonds matching theorem*, is given in the notes at the end of this chapter.

2.3 The Erdős-Pósa theorem

Much of the charm of König's and Hall's theorems in Section 2.1 lies in the fact that they guarantee the existence of the desired matching as soon as some obvious obstruction does not occur. In König's theorem, we can find k independent edges in our graph unless we can cover all its edges by fewer than k vertices (in which case it is obviously impossible).

More generally, if G is an arbitrary graph, not necessarily bipartite, and $\mathcal{H}$ is any class of graphs, we might compare the largest number k of graphs from $\mathcal{H}$ (not necessarily distinct) that we can pack disjointly into G with the smallest number s of vertices of G that will cover all its subgraphs in $\mathcal{H}$. If s can be bounded by a function of k, i.e. independently of G, we say that $\mathcal{H}$ has the *Erdős-Pósa property*. (Thus, formally, $\mathcal{H}$ has this property if there exists an $\mathbb{N} \to \mathbb{N}$ function $k \mapsto f(k)$ such that, for every k and G, either G contains k disjoint subgraphs each isomorphic to a graph in $\mathcal{H}$, or there is a set $U \subseteq V(G)$ of at most $f(k)$ vertices such that $G - U$ has no subgraph in $\mathcal{H}$.)

Our aim in this section is to prove the theorem of Erdős and Pósa that the class of all cycles has this property: we shall find a function f (about $4k \log k$) such that every graph contains either k disjoint cycles or a set of at most $f(k)$ vertices covering all its cycles.

We begin by proving a stronger assertion for cubic graphs. For $k \in \mathbb{N}$, put

r_k, s_k

$$s_k := \begin{cases} 4kr_k & \text{if } k \geqslant 2 \\ 1 & \text{if } k \leqslant 1 \end{cases} \quad \text{where} \quad r_k := \log k + \log \log k + 4.$$

Lemma 2.3.1. *Let $k \in \mathbb{N}$, and let H be a cubic multigraph. If $|H| \geqslant s_k$, then H contains k disjoint cycles.*

(1.3.5)

Proof. We apply induction on k. For $k \leqslant 1$ the assertion is trivial, so let $k \geqslant 2$ be given for the induction step. Let C be a shortest cycle in H.

We first show that $H - C$ contains a subdivision of a cubic multi-

m

graph H' with $|H'| \geqslant |H| - 2|C|$. Let m be the number of edges between C and $H - C$. Since H is cubic and $d(C) = 2$, we have $m \leqslant |C|$. We now consider bipartitions $\{V_1, V_2\}$ of $V(H)$, beginning with $V_1 := V(C)$ and allowing $V_2 = \emptyset$. If $H[V_2]$ has a vertex of degree at most 1 we move this vertex to V_1, obtaining a new partition $\{V_1, V_2\}$ crossed by fewer edges. Suppose we can perform a sequence of n such moves, but

n

no more. (Our assumptions imply $n \leqslant 3$, but we do not formally need this.) Then the resulting partition $\{V_1, V_2\}$ is crossed by at most $m - n$ edges. And $H[V_2]$ has at most $m - n$ vertices of degree less than 3, because each of these is incident with a crossing edge. These vertices have degree exactly 2 in $H[V_2]$, since we could not move them to V_1. Let H' be the cubic multigraph obtained from $H[V_2]$ by suppressing these vertices. Then

$$|H'| \geqslant |H| - |C| - n - (m - n) \geqslant |H| - 2|C|,$$

as desired.

To complete the proof, it suffices to show that $|H'| \geqslant s_{k-1}$. Since $|C| \leqslant 2 \log |H|$ by Corollary 1.3.5 (or by $|H| \geqslant s_k$, if $|C| = g(H) \leqslant 2$), and $|H| \geqslant s_k \geqslant 6$, we have

$$|H'| \geqslant |H| - 2|C| \geqslant |H| - 4 \log |H| \geqslant s_k - 4 \log s_k.$$

(In the last inequality we use that the function $x \mapsto x - 4 \log x$ increases for $x \geqslant 6$.)

It thus remains to show that $s_k - 4\log s_k \geqslant s_{k-1}$. For $k = 2$ this is clear, so we assume that $k \geqslant 3$. Then $r_k \leqslant 4\log k$ (which is obvious for $k \geqslant 4$, while the case of $k = 3$ has to be calculated), and hence

$$
\begin{aligned}
s_k - 4\log s_k &= 4(k-1)r_k + 4\log k + 4\log\log k + 16 \\
&\qquad - \big(8 + 4\log k + 4\log r_k\big) \\
&\geqslant s_{k-1} + 4\log\log k + 8 - 4\log(4\log k) \\
&= s_{k-1}\,.
\end{aligned}
$$
$\square$

Theorem 2.3.2. (Erdős & Pósa 1965)
There is a function $f: \mathbb{N} \to \mathbb{N}$ such that, given any $k \in \mathbb{N}$, every graph contains either k disjoint cycles or a set of at most $f(k)$ vertices meeting all its cycles.

Proof. We show the result for $f(k) := \lfloor s_k + k - 1 \rfloor$. Let k be given, and let G be any graph. We may assume that G contains a cycle, and so it has a maximal subgraph H in which every vertex has degree 2 or 3. Let U be its set of degree 3 vertices. $\qquad\qquad\qquad\qquad\qquad\qquad\qquad\qquad U$

Let $\mathcal{C}$ be the set of all cycles in G that avoid U and meet H in exactly one vertex. Let $Z \subseteq V(H) \smallsetminus U$ be the set of those vertices. For each $\qquad Z$
$z \in Z$ pick a cycle $C_z \in \mathcal{C}$ that meets H in z, and put $\mathcal{C}' := \{\, C_z \mid z \in Z \,\}$. By the maximality of H, the cycles in $\mathcal{C}'$ are disjoint.

Let $\mathcal{D}$ be the set of the 2-regular components of H that avoid Z. Then $\mathcal{C}' \cup \mathcal{D}$ is another set of disjoint cycles. If $|\mathcal{C}' \cup \mathcal{D}| \geqslant k$, we are done. Otherwise we can add to Z one vertex from each cycle in $\mathcal{D}$ to obtain a set X of at most $k-1$ vertices that meets all the cycles in $\mathcal{C}$ and all the $\qquad X$
2-regular components of H. Now consider any cycle of G that avoids X. By the maximality of H it meets H. But it is not a component of H, it does not lie in $\mathcal{C}$, and it does not contain an H-path between distinct vertices outside U (by the maximality of H). So this cycle meets U.

We have shown that every cycle in G meets $X \cup U$. As $|X| \leqslant k-1$, it thus suffices to show that $|U| < s_k$ unless H contains k disjoint cycles. But this follows from Lemma 2.3.1 applied to the multigraph obtained from H by suppressing its vertices of degree 2. $\qquad\qquad\qquad\square$

The proof of Theorem 2.3.2 can be adapted to give an analogous result for packing cycles edge-disjointly and covering them by edges; this is outlined in Exercise 22 of Chapter 7. A simpler proof of the edge version using Ramsey's theorem is indicated in Exercise 6 of Chapter 9.

We shall also meet the Erdős-Pósa property again in Chapter 12. There, a considerable extension of Theorem 2.3.2 will appear as an unexpected and easy corollary of the theory of graph minors.

2.4 Tree packing and arboricity

In this section we consider packing and covering in terms of edges rather than vertices. How many edge-disjoint spanning trees can we find in a given connected graph? And how few trees, not necessarily edge-disjoint, suffice to cover all its edges? These two questions have two classical theorems answering them. But rather than proving these theorems directly, we shall obtain them both as corollaries of a beautiful recent unification due to Bowler and Carmesin: the *packing-covering* theorem.

To motivate the tree packing problem, assume for a moment that our graph represents a communication network, and that for every choice of two vertices we want to be able to find k edge-disjoint paths between them. Menger's theorem (3.3.6) in the next chapter will tell us that such paths exist as soon as our graph is k-edge-connected, which is clearly also necessary. This is a good theorem, but it does not tell us how to find those paths; in particular, having found them for one pair of endvertices we are not necessarily better placed to find them for another pair. If our graph has k edge-disjoint spanning trees, however, there will always be k canonical such paths, one in each tree. Once we have stored those trees in our computer, we shall always be able to find the k paths quickly, between any given pair of vertices.

When does a graph G have k edge-disjoint spanning trees? If it does, it clearly must be k-edge-connected. The converse, however, is easily seen to be false (try $k = 2$); indeed it is not even clear that any edge-connectivity will imply the existence of k edge-disjoint spanning trees. (But see Corollary 2.4.2 below.)

Here is another necessary condition. If G has k edge-disjoint spanning trees, then with respect to any partition of $V(G)$ into r sets, every spanning tree of G has at least $r - 1$ *cross-edges*, edges whose ends lie in different partition sets. (Why?) Thus if G has k edge-disjoint spanning trees, it has at least $k(r - 1)$ cross-edges. This condition is also sufficient:

cross-edges

tree
packing
theorem
Theorem 2.4.1. (Nash-Williams 1961; Tutte 1961)
A multigraph contains k edge-disjoint spanning trees if and only if for every partition P of its vertex set it has at least $k(|P| - 1)$ cross-edges.
[8.6.9]

Theorem 2.4.1 has a striking corollary: $2k$-edge-connectedness is enough to ensure the existence of k edge-disjoint spanning trees.

[6.4.4]
Corollary 2.4.2. *Every $2k$-edge-connected multigraph G has k edge-disjoint spanning trees.*

Proof. Every class in a vertex partition of G is joined to other partition classes by at least $2k$ edges. Hence, for any partition into r sets, G has at least $\frac{1}{2}\sum_{i=1}^{r} 2k = kr$ cross-edges. The assertion thus follows from Theorem 2.4.1.   □

Note that the quantitative condition on cross-edges in Theorem 2.4.1 is equivalent to asking the same only for partitions into connected vertex sets: any other partition is refined by such a partition, and if the latter has enough cross-edges (even though it has more classes) then clearly so does the former. The tree packing theorem thus says that a multigraph has k edge-disjoint spanning trees as soon as all its contraction minors have enough edges to support k edge-disjoint spanning trees.

We shall meet Theorem 2.4.1 again in Chapter 8.6, where we prove an infinite analouge. This is based not on ordinary spanning trees (for which the result is false) but on 'topological spanning trees': the analogous structures in a topological space formed by the graph together with its ends, points at infinity that make it compact.

Let us now turn to the covering problem. To bring out its duality to the packing problem, we begin by rephrasing the latter. Let us say that some given subgraphs of a multigraph G form an *edge-decomposition* of G if their edge sets partition $E(G)$. Our spanning tree problem can now be recast as follows: into how many *connected* spanning subgraphs can we edge-decompose G? Since a spanning subgraph is connected if and only if it has an edge in every bond, the packing problem in this new guise has a 'dual' reminiscent of Theorems 1.5.1 and 1.9.4: into how *few* acyclic subgraphs—those whose complement meets all their circuits— can we edge-decompose G?

Let us say that some given graphs, not necessarily subgraphs of G, *cover its edges* if every edge of G lies in at least one of them. Our dual problem, then, is for which multigraphs G can we cover their edges by at most k trees.

cover

An obvious necessary condition is that every set $U \subseteq V(G)$ induces at most $k(|U| - 1)$ edges, no more than $|U| - 1$ for each tree. Or, to phrase it dually to the tree packing condition, that no 'deletion minor' (subgraph) of G has too many edges to be covered by k trees.

Once more, this condition turns out to be sufficient too:

Theorem 2.4.3. (Nash-Williams 1964)
The edges of a multigraph $G = (V, E)$ can be covered by at most k trees if and only if $\|G[U]\| \leqslant k(|U| - 1)$ for every non-empty set $U \subseteq V$.

tree covering theorem

The least number of trees that can cover the edges of a graph is its *arboricity*. By Theorem 2.4.3, the arboricity of a graph is a measure for its maximum local density: it has small arboricity if and only if it is 'nowhere dense' in the sense that it has no subgraph H with $\varepsilon(H)$ large.

arboricity

We finally come to the packing-covering theorem. Recall from Chapter 1.10 that when we form a contraction minor G/P of a multigraph G, we keep all the edges of G between different partition classes: edges between the same two classes $U, U' \in P$ become parallel edges of G/P.

Theorem 2.4.4. (Bowler & Carmesin 2015)

For every connected multigraph $G = (V, E)$ and every $k \in \mathbb{N}$ there is a partition P of V such that every $G[U]$ with $U \in P$ has k edge-disjoint spanning trees and the edges of G/P can be covered by k spanning trees.

Before we prove the packing-covering theorem, let us deduce Theorems 2.4.1 and 2.4.3.

Proof of Theorem 2.4.1 *from Theorem 2.4.4.*
Suppose a multigraph G has at least $k(|P| - 1)$ cross-edges for every partition P of $V(G)$. Let P be the partition provided by Theorem 2.4.4. By the theorem, G/P has k spanning trees covering its edges. Since $\|G/P\| \geqslant k(|P| - 1)$, they must be edge-disjoint. Combining them with the edge-disjoint spanning trees in the $G[U]$ that are also provided by Theorem 2.4.4, we obtain the desired k spanning trees of G.   □

Proof of Theorem 2.4.3 *from Theorem 2.4.4.*
Suppose every $U \subseteq V$ induces at most $k(|U| - 1)$ edges in G. Let C be a component of G, and P the partition of $V(C)$ provided by Theorem 2.4.4. For each $U \in P$, each of the k edge-disjoint spanning trees of $G[U]$ that the theorem provides has $|U| - 1$ edges, so all the edges of $G[U]$ lie in these trees. Combining these trees with the spanning trees of C/P that cover its edges, also provided by Theorem 2.4.4, we obtain k spanning trees of C covering its edges. These can be combined to k forests covering the edges of G. Add edges to turn these into the desired k trees.   □

Given the power of the packing-covering theorem, its proof is strikingly short and elegant. To prepare some notation, consider a spanning tree T of G, a chord e, and an edge $f \in T$ on its fundamental cycle C_e. Then $T' = T + e - f$ is another spanning tree: this is immediate from Corollary 1.5.3, because T' is still connected and has the same number of edges as T. One says that T' is obtained from T by *exchanging* f for e.

Now let $\mathcal{T} = (T_1, \ldots, T_k)$ be a family of spanning trees of G. Call a sequence $e_0, \ldots, e_n$ of edges an *exchange chain* for $\mathcal{T}$ *started by* e_0 if e_n lies on none of these trees but for every $i < n$ there exists $j =: j(i)$ such that $e_i \in T_j$ while e_{i+1} is a chord of T_j whose fundamental cycle with respect to T_j contains e_i.

Let us write $E(\mathcal{T}) := \bigcup \{ E(T) \mid T \in \mathcal{T} \}$ for any such family.

Lemma 2.4.5. *If e_0 starts an exchange chain for $\mathcal{T}$ and lies in two of its trees, then there is a family $\mathcal{T}'$ of k spanning trees of G such that $E(\mathcal{T}) \subsetneq E(\mathcal{T}')$.*

Proof. Choose $e_0, \ldots, e_n$ of minimal length among the exchange chains for $\mathcal{T}$ that start with e_0. Then no e_i lies on the fundamental cycle, with respect to any tree in $\mathcal{T}$, of any e_ℓ with $\ell > i + 1$: otherwise we could shorten the sequence by skipping from e_i straight to e_ℓ or $e_{\ell+1}$.

Starting with $T^0 = T$, define T^{i+1} inductively for $i = 0, \ldots, n-1$ by replacing in $T^i = (T_1^i, \ldots, T_k^i)$ the tree T_j^i with $T_j^i + e_{i+1} - e_i =: T_j^{i+1}$ for $j = j(i)$ and letting $T_j^{i+1} := T_j^i$ for all other j. Note that, for $j = j(i)$, the minimality of our sequence implies that every edge e of T_j on its fundamental cycle for e_{i+1} is still in T_j^i: otherwise $e = e_{i'}$ for some $i' < i$, with a contradiction for $\ell := i + 1 > i' + 1$. Thus, if T_j^i is a spanning tree of G, as we may assume inductively, then so is T_j^{i+1}.

Clearly, $T' := (T_1^n, \ldots, T_k^n)$ satisfies $E(T') = E(T) \cup \{e_n\}$. □

Proof of Theorem 2.4.4. Let $T = (T_1, \ldots, T_k)$ be a family of k spanning trees of G, chosen with $E(T)$ maximal. Let D be the set of all edges of G that start an exchange chain for T. These include all edges not in $E(T)$, since they form singleton exchange chains. Let P be the partition of V into the vertex sets of the components of (V, D).

For the theorem's packing assertion, let $U \in P$ be given. For all $j = 1, \ldots, k$ let S_j be the subgraph of $T_j[U]$ formed by its edges in D. These forests S_j are edge-disjoint, since by the maximality of T and Lemma 2.4.5 no edge in D lies in more than one T_j. Let us show that the S_j are connected.

Since the edges of D form a connected submultigraph on U, it suffices to show that for every edge $uu' \in D$ with $u, u' \in U$ there is a u–u' path in S_j. This is clear if uu' lies in T_j, and hence in S_j. If it does not, then the path uT_ju' still has all its edges e in D, and hence lies in S_j: if $e_0, \ldots, e_n$ is an exchange chain witnessing that $e_0 = uu' \in D$, then $e, e_0, \ldots e_n$ is an exchange chain putting e in D, because e lies on the fundamental cycle of e_0 with respect to T_j.

As every T_j induces connected subgraphs S_j on the partition classes of P, contracting these S_j turns the T_j into spanning trees T_j' of G/P. These T_j' cover all the edges of G/P, since $E \smallsetminus E(T) \subseteq D$. □

The packing-covering theorem differs from both the tree packing and the tree covering theorem in a fundamental way. The non-trivial directions of the latter two theorems each obtain a structural assertion about a graph, the existence of a packing or covering, as a consequence of quantitative assumptions about all their minors of a certain type: contraction minors for the packing theorem, and 'deletion minors'—i.e., subgraphs—for the covering theorem. This format makes them interesting: they offer valuable structural information for one graph in exchange for less valuable quantitative information about many smaller graphs.

The packing-covering theorem, by contrast, makes a structural assertion about every graph: with no need for any assumptions at all, neither quantitative nor qualitative.

The packing-covering theorem extends to infinite graphs in two interestingly different ways; see Exercises 18 and 126 in Chapter 8.

2.5 Path covers

Let us return once more to König's duality theorem for bipartite graphs, Theorem 2.1.1. If we orient every edge of G from A to B, the theorem tells us how many disjoint directed paths we need in order to cover all the vertices of G: every directed path has length 0 or 1, and clearly the number of paths in such a 'path cover' is smallest when it contains as many paths of length 1 as possible—in other words, when it contains a maximum-cardinality matching.

In this section we put the above question more generally: how many paths in a given directed graph will suffice to cover its entire vertex set? Of course, this could be asked just as well for undirected graphs. As it turns out, however, the result we shall prove is rather more trivial in the undirected case (exercise), and the directed case will also have an interesting corollary.

path

ter(P)

path cover

A *directed path* is a directed graph $P \neq \emptyset$ with distinct vertices $x_0, \ldots, x_k$ and edges $e_0, \ldots, e_{k-1}$ such that e_i is an edge directed from x_i to x_{i+1}, for all $i < k$. In this section, *path* will always mean 'directed path'. The vertex x_k above is the *last vertex* of the path P, and when $\mathcal{P}$ is a set of paths we write $\mathrm{ter}(\mathcal{P})$ for the set of their last vertices. A *path cover* of a directed graph G is a set of disjoint paths in G which together contain all the vertices of G.

Theorem 2.5.1. (Gallai & Milgram 1960)
Every directed graph G has a path cover $\mathcal{P}$ and an independent set $\{v_P \mid P \in \mathcal{P}\}$ of vertices such that $v_P \in P$ for every $P \in \mathcal{P}$.

P, P_i

v_i

Proof. Clearly, G has a path cover, e.g. by trivial paths. We prove by induction on $|G|$ that for every path cover $\mathcal{P} = \{P_1, \ldots, P_m\}$ with $\mathrm{ter}(\mathcal{P})$ minimal there is a set $\{v_P \mid P \in \mathcal{P}\}$ as claimed. For each i, let v_i denote the last vertex of P_i.

v

P', G'

If $\mathrm{ter}(\mathcal{P}) = \{v_1, \ldots, v_m\}$ is independent there is nothing more to show, so we assume that G has an edge from v_2 to v_1. Since $P_2v_2v_1$ is again a path, the minimality of $\mathrm{ter}(\mathcal{P})$ implies that v_1 is not the only vertex of P_1; let v be the vertex preceding v_1 on P_1. Then $\mathcal{P}' := \{P_1v, P_2, \ldots, P_m\}$ is a path cover of $G' := G - v_1$ (Fig. 2.5.1). Clearly, any independent set of representatives for $\mathcal{P}'$ in G' will also work for $\mathcal{P}$ in G, so all we have to check is that we may apply the induction hypothesis to $\mathcal{P}'$. It thus remains to show that $\mathrm{ter}(\mathcal{P}') = \{v, v_2, \ldots, v_m\}$ is minimal among the sets of last vertices of path covers of G'.

Suppose then that G' has a path cover $\mathcal{P}''$ with $\mathrm{ter}(\mathcal{P}'') \subsetneq \mathrm{ter}(\mathcal{P}')$. If a path $P \in \mathcal{P}''$ ends in v, we may replace P in $\mathcal{P}''$ by Pvv_1 to obtain a path cover of G whose set of last vertices is a proper subset of $\mathrm{ter}(\mathcal{P})$, contradicting the choice of $\mathcal{P}$. If a path $P \in \mathcal{P}''$ ends in v_2 (but none in v), we similarly replace P in $\mathcal{P}''$ by Pv_2v_1 to obtain a contradiction to the minimality of $\mathrm{ter}(\mathcal{P})$. Hence $\mathrm{ter}(\mathcal{P}'') \subseteq \{v_3, \ldots, v_m\}$. But now $\mathcal{P}''$ and

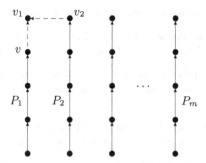

Fig. 2.5.1. Path covers of G and G'

the trivial path $\{v_1\}$ together form a path cover of G that contradicts
the minimality of $\mathrm{ter}(\mathcal{P})$. $\square$

As a corollary to Theorem 2.5.1 we obtain a classical result from
the theory of partial orders. Recall that a subset of a partially ordered
set $(P, \leqslant)$ is a *chain* in P if its elements are pairwise comparable; it is   *chain*
an *antichain* if they are pairwise incomparable.   *antichain*

Corollary 2.5.2. (Dilworth 1950)
*In every finite partially ordered set $(P, \leqslant)$, the minimum number of
chains with union P is equal to the maximum cardinality of an antichain
in P.*

Proof. If A is an antichain in P of maximum cardinality, then clearly
P cannot be covered by fewer than $|A|$ chains. The fact that $|A|$ chains
will suffice follows from Theorem 2.5.1 applied to the directed graph on
P with the edge set $\{\, (x, y) \mid x < y \,\}$. $\square$

Exercises

1. Let M be a matching in a bipartite graph G. Show that if M is sub-
    optimal, i.e. contains fewer edges than some other matching in G, then
    G contains an augmenting path with respect to M. Does this fact
    generalize to matchings in non-bipartite graphs?

2. Describe an algorithm that finds, as efficiently as possible, a matching
    of maximum cardinality in any bipartite graph.

3. Show that if there exist injective functions $A \to B$ and $B \to A$ between
    two infinite sets A and B then there exists a bijection $A \to B$.

4. Moving alternately, two players jointly construct a path in some fixed
    graph G. If $v_1 \ldots v_n$ is the path constructed so far, the player to move
    next has to find a vertex v_{n+1} such that $v_1 \ldots v_{n+1}$ is again a path.
    Whichever player cannot move loses. For which graphs G does the first
    player have a winning strategy, for which the second?

5.   Derive the marriage theorem from König's theorem.

6.   Let G and H be defined as for the third proof of Hall's theorem. Show
     that $d_H(b) \leqslant 1$ for every $b \in B$, and deduce the marriage theorem.

7.   Does our first proof of the marriage theorem use the assumption that
     the graph is finite? If so, can it be adapted so that it works for infinite
     graphs too?

8.   Let k be an integer. Show that any two partitions of a finite set into
     k-sets admit a common choice of representatives.

9.   Let A be a finite set with subsets $A_1, \ldots, A_n$, and let $d_1, \ldots, d_n \in \mathbb{N}$.
     Show that there are disjoint subsets $D_k \subseteq A_k$, with $|D_k| = d_k$ for all
     $k \leqslant n$, if and only if
     $$\left| \bigcup_{i \in I} A_i \right| \geqslant \sum_{i \in I} d_i$$
     for all $I \subseteq \{1, \ldots, n\}$.

10.+ Prove that in an n-set X there are never more than $\binom{n}{\lfloor n/2 \rfloor}$ subsets
     such that none of these contains another.

     (Hint. Construct $\binom{n}{\lfloor n/2 \rfloor}$ chains covering the power set lattice of X.)

11. Let G be a bipartite graph with bipartition $\{A, B\}$. Assume that
     $\delta(G) \geqslant 1$, and that $d(a) \geqslant d(b)$ for every edge ab with $a \in A$. Show
     that G contains a matching of A.

12.− Find a bipartite graph with a set of preferences such that no matching
     of maximum size is stable and no stable matching has maximum size.
     Find a non-bipartite graph with a set of preferences that has no stable
     matching.

13.− Consider the algorithm described in the proof of the stable marriage
     theorem. Observe that once a vertex of B is matched, she remains
     matched and gets happier with every change of her matching edge.
     On the other hand, show that the sequence of matching edges incident
     with a given vertex of A makes this vertex unhappier with every change
     (disregarding the interim periods when he is unmatched).

14. Show that all stable matchings of a given graph cover the same vertices.
     (In particular, they have the same size.)

15.+ Show that the algorithm in our proof of Theorem 2.1.4 produces a
     matching M such that no other stable matching makes any vertex in A
     happier or any vertex in B unhappier than he or she is in M. Consider
     only matched vertices for happiness.

16.+ Show that the following 'obvious' algorithm need not produce a stable
     matching in a bipartite graph. Start with any matching. If the current
     matching is not maximal, add an edge. If it is maximal but not stable,
     insert an edge that creates instability, deleting any current matching
     edges at its ends.

17. Show that the union of two partial orderings $\leqslant_1, \leqslant_2$ of a finite set P has a 'dominating antichain', a set $A \subseteq P$ such that no two elements of A are related in either $\leqslant_1$ or $\leqslant_2$ and for every $x \in P$ there exists an $a \in A$ such that $x \leqslant_1 a$ or $x \leqslant_2 a$. Deduce Theorem 2.1.4.

18. Find a set S for Theorem 2.2.3 when G is a forest.

19. A graph G is called (vertex-) *transitive* if, for any two vertices $v, w \in G$, there is an automorphism of G mapping v to w. Using the observations following the proof of Theorem 2.2.3, show that every transitive connected graph of even order contains a 1-factor.

20.$^{+}$ Show that a graph G contains k independent edges if and only if $q(G - S) \leqslant |S| + |G| - 2k$ for all sets $S \subseteq V(G)$.

21.$^{-}$ Find a cubic graph without a 1-factor.

22.$^{+}$ Derive the marriage theorem from Tutte's theorem.

23.$^{-}$ Disprove the analogue of König's theorem (2.1.1) for non-bipartite graphs, but show that $\mathcal{H} = \{K^2\}$ has the Erdős-Pósa property.

24. Let T be a tree and $\mathcal{T}$ a set of subtrees of T. Show that the maximum number of disjoint trees in $\mathcal{T}$ equals the least cardinality of a set X of vertices such that $T - X$ contains no tree from $\mathcal{T}$.

25. For cubic graphs, Lemma 2.3.1 is considerably stronger than the Erdős-Pósa theorem. Extend the lemma to arbitrary multigraphs of minimum degree $\geqslant 3$, by finding a function $g: \mathbb{N} \to \mathbb{N}$ such that every multigraph of minimum degree $\geqslant 3$ and order at least $g(k)$ contains k disjoint cycles, for all $k \in \mathbb{N}$. Alternatively, show that no such function g exists.

26. Given a graph G, let $\alpha(G)$ denote the largest size of a set of independent vertices in G. Prove that the vertices of G can be covered by at most $\alpha(G)$ disjoint subgraphs each isomorphic to a cycle or a K^2 or K^1.

27. Show that if G has two edge-disjoint spanning trees, it has a connected spanning subgraph all whose degrees are even.

28. In the proofs of Theorems 2.4.1, 2.4.3 and 2.4.4, there is exactly one place where we use that we are working with multigraphs. Where is it?

29. Find the error in the following short 'proof' of Theorem 2.4.1. Call a partition *non-trivial* if it has at least two classes and at least one of the classes has more than one element. We show by induction on $|V| + |E|$ that $G = (V, E)$ has k edge-disjoint spanning trees if every non-trivial partition of V into r sets (say) has at least $k(r - 1)$ cross-edges. The induction starts trivially with $G = K^1$ if we allow k copies of K^1 as a family of k edge-disjoint spanning trees of K^1. We now consider the induction step. If every non-trivial partition of V into r sets (say) has more than $k(r - 1)$ cross-edges, we delete any edge of G and are done by induction. So V has a non-trivial partition $\{V_1, \ldots, V_r\}$ with exactly $k(r - 1)$ cross-edges. Assume that $|V_1| \geqslant 2$. If $G' := G[V_1]$ has k disjoint spanning trees, we may combine these with k disjoint spanning trees that exist in G/V_1 by induction. We may thus assume that G' has

no k disjoint spanning trees. Then by induction it has a non-trivial vertex partition $\{V'_1, \ldots, V'_s\}$ with fewer than $k(s-1)$ cross-edges. Then $\{V'_1, \ldots, V'_s, V_2, \ldots, V_r\}$ is a non-trivial vertex partition of G into $r+s-1$ sets with fewer than $k(r-1) + k(s-1) = k((r+s-1)-1)$ cross-edges, a contradiction.

30. A graph G is called *balanced* if $\varepsilon(H) \leqslant \varepsilon(G)$ for every subgraph $H \subseteq G$.

   (i)   Find a few natural classes of balanced graphs.

   (ii) Show that the arboricity of a balanced graph is bounded above by its average degree. Is it even bounded by ε? Or by $\varepsilon + 1$?

   (iii) Characterize, in terms of the balanced graphs or otherwise, the graphs G such that $\varepsilon(H) \geqslant \varepsilon(G)$ for every induced subgraph $H \subseteq G$.

31. Rephrase König's and Dilworth's theorems as pure existence statements without any inequalities.

32.⁻ Prove the undirected version of the theorem of Gallai & Milgram (without using the directed version).

33. Derive the marriage theorem from the theorem of Gallai & Milgram.

34.⁻ Show that a partially ordered set of at least $rs+1$ elements contains either a chain of size $r+1$ or an antichain of size $s+1$.

35. Prove the following dual version of Dilworth's theorem: in every finite partially ordered set $(P, \leqslant)$, the minimum number of antichains with union P is equal to the maximum cardinality of a chain in P.

36. Derive König's theorem from Dilworth's theorem.

37. Find a partially ordered set that has no infinite antichain but is not a union of finitely many chains.

Notes

There is a very readable and comprehensive monograph about matching in finite graphs: L. Lovász & M.D. Plummer, *Matching Theory*, Annals of Discrete Math. **29**, North Holland 1986. Two other very comprehensive sources are A. Schrijver, *Combinatorial optimization*, Springer 2003, and A. Frank, *Connections in combinatorial optimization*, Oxford University Press 2011. All the references for the results in this chapter can be found in these books.

As we shall see in Chapter 3, König's Theorem of 1931 is no more than the bipartite case of a more general theorem due to Menger, of 1929. At the time, neither of these results was nearly as well known as Hall's marriage theorem, which he proved even later, in 1935. To this day, Hall's theorem remains one of the most applied graph-theoretic results. The first two of our proofs are folklore. The edge-minimal subgraph approach of our third proof can be traced back to a paper of Rado (1967); our version and its dual, Exercise 6, are due to Kriesell.

More on the stable marriage theorem can be found in D. Gusfield & R.W. Irving, *The Stable Marriage Problem: Structure and Algorithms*, MIT Press 1989, and in A. Tamura, Transformation from arbitrary matchings to stable matchings, *J. Comb. Theory A* **62** (1993), 310–323. Some particularly rewarding applications are listed under 'Advanced Information' on http://www.nobelprize.org/nobel_prizes/economic-sciences/laureates/2012.

Our proof of Tutte's 1-factor theorem is based on a proof by Lovász (1975). Our extension of Tutte's theorem, Theorem 2.2.3 (including the informal discussion following it) is a lean version of a comprehensive structure theorem for matchings, due to Gallai (1964) and Edmonds (1965). See Lovász & Plummer for a detailed statement and discussion of this theorem.

Theorem 2.3.2 is due to P. Erdős & L. Pósa, On independent circuits contained in a graph, *Canad. J. Math.* **17** (1965), 347–352. Our proof is essentially due to M. Simonovits, A new proof and generalization of a theorem of Erdős and Pósa on graphs without $k + 1$ independent circuits, *Acta Sci. Hungar* **18** (1967), 191–206. Calculations such as in Lemma 2.3.1 are standard for proofs where one aims to bound one numerical invariant in terms of another. This book does not emphasize this aspect of graph theory, but it is not atypical.

There is also an analogue of the Erdős-Pósa theorem for directed graphs, due to B. Reed, N. Robertson, P.D. Seymour and R. Thomas, Packing directed circuits, *Combinatorica* **16** (1996), 535–554. Its proof is more difficult than the undirected case; see Chapter 12.6, and in particular Theorem 12.6.5, for a glimpse of the techniques used.

The tree packing theorem, Theorem 2.4.1, was proved independently by Nash-Williams and Tutte; both papers are contained in *J. Lond. Math. Soc.* **36** (1961). The tree covering theorem, Theorem 2.4.3, is due to C.St.J.A. Nash-Williams, Decompositions of finite graphs into forests, *J. Lond. Math. Soc.* **39** (1964), 12. The partitions whose existence is asserted by the packing-covering theorem, Theorem 2.4.4, were first constructed explicitly by B. Jackson and T. Jordán, Brick partitions of graphs, *Discrete Math.* **310** (2010), 270–275. They may not be unique, and are interesting in their own right; see the paper, and Frank's monograph cited earlier, for more.

The packing-covering theorem itself, together with its direct proof that does not rely on the classical tree packing and covering theorems but implies them, is from N. Bowler and J. Carmesin, Matroid intersection, base packing and base covering for infinite matroids, *Combinatorica* **35** (2015), 153–180, arXiv:1202.3409.

It has long been known that the tree packing and covering theorems can be naturally expressed in terms of matroids; see Schrijver's book cited earlier. However it was only recently when infinite matroids were axiomatized and thus made accessible to systematic study, forcing the translation of quantitative assertions about finite matroids into structural ones to make them meaningful also for infinite matroids, that Bowler and Carmesin found the packing-covering theorem. The main focus of their paper is to show how the unproved infinite version of the packing-covering theorem for matroids, the *packing-covering conjecture*, plays a central role in infinite matroid theory. The conjecture implies, among other things, the Aharoni-Berger theorem for infinite graphs (Theorem 8.4.2), one of the deepest theorems in graph theory.

The packing-covering theorem extends to infinite graphs in two ways:

with ordinary spanning trees (Exercise 18, Ch. 8), and with 'topological' spanning trees (Exercise 126, Ch. 8). These infinite versions also follow from two cases of the infinite packing-covering conjecture that Bowler and Carmesin prove in their paper, those for finitary and for cofinitary matroids.

An interesting vertex analogue of Corollary 2.4.2 is to ask which connectivity forces the existence of k spanning trees $T_1, \ldots, T_k$, all rooted at a given vertex r, such that for every vertex v the k paths vT_ir are independent. For example, if G is a cycle then deleting the edge left or right of r produces two such spanning trees. A. Itai and A. Zehavi, Three tree-paths, *J. Graph Theory* **13** (1989), 175–187, conjectured that $\kappa \geqslant k$ should suffice. This conjecture has been proved for $k \leqslant 4$; see S. Curran, O. Lee & X. Yu, Chain decompositions and independent trees in 4-connected graphs, *Proc. 14th Ann. ACM SIAM symposium on Discrete algorithms* (Baltimore 2003), 186–191.

Theorem 2.5.1 is due to T. Gallai & A. N. Milgram, Verallgemeinerung eines graphentheoretischen Satzes von Rédei, *Acta Sci. Math. (Szeged)* **21** (1960), 181–186.

3 Connectivity

Our definition of k-connectedness, given in Chapter 1.4, is somewhat unintuitive. It does not tell us much about 'connections' in a k-connected graph: all it says is that we need at least k vertices to disconnect it. The following definition—which, incidentally, implies the one above—might have been more descriptive: 'a graph is k-*connected* if any two of its vertices can be joined by k independent paths'.

It is one of the classic results of graph theory that these two definitions are in fact equivalent, are dual aspects of the same property. We shall study this theorem of Menger (1927) in some depth in Section 3.3.

In Sections 3.1 and 3.2, we investigate the structure of the 2-connected and the 3-connected graphs. For these small values of k it is still possible to give a simple general description of how these graphs can be constructed.

In Sections 3.4 and 3.5 we look at other concepts of connectedness, more recent than the standard one but no less important: the number of H-paths in G for a subgraph H of G, and the existence of disjoint paths in G linking up specified pairs of vertices.

3.1 2-Connected graphs and subgraphs

The simplest 2-connected graphs are the cycles. All the others can be constructed inductively from a cycle by adding paths:

Proposition 3.1.1. *A graph is 2-connected if and only if it can be* [4.2.6]
constructed from a cycle by successively adding H-paths to graphs H
already constructed (Fig. 3.1.1).

Proof. Clearly, every graph constructed as described is 2-connected. Conversely, let a 2-connected graph G be given. Then G contains a

© Reinhard Diestel 2017

R. Diestel, *Graph Theory*, Graduate Texts in Mathematics 173,
DOI 10.1007/978-3-662-53622-3_3

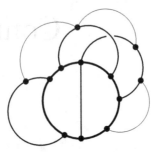

Fig. 3.1.1. The construction of 2-connected graphs

H

cycle, and hence has a maximal subgraph H constructible as above. Since any edge $xy \in E(G) \smallsetminus E(H)$ with $x, y \in H$ would define an H-path, H is an induced subgraph of G. Thus if $H \neq G$, then by the connectedness of G there is an edge vw with $v \in G - H$ and $w \in H$. As G is 2-connected, $G - w$ contains a v–H path P. Then wvP is an H-path in G, and $H \cup wvP$ is a constructible subgraph of G larger than H. This contradicts the maximality of H.   □

Just as an arbitrary graph can be decomposed into its maximal connected subgraphs, or *components*, we can try to decompose a connected graph G into its maximal 2-connected subgraphs. These may not quite be disjoint, and they may not quite cover all of G. However, it is easy to weaken the notion of 'maximal 2-connected subgraph' slightly so that the subgraphs fitting the weaker notion do cover G and are still nearly disjoint. These 'blocks' fit together nicely in a tree-like fashion, which captures precisely the overall structure of G in terms of those blocks.

block

Formally, a *block* is a maximal connected subgraph without a cutvertex.[1] Thus, every block is either a maximal 2-connected subgraph, or a bridge (with its ends), or an isolated vertex. Conversely, every such subgraph is a block. By their maximality, different blocks of G overlap in at most one vertex, which is then a cutvertex of G. Hence every edge of G lies in a unique block, and G is the union of its blocks.

Cycles and bonds are confined to a single block:

[4.6]

Lemma 3.1.2. *Let G be any graph.*

  (i) *The cycles of G are precisely the cycles of its blocks.*

  (ii) *The bonds of G are precisely the minimal cuts of its blocks.*

Proof. (i) Any cycle in G is a connected subgraph without a cutvertex, and hence lies in some maximal such subgraph. By definition, this is a block of G.

(ii) The proof follows easily by repeated application of the following observation. Consider any cut in G. Let xy be one of its edges, and B

---

[1] . . . of the subgraph; it may contain cutvertices of G.

the block containing it. By the maximality of B in the definition of a block, G contains no B-path. Hence every x–y path of G lies in B, so those edges of our cut that lie in B separate x from y even in G.   □

As every edge lies in a unique block, belonging to a common block is an equivalence relation on the edge set of a graph. This equivalence can be expressed in two other interesting ways:

Lemma 3.1.3. *The following statements are equivalent for distinct* [4.6]
edges e, f *of a graph* G:
  (i) *The edges* e, f *belong to a common block of* G.
  (ii) *The edges* e, f *belong to a common cycle in* G.
  (iii) *The edges* e, f *belong to a common bond of* G.

Proof. (i)→(ii) It clearly suffices to prove that in a 2-connected graph any two 2-sets of vertices can be joined by two disjoint paths. This follows easily by induction based on Proposition 3.1.1.[2]

(ii)→(iii) Deleting e and f from a cycle $C \ni e, f$ leaves a partition of $V(C)$ into two connected sets. Extend this to a partition into two connected sets of the vertex set of the component of G containing C. (How?) The edges between these sets form a bond of G containing e and f.

(iii)→(i) By Lemma 3.1.2 (ii), two edges can lie in a common bond only if they belong to the same block.   □

Our last lemma on blocks shows how they fit together to form the coarse structure of G. Let A denote the set of cutvertices of G, and $\mathcal{B}$ the set of its blocks. We then have a natural bipartite graph on $A \cup \mathcal{B}$ formed by the edges aB with $a \in B$. This *block graph* of G is shown in Figure 3.1.2.   *block graph*

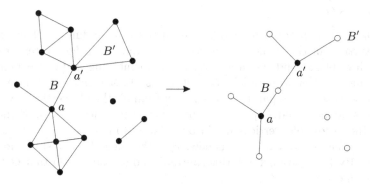

Fig. 3.1.2. A graph and its block graph

Lemma 3.1.4. *The block graph of a connected graph is a tree.*   □

---

[2] See Exercise 5. Note that this is the case $k = 2$ of Menger's theorem (3.3.1).

Lemma 3.1.4 generalizes to graphs of higher connectivity: every $(k-1)$-connected graph has a canonical tree-like decomposition that separates all its 'k-blocks'. See Theorem 12.3.7 for the precise statement, and Exercise 17 in Chapter 12 for the case of $k = 3$.

3.2 The structure of 3-connected graphs

In this section we describe how every 3-connected graph can be obtained from a K^4 by a succession of elementary operations preserving 3-connectedness. We then prove a theorem of Tutte about the algebraic structure of the cycle space of 3-connected graphs; this will play an important role again in Chapter 4.5.

Proposition 3.1.1 describes how the 2-connected graphs can be constructed inductively, starting from a cycle. All the graphs constructed in the process were themselves 2-connected, so the graphs constructible in this way are precisely the 2-connected graphs. We shall now do something similar for 3-connected graphs. We shall prove that every 3-connected graph $G \neq K^4$ can be turned into a smaller 3-connected graph in two ways: by deleting an edge (and suppressing any vertices of degree 2 that may arise), and by contracting an edge. Inverting these processes will give us two independent ways of building all 3-connected graphs from a K^4.

$G \doteq e$ Given an edge e in a graph G, let us write $G \doteq e$ for the *multigraph* obtained from $G - e$ by suppressing any end of e that has degree 2 in $G - e$.[3]

Lemma 3.2.1. *Let e be an edge in a graph G. If $G \doteq e$ is 3-connected, then so is G.*

(1.4.2)   *Proof.* Thinking of G as obtained from $G \doteq e$ by adding e, let us call the vertices of $G \doteq e$ the *old* vertices of G, and any other vertex of G (which will be an end of e) a *new* vertex. Remembering that $G \doteq e$, being 3-connected, has no parallel edges, it is easy to see that, in G, no two vertices x_1, x_2 can separate a new vertex from all the old vertices. So it suffices to show that $\{x_1, x_2\}$ cannot separate two old vertices. If they did, then those old vertices would be separated in $G \doteq e$ by x_1' and x_2', where either $x_i' = x_i$ or, if x_i is new, x_i' is the edge of $G \doteq e$ subdivided by x_i. By Proposition 1.4.2, this contradicts our assumption that $G \doteq e$ is 3-connected. $\qquad\qquad\square$

---

[3] See Chapter 1.10 for the formal definition of suppressing vertices in a multigraph. Recall also that 3-connected multigraphs cannot have multiple edges. Since parallel edges arising when a vertex is suppressed are not deleted, our assumption in Lemma 3.2.1 that the multigraph $G \doteq e$ is 3-connected implies that no parallel edges arise when it is formed from the graph G. Thus $G \doteq e$, too, is in fact a graph.

Lemma 3.2.2. *Every 3-connected graph $G \neq K^4$ has an edge e such that $G \doteq e$ is another 3-connected graph.*

Proof. We start by showing that G contains a TK^4. Let C be a shortest cycle and $P = u \ldots v$ a C-path in G. Then $\mathring{P} \neq \emptyset$ since C is induced, so $G - \{u, v\}$ contains a C–P path Q. Now $C \cup P \cup Q = TK^4$.

As $G \neq K^4$, there is a 3-connected graph $J \not\simeq G$ such that G contains a TJ. Choose J with $\|J\|$ maximum, and then $H = TJ \subseteq G$ with $\|H\|$ maximum. We shall find an edge e such that $G \doteq e \simeq J$. $\qquad J$
$\qquad\qquad\qquad\qquad\qquad\qquad\qquad\qquad\qquad\qquad\qquad\qquad\qquad\qquad H$

Clearly $H \neq G$. Let $P = u \ldots v$ be an H-path in G, chosen if possible so that

$$u \text{ and } v \text{ do not lie on the same (subdivided) edge of } J. \qquad (*) \qquad P = u \ldots v$$

If P does *not* satisfy $(*)$ then $H = J$; for since G is 3-connected, the vertices subdividing an edge of J could be joined by an H-path to a vertex not on the same subdivided edge of J. Our assumption that P does not satisfy $(*)$ thus implies that $uv \in E(J) = E(H)$. Since G has no parallel edges, P has an inner vertex. Now $(H - uv) \cup P$ is another TJ with more edges than H, contradicting our choice of H.

Therefore P does satisfy $(*)$. Suppressing any vertices of degree 2 in $H \cup P$ we obtain a multigraph J' such that $J' \doteq e = J$, where e is the edge corresponding to P. By $(*)$ the edge e is not parallel to an edge of J, so J' is in fact a graph. By Lemma 3.2.1, J' is 3-connected. Hence $J' \simeq G$ by the maximality of J, completing the proof. $\qquad\square$

Theorem 3.2.3. (Tutte 1966)
A graph G is 3-connected if and only if there exists a sequence $G_0, \ldots, G_n$ of graphs such that

   (i) $G_0 = K^4$ *and* $G_n = G$;

   (ii) G_{i+1} *has an edge e such that* $G_i = G_{i+1} \doteq e$, *for every* $i < n$.

Moreover, the graphs in any such sequence are all 3-connected.

Proof. If G is 3-connected, use Lemma 3.2.2 to find $G_n, \ldots, G_0$ in turn. Conversely, if $G_0, \ldots, G_n$ is any sequence of graphs satisfying (i) and (ii), then all these graphs, and in particular $G = G_n$, are 3-connected by Lemma 3.2.1. $\qquad\square$

Theorem 3.2.3 enables us to construct, recursively, the entire class of 3-connected graphs. Starting from K^4, we simply add to every graph already constructed a new edge in every way compatible with (ii): between two already existing vertices, between newly inserted subdividing vertices (not on the same edge), or between one old vertex and one new subdividing vertex.

We now turn to our second method of reducing 3-connected graphs to K^4, by contracting edges. In what follows we only consider graphs, not multigraphs.

[4.4.3]

Lemma 3.2.4. *Every 3-connected graph $G \neq K^4$ has an edge e such that G/e is again 3-connected.*

Proof. Suppose there is no such edge e. Then, for every edge $xy \in G$, the graph G/xy contains a separator S of at most 2 vertices. Since $\kappa(G) \geqslant 3$, the contracted vertex v_{xy} of G/xy (see Chapter 1.7) lies in S and $|S| = 2$, i.e. G has a vertex $z \notin \{x, y\}$ such that $\{v_{xy}, z\}$ separates G/xy. Then any two vertices separated by $\{v_{xy}, z\}$ in G/xy are separated in G by $T := \{x, y, z\}$. Since no proper subset of T separates G, every vertex in T has a neighbour in every component C of $G - T$.

We choose the edge xy, the vertex z, and the component C so that $|C|$ is as small as possible, and pick a neighbour v of z in C (Fig. 3.2.1). By assumption, G/zv is again not 3-connected, so again there is a vertex w such that $\{z, v, w\}$ separates G, and as before every vertex in $\{z, v, w\}$ has a neighbour in every component of $G - \{z, v, w\}$.

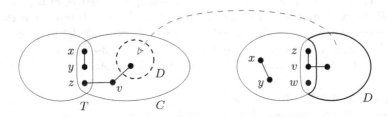

Fig. 3.2.1. Separating vertices in the proof of Lemma 3.2.4

As x and y are adjacent, $G - \{z, v, w\}$ has a component D such that $D \cap \{x, y\} = \emptyset$. Then every neighbour of v in D lies in C (since $v \in C$), so $D \cap C \neq \emptyset$ and hence $D \subsetneq C$ by the choice of D. This contradicts the choice of xy, z and C. $\square$

Theorem 3.2.5. (Tutte 1961)
A graph G is 3-connected if and only if there exists a sequence $G_0, \ldots, G_n$ of graphs with the following two properties:

(i) *$G_0 = K^4$ and $G_n = G$;*

(ii) *G_{i+1} has an edge xy such that $d(x), d(y) \geqslant 3$ and $G_i = G_{i+1}/xy$, for every $i < n$.*

Moreover, the graphs in any such sequence are all 3-connected.

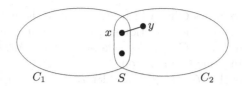

Fig. 3.2.2. The position of $xy \in G_{i+1}$ in the proof of Theorem 3.2.5

Proof. If G is 3-connected, then by Lemma 3.2.4 there is a sequence $G_n, \ldots, G_0$ of 3-connected graphs satisfying (i) and (ii).

Conversely, and to show the final statement of the theorem, let $G_0, \ldots, G_n$ be a sequence of graphs satisfying (i) and (ii); we show that if G_i is 3-connected then so is G_{i+1}, for every $i < n$. Suppose not, let S be a separator of at most 2 vertices in G_{i+1}, and let C_1, C_2 be two components of $G_{i+1} - S$. As x and y are adjacent, we may assume that $\{x, y\} \cap V(C_1) = \emptyset$ (Fig. 3.2.2). Then C_2 contains neither both vertices x, y nor a vertex $v \notin \{x, y\}$: otherwise v_{xy} or v would be separated from C_1 in G_i by at most two vertices, a contradiction. But now C_2 contains only one vertex: either x or y. This contradicts our assumption of $d(x), d(y) \geqslant 3$. $\square$

Like Theorem 3.2.3, Theorem 3.2.5 enables us to construct all 3-connected graphs inductively from K^4, by simple local alterations and without ever leaving the class of 3-connected graphs. Given a 3-connected graph already constructed, pick any vertex v and split it into two adjacent vertices v', v''; then join these to all the former neighbours of v, each to at least two. This is the essential core of a result of Tutte known as his *wheel theorem*.[4]

For larger integers k it is no longer true that in any k-connected graph we can contract an edge so as to obtain another k-connected graph. However, for every k there is a constant n_k such that in every k-connected graph we can either delete or contract an edge so that the resulting graph has no separation of order less than k in which both sides have at least n_k vertices. See the notes.

Theorem 3.2.6. (Tutte 1963) [4.5.2]
The cycle space of a 3-connected graph is generated by its non-separating induced cycles.

Proof. Let G be a fixed 3-connected graph, of order n say. We prove that each of its cycles C is a sum of non-separating induced cycles, applying induction on $k(C) := n - b$, where b denotes the largest order of a component of $G - C$ if there is one, and $b = 0$ if $V(C) = V(G)$. k

---

[4] Graphs of the form $C^n * K^1$ are called *wheels*; thus, K^4 is the smallest wheel. *wheel*

C

B

There are no cycles C for which $k(C) = 0$, so the induction starts. Now let C be given for the induction step. If C is a spanning cycle, it is the sum of two cycles $C_1, C_2 \subseteq C + e$, where e a chord. As $k(C_1), k(C_2) < n = k(C)$, we are home by induction.

Assume now that $G - C \neq \emptyset$, and let B be a largest component of $G - C$. Suppose first that

> $G - B$ contains a C-path $P = u \ldots v$ such that each of the two u–v paths on C has an inner vertex in $N(B)$. $(*)$

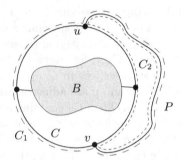

Fig. 3.2.3. C_1 and C_2 are drawn in broken lines

Then C is the sum of the two cycles $C_1, C_2 \subseteq C \cup P$ containing P, and for each of these C_i there is a component of $G - C_i$ that contains B properly (Fig. 3.2.3). Hence $k(C_i) < k(C)$, and we are again home by induction.

Suppose finally that $(*)$ fails. Then every vertex of C sends an edge to B. (Indeed, if not then C contains an $N(B)$-path $Q = x \ldots y$ with $\mathring{Q} \neq \emptyset$. As G is 3-connected, $C - Q \neq \emptyset$, and there is a $\mathring{Q}$–$(C - Q)$ path in $G - \{x, y\}$. Such a path P would satisfy $(*)$.) Since $V(C) = N(B)$, any chord of C would also be a path P as in $(*)$, so C has no chord. Hence unless C itself is induced and non-separating, $G - C$ has a component $B' \neq B$. Let $P = u \ldots v$ be a C-path through B', and let Q be a C–P

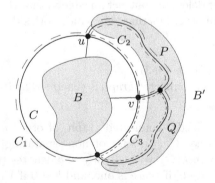

Fig. 3.2.4. Three cycles C_1, C_2, C_3 summing to C, and each missing a vertex of C that sends an edge to B

path in $G - \{u, v\}$. Note that Q too avoids B. Now $C \cup P \cup Q$ contains three cycles C_1, C_2, C_3 summing to C and each missing a vertex of C (Fig. 3.2.4). As every vertex of C sends an edge to B, we therefore have $k(C_i) < k(C)$ for every i, completing the induction step. $\qquad \square$

3.3 Menger's theorem

The following theorem is one of the cornerstones of graph theory.

Theorem 3.3.1. (Menger 1927)
Let $G = (V, E)$ be a graph and $A, B \subseteq V$. Then the minimum number of vertices separating A from B in G is equal to the maximum number of disjoint A–B paths in G.

[3.5.2]
[8.2.5]
[8.4.1]
[12.4.3]
[12.6.3]

We offer three proofs. Whenever G, A, B are given as in the theorem, we denote by $k = k(G, A, B)$ the minimum number of vertices separating A from B in G. Clearly, G cannot contain more than k disjoint A–B paths; our task will be to show that k such paths exist.

k

First proof. We apply induction on $\|G\|$. If G has no edge, then $|A \cap B| = k$ and we have k trivial A–B paths. So we assume that G has an edge $e = xy$. If G has no k disjoint A–B paths, then neither does G/e; here, we count the contracted vertex v_e as an element of A (resp. B) in G/e if in G at least one of x, y lies in A (resp. B). By the induction hypothesis, G/e contains an A–B separator Y of fewer than k vertices. Among these must be the vertex v_e, since otherwise $Y \subseteq V$ would be an A–B separator in G. Then $X := (Y \smallsetminus \{v_e\}) \cup \{x, y\}$ is an A–B separator in G of exactly k vertices.

We now consider the graph $G - e$. Since $x, y \in X$, every A–X separator in $G - e$ is also an A–B separator in G and hence contains at least k vertices. So by induction there are k disjoint A–X paths in $G - e$, and similarly there are k disjoint X–B paths in $G - e$. As X separates A from B, these two path systems do not meet outside X, and can thus be combined to k disjoint A–B paths. $\qquad \square$

Let $\mathcal{P}$ be a set of disjoint A–B paths, and let $\mathcal{Q}$ be another such set. We say that $\mathcal{Q}$ *exceeds* $\mathcal{P}$ if the set of vertices in A that lie on a path in $\mathcal{P}$ is a proper subset of the set of vertices in A that lie on a path in $\mathcal{Q}$, and likewise for B. Then, in particular, $|\mathcal{Q}| \geqslant |\mathcal{P}| + 1$.

exceeds

Second proof. We prove the following stronger statement:

If $\mathcal{P}$ is any set of fewer than k disjoint A–B paths in G, then G contains a set of $|\mathcal{P}| + 1$ disjoint A–B paths exceeding $\mathcal{P}$.

Keeping G and A fixed, we let B vary and apply induction on $|\bigcup \mathcal{P}|$. Let R be an A–B path that avoids the (fewer than k) vertices of B that lie on a path in $\mathcal{P}$. If R avoids all the paths in $\mathcal{P}$, then $\mathcal{P} \cup \{R\}$ exceeds $\mathcal{P}$, as desired. (This will happen when $\mathcal{P} = \emptyset$, so the induction starts.) If not, let x be the last vertex of R that lies on some $P \in \mathcal{P}$ (Fig. 3.3.1).

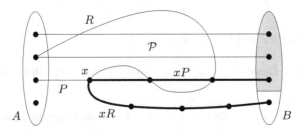

Fig. 3.3.1. Paths in the second proof of Menger's theorem

Put

$$B' := B \cup V(xP \cup xR) \quad \text{and} \quad \mathcal{P}' := (\mathcal{P} \smallsetminus \{P\}) \cup \{Px\}.$$

Then $|\mathcal{P}'| = |\mathcal{P}|$ (but $|\bigcup \mathcal{P}'| < |\bigcup \mathcal{P}|$) and $k(G, A, B') \geqslant k(G, A, B)$, so by the induction hypothesis there is a set $\mathcal{Q}'$ of $|\mathcal{P}'| + 1$ disjoint A–B' paths exceeding $\mathcal{P}'$. Then $\mathcal{Q}'$ contains a path Q ending in x, and a unique path Q' whose last vertex y is not among the last vertices of the paths in $\mathcal{P}'$.

If $y \notin xP$, we let $\mathcal{Q}$ be obtained from $\mathcal{Q}'$ by adding xP to Q, and adding yR to Q' if $y \notin B$. Otherwise $y \in \mathring{x}P$, and we let $\mathcal{Q}$ be obtained from $\mathcal{Q}'$ by adding xR to Q and adding yP to Q'. In all cases $\mathcal{Q}$ exceeds $\mathcal{P}$, as desired. $\qquad \square$

Applied to a bipartite graph, Menger's theorem specializes to the assertion of König's theorem (2.1.1). For our third proof, we shall adapt the alternating path proof of König's theorem to the more general set- up of Theorem 3.3.1. Let again G, A, B be given, and let $\mathcal{P}$ be a set of disjoint A–B paths in G. Let us say that an A–B separator $X \subseteq V$ lies *on* $\mathcal{P}$ if it consists of a choice of exactly one vertex from each path in $\mathcal{P}$. If we can find such a separator X, then clearly $k \leqslant |X| = |\mathcal{P}|$, and Menger's theorem will be proved.

Put

$$V[\mathcal{P}] := \bigcup \{ V(P) \mid P \in \mathcal{P} \}$$

$$E[\mathcal{P}] := \bigcup \{ E(P) \mid P \in \mathcal{P} \}.$$

Let a walk $W = x_0 e_0 x_1 e_1 \ldots e_{n-1} x_n$ in G with $e_i \neq e_j$ for $i \neq j$ be said to *alternate* with respect to $\mathcal{P}$ (Fig. 3.3.2) if it starts in $A \smallsetminus V[\mathcal{P}]$ and the following three conditions hold for all $i < n$ (with $e_{-1} := e_0$ in (iii)):

Margin notes:
$\mathcal{P}$

on

W, x_i, e_i

alternating walk

(i) if $e_i = e \in E[\mathcal{P}]$, then W traverses the edge e backwards, i.e. $x_{i+1} \in P\mathring{x}_i$ for some $P \in \mathcal{P}$;

(ii) if $x_i = x_j$ with $i \neq j$, then $x_i \in V[\mathcal{P}]$;

(iii) if $x_i \in V[\mathcal{P}]$, then $\{e_{i-1}, e_i\} \cap E[\mathcal{P}] \neq \emptyset$.

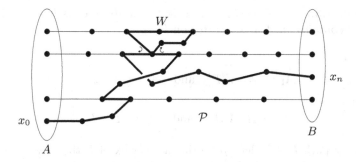

Fig. 3.3.2. An alternating walk from A to B

Note that, by (ii), any vertex outside $V[\mathcal{P}]$ occurs at most once on W. And since the edges e_i of W are all distinct, (iii) implies that any vertex $v \in V[\mathcal{P}]$ occurs at most twice on W. For $v \neq x_n$, this can happen in exactly the following two ways. If $x_i = x_j$ with $0 < i < j < n$, then

either $e_{i-1}, e_j \in E[\mathcal{P}]$ and $e_i, e_{j-1} \notin E[\mathcal{P}]$

or $e_i, e_{j-1} \in E[\mathcal{P}]$ and $e_{i-1}, e_j \notin E[\mathcal{P}]$.

Unless otherwise stated, any use of the word 'alternate' below will refer to our fixed path system $\mathcal{P}$.

The next two lemmas together make up our third proof of Menger's theorem. We state and prove them in a way that makes them reusable in Chapter 8, when we prove Menger's theorem for infinite graphs.

Lemma 3.3.2. *If an alternating walk W as above ends in $B \smallsetminus V[\mathcal{P}]$,* [8.4.5] *then G contains a set of disjoint A–B paths exceeding $\mathcal{P}$.*

Proof. We may assume that W has only its first vertex in $A \smallsetminus V[\mathcal{P}]$ and only its last vertex in $B \smallsetminus V[\mathcal{P}]$. Let H be the graph on $V(G)$ whose edge set is the symmetric difference of $E[\mathcal{P}]$ with $\{e_0, \dots, e_{n-1}\}$. In H, the ends of the paths in $\mathcal{P}$ and of W have degree 1 (or 0, if the path or W is trivial), and all other vertices have degree 0 or 2.

For each vertex $a \in (A \cap V[\mathcal{P}]) \cup \{x_0\}$, therefore, the component of H containing a is a path, $P = v_0 \dots v_k$ say, which starts in a and ends in A or B. Using conditions (i) and (iii), one easily shows by induction on $i = 0, \dots, k-1$ that P traverses each of its edges $e = v_i v_{i+1}$ in the forward direction with respect to $\mathcal{P}$ or W. (Formally: if $e \in P'$ with $P' \in \mathcal{P}$, then $v_i \in P'\mathring{v}_{i+1}$; if $e = e_j \in W$, then $v_i = x_j$ and $v_{i+1} = x_{j+1}$.) Hence, P is an A–B path. (When G is infinite, this last conclusion uses the fact that W meets only finitely many paths in $\mathcal{P}$, and hence every component of H is finite.)

Similarly, for every $b \in (B \cap V[\mathcal{P}]) \cup \{x_n\}$ there is an A–B path in H that ends in b. The set of A–B paths in H therefore exceeds $\mathcal{P}$.   □

[8.4.5]

Lemma 3.3.3. *If no alternating walk W as above ends in $B \setminus V[\mathcal{P}]$, then G contains an A–B separator on $\mathcal{P}$.*

Proof. Let

A_1, A_2

$$A_1 := A \cap V[\mathcal{P}] \quad \text{and} \quad A_2 := A \setminus A_1,$$

and

B_1, B_2

$$B_1 := B \cap V[\mathcal{P}] \quad \text{and} \quad B_2 := B \setminus B_1.$$

x_P

For every path $P \in \mathcal{P}$, let x_P be the last vertex of P that lies on some alternating walk; if no such vertex exists, let x_P be the first vertex of P. Our aim is to show that

X

$$X := \{ x_P \mid P \in \mathcal{P} \}$$

meets every A–B path in G; then X is an A–B separator on $\mathcal{P}$.

Q

Suppose there is an A–B path Q that avoids X. We know that Q meets $V[\mathcal{P}]$, as otherwise it would be an alternating walk ending in B_2. Now the A–$V[\mathcal{P}]$ path in Q is either an alternating walk or consists only of the first vertex of some path in $\mathcal{P}$. Therefore Q also meets the vertex set $V[\mathcal{P}']$ of

$\mathcal{P}'$

$$\mathcal{P}' := \{ Px_P \mid P \in \mathcal{P} \}.$$

y, P, x

Let y be the last vertex of Q in $V[\mathcal{P}']$, say $y \in P \in \mathcal{P}$, and let $x := x_P$.

W

As Q avoids X and hence x, we have $y \in P\mathring{x}$. In particular, $x = x_P$ is not the first vertex of P, and so there is an alternating walk W ending at x. Then $W \cup xPyQ$ is a walk from A_2 to B (Fig. 3.3.3). If this walk alternates and ends in B_2, we have our desired contradiction.

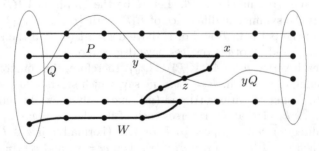

Fig. 3.3.3. Alternating walks in the proof of Lemma 3.3.3.

How could $W \cup xPyQ$ fail to alternate? For example, W might
already use an edge of xPy. But if x' is the first vertex of W on $xP\mathring{y}$, x', W'
then $W' := Wx'Py$ is an alternating walk from A_2 to y. (By Wx' we
mean the initial segment of W ending at the first occurrence of x' on W;
from there, W' follows P back to y.) Even our new walk $W'yQ$ need not
yet alternate: for example, W' might still meet $\mathring{y}Q$. By definition of $\mathcal{P}'$
and W, however, and the choice of y on Q, we have

$$V(W') \cap V[\mathcal{P}] \subseteq V[\mathcal{P}'] \quad \text{and} \quad V(\mathring{y}Q) \cap V[\mathcal{P}'] = \emptyset.$$

Thus, W' and $\mathring{y}Q$ can meet only outside $\mathcal{P}$.

If W' does indeed meet $\mathring{y}Q$, we let z be the first vertex of W' on $\mathring{y}Q$ z
and set $W'' := W'zQ$. Otherwise we set $W'' := W' \cup yQ$. In both cases W''
W'' alternates with respect to $\mathcal{P}'$, because W' does and $\mathring{y}Q$ avoids $V[\mathcal{P}']$.
(W'' satisfies condition (iii) at y in the second case even if y occurs
twice on W', because W'' then contains the entire walk W' and not
just its initial segment $W'y$.) By definition of $\mathcal{P}'$, therefore, W'' avoids
$V[\mathcal{P}] \smallsetminus V[\mathcal{P}']$. Thus W'' also alternates with respect to $\mathcal{P}$ and ends in B_2,
contrary to our assumptions. □

Third proof of Menger's theorem. Let $\mathcal{P}$ contain as many disjoint
A–B paths in G as possible. Then by Lemma 3.3.2, no alternating walk
ends in $B \smallsetminus V[\mathcal{P}]$. By Lemma 3.3.3, this implies that G has an A–B
separator X on $\mathcal{P}$, giving $k \leqslant |X| = |\mathcal{P}|$ as desired. □

A set of a–B paths is called an a–B *fan* if any two of the paths have *fan*
only a in common.

Corollary 3.3.4. *For $B \subseteq V$ and $a \in V \smallsetminus B$, the minimum number of* [10.1.2]
*vertices separating a from B in G is equal to the maximum number of
paths forming an a–B fan in G.*

Proof. Apply Theorem 3.3.1 to $G - a$ with $A := N_G(a)$. □

Corollary 3.3.5. *Let a and b be two distinct vertices of G.*

  (i) *If $ab \notin E$, then the minimum number of vertices separating a
     from b in G is equal to the maximum number of independent a–b
     paths in G.*

  (ii) *The minimum number of edges separating a from b in G is equal
     to the maximum number of edge-disjoint a–b paths in G.*

Proof. (i) Apply Theorem 3.3.1 to $G - \{a, b\}$, with $A := N_G(a)$ and
$B := N_G(b)$.

  (ii) Apply Theorem 3.3.1 to the line graph of G, with $A := E(a)$
and $B := E(b)$. □

[4.2.7]
[6.6.1]
[9.4.2]

Theorem 3.3.6. (Global Version of Menger's Theorem)

(i) *A graph is k-connected if and only if it contains k independent paths between any two vertices.*

(ii) *A graph is k-edge-connected if and only if it contains k edge-disjoint paths between any two vertices.*

Proof. (i) If a graph G contains k independent paths between any two vertices, then $|G| > k$ and G cannot be separated by fewer than k vertices; thus, G is k-connected.

Conversely, suppose that G is k-connected (and, in particular, has more than k vertices) but contains vertices a, b not linked by k independent paths. By Corollary 3.3.5 (i), a and b are adjacent; let $G' := G - ab$. Then G' contains at most $k - 2$ independent a–b paths. By Corollary 3.3.5 (i), we can separate a and b in G' by a set X of at most $k - 2$ vertices. As $|G| > k$, there is at least one further vertex $v \notin X \cup \{a, b\}$ in G. Now X separates v in G' from either a or b—say, from a. But then $X \cup \{b\}$ is a set of at most $k - 1$ vertices separating v from a in G, contradicting the k-connectedness of G.

(ii) follows straight from Corollary 3.3.5 (ii).   □

a, b

G'

X

v

3.4 Mader's theorem

In analogy to Menger's theorem we may consider the following question: given a graph G with an induced subgraph H, up to how many independent H-paths can we find in G?

In this section, we present without proof a deep theorem of Mader, which solves the above problem in a fashion similar to Menger's theorem. Again, the theorem says that an upper bound on the number of such paths that arises naturally from the size of certain separators is indeed attained by some suitable set of paths.

What could such an upper bound look like? Clearly, if $X \subseteq V(G - H)$ and $F \subseteq E(G - H)$ are such that every H-path in G has a vertex or an edge in $X \cup F$, then G cannot contain more than $|X \cup F|$ independent H-paths. Hence, the least cardinality of such a set $X \cup F$ is a natural upper bound for the maximum number of independent H-paths. (Note that every H-path meets $G - H$, because H is induced in G and edges of H do not count as H-paths.)

In contrast to Menger's theorem, this bound can still be improved. The minimality of $|X \cup F|$ implies that no edge in F has an end in X: otherwise this edge would not be needed in the separator. Let $Y :=$ $V(G - H) \setminus X$, and denote by $\mathcal{C}_F$ the set of components of the graph (Y, F). Since every H-path avoiding X contains an edge from F, it has

X

F

$\mathcal{C}_F$

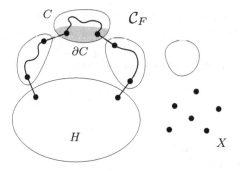

Fig. 3.4.1. An H-path in $G - X$

at least two vertices in ∂C for some $C \in \mathcal{C}_F$, where ∂C denotes the set $\qquad$ ∂C
of vertices in C that send an edge of G to $G - X - C$ (Fig. 3.4.1). The
number of independent H-paths in G is therefore bounded above by

$$M_H(G) := \min \left(|X| + \sum_{C \in \mathcal{C}_F} \lfloor \tfrac{1}{2} |\partial C| \rfloor \right), \qquad\qquad M_H(G)$$

where the minimum is taken over all X and F as described above: $X \subseteq$ $\qquad$ X
$V(G - H)$ and $F \subseteq E(G - H - X)$ such that every H-path in G has a
vertex or an edge in $X \cup F$.

Now Mader's theorem says that this upper bound is always attained
by some set of independent H-paths:

Theorem 3.4.1. (Mader 1978)
*Given a graph G with an induced subgraph H, there are always $M_H(G)$
independent H-paths in G.*

In order to obtain direct analogues to the vertex and edge version
of Menger's theorem, let us consider the two special cases of the above
problem where either F or X is required to be empty. Given an induced
subgraph $H \subseteq G$, we denote by $\kappa_H(G)$ the least cardinality of a vertex $\qquad$ $\kappa_H(G)$
set $X \subseteq V(G - H)$ that meets every H-path in G. Similarly, we let
$\lambda_H(G)$ denote the least cardinality of an edge set $F \subseteq E(G)$ that meets $\qquad$ $\lambda_H(G)$
every H-path in G.

Corollary 3.4.2. *Given a graph G with an induced subgraph H, there
are at least $\tfrac{1}{2}\kappa_H(G)$ independent H-paths and at least $\tfrac{1}{2}\lambda_H(G)$ edge-
disjoint H-paths in G.*

Proof. To prove the first assertion, let k be the maximum number of inde- $\qquad$ k
pendent H-paths in G. By Theorem 3.4.1, there are sets $X \subseteq V(G - H)$
and $F \subseteq E(G - H - X)$ with

$$k = |X| + \sum_{C \in \mathcal{C}_F} \lfloor \tfrac{1}{2} |\partial C| \rfloor$$

such that every H-path in G has a vertex in X or an edge in F. For every $C \in \mathcal{C}_F$ with $\partial C \neq \emptyset$, pick a vertex $v \in \partial C$ and let $Y_C := \partial C \smallsetminus \{v\}$; if $\partial C = \emptyset$, let $Y_C := \emptyset$. Then $\lfloor \frac{1}{2} |\partial C| \rfloor \geqslant \frac{1}{2} |Y_C|$ for all $C \in \mathcal{C}_F$. Moreover, for $Y := \bigcup_{C \in \mathcal{C}_F} Y_C$ every H-path has a vertex in $X \cup Y$. Hence

Y

$$k \geqslant |X| + \sum_{C \in \mathcal{C}_F} \tfrac{1}{2} |Y_C| \geqslant \tfrac{1}{2} |X \cup Y| \geqslant \tfrac{1}{2} \kappa_H(G)$$

as claimed.

The second assertion follows from the first by considering the line graph of G (Exercise 25). $\square$

It may come as a surprise to see that the bounds in Corollary 3.4.2 are best possible (as general bounds): one can find examples for G and H where G contains no more than $\frac{1}{2} \kappa_H(G)$ independent H-paths or no more than $\frac{1}{2} \lambda_H(G)$ edge-disjoint H-paths (Exercises 26 and 27).

3.5 Linking pairs of vertices

Let G be a graph, and let $X \subseteq V(G)$ be a set of vertices. We call X
linked
linked in G if whenever we pick distinct vertices $s_1, \ldots, s_\ell, t_1, \ldots, t_\ell$ in X we can find disjoint paths $P_1, \ldots, P_\ell$ in G such that each P_i links s_i to t_i and has no inner vertex in X. Thus, unlike in Menger's theorem, we are not merely asking for disjoint paths between two *sets* of vertices: we insist that each of these paths shall link a specified pair of endvertices.

If $|G| \geqslant 2k$ and every set X of at most $2k$ vertices is linked in G,
k-linked
then G is *k-linked*. Clearly, this is equivalent to requiring merely that $|G| \geqslant 2k$ and disjoint paths $P_i = s_i \ldots t_i$ exist for every choice of *exactly* $2k$ distinct vertices $s_1, \ldots, s_k, t_1, \ldots, t_k$: just add dummy vertices to X to bring it up to size $2k$. In practice, the latter is easier to prove, because we need not worry about inner vertices in X.

Clearly, every k-linked graph is k-connected. The converse, however, seems far from true: being k-linked is clearly a much stronger property than k-connectedness. Still, we shall prove in this section that we can force a graph to be k-linked by assuming that it is $f(k)$-connected, for some function $f: \mathbb{N} \to \mathbb{N}$. We first borrow a lemma from Chapter 7 to give a nice and simple proof that such a function f exists at all. In the remainder of the section we then prove that f can even be chosen linear.

The basic idea in the simple proof is as follows. If we can prove that G contains a subdivision K of a large complete graph, we can use Menger's theorem to link the vertices of X disjointly to branch vertices of K, and then hope to pair them up as desired through the subdivided edges of K. This requires, of course, that our paths do not hit too many of the subdivided edges before reaching the branch vertices of K.

The lemma saying that large enough connectivity does indeed force the existence of such a complete topological minor K will be proved in Chapter 7.2, where we consider several results of this type. By Theorem 1.4.3 it suffices to assume that G has large average degree:

Lemma 3.5.1. *There is a function $h: \mathbb{N} \to \mathbb{N}$ such that, for every $r \in \mathbb{N}$, every graph of average degree at least $h(r)$ contains K^r as a topological minor.*

Theorem 3.5.2. (Jung 1970; Larman & Mani 1970)
There is a function $f: \mathbb{N} \to \mathbb{N}$ such that every $f(k)$-connected graph is k-linked, for all $k \in \mathbb{N}$.

Proof. We prove the assertion for $f(k) = h(3k) + 2k$, where h is a function as in Lemma 3.5.1. Let G be an $f(k)$-connected graph. Then $d(G) \geqslant \delta(G) \geqslant \kappa(G) \geqslant h(3k)$; let K be a TK^{3k} in G as provided by Lemma 3.5.1, and let U denote its set of branch vertices.

For the proof that G is k-linked, let distinct vertices $s_1, \ldots, s_k$ and $t_1, \ldots, t_k$ be given. By definition of $f(k)$, we have $\kappa(G) \geqslant 2k$. Hence by Menger's theorem (3.3.1), G contains disjoint paths $P_1, \ldots, P_k$, $Q_1, \ldots, Q_k$, such that each P_i starts in s_i, each Q_i starts in t_i, and all these paths end in U but have no inner vertices in U. Let the set $\mathcal{P}$ of these paths be chosen so that their total number of edges outside $E(K)$ is as small as possible.

(3.3.1)

G

K

U

s_i, t_i

P_i, Q_i

$\mathcal{P}$

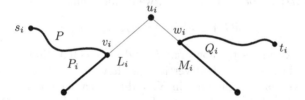

Fig. 3.5.1. Constructing an s_i–t_i path via u_i

Let $u_1, \ldots, u_k$ be those k vertices in U that are not an end of a path in $\mathcal{P}$. For each $i = 1, \ldots, k$, let L_i be the U-path in K (i.e., the subdivided edge of the K^{3k}) from u_i to the end of P_i in U, and let v_i be the first vertex of L_i on any path $P \in \mathcal{P}$. By definition of $\mathcal{P}$, P has no more edges outside $E(K)$ than $Pv_iL_iu_i$ does, so $v_iP = v_iL_i$ and hence $P = P_i$ (Fig. 3.5.1). Similarly, if M_i denotes the U-path in K from u_i to the end of Q_i in U, and w_i denotes the first vertex of M_i on any path in $\mathcal{P}$, then this path is Q_i. Then the paths $s_iP_iv_iL_iu_iM_iw_iQ_it_i$ are disjoint for different i and show that G is k-linked. $\square$

The function h of Lemma 3.5.1 which our proof in Chapter 7.2 will yield will be exponential in r, and will therefore give only an exponential upper bound for the function $f(k)$ in Theorem 3.5.2. As

$2\varepsilon(G) \geqslant \delta(G) \geqslant \kappa(G)$, the following result implies the linear bound of
$f(k) = 16k$:

[7.2.3]
Theorem 3.5.3. (Thomas & Wollan 2005)
*Let G be a graph and $k \in \mathbb{N}$. If G is $2k$-connected and $\varepsilon(G) \geqslant 8k$, then
G is k-linked.*

We begin our proof of Theorem 3.5.3 with a lemma.

Lemma 3.5.4. *Any graph H with $\delta(H) \geqslant 8k \geqslant |H|/2$ has a k-linked
subgraph.*

Proof. If H itself is k-linked there is nothing to show, so suppose not.
Then we can find a set X of $2k$ vertices $s_1, \ldots, s_k, t_1, \ldots, t_k$ that cannot
be linked in H by disjoint paths $P_i = s_i \ldots t_i$. Let $\mathcal{P}$ be a set of as many
such paths as possible, without inner vertices in X and all of length
at most 7. If there are several such sets $\mathcal{P}$, we choose one with $|\bigcup\mathcal{P}|$
minimum. We may assume that $\mathcal{P}$ contains no path from s_1 to t_1. Let
J be the subgraph of H induced by X and all the vertices on the paths
in $\mathcal{P}$, and let $K := H - J$.

Note that each vertex $v \in K$ has at most three neighbours on any
given $P_i \in \mathcal{P}$: if it had four, then replacing the segment uP_iw between
its first and its last neighbour on P_i by the path uvw would reduce $|\bigcup\mathcal{P}|$
and thus contradict our choice of $\mathcal{P}$. So v has at most 3 neighbours in J
for every $i = 1, \ldots, k$, at most $3k$ in total. As $\delta(H) \geqslant 8k$ by assumption,
as well as $|H| \leqslant 16k$ and $|X| = 2k$, we deduce that

$$\delta(K) \geqslant 5k \quad \text{and} \quad |K| \leqslant 14k \,. \tag{1}$$

Our next aim is to show that K is disconnected. Since each of the
paths in $\mathcal{P}$ has at most eight vertices, we have $|J - \{s_1, t_1\}| \leqslant 8(k-1)$.
Therefore s_1 has a neighbour s in K, and t_1 has a neighbour t in K. Put
$S := \{ s' \in K \mid d_K(s, s') \leqslant 2 \}$ and $T := \{ t' \in K \mid d_K(t, t') \leqslant 2 \}$. Since
$H - \bigcup\mathcal{P}$ contains no s_1–t_1 path of length at most 7, we have $S \cap T = \emptyset$,
and there is no S–T edge in K. To prove that K is disconnected, it thus
suffices to show that $V(K) = S \cup T$. But for any vertex $v \in K - (S \cup T)$
the sets $N_K(s)$, $N_K(t)$ and $N_K(v)$ are disjoint and each have size at
least $5k$, contradicting (1).

So K is disconnected; let C be its smaller component. By (1),

$$2\delta(C) \geqslant 2\delta(K) \geqslant 7k + 3k \geqslant \tfrac{1}{2}|K| + 3k \geqslant |C| + 3k \,. \tag{2}$$

We complete the proof by showing that C is k-linked. As $\delta(C) \geqslant 5k$,
we have $|C| \geqslant 2k$. Let Y be a set of at most $2k$ vertices in C. By (2),
every two vertices in Y have at least $3k$ common neighbours, at least k
of which lie outside Y. We can therefore link any desired $\ell \leqslant k$ pairs
of vertices in Y inductively by paths of length 2 whose inner vertex lies
outside Y. $\qquad\square$

Before we launch into the proof of Theorem 3.5.3, let us look at its main ideas. To prove that G is k-linked, we have to consider a given set X of up to $2k$ vertices and show that X is linked in G. Ideally, we would like to use Lemma 3.5.4 to find a linked subgraph L somewhere in G, and then use our assumption of $\kappa(G) \geqslant 2k$ to obtain a set of $|X|$ disjoint X–L paths by Menger's theorem (3.3.1). Then X could be linked via these paths and L, completing the proof.

Unfortunately, we cannot expect to find a subgraph H such that $\delta(H) \geqslant 8k$ and $|H| \leqslant 16k$ (in which L could be found by Lemma 3.5.4); cf. Corollary 11.2.3. However, it is not too difficult to find a minor $H \preccurlyeq G$ that has such a subgraph (Ex. 21, Ch. 7), even so that the vertices of X come to lie in distinct branch sets of H. We may then regard X as a subset of $V(H)$, and Lemma 3.5.4 provides us with a linked subgraph L of H. The only problem now is that H need no longer be $2k$-connected, that is, our assumption of $\kappa(G) \geqslant 2k$ will not ensure that we can link X to L by $|X|$ disjoint paths in H.

And here comes the clever bit of the proof: it relaxes the assumption of $\kappa \geqslant 2k$ to a weaker assumption that does get passed on to H. This weaker assumption is that if we can separate X from another part of G (or H) by fewer than $|X|$ vertices, then this other part must be 'light': roughly, its own value of ε must not exceed $8k$. If X then fails to link to L by $|X|$ disjoint paths, and hence H has a separation $\{A, B\}$ with $X \subseteq A$ and $L \subseteq B$ and $|A \cap B| < |X|$, we know that ε is still at least $8k$ on $H[A]$, because the B-part of H was light.

The idea now is to continue the proof inside $H' := H[A]$ by induction. This still needs some ingenuity, since it is not enough that ε is large on H': we also need that for every low-order separation (A', B') of H' with $X \subseteq A'$ the B'-part is light. That need not be true. But when it fails, we shall be able to use induction on $H'[B']$ to show that $A' \cap B'$ is linked in $H'[B']$, and use this for our proof that X is linked in H.

Given $k \in \mathbb{N}$, a graph G, and $A, B, X \subseteq V(G)$, call the ordered pair (A, B) an *X-separation* of G if $\{A, B\}$ is a proper separation of G of order at most $|X|$ and $X \subseteq A$. An X-separation (A, B) is *small* if $|A \cap B| < |X|$, and *linked* if $A \cap B$ is linked in $G[B]$.

Call a set $U \subseteq V(G)$ *light* in G if $\|U\|^+ \leqslant 8k\,|U|$, where $\|U\|^+$ denotes the number of edges of G with at least one end in U. A set of vertices is *heavy* if it is not light.

Proof of Theorem 3.5.3. We shall prove the following, for fixed $k \in \mathbb{N}$:

> Let $G = (V, E)$ be a graph and $X \subseteq V$ a set of at most $2k$ vertices. If $V \smallsetminus X$ is heavy and for every small X-separation (A, B) the set $B \smallsetminus A$ is light, then X is linked in G. $\quad(*)$

To see that $(*)$ implies the theorem, assume that $\kappa(G) \geqslant 2k$ and $\varepsilon(G) \geqslant 8k$, and let X be a set of exactly $2k$ vertices. Then G has no

X-separation

small/linked

$\|\;\|^+$
light
heavy

k

$G = (V, E)$
X

small X-separation. And $V \smallsetminus X$ is heavy, since

$$\|V \smallsetminus X\|^+ \geqslant \|G\| - \binom{2k}{2} > 8k\,|V| - 16k^2 = 8k\,|V \smallsetminus X|.$$

By $(*)$, X is linked in G, completing the proof that G is k-linked.

We prove $(*)$ by induction on $|G|$, and for each value of $|G|$ by induction on $\|V \smallsetminus X\|^+$. If $|G| = 1$ then X is linked in G. For the induction step, let G and X be given as in $(*)$. We first prove the following:

> We may assume that G has no linked X-separation.   (1)

For our proof of (1), suppose that G has a linked X-separation (A, B). Let us choose one with A minimal, and put $S := A \cap B$.

We first consider the case that $|S| = |X|$. If $G[A]$ contains $|X|$ disjoint X–S paths, then X is linked in G because (A, B) is linked, completing the proof of $(*)$. If not, then by Menger's theorem (3.3.1) $G[A]$ has a small X-separation (A', B') such that $B' \supseteq S$. If we choose this of minimum order, i.e. with $|A' \cap B'|$ minimum, we can link $A' \cap B'$ to S in $G[B']$ by $|A' \cap B'|$ disjoint paths, again by Menger's theorem. But then $(A', B' \cup B)$ is a linked X-separation of G that contradicts the choice of (A, B).

So $|S| < |X|$. Let G' be obtained from $G[A]$ by adding any missing edges on S, so that $G'[S]$ is a complete subgraph of G'. As (A, B) is now a small X-separation, our assumption in $(*)$ says that $B \smallsetminus A$ is light in G. Thus, G' arises from G by deleting $|B \smallsetminus A|$ vertices outside X and at most $8k\,|B \smallsetminus A|$ edges, and possibly adding some edges. As $V \smallsetminus X$ is heavy in G, this implies that

$$A \smallsetminus X \text{ is heavy in } G'.$$

In order to be able to apply the induction hypothesis to G', let us show next that for every small X-separation (A', B') of G' the set $B' \smallsetminus A'$ is light in G'. Suppose not, and choose a counterexample (A', B') with B' minimal. As $G'[S]$ is complete, we have $S \subseteq A'$ or $S \subseteq B'$.

If $S \subseteq A'$ then $B \cap B' \subseteq S \subseteq A'$, so $(A' \cup B, B')$ is a small X-separation of G. Moreover,

$$B' \smallsetminus (A' \cup B) = B' \smallsetminus A',$$

and no edge of $G' - E$ on S is incident with this set (Fig 3.5.2). Our assumption that this set is heavy in G', by the choice of (A', B'), therefore implies that it is heavy also in G. As $(A' \cup B, B')$ is a small X-separation of G, this contradicts our assumptions in $(*)$.

Hence $S \subseteq B'$. By our choice of (A', B'), the graph $G'' := G'[B']$ satisfies the premise of $(*)$ for $X'' := A' \cap B'$. Indeed, $B' \smallsetminus X'' = B' \smallsetminus A'$

(A, B)
S

G'

(A', B')

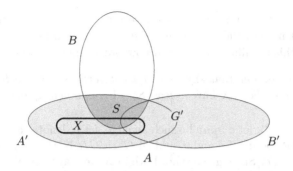

Fig. 3.5.2. If $S \subseteq A'$, then $(A' \cup B, B')$ is an X-separation of G

is heavy, and by the minimality of B' any small X''-separation (A'', B'') of G'' will be such that $B'' \smallsetminus A''$ is light, because $(A' \cup A'', B'')$ will be a small X-separation of G', and $B'' \smallsetminus A'' = B'' \smallsetminus (A' \cup A'')$.

By the induction hypothesis, therefore, X'' is linked in G''. But then X'' is also linked in $G[B' \cup B]$: as S was linked in $G[B]$, we simply replace any edges added on S in the definition of G' by disjoint paths through B (Fig. 3.5.3). But now $(A', B' \cup B)$ is a linked X-separation of G that violates the minimality of A in the choice of (A, B).

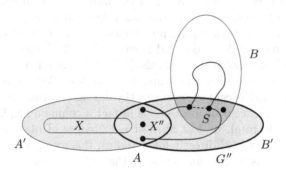

Fig. 3.5.3. If $S \subseteq B'$, then $(A', B' \cup B)$ is linked in G

We have thus shown that G' satisfies the premise of $(*)$ with respect to X. Since $\{A, B\}$ is a proper separation, G' has fewer vertices than G. By the induction hypothesis, therefore, X is linked in G'. Replacing edges of $G' - E$ on S by paths through B as before, we can turn any linkage of X in G' into one in G, completing the proof of $(*)$. This completes the proof of (1).

Our next goal is to show that, by the induction hypothesis, we may assume that G has not only large average degree but even large minimum degree. For our proof that X is linked in G, let $s_1, \ldots, s_\ell, t_1, \ldots, t_\ell$ be the distinct vertices in X which we wish to link by disjoint paths $P_i = s_i \ldots t_i$. Let us add to G any missing edges on X except those of the form $s_i t_i$;　　$G[X]$

as the paths P_i are not allowed to have inner vertices in X, these new edges affect neither the premise nor the conclusion in $(*)$.

After this modification, we can now prove the following:

> We may assume that any two adjacent vertices u, v which do not both lie in X have at least $8k - 1$ common neighbours. $\qquad(2)$

To prove (2), let $e = uv$ be such an edge, let n denote the number of common neighbours of u and v, and let $G' := G/e$ be the graph obtained by contracting e. Since u, v are not both in X we may view X as a subset also of $V' := V(G')$, replacing u or v in X with the contracted vertex v_e if $X \cap \{u, v\} \neq \emptyset$. Our aim is to show that unless $n \geqslant 8k - 1$ as desired in (2), G' satisfies the premise of $(*)$. Then X will be linked in G' by the induction hypothesis, so the desired paths $P_1, \ldots, P_\ell$ exist in G'. If one of them contains v_e, replacing v_e by u or v or uv turns it into a path in G, completing the proof of $(*)$.

In order to show that G' satisfies the premise of $(*)$ with respect to X, let us show first that $V' \smallsetminus X$ is heavy. Since $V \smallsetminus X$ was heavy and $|V' \smallsetminus X| = |V \smallsetminus X| - 1$, it suffices to show that the contraction of e resulted in the loss of at most $8k$ edges incident with a vertex outside X. If u and v are both outside X, then the number of such edges lost is only $n + 1$: one edge at every common neighbour of u and v, as well as e. But if $u \in X$, then $v \notin X$, and we lost all the X–v edges xv of G with $x \neq u$, too: while xv counted towards $\|V \smallsetminus X\|^+$, the edge xv_e lies in $G'[X]$ and does not count towards $\|V' \smallsetminus X\|^+$. If x is not a common neighbour of u and v, then this is an additional loss. But u is adjacent to every $x \in X \smallsetminus \{u\}$ except at most one (by our assumption about $G[X]$), so every such x except at most one is in fact a common neighbour of u and v. Thus in total, we lost at most $n + 2$ edges. Unless $n \geqslant 8k - 1$ (which would prove (2) directly for u and v), this means that we lost at most $8k$ edges, as desired for our proof that $V' \smallsetminus X$ is heavy.

It remains to show that for every small X-separation (A', B') of G' the set $B' \smallsetminus A'$ is light. Let (A', B') be a counterexample, chosen with B' minimal. Then $G'[B']$, as in the proof of (1), satisfies the premise of $(*)$ with respect to $X' := A' \cap B'$. Hence X' is linked in $G'[B']$ by induction. Let A and B be obtained from A' and B' by replacing v_e, where applicable, with both u and v, and put $X'' := A \cap B$. We shall prove that the separation (A, B) of G contradicts our assumption (1).

Let us consider two possible positions of v_e in turn. If v_e lies in $A' \smallsetminus B'$ or $B' \smallsetminus A'$, then $u, v \in A \smallsetminus B$ or $u, v \in B \smallsetminus A$. Then $X'' = X'$ is linked in $G[B]$, because it is linked in $G'[B']$: if v_e occurs on one of the linking paths for X', just replace it by u or v or uv as earlier. This contradicts (1). The other possibility is that $v_e \in X'$. We show that $G[B]$ satisfies the premise of $(*)$ with respect to X''; then X'' will be linked in $G[B]$ by induction, again contradicting (1). Since (A', B') is a

$e = uv$
n
G'
V'

(A', B')

X'

(A, B), X''

small X-separation, we have

$$|X''| \leqslant |X'| + 1 \leqslant |X| \leqslant 2k.$$

Moreover, $B \smallsetminus X'' = B' \smallsetminus A'$ is heavy in G, because it is heavy in G' by the choice of (A', B'). Now consider a small X''-separation (A'', B'') of $G[B]$. Then $(A \cup A'', B'')$ is a small X-separation of G, because $|X''| \leqslant |X|$. Therefore $B'' \smallsetminus A'' = B'' \smallsetminus (A \cup A'')$ is light by the assumption in $(*)$. Hence $G[B]$ does satisfy the premise of $(*)$ for X'', completing the proof of (2).

Using induction by contracting an edge, we have just shown that the vertices in $V \smallsetminus X$ may be assumed to have large degree. Using induction by deleting an edge, we now show that their degrees cannot be too large. Since $(*)$ holds if $V = X$, we may assume that $V \smallsetminus X \neq \emptyset$; let d^* denote $\qquad$ d^* the smallest degree in G of a vertex in $V \smallsetminus X$. Let us prove the following:

> We may assume that $8k \leqslant d^* \leqslant 16k - 1$. $\qquad$ (3)

The lower bound in (3) follows from (2) if we assume that G has no isolated vertex outside X, which we may clearly assume by induction. For the upper bound, let us see what happens if we delete an edge e $\qquad$ $e = uv$ whose ends u, v are not both in X. If $G - e$ satisfies the premise of $(*)$ with respect to X, then X is linked in $G - e$ by induction, and hence in G. If not, then either $V \smallsetminus X$ is light in $G - e$, or $G - e$ has a small X-separation (A, B) such that $B \smallsetminus A$ is heavy. If the latter happens then e must be an $(A \smallsetminus B)$–$(B \smallsetminus A)$ edge: otherwise, (A, B) would be a small X-separation also of G, and $B \smallsetminus A$ would be heavy also in G, in contradiction to our assumptions in $(*)$. But if e is such an edge then any common neighbours of u and v lie in $A \cap B$, so there are fewer than $|X| \leqslant 2k$ such neighbours. This contradicts (2).

So $V \smallsetminus X$ must be light in $G - e$. For G, this yields

$$\|V \smallsetminus X\|^+ \leqslant 8k\,|V \smallsetminus X| + 1. \qquad (4)$$

In order to show that this implies the desired upper bound for d^*, let us estimate the number $f(x)$ of edges that a vertex $x \in X$ sends to $V \smallsetminus X$. $\qquad$ $f(x)$ There must be at least one such edge, xy say, as otherwise $(X, V \smallsetminus \{x\})$ would be a small X-separation of G that contradicts our assumptions in $(*)$. But then, by (2), x and y have at least $8k - 1$ common neighbours, at most $2k - 1$ of which lie in X. Hence $f(x) \geqslant 6k$. As

$$2\,\|V \smallsetminus X\|^+ = \sum_{v \in V \smallsetminus X} d_G(v) + \sum_{x \in X} f(x),$$

an assumption of $d^* \geqslant 16k$ would thus imply that

$$2\,(8k\,|V \smallsetminus X| + 1) \underset{(4)}{\geqslant} 2\,\|V \smallsetminus X\|^+ \geqslant 16k\,|V \smallsetminus X| + 6k\,|X|\,,$$

yielding the contradiction of $2 \geqslant 6k\,|X|$. This completes the proof of (3).

    To complete our proof of $(*)$, pick a vertex $v_0 \in V \smallsetminus X$ of degree d^*, and consider the subgraph H induced in G by v_0 and its neighbours. By (2) we have $\delta(H) \geqslant 8k$, and by (3) and the choice of v_0 we have $|H| \leqslant 16k$. By Lemma 3.5.4, then, H has a k-linked subgraph; let L be its vertex set. By definition of 'k-linked', we have $|L| \geqslant 2k \geqslant |X|$. If G contains $|X|$ disjoint X–L paths, then X is linked in G, as desired. If not, then G has a small X-separation (A, B) with $L \subseteq B$. If we choose (A, B) of minimum order, then $G[B]$ contains $|A \cap B|$ disjoint $(A \cap B)$–L paths by Menger's theorem (3.3.1). But then (A, B) is a linked X-separation that contradicts (1). $\qquad\square$

Exercises

For the first three exercises let G be a graph with vertices a and b, and let $X \subseteq V(G) \smallsetminus \{a, b\}$ be an a–b separator in G.

1.⁻ Show that X is minimal as an a–b separator if and only if every vertex in X has a neighbour in the component C_a of $G - X$ containing a, and another in the component C_b of $G - X$ containing b.

2.⁻ (continued)

    Let $X' \subseteq V(G) \smallsetminus \{a, b\}$ be another a–b separator, and define C_a' and C_b' correspondingly. Show that both

and

$$Y_a := (X \cap C_a') \cup (X \cap X') \cup (X' \cap C_a)$$

$$Y_b := (X \cap C_b') \cup (X \cap X') \cup (X' \cap C_b)$$

separate a from b (Figure 3.6.1).

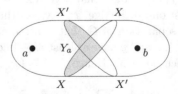

Fig. 3.6.1. The separators in Exercise 2

3.   (continued)

Are Y_a and Y_b minimal a–b separators if X and X' are? Are $|Y_a|$ and $|Y_b|$ minimum for a–b separators if $|X|$ and $|X'|$ are?

4.   Let X and X' be minimal separators in G such that X meets at least two components of $G - X'$. Show that X' meets all the components of $G - X$, and that X meets all the components of $G - X'$.

5.⁻ Deduce the $k = 2$ case of Menger's theorem (3.3.1) from Proposition 3.1.1.

6.⁻ Prove the elementary properties of blocks mentioned after their formal definition.

7.   Show that the block graph of any connected graph is a tree.

8.⁻ Let G be a k-connected graph, and let xy be an edge of G. Show that G/xy is k-connected if and only if $G - \{x, y\}$ is $(k-1)$-connected.

9.   (i) Let e be an edge in a 2-connected graph $G \neq K^3$. Show that either $G - e$ or G/e is again 2-connected.

(ii) Does every 2-connected graph $G \neq K^3$ have an edge e such that G/e is still 2-connected?

10.⁺ Let e be an edge in a 3-connected graph $G \neq K^4$. Show that either $G - e$ or G/e is again 3-connected.

11. Show without using Theorem 3.2.6 that every edge of a 3-connected graph lies on some non-separating induced cycle.

12. Give an inductive proof of Theorem 3.2.6 based on Lemma 3.2.2. You may use the previous exercise.

13.⁺ Give an inductive proof of Theorem 3.2.6 based on Lemma 3.2.4.

14.⁺ Show that every transitive graph G with $\kappa(G) = 2$ is a cycle.

15.⁻ At which point does the first proof of Menger's theorem fail if we assign the contracted vertex v_e to A in G/e only if A in G contains both ends of e, and similarly for B?

16. When one tries to prove an unknown implication $a \Rightarrow b$, it can be dangerous to attempt to prove $a \Rightarrow c \Rightarrow b$ for some assertion c that clearly implies b: if c is too strong, then $a \Rightarrow c$ may fail even if $a \Rightarrow b$ is true. But the first proof of Menger's theorem appears to be doing just that: it proves the seemingly very strong assertion that, given G, A, B and any edge e of G, we can contract or delete e without decreasing $k(G, A, B)$—from which the existence of k disjoint A–B paths follows easily by induction. Can one see already at the outset of the proof that this route will in fact not be dangerous?

17.   (i) Find the error in the following 'simple proof' of Menger's theorem
      (3.3.1). Let X be an A–B separator of minimum size. Denote by G_A
      the subgraph of G induced by X and all the components of $G - X$ that
      meet A, and define G_B correspondingly. By the minimality of X, there
      can be no A–X separator in G_A with fewer than $|X|$ vertices, so G_A
      contains k disjoint A–X paths by induction. Similarly, G_B contains k
      disjoint X–B paths. Together, all these paths form the desired A–B
      paths in G.

      (ii) Fill the gap in the proof of (i) by considering, if possible, a vertex
      or edge outside A and B.

18.   Prove Menger's theorem by induction on $\|G\|$, as follows. Given an
      edge $e = xy$, consider a smallest A–B separator S in $G - e$. Show that
      the induction hypothesis implies a solution for G unless $S \cup \{x\}$ and
      $S \cup \{y\}$ are smallest A–B separators in G. Then show that if choosing
      neither of these separators as X in the previous exercise gives a valid
      proof, there is only one easy case left to do.

19.   Work out the details of the proof of Corollary 3.3.5 (ii).

20.⁻ Show that the least number of edges separating two disjoint sets A, B
      of vertices in a graph G equals the maximum number of edge-disjoint
      A–B paths in G.

21.   Let $k \geqslant 2$. Show that every k-connected graph of order at least $2k$
      contains a cycle of length at least $2k$.

22.   Let $k \geqslant 2$. Show that in a k-connected graph any k vertices lie on a
      common cycle.

23.   Find a subset D of the plane and two infinite subsets $A, B \subseteq D$ such
      that for every finite set $X \subseteq D$ there is an A–B arc in $D \smallsetminus X$ but D
      contains no infinite set of disjoint A–B arcs.

24.⁺ Given a collection $\mathcal{S}$ of disjoint vertex sets in a graph $G = (V, E)$, a path
      in G is an $\mathcal{S}$-path if it joins distinct sets in $\mathcal{S}$ and has no inner vertices
      in $S := \bigcup \mathcal{S}$. Show that the following version of Mader's theorem is
      equivalent to Theorem 3.4.1: The maximum number of disjoint $\mathcal{S}$-paths
      in G is always equal to the minimum value of $|V_0| + \sum_{i=1}^{r} \lfloor \frac{1}{2} |\partial V_i| \rfloor$ taken
      over all partitions $\{V_0, \ldots, V_r\}$ of V such that every $\mathcal{S}$-path in $G - V_0$
      has an edge spanned by some V_i, where ∂V_i is the set of vertices in V_i
      that lie in S or have a neighbour outside $V_0 \cup V_i$.

25.   Derive the edge part of Corollary 3.4.2 from the vertex part.

      (Hint. Consider the H-paths in the graph obtained from the disjoint
      union of H and the line graph $L(G)$ by adding all the edges he such
      that h is a vertex of H and $e \in E(G) \smallsetminus E(H)$ is an edge at h.)

26.⁻ To the disjoint union of the graph $H = \overline{K^{2m+1}}$ with k copies of K^{2m+1}
      add edges joining H bijectively to each of the K^{2m+1}. Show that the
      resulting graph G contains at most $km = \frac{1}{2}\kappa_H(G)$ independent H-
      paths.

27. Find a bipartite graph G, with partition classes A and B say, such that for $H := G[A]$ there are at most $\frac{1}{2}\lambda_H(G)$ edge-disjoint H-paths in G.

28.[+] Derive Tutte's 1-factor theorem (2.2.1) from Mader's theorem.

(Hint. Extend the given graph G to a graph G' by adding, for each vertex $v \in G$, a new vertex v' and joining v' to v. Choose $H \subseteq G'$ so that the 1-factors in G correspond to the large enough sets of independent H-paths in G'.)

29.[−] Show that $2k$-edge-connected graphs are k-edge-linked in the sense that for all distinct vertices $s_1, \ldots, s_k, t_1, \ldots, t_k$ there are edge-disjoint paths $P_i = s_i \ldots t_i$ for $i = 1, \ldots, k$.

30.[−] Show that k-linked graphs are $(2k-1)$-connected. Are they even $2k$-connected?

31. For every $k \in \mathbb{N}$ find an $\ell = \ell(k)$, as large as possible, such that not every ℓ-connected graph is k-linked.

32. Show that if G is k-linked and $s_1, \ldots, s_k, t_1, \ldots, t_k$ are not necessarily distinct vertices such that $s_i \neq t_i$ for all i, then G contains independent paths $P_i = s_i \ldots t_i$ for $i = 1, \ldots, k$.

33. Go through the proof of Theorem 3.5.3 monitoring the use of $\|V \setminus X\|^+$. How would the proof fail if $\|G[V \setminus X]\|$ was used instead? Which arguments would become simpler?

34. Use Theorem 3.5.3 to show that the function h in Lemma 3.5.1 can be chosen as $h(r) = cr^2$, for some $c \in \mathbb{N}$.

Notes

Although connectivity theorems are doubtless among the most natural, and also the most applicable, results in graph theory, there is still no monograph on this subject. The most comprehensive sources to date are A. Schrijver, *Combinatorial optimization*, Springer 2003, and A. Frank, *Connections in combinatorial optimization*, Oxford University Press 2011. Some areas are covered in B. Bollobás, *Extremal Graph Theory*, Academic Press 1978, in R. Halin, *Graphentheorie*, Wissenschaftliche Buchgesellschaft 1980, and in A. Frank's chapter of the *Handbook of Combinatorics* (R.L. Graham, M. Grötschel & L. Lovász, eds.), North-Holland 1995. A survey specifically of techniques and results on minimally k-connected graphs (see below) is given by W. Mader, On vertices of degree n in minimally n-connected graphs and digraphs, in (D. Miklós, V.T. Sós & T. Szőnyi, eds.) *Paul Erdős is 80*, Vol. 2, Proc. Colloq. Math. Soc. János Bolyai, Budapest 1996.

Theorem 3.2.3 is often attributed to Barnette and Grünbaum (1969). It can also be extracted from W.T. Tutte, *Connectivity in graphs*, Oxford University Press 1966. Tutte's *wheel theorem*, proved in W.T. Tutte, A theory of 3-connected graphs, *Nederl. Akad. Wet. Proc. Ser. A* **64** (1961), 441–455, differs from our Theorem 3.2.5 as follows. As an alternative to the contraction of an edge in the reduction step, the wheel theorem also allows its deletion. Of edges to be contracted, however, it requires that they do not lie in any triangle. The starting set for the construction of all 3-connected graphs therefore consists of all wheels, not only K^4.

Tutte's wheel theorem has been extended to 3-connected graphs H other than K^4: from any 3-connected graph $G \succcurlyeq H$ that is not a wheel we can obtain H by contracting or deleting edges step by step, remaining 3-connected at every step. (As in Tutte's theorem, one is not allowed to contract edges that lie in a triangle.) This was proved by S. Negami, A characterization of 3-connected graphs containing a given graph, *J. Comb. Theory B* **32** (1982), 69–74. It also follows from an earlier theorem of Seymour on matroid decompositions, and is sometimes called Seymour's *splitter theorem* for 3-connected graphs.

The fact, mentioned after the proof of Theorem 3.2.5, that in k-connected graphs we can either delete or contract an edge so that the resulting graphs have no separations of order $< k$ with unboundedly large sides follows from Lemma 3.1 of J. Geelen, B. Gerards, N. Robertson and G. Whittle, On the excluded minors for the matroids of branch-width k, *J. Comb. Theory B* **88** (2003), 261–265.

Our proof of Theorem 3.2.6 is the original from W.T. Tutte, How to draw a graph, *Proc. Lond. Math. Soc.* **13** (1963), 743–767. Alternative proofs are indicated in Exercises 12 and 13.

An approach to the study of connectivity not touched upon in this chapter is the investigation of *minimal* k-connected graphs, those that lose their k-connectedness as soon as we delete an edge. Like all k-connected graphs, these have minimum degree at least k, and by a fundamental result of Halin (1969), their minimum degree is exactly k. The existence of a vertex of small degree can be particularly useful in induction proofs about k-connected graphs. Halin's theorem was the starting point for a series of more and more sophisticated studies of minimal k-connected graphs; see the books of Bollobás and Halin cited above, and in particular Mader's survey.

Menger's theorem goes back to his paper, Zur allgemeinen Kurventheorie, *Fundamenta Math.* **10** (1927), 96–115. It is probably the most-used classical result in graph theory. Our first proof is extracted from Halin's book. The second is due to T. Böhme, F. Göring and J. Harant, Menger's theorem, *J. Graph Theory* **37** (2001), 35–36, the third to T. Grünwald (later Gallai), Ein neuer Beweis eines Mengerschen Satzes, *J. Lond. Math. Soc.* **13** (1938), 188–192. A fourth proof is sketched in Exercise 18, and in Chapter 6 we shall obtain a fifth proof as an application of a theorem about network flows (Ch. 6, Ex. 3.) The global version of Menger's theorem (Theorem 3.3.6) was first stated and proved by Whitney (1932). Topological generalizations of Menger's theorem have been known since the 1930s; see C. Thomassen and A. Vella, Graph-like continua and Menger's theorem, *Combinatorica* **28** (2008), 595–623.

Mader's Theorem 3.4.1 is taken from W. Mader, Über die Maximalzahl kreuzungsfreier H-Wege, *Arch. Math.* **31** (1978), 387–402; our formulation is easily seen to be equivalent to the original. The shortest proof known to me is given by Schrijver in his book. The theorem may be viewed as a common generalization of Menger's theorem and Tutte's 1-factor theorem (Exercise 28).

Theorem 3.5.3 is due to R. Thomas and P. Wollan, An improved linear bound for graph linkages, *Eur. J. Comb.* **26** (2005), 309–324. Using a more involved version of Lemma 3.5.4, they prove that $2k$-connected graphs even with only $\varepsilon \geqslant 5k$ must be k-linked. And for graphs of large enough girth the condition on ε can be dropped altogether: as shown by W. Mader, Topological subgraphs in graphs of large girth, *Combinatorica* **18** (1998), 405–412, such graphs are k-linked as soon as they are $2k$-connected, which is best possible. (Mader assumes a lower bound on the girth that depends on k, but this is not necessary; see D. Kühn & D. Osthus, Topological minors in graphs of large girth, *J. Comb. Theory B* **86** (2002), 364–380.) In fact, for every $s \in \mathbb{N}$ there exists a k_s such that if $G \not\supseteq K_{s,s}$ and $\kappa(G) \geqslant 2k \geqslant k_s$ then G is k-linked; see D. Kühn & D. Osthus, Complete minors in $K_{s,s}$-free graphs, *Combinatorica* **25** (2005) 49–64.

4   Planar Graphs

When we draw a graph on a piece of paper, we naturally try to do this as transparently as possible. One obvious way to limit the mess created by all the lines is to avoid intersections. For example, we may ask if we can draw the graph in such a way that no two edges meet in a point other than a common end.

Graphs drawn in this way are called *plane graphs*; abstract graphs that *can* be drawn in this way are called *planar*. In this chapter we study both plane and planar graphs, as well as the relationship between the two: the question of how an abstract planar graph might be drawn in fundamentally different ways. After collecting together in Section 4.1 the few basic topological facts that will enable us later to prove all results rigorously without too much technical ado, we begin in Section 4.2 by studying the structural properties of plane graphs. In Section 4.3, we investigate how two drawings of the same graph can differ. The main result of that section is that 3-connected planar graphs have essentially only one drawing, in some very strong and natural topological sense. The next two sections are devoted to the proofs of all the classical planarity criteria, conditions telling us when an abstract graph is planar. We complete the chapter with a section on *plane duality*, a notion with fascinating links to algebraic, colouring, and flow properties of graphs (Chapters 1.9 and 6.5).

The traditional notion of a graph drawing is that its vertices are represented by points in the Euclidean plane, its edges are represented by curves between these points, and different curves meet only in common endpoints. To avoid unnecessary topological complication, however, we shall only consider curves that are piecewise linear; it is not difficult to show that any drawing can be straightened out in this way, so the two notions come to the same thing.

© Reinhard Diestel 2017
R. Diestel, *Graph Theory*, Graduate Texts in Mathematics 173,
DOI 10.1007/978-3-662-53622-3_4

4.1 Topological prerequisites

In this section we briefly review some basic topological definitions and facts needed later. All these facts have (by now) easy and well-known proofs; see the notes for sources. Since those proofs contain no graph theory, we do not repeat them here: indeed our aim is to collect precisely those topological facts that we need but do *not* want to prove. Later, all proofs will follow strictly from the definitions and facts stated here (and be guided by but not rely on geometric intuition), so the material presented now will help to keep elementary topological arguments in those proofs to a minimum.

polygon

A *straight line segment* in the Euclidean plane is a subset of $\mathbb{R}^2$ that has the form $\{\, p + \lambda(q - p) \mid 0 \leqslant \lambda \leqslant 1 \,\}$ for distinct points $p, q \in \mathbb{R}^2$. A *polygon* is a subset of $\mathbb{R}^2$ which is the union of finitely many straight line segments and is homeomorphic to the unit circle S^1, the set of points in $\mathbb{R}^2$ at distance 1 from the origin. Here, as later, any subset of a topological space is assumed to carry the subspace topology. A *polygonal arc* is a subset of $\mathbb{R}^2$ which is the union of finitely many straight line segments and is homeomorphic to the closed unit interval $[0, 1]$. The images of 0 and of 1 under such a homeomorphism are the *endpoints* of this polygonal arc, which *links* them and runs *between* them. Instead of 'polygonal

arc
$\mathring{P}$, $\mathring{e}$

arc' we shall simply say *arc* in this chapter. If P is an arc between x and y, we denote the point set $P \smallsetminus \{x, y\}$, the *interior* of P, by $\mathring{P}$. As continuous images of $[0, 1]$, arcs, and finite unions of arcs, are compact, and hence closed in $\mathbb{R}^2$. Their complements in $\mathbb{R}^2$, therefore, are open.

region
separate
frontier

Let $O \subseteq \mathbb{R}^2$ be any open set. Being linked by an arc in O defines an equivalence relation on O. The corresponding equivalence classes are again open; they are the *regions* of O. A closed set $X \subseteq \mathbb{R}^2$ is said to *separate* a region O' of O if $O' \smallsetminus X$ has more than one region. The *frontier* of a set $X \subseteq \mathbb{R}^2$ is the set Y of all points $y \in \mathbb{R}^2$ such that every neighbourhood of y meets both X and $\mathbb{R}^2 \smallsetminus X$. Note that if X is open then its frontier lies in $\mathbb{R}^2 \smallsetminus X$.

The frontier of a region O of $\mathbb{R}^2 \smallsetminus X$, where X is a finite union of points and arcs, has two important properties. The first is accessibility: if $x \in X$ lies on the frontier of O, then x can be linked to some point in O by a straight line segment whose interior lies wholly inside O. As a consequence, any two points on the frontier of O can be linked by an arc whose interior lies in O (why?). The second notable property of the frontier of O is that it separates O from the rest of $\mathbb{R}^2$. Indeed, if $\varphi \colon [0, 1] \to P \subseteq \mathbb{R}^2$ is continuous, with $\varphi(0) \in O$ and $\varphi(1) \notin O$, then P meets the frontier of O at least in the point $\varphi(y)$ for $y := \inf \{\, x \mid \varphi(x) \notin O \,\}$, the *first point* of P in $\mathbb{R}^2 \smallsetminus O$.

Theorem 4.1.1. (Jordan Curve Theorem for Polygons)
For every polygon $P \subseteq \mathbb{R}^2$, the set $\mathbb{R}^2 \smallsetminus P$ has exactly two regions. Each of these has the entire polygon P as its frontier.

With the help of Theorem 4.1.1, it is not difficult to prove the following lemma.

Lemma 4.1.2. *Let P_1, P_2, P_3 be three arcs, between the same two endpoints but otherwise disjoint.*

(i) $\mathbb{R}^2 \smallsetminus (P_1 \cup P_2 \cup P_3)$ *has exactly three regions, with frontiers $P_1 \cup P_2$, $P_2 \cup P_3$ and $P_1 \cup P_3$.*

(ii) *If P is an arc between a point in $\mathring{P}_1$ and a point in $\mathring{P}_3$ whose interior lies in the region of $\mathbb{R}^2 \smallsetminus (P_1 \cup P_3)$ that contains $\mathring{P}_2$, then $\mathring{P} \cap \mathring{P}_2 \neq \emptyset$.*

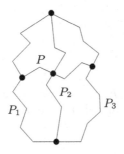

Fig. 4.1.1. The arcs in Lemma 4.1.2 (ii)

Our next lemma complements the Jordan curve theorem by saying that an arc does *not* separate the plane. For easier application later, we phrase this a little more generally:

Lemma 4.1.3. *Let $X_1, X_2 \subseteq \mathbb{R}^2$ be disjoint sets, each the union of finitely many points and arcs, and let P be an arc between a point in X_1 and one in X_2 whose interior lies in a region O of $\mathbb{R}^2 \smallsetminus (X_1 \cup X_2)$. Then $O \smallsetminus P$ is a region of $\mathbb{R}^2 \smallsetminus (X_1 \cup P \cup X_2)$.*

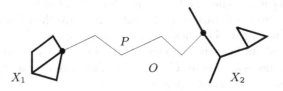

Fig. 4.1.2. P does not separate the region O of $\mathbb{R}^2 \smallsetminus (X_1 \cup X_2)$

It remains to introduce a few terms and facts that will be used only once, when we consider notions of equivalence for graph drawings in Chapter 4.3.

S^n As usual, we denote by S^n the n-dimensional sphere, the set of points in $\mathbb{R}^{n+1}$ at distance 1 from the origin. The 2-sphere minus its 'north pole' $(0,0,1)$ is homeomorphic to the plane; let us choose a fixed

π such homeomorphism $\pi\colon S^2 \smallsetminus \{(0,0,1)\} \to \mathbb{R}^2$ (for example, stereographic projection). If $P \subseteq \mathbb{R}^2$ is a polygon and O is the bounded region of $\mathbb{R}^2 \smallsetminus P$, let us call $C := \pi^{-1}(P)$ a *circle on* S^2, and the sets $\pi^{-1}(O)$ and $S^2 \smallsetminus \pi^{-1}(P \cup O)$ the *regions* of $S^2 \smallsetminus C$.

Our last tool is the theorem of Jordan and Schoenflies, again adapted slightly for our purposes:

[4.3.1]

Theorem 4.1.4. *Let* $\varphi\colon C_1 \to C_2$ *be a homeomorphism between two circles on* S^2, *let* O_1 *be a region of* $S^2 \smallsetminus C_1$, *and let* O_2 *be a region of* $S^2 \smallsetminus C_2$. *Then* φ *can be extended to a homeomorphism* $C_1 \cup O_1 \to C_2 \cup O_2$.

4.2 Plane graphs

plane
graph   A *plane graph* is a pair (V, E) of finite sets with the following properties (the elements of V are again called *vertices*, those of E *edges*):

(i) $V \subseteq \mathbb{R}^2$;

(ii) every edge is an arc between two vertices;

(iii) different edges have different sets of endpoints;

(iv) the interior of an edge contains no vertex and no point of any other edge.

A plane graph (V, E) defines a graph G on V in a natural way. As long as no confusion can arise, we shall use the name G of this abstract graph also for the plane graph (V, E), or for the point set $V \cup \bigcup E$; similar notational conventions will be used for abstract versus plane edges, for subgraphs, and so on.[1]

faces   When G is a plane graph, we call the regions of $\mathbb{R}^2 \smallsetminus G$ the *faces* of G. These are open subsets of $\mathbb{R}^2$ and hence have their frontiers in G. Since G is bounded—i.e., lies inside some sufficiently large disc D— exactly one of its faces is unbounded, the face that contains $\mathbb{R}^2 \smallsetminus D$. This face is the *outer face* of G; the other faces are its *inner faces*. We

$F(G)$ denote the set of faces of G by $F(G)$.

The faces of plane graphs and their subgraphs are related in the obvious way:

---

[1] However, we shall continue to use $\smallsetminus$ for differences of point sets and $-$ for graph differences—which may help a little to keep the two apart.

Lemma 4.2.1. *Let G be a plane graph, $f \in F(G)$ a face, and $H \subseteq G$ a subgraph.*

(i) *H has a face f' containing f.*

(ii) *If the frontier of f lies in H, then $f' = f$.*

Proof. (i) Clearly, the points in f are equivalent also in $\mathbb{R}^2 \setminus H$; let f' be the equivalence class of $\mathbb{R}^2 \setminus H$ containing them.

(ii) Recall from Section 4.1 that any arc between f and $f' \setminus f$ meets the frontier X of f. If $f' \setminus f \neq \emptyset$ then there is such an arc inside f', whose points in X do not lie in H. Hence $X \not\subseteq H$. $\square$

In order to lay the foundations for the (easy but) rigorous introduction to plane graphs that this section aims to provide, let us descend once now into the realm of truly elementary topology of the plane, and prove what seems entirely obvious:[2] that the frontier of a face of a plane graph G is always a subgraph of G—not, say, half an edge.

The following lemma states this formally, together with two similarly 'obvious' properties of plane graphs:

[4.5.1]
[4.5.2]
[12.7.4]

Lemma 4.2.2. *Let G be a plane graph and e an edge of G.*

(i) *If X is the frontier of a face of G, then either $e \subseteq X$ or $X \cap \mathring{e} = \emptyset$.*

(ii) *If e lies on a cycle $C \subseteq G$, then e lies on the frontier of exactly two faces of G, and these are contained in distinct faces of C.*

(iii) *If e lies on no cycle, then e lies on the frontier of exactly one face of G.*

Proof. We prove all three assertions together. Let us start by considering one point $x_0 \in \mathring{e}$. We show that x_0 lies on the frontier of either exactly two faces or exactly one, according as e lies on a cycle in G or not. We then show that every other point in $\mathring{e}$ lies on the frontier of exactly the same faces as x_0. Then the endpoints of e will also lie on the frontier of these faces—simply because every neighbourhood of an endpoint of e is also the neighbourhood of an inner point of e.

(4.1.1)
(4.1.3)

Since $G \setminus \mathring{e}$ is compact, we can find around every point $x \in \mathring{e}$ an open disc D_x that meets G only in those (one or two) straight line segments that contain x.

D_x

Let us pick an inner point x_0 from a straight line segment $S \subseteq e$. Then $D_{x_0} \cap G = D_{x_0} \cap S$, so $D_{x_0} \setminus G$ is the union of two open half-discs. Since these half-discs do not meet G, they each lie in a face of G. Let

x_0, S

---

[2] Note that even the best intuition can only ever be 'accurate', i.e., coincide with what the technical definitions imply, inasmuch as those definitions do indeed formalize what is intuitively intended. Given the complexity of definitions in elementary topology, this can hardly be taken for granted.

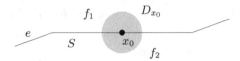

Fig. 4.2.1. Faces f_1, f_2 of G in the proof of Lemma 4.2.2

f_1, f_2 us denote these faces by f_1 and f_2; they are the only faces of G with x_0 on their frontier, and they may coincide (Fig. 4.2.1).

If e lies on a cycle $C \subseteq G$, then D_{x_0} meets both faces of C (Theorem 4.1.1). Since f_1 and f_2 are contained in faces of C by Lemma 4.2.1, this implies $f_1 \neq f_2$. If e does not lie on any cycle, then e is a bridge and thus links two disjoint point sets X_1, X_2 as in Lemma 4.1.3, with $X_1 \cup X_2 = G \smallsetminus \mathring{e}$. Clearly, $f_1 \cup \mathring{e} \cup f_2$ is the subset of a face f of $G - e$. By Lemma 4.1.3, $f \smallsetminus \mathring{e}$ is a face of G, while $f_1, f_2 \subseteq f \smallsetminus \mathring{e}$ by definition of f. Since f_1 and f_2 are also faces of G, this implies $f_1 = f \smallsetminus \mathring{e} = f_2$.

x_1
P
$D_0, \ldots, D_n$

y
z

     Now consider any other point $x_1 \in \mathring{e}$. Let P be the arc from x_0 to x_1 contained in e. Since P is compact, finitely many of the discs D_x with $x \in P$ cover P. Let us enumerate these discs as $D_0, \ldots, D_n$ in the natural order of their centres along P; adding D_{x_0} or D_{x_1} as necessary, we may assume that $D_0 = D_{x_0}$ and $D_n = D_{x_1}$. By induction on n, one easily proves that every point $y \in D_n \smallsetminus e$ can be linked by an arc inside $(D_0 \cup \ldots \cup D_n) \smallsetminus e$ to a point $z \in D_0 \smallsetminus e$ (Fig. 4.2.2); then y and z are equivalent in $\mathbb{R}^2 \smallsetminus G$. Hence, every point of $D_n \smallsetminus e$ lies in f_1 or in f_2, so x_1 cannot lie on the frontier of any other face of G. Since both half-discs of $D_0 \smallsetminus e$ can be linked to $D_n \smallsetminus e$ in this way (swap the roles of D_0 and D_n), we find that x_1 lies on the frontier of both f_1 and f_2. $\quad\square$

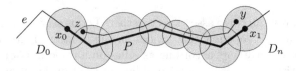

Fig. 4.2.2. An arc from y to D_0, close to P

Corollary 4.2.3. *The frontier of a face is always the point set of a subgraph.* $\quad\square$

boundary
$G[f]$

     The subgraph of G whose point set is the frontier of a face f is said to *bound* f and is called its *boundary*; we denote it by $G[f]$. A face is said to be *incident* with the vertices and edges of its boundary. By Lemma 4.2.1 (ii), every face of G is also a face of its boundary; we shall use this fact frequently in the proofs to come.

[4.6.1]   **Proposition 4.2.4.** *A plane forest has exactly one face.*

(4.1.3)   *Proof.* Use induction on the number of edges and Lemma 4.1.3. $\quad\square$

With just one exception, different faces of a plane graph have different boundaries:

Lemma 4.2.5. *If a plane graph has different faces with the same boundary, then the graph is a cycle.*   [4.3.1]

Proof. Let G be a plane graph, and let $H \subseteq G$ be the boundary of distinct   (4.1.1)
faces f_1, f_2 of G. Since f_1 and f_2 are also faces of H, Proposition 4.2.4
implies that H contains a cycle C. By Lemma 4.2.2 (ii), f_1 and f_2 are
contained in different faces of C. Since f_1 and f_2 both have all of H
as boundary, this implies that $H = C$: any further vertex or edge of H
would lie in one of the faces of C and hence not on the boundary of the
other. Thus, f_1 and f_2 are distinct faces of C. As C has only two faces,
it follows that $f_1 \cup C \cup f_2 = \mathbb{R}^2$ and hence $G = C$.   □

                                                                                  [4.3.1]
Proposition 4.2.6. *In a 2-connected plane graph, every face is bounded*   [4.4.3]
by a cycle.   [4.5.1]
                                                                                  [4.5.2]
Proof. Let f be a face in a 2-connected plane graph G. We show by   (3.1.1)
induction on $\|G\|$ that $G[f]$ is a cycle. If G is itself a cycle, this holds   (4.1.1)
by Theorem 4.1.1; we therefore assume that G is not a cycle.   (4.1.2)
   By Proposition 3.1.1, there exist a 2-connected plane graph $H \subseteq G$ H
and a plane H-path P such that $G = H \cup P$. The interior of P lies in a P
face f' of H, which by the induction hypothesis is bounded by a cycle C. f', C
   If $G[f] \subseteq H$, then f is also a face of H (Lemma 4.2.1 (ii)), and we are
home by the induction hypothesis. If $G[f] \not\subseteq H$, then $G[f]$ meets $P \smallsetminus H$,
so $f \subseteq f'$ and $G[f] \subseteq C \cup P$ (why?). By Lemma 4.2.1 (ii), then, f is a
face of $C \cup P$ and hence bounded by a cycle (Lemma 4.1.2 (i)).   □

   In a 3-connected graph, we can identify the face boundaries among
the other cycles in purely combinatorial terms:

Proposition 4.2.7. *The face boundaries in a 3-connected plane graph*   [4.3.2]
are precisely its non-separating induced cycles.   [4.5.2]

                                                                                  (3.3.6)
Proof. Let G be a 3-connected plane graph, and let $C \subseteq G$. If C is a non-   (4.1.1)
separating induced cycle, then by the Jordan curve theorem its two faces   (4.1.2)
cannot both contain points of $G \smallsetminus C$. Therefore it bounds a face of G.
   Conversely, suppose that C bounds a face f. By Proposition 4.2.6, C, f
C is a cycle. If C has a chord $e = xy$, then the components of $C - \{x, y\}$
are linked by a C-path in G, because G is 3-connected. This path and
e both run through the other face of C (not f) but do not intersect,
a contradiction to Lemma 4.1.2 (ii).
   It remains to show that C does not separate any two vertices $x, y \in$
$G - C$. By Menger's theorem (3.3.6), x and y are linked in G by three
independent paths. Clearly, f lies inside a face of their union, and by
Lemma 4.1.2 (i) this face is bounded by only two of the paths. The third
therefore avoids f and its boundary C.   □

maximal
plane graph   A plane graph G is called *maximally plane*, or just *maximal*, if we
                  cannot add a new edge to form a plane graph $G' \supsetneq G$ with $V(G') = V(G)$.
plane   We call G a *plane triangulation* if every face of G (including the outer
triangulation face) is bounded by a triangle.

[4.4.1]
[5.4.2]   **Proposition 4.2.8.** *A plane graph of order at least 3 is maximally plane*
                  if and only if it is a plane triangulation.

(4.1.2)   *Proof.* Let G be a plane graph of order at least 3. It is easy to see that
                  if every face of G is bounded by a triangle, then G is maximally plane.
                  Indeed, any additional edge e would have its interior inside a face of G
                  and its ends on the boundary of that face. Hence these ends are already
                  adjacent in G, so $G \cup e$ cannot satisfy condition (iii) in the definition of
                  a plane graph.

f Conversely, assume that G is maximally plane and let $f \in F(G)$ be
H a face; let us write $H := G[f]$. Since G is maximal as a plane graph,
                  $G[H]$ is complete: any two vertices of H that are not already adjacent
                  in G could be linked by an arc through f, extending G to a larger plane
n graph. Thus $G[H] = K^n$ for some n—but we do not know yet which
                  edges of $G[H]$ lie in H.
                          Let us show first that H contains a cycle. If not, then $G \setminus H \neq \emptyset$:
                  by $G \supseteq K^n$ if $n \geqslant 3$, or else by $|G| \geqslant 3$. On the other hand we have
                  $f \cup H = \mathbb{R}^2$ by Proposition 4.2.4 and hence $G = H$, a contradiction.
                          Since H contains a cycle, it suffices to show that $n \leqslant 3$: then $H = K^3$
C, v_i as claimed. Suppose $n \geqslant 4$, and let $C = v_1 v_2 v_3 v_4 v_1$ be a cycle in $G[H]$
                  $(= K^n)$. By $C \subseteq G$, our face f is contained in a face f_C of C; let f'_C
f_C, f'_C be the other face of C. Since the vertices v_1 and v_3 lie on the boundary
                  of f, they can be linked by an arc whose interior lies in f_C and avoids G.
                  Hence by Lemma 4.1.2 (ii), the plane edge $v_2 v_4$ of $G[H]$ runs through f'_C
                  rather than f_C (Fig. 4.2.3). Analogously, since $v_2, v_4 \in G[f]$, the edge
                  $v_1 v_3$ runs through f'_C. But the edges $v_1 v_3$ and $v_2 v_4$ are disjoint, so this
                  contradicts Lemma 4.1.2 (ii). $\square$

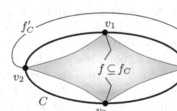

Fig. 4.2.3. The edge $v_2 v_4$ of G runs through the face f'_C

        The following classic result of Euler (1752)—here stated in its sim-
plest form, for the plane—marks one of the common origins of graph
theory and topology. The theorem relates the number of vertices, edges

and faces in a plane graph: taken with the correct signs, these numbers always add up to 2. The general form of Euler's theorem asserts the same for graphs suitably embedded in other surfaces, too: the sum obtained is always a fixed number depending only on the surface, not on the graph, and this number differs for distinct (orientable closed) surfaces. Hence, any two such surfaces can be distinguished by a simple arithmetic invariant of the graphs embedded in them![3]

Let us then prove Euler's theorem in its simplest form:

Theorem 4.2.9. (Euler's Formula)
Let G be a connected plane graph with n vertices, m edges, and ℓ faces. Then

$$n - m + \ell = 2\,.$$

Proof. We fix n and apply induction on m. For $m \leqslant n - 1$, G is a tree and $m = n - 1$ (why?), so the assertion follows from Proposition 4.2.4.

Now let $m \geqslant n$. Then G has an edge e that lies on a cycle; let $G' := G - e$. By Lemma 4.2.2 (ii), e lies on the boundary of exactly two faces f_1, f_2 of G, and as the points in $\mathring{e}$ are all equivalent in $\mathbb{R}^2 \smallsetminus G'$, there is a face f_e of G' containing $\mathring{e}$. We show that

$$F(G) \smallsetminus \{f_1, f_2\} = F(G') \smallsetminus \{f_e\}\,; \qquad\qquad (*)$$

then G' has exactly one face and one edge less than G, and so the assertion follows from the induction hypothesis for G'.

For a proof of $(*)$ let first $f \in F(G) \smallsetminus \{f_1, f_2\}$ be given. By Lemma 4.2.2 (i) we have $G[f] \subseteq G \smallsetminus \mathring{e} = G'$, and hence $f \in F(G')$ by Lemma 4.2.1 (ii). As clearly $f \neq f_e$, this establishes the forward inclusion in $(*)$.

Conversely, consider any face $f' \in F(G') \smallsetminus \{f_e\}$. Clearly $f' \neq f_1, f_2$, and $f' \cap \mathring{e} = \emptyset$. Hence every two points of f' lie in $\mathbb{R}^2 \smallsetminus G$ and are equivalent there, so G has a face f containing f'. By Lemma 4.2.1 (i), however, f lies inside a face f'' of G'. Thus $f' \subseteq f \subseteq f''$ and hence $f' = f = f''$, since both f' and f'' are faces of G'. $\qquad \square$

Corollary 4.2.10. *A plane graph with $n \geqslant 3$ vertices has at most $3n - 6$ edges. Every plane triangulation with n vertices has $3n - 6$ edges.*

Proof. By Proposition 4.2.8 it suffices to prove the second assertion. In a plane triangulation G, every face boundary contains exactly three edges, and every edge lies on the boundary of exactly two faces (Lemma 4.2.2). The bipartite graph on $E(G) \cup F(G)$ with edge set $\{ ef \mid e \subseteq G[f] \}$ thus has exactly $2\,|E(G)| = 3\,|F(G)|$ edges. According to this identity we may replace ℓ with $2m/3$ in Euler's formula, and obtain $m = 3n - 6$. $\qquad \square$

Right margin notes:
(1.5.1)
(1.5.3)

e
G'
f_1, f_2
f_e

[4.4.1]
[5.1.2]
[7.3.5]

Euler's formula can be useful for showing that certain graphs cannot occur as plane graphs. The graph K^5, for example, has $10 > 3 \cdot 5 - 6$ edges, more than allowed by Corollary 4.2.10. Similarly, $K_{3,3}$ cannot be a plane graph. For since $K_{3,3}$ is 2-connected but contains no triangle, every face of a plane $K_{3,3}$ would be bounded by a cycle of length $\geqslant 4$ (Proposition 4.2.6). As in the proof of Corollary 4.2.10 this implies $2m \geqslant 4\ell$, which yields $m \leqslant 2n - 4$ when substituted in Euler's formula. But $K_{3,3}$ has $9 > 2 \cdot 6 - 4$ edges.

Clearly, along with K^5 and $K_{3,3}$ themselves, their subdivisions cannot occur as plane graphs either:

[4.4.5]
[4.4.6]

Corollary 4.2.11. *A plane graph contains neither K^5 nor $K_{3,3}$ as a topological minor.*   □

Surprisingly, it turns out that this simple property of plane graphs identifies them among all other graphs: as Section 4.4 will show, an arbitrary graph can be drawn in the plane if and only if it has no (topological) K^5 or $K_{3,3}$ minor.

4.3 Drawings

planar embedding

An embedding in the plane, or *planar embedding*, of an (abstract) graph G is an isomorphism between G and a plane graph H. The latter will be called a *drawing* of G. We shall not always distinguish notationally between the vertices and edges of G and of H. In this section we investigate how two planar embeddings of a graph can differ.

drawing

How should we measure the likeness of two embeddings $\rho \colon G \to H$ and $\rho' \colon G \to H'$ of a planar graph G? An obvious way to do this is to consider the canonical isomorphism $\sigma := \rho' \circ \rho^{-1}$ between H and H' as abstract graphs, and ask how much of their position in the plane this isomorphism respects or preserves. For example, if σ is induced by a simple rotation of the plane, we would hardly consider ρ and ρ' as genuinely different ways of drawing G.

σ
$H; V, E, F$
$H'; V', E', F'$

So let us begin by considering any abstract isomorphism $\sigma \colon V \to V'$ between two plane graphs $H = (V, E)$ and $H' = (V', E')$, with face sets $F(H) =: F$ and $F(H') =: F'$ say, and try to measure to what degree σ respects or preserves the features of H and H' as plane graphs. In what follows we shall propose three criteria for this in decreasing order of strictness (and increasing order of ease of handling), and then prove that for most graphs these three criteria turn out to agree. In particular, applied to the isomorphism $\sigma = \rho' \circ \rho^{-1}$ considered earlier, all three criteria will say that there is essentially only one way to draw a 3-connected graph.

Our first criterion for measuring how well our abstract isomorphism σ preserves the plane features of H and H' is perhaps the most natural one. Intuitively, we would like to call σ 'topological' if it is induced by a homeomorphism from the plane $\mathbb{R}^2$ to itself. To avoid having to grant the outer faces of H and H' a special status, however, we take a detour via the homeomorphism $\pi\colon S^2 \smallsetminus \{(0,0,1)\} \to \mathbb{R}^2$ chosen in Section 4.1: π
we call σ a *topological isomorphism* between the plane graphs H and H'
if there exists a homeomorphism $\varphi\colon S^2 \to S^2$ such that $\psi := \pi \circ \varphi \circ \pi^{-1}$ *topological isomorphism*
induces σ on $V \cup E$. (More formally: we ask that ψ agree with σ on V, and that it map every plane edge $xy \in H$ onto the plane edge $\sigma(x)\sigma(y) \in H'$. Unless φ fixes the point $(0,0,1)$, the map ψ will be undefined at $\pi(\varphi^{-1}(0,0,1))$.)

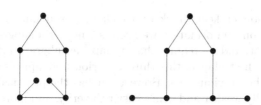

Fig. 4.3.1. Two drawings of a graph that are not topologically isomorphic—why not?

It can be shown that, up to topological isomorphism, inner and outer faces are indeed no longer different: if we choose as φ a rotation of S^2 mapping the π^{-1}-image of a point of some inner face of H to the north pole $(0,0,1)$ of S^2, then ψ maps the rest of this face to the outer face of $\psi(H)$. (To ensure that the edges of $\psi(H)$ are again piecewise linear, however, one may have to adjust φ a little.)

If σ is a topological isomorphism as above, then—except possibly for a pair of missing points where ψ or ψ^{-1} is undefined—ψ maps the faces of H onto those of H' (proof?). In this way, σ extends naturally to a bijection $\sigma\colon V \cup E \cup F \to V' \cup E' \cup F'$ which preserves incidence of vertices, edges and faces.

Let us single out this last property of a topological isomorphism as the second criterion for how well an abstract isomorphism between plane graphs respects their position in the plane: let us call σ a *combinatorial* *combinatorial isomorphism* of the plane graphs H and H' if it can be extended to a *isomorphism*
bijection $\sigma\colon V \cup E \cup F \to V' \cup E' \cup F'$ that preserves incidence not only of vertices with edges but also of vertices and edges with faces. (Formally: we require that a vertex or edge $x \in H$ shall lie on the boundary of a face $f \in F$ if and only if $\sigma(x)$ lies on the boundary of the face $\sigma(f)$.)

If σ is a combinatorial isomorphism of the plane graphs H and H', it maps the face boundaries of H to those of H'. Let us pick out this prop- *graph-* erty as our third criterion, and call σ a *graph-theoretical isomorphism* of *theoretical* *isomorphism*

Fig. 4.3.2. Two drawings of a graph that are combinatorially
isomorphic but not topologically—why not?

the plane graphs H and H' if

$$\{ \sigma(H[f]) : f \in F \} = \{ H'[f'] : f' \in F' \}.$$

Thus, we no longer keep track of which face is bounded by a given
subgraph: the only information we keep is whether a subgraph bounds
some face or not, and we require that σ map the subgraphs that do onto
each other. At first glance, this third criterion may appear a little less
natural than the previous two. However, it has the practical advantage
of being formally weaker and hence easier to verify, and moreover, it will
turn out to be equivalent to the other two in most cases.

As we have seen, every topological isomorphism between two plane
graphs is also combinatorial, and every combinatorial isomorphism is also
graph-theoretical. The following theorem shows that, for most graphs,
the converse is true as well:

Theorem 4.3.1.

(i) *Every graph-theoretical isomorphism between two plane graphs is
combinatorial. Its extension to a face bijection is unique if and
only if the graph is not a cycle.*

(ii) *Every combinatorial isomorphism between two 2-connected plane
graphs is topological.*

(4.1.1)
(4.1.4)
(4.2.5)
(4.2.6)

Proof. Let $H = (V, E)$ and $H' = (V', E')$ be two plane graphs,
put $F(H) =: F$ and $F(H') =: F'$, and let $\sigma : V \to V'$ be an isomorphism between the underlying abstract graphs. Extend σ to a map
$V \cup E \to V' \cup E'$ by letting $\sigma(xy) := \sigma(x)\sigma(y)$.

(i) If H is a cycle, the assertion follows from the Jordan curve theo-
rem. We now assume that H is not a cycle. Let $\mathcal{B}$ and $\mathcal{B}'$ be the sets of
all face boundaries in H and H', respectively. If σ is a graph-theoretical
isomorphism, then the map $B \mapsto \sigma(B)$ is a bijection between $\mathcal{B}$ and $\mathcal{B}'$.
By Lemma 4.2.5, the map $f \mapsto H[f]$ is a bijection between F and $\mathcal{B}$,
and likewise for F' and $\mathcal{B}'$. The composition of these three bijections is
a bijection between F and F', which we choose as $\sigma : F \to F'$. By construction, this extension of σ to $V \cup E \cup F$ preserves incidences (and is
unique with this property), so σ is indeed a combinatorial isomorphism.

(ii) Let us assume that H is 2-connected, and that σ is a combinatorial isomorphism. We have to construct a homeomorphism $\varphi\colon S^2 \to S^2$
which, for every vertex or plane edge $x \in H$, maps $\pi^{-1}(x)$ to $\pi^{-1}(\sigma(x))$.
Since σ is a combinatorial isomorphism, $\tilde{\sigma}\colon \pi^{-1} \circ \sigma \circ \pi$ is an incidence
preserving bijection from the vertices, edges and faces[4] of $\tilde{H} := \pi^{-1}(H)$
to the vertices, edges and faces of $\tilde{H}' := \pi^{-1}(H')$.

σ

$\tilde{\sigma}$

$\tilde{H}, \tilde{H}'$

$$
\begin{array}{ccc}
S^2 \supseteq \tilde{H} & \xrightarrow{\ \tilde{\sigma}\ } & \tilde{H}' \subseteq S^2 \\[4pt]
\pi \Big\downarrow & & \Big\downarrow \pi \\[4pt]
\mathbb{R}^2 \supseteq H & \xrightarrow[\ \sigma\ ]{} & H' \subseteq \mathbb{R}^2
\end{array}
$$

Fig. 4.3.3. Defining $\tilde{\sigma}$ via σ

We construct φ in three steps. Let us first define φ on the vertex
set of $\tilde{H}$, setting $\varphi(x) := \tilde{\sigma}(x)$ for all $x \in V(\tilde{H})$. This is trivially a
homeomorphism between $V(\tilde{H})$ and $V(\tilde{H}')$.

As the second step, we now extend φ to a homeomorphism between
$\tilde{H}$ and $\tilde{H}'$ that induces $\tilde{\sigma}$ on $V(\tilde{H}) \cup E(\tilde{H})$. We may do this edge by
edge, as follows. Every edge xy of $\tilde{H}$ is homeomorphic to the edge
$\tilde{\sigma}(xy) = \varphi(x)\varphi(y)$ of $\tilde{H}'$, by a homeomorphism mapping x to $\varphi(x)$ and
y to $\varphi(y)$. Then the union of all these homeomorphisms, one for every
edge of $\tilde{H}$, is indeed a homeomorphism between $\tilde{H}$ and $\tilde{H}'$—our desired
extension of φ to $\tilde{H}$: all we have to check is continuity at the vertices
(where the edge homeomorphisms overlap), and this follows at once from
our assumption that the two graphs and their individual edges all carry
the subspace topology in $\mathbb{R}^3$.

In the third step we now extend our homeomorphism $\varphi\colon \tilde{H} \to \tilde{H}'$ to
all of S^2. This can be done analogously to the second step, face by face.
By Proposition 4.2.6, all face boundaries in $\tilde{H}$ and $\tilde{H}'$ are cycles. Now if
f is a face of $\tilde{H}$ and C its boundary, then $\tilde{\sigma}(C) := \bigcup \{\, \tilde{\sigma}(e) \mid e \in E(C)\,\}$
bounds the face $\tilde{\sigma}(f)$ of $\tilde{H}'$. By Theorem 4.1.4, we may therefore extend
the homeomorphism $\varphi\colon C \to \tilde{\sigma}(C)$ defined so far to a homeomorphism
from $C \cup f$ to $\tilde{\sigma}(C) \cup \tilde{\sigma}(f)$. We finally take the union of all these homeomorphisms, one for every face f of $\tilde{H}$, as our desired homeomorphism
$\varphi\colon S^2 \to S^2$; as before, continuity is easily checked. $\square$

Let us return now to our original goal, the definition of equivalence
for planar embeddings. Let us call two planar embeddings ρ, ρ' of a graph

---

[4] By the 'vertices, edges and faces' of $\tilde{H}$ and $\tilde{H}'$ we mean the images under π^{-1}
of the vertices, edges and faces of H and H' (plus $(0,0,1)$ in the case of the outer
face). Their sets will be denoted by $V(\tilde{H})$, $E(\tilde{H})$, $F(\tilde{H})$ and $V(\tilde{H}')$, $E(\tilde{H}')$, $F(\tilde{H}')$,
and incidence is defined as inherited from H and H'.

equivalent embeddings G *topologically* (respectively, *combinatorially*) *equivalent* if $\rho' \circ \rho^{-1}$ is a topological (respectively, combinatorial) isomorphism between $\rho(G)$ and $\rho'(G)$. If G is 2-connected, the two definitions coincide by Theorem 4.3.1, and we simply speak of *equivalent* embeddings. Clearly, this is indeed an equivalence relation on the set of planar embeddings of any given graph.

Note that two drawings of G resulting from inequivalent embeddings may well be topologically isomorphic (exercise): for the equivalence of two embeddings we ask not only that some (topological or combinatorial) isomorphism exist between the their images, but that the canonical isomorphism $\rho' \circ \rho^{-1}$ be a topological or combinatorial one.

Even in this strong sense, 3-connected graphs have only one embedding up to equivalence:

[12.7.4]   **Theorem 4.3.2.** (Whitney 1933)
Any two planar embeddings of a 3-connected graph are equivalent.

(4.2.7)   *Proof.* Let G be a 3-connected graph with planar embeddings $\rho \colon G \to H$ and $\rho' \colon G \to H'$. By Theorem 4.3.1 it suffices to show that $\rho' \circ \rho^{-1}$ is a graph-theoretical isomorphism, i.e. that $\rho(C)$ bounds a face of H if and only if $\rho'(C)$ bounds a face of H', for every subgraph $C \subseteq G$. This follows at once from Proposition 4.2.7. $\square$

4.4 Planar graphs: Kuratowski's theorem

planar   A graph is called *planar* if it can be embedded in the plane: if it is isomorphic to a plane graph. A planar graph is *maximal*, or *maximally planar*, if it is planar but cannot be extended to a larger planar graph by adding an edge (but no vertex).

Drawings of maximal planar graphs are clearly maximally plane. The converse, however, is not obvious: when we start to draw a planar graph, could it happen that we get stuck half-way with a proper subgraph that is already maximally plane? Our first proposition says that this can never happen, that is, a plane graph is never maximally plane just because it is badly drawn:

Proposition 4.4.1.

    (i) *Every maximal plane graph is maximally planar.*

    (ii) *A planar graph with $n \geqslant 3$ vertices is maximally planar if and only if it has $3n - 6$ edges.*

(4.2.8)
(4.2.10)   *Proof.* Apply Proposition 4.2.8 and Corollary 4.2.10. $\square$

Which graphs are planar? As we saw in Corollary 4.2.11, no planar graph contains K^5 or $K_{3,3}$ as a topological minor. Our aim in this section is to prove the surprising converse, a classic theorem of Kuratowski: any graph without a topological K^5 or $K_{3,3}$ minor is planar.

Before we prove Kuratowski's theorem, let us note that it suffices to consider ordinary minors rather than topological ones:

Lemma 4.4.2. *A graph contains K^5 or $K_{3,3}$ as a minor if and only if it contains K^5 or $K_{3,3}$ as a topological minor.*

Proof. By Proposition 1.7.3 it suffices to show that every graph G with a K^5 minor contains either K^5 as a topological minor or $K_{3,3}$ as a minor. So suppose that $G \succcurlyeq K^5$, and let K be a minimal model of K^5 in G. Then every branch set of K induces a tree in K, and between any two branch sets K has exactly one edge. If we take the tree induced by a branch set V_x and add to it the four edges joining it to other branch sets, we obtain another tree, T_x say. By the minimality of K, the tree T_x has exactly 4 leaves, the 4 neighbours of V_x in other branch sets (Fig. 4.4.1).

(1.7.3)

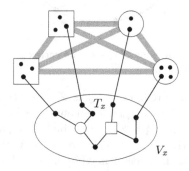

Fig. 4.4.1. Every IK^5 contains a TK^5 or $IK_{3,3}$

If each of the five trees T_x is a $TK_{1,4}$ then K is a TK^5, and we are done. If one of the T_x is not a $TK_{1,4}$ then it has exactly two vertices of degree 3. Contracting V_x onto these two vertices, and every other branch set to a single vertex, we obtain a graph on 6 vertices containing a $K_{3,3}$. Thus, $G \succcurlyeq K_{3,3}$ as desired.   □

We first prove Kuratowski's theorem for 3-connected graphs. This is the heart of the proof: the general case will then follow easily.

Lemma 4.4.3. *Every 3-connected graph G without a K^5 or $K_{3,3}$ minor is planar.*

(3.2.4)
(4.2.6)

xy

Proof. We apply induction on $|G|$. For $|G| = 4$ we have $G = K^4$, and the assertion holds. Now let $|G| > 4$, and assume the assertion is true for smaller graphs. By Lemma 3.2.4, G has an edge xy such that G/xy is again 3-connected. Since the minor relation is transitive, G/xy has no K^5 or $K_{3,3}$ minor either. Thus, by the induction hypothesis, G/xy has

$\tilde{G}$

a drawing $\tilde{G}$ in the plane. Let f be the face of $\tilde{G} - v_{xy}$ containing the

f, C

point v_{xy}, and let C be the boundary of f. Let $X := N_G(x) \smallsetminus \{y\}$ and

X, Y

$Y := N_G(y) \smallsetminus \{x\}$; then $X \cup Y \subseteq V(C)$, because $v_{xy} \in f$. Clearly,

$\tilde{G}'$

$$\tilde{G}' := \tilde{G} - \{\, v_{xy}v \mid v \in Y \smallsetminus X \,\}$$

may be viewed as a drawing of $G - y$, in which the vertex x is represented by the point v_{xy} (Fig. 4.4.2). Our aim is to add y to this drawing to obtain a drawing of G.

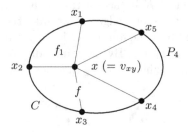

Fig. 4.4.2. $\tilde{G}'$ as a drawing of $G - y$: the vertex x is represented by the point v_{xy}

$x_1, \ldots, x_k$

P_i

Since $\tilde{G}$ is 3-connected, $\tilde{G} - v_{xy}$ is 2-connected, so C is a cycle (Proposition 4.2.6). Let $x_1, \ldots, x_k$ be an enumeration along this cycle of the vertices in X, and let $P_i = x_i \ldots x_{i+1}$ be the X-paths on C between them ($i = 1, \ldots, k$; with $x_{k+1} := x_1$). Let us show that $Y \subseteq V(P_i)$ for some i. Suppose not. If y has a neighbour $y' \in \mathring{P}_i$ for some i, it has another neighbour $y'' \in C - P_i$, and these are separated in C by $x' := x_i$ and $x'' := x_{i+1}$. If $Y \subseteq X$ and $|Y \cap X| \leqslant 2$, then y has exactly two neighbours y', y'' on C but not in the same P_i, so again y' and y'' are separated on C by two vertices $x', x'' \in X$. In either case, x, y', y'' and y, x', x'' are the branch vertices of a $TK_{3,3}$ in G, a contradiction. The only remaining case is that y and x have three common neighbours on C. Then these form a TK^5 with x and y, again a contradiction.

C_i

f_i

Fix i so that $Y \subseteq P_i$. The set $C \smallsetminus P_i$ is contained in one of the two faces of the cycle $C_i := x x_i P_i x_{i+1} x$; we denote the *other* face of C_i by f_i. Since f_i contains points of f (close to x) but no points of its boundary C, we have $f_i \subseteq f$. Moreover, the plane edges $x x_j$ with $j \notin \{i, i+1\}$ meet C_i only in x and end outside f_i in $C \smallsetminus P_i$, so f_i meets none of those edges. Hence $f_i \subseteq \mathbb{R}^2 \smallsetminus \tilde{G}'$, that is, f_i is contained in (and hence equal to) a face of $\tilde{G}'$. We may therefore extend $\tilde{G}'$ to a drawing of G by placing y and its incident edges in f_i. $\qquad\square$

Compared with other proofs of Kuratowski's theorem, the above proof has the attractive feature that it can easily be adapted to produce a drawing in which every inner face is convex (exercise); in particular, every edge can be drawn straight. Note that 3-connectedness is essential here: a 2-connected planar graph need not have a drawing with all inner faces convex (example?), although it always has a straight-line drawing (Exercise 15).

It is not difficult, in principle, to reduce the general Kuratowski theorem to the 3-connected case by manipulating and combining partial drawings assumed to exist by induction. For example, if $\kappa(G) = 2$ and $G = G_1 \cup G_2$ with $V(G_1 \cap G_2) = \{x, y\}$, and if G has no TK^5 or $TK_{3,3}$ subgraph, then neither $G_1 + xy$ nor $G_2 + xy$ has such a subgraph, and we may try to combine drawings of these graphs to one of $G + xy$. (If xy is already an edge of G, the same can be done with G_1 and G_2.) For $\kappa(G) \leqslant 1$, things become even simpler. However, the geometric operations involved require some cumbersome shifting and scaling, even if all the plane edges occurring are assumed to be straight.

The following more combinatorial route offers an ingenious alternative. In order to show that a given graph $G \not\supseteq TK^5, TK_{3,3}$ is planar we start by adding edges to G until it is edge-maximal with the property of not containing a TK^5 or $TK_{3,3}$. In Lemma 4.4.5 we show that this makes our graph 3-connected, so by Lemma 4.4.3 it is planar.

For the proof of Lemma 4.4.5 we need another lemma. We state this a little more generally, so that we can use it again in another context in Chapter 7. For our application here put $\mathcal{X} := \{K^5, K_{3,3}\}$.

Lemma 4.4.4. *Let $\mathcal{X}$ be a set of 3-connected graphs. Let G be a graph with a proper separation $\{V_1, V_2\}$ of order $\kappa(G) \leqslant 2$. If G is edge-maximal without a topological minor in $\mathcal{X}$, then so are $G_1 := G[V_1]$ and $G_2 := G[V_2]$, and $G_1 \cap G_2 = K^2$.* [7.3.1]

Proof. Note first that every vertex $v \in S := V_1 \cap V_2$ has a neighbour in every component of $G_i - S$, $i = 1, 2$: otherwise $S \setminus \{v\}$ would separate G, contradicting $|S| = \kappa(G)$. By the maximality of G, every edge e added to G lies in a $TX \subseteq G + e$ with $X \in \mathcal{X}$. For all the choices of e considered below, the 3-connectedness of X will imply that the branch vertices of this TX all lie in the same V_i, say in V_1. (The position of e will always be symmetrical with respect to V_1 and V_2, so this assumption entails no loss of generality.) Then the TX meets V_2 at most in a path P corresponding to an edge of X. S X P

If $S = \emptyset$, we obtain an immediate contradiction by choosing e with one end in V_1 and the other in V_2. If $S = \{v\}$ is a singleton, let e join a neighbour v_1 of v in $V_1 \setminus S$ to a neighbour v_2 of v in $V_2 \setminus S$ (Fig. 4.4.3). Then P contains both v and the edge $e = v_1 v_2$; replacing its segment $vPv_2 v_1$ with the edge vv_1 we obtain a TX in $G_1 \subseteq G$, a contradiction.

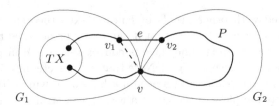

Fig. 4.4.3. If $G + e$ contains a TX, then so does G_1 or G_2

x, y

So $|S| = 2$, say $S = \{x, y\}$. If $xy \notin G$, we let $e := xy$, and in the arising TX replace e by an x–y path through G_2. This yields a TX in G, a contradiction. Hence $xy \in G$, and $G[S] = K^2$ as claimed.

It remains to show that G_1 and G_2 are edge-maximal without a topological minor in $\mathcal{X}$. So let e' be an additional edge for G_1, say. Replacing xPy with the edge xy if necessary, we obtain a TX either in $G_1 + e'$ (which shows the edge-maximality of G_1, as desired) or in G_2 (which contradicts $G_2 \subseteq G$). $\qquad\square$

Lemma 4.4.5. *If $|G| \geqslant 4$ and G is edge-maximal with $TK^5, TK_{3,3} \nsubseteq G$, then G is 3-connected.*

(4.2.11)

G_1, G_2

x, y

f_i
z_i
K

Proof. We apply induction on $|G|$. For $|G| = 4$, we have $G = K^4$ and the assertion holds. Now let $|G| > 4$, and let G be edge-maximal without a TK^5 or $TK_{3,3}$. Suppose $\kappa(G) \leqslant 2$, and choose G_1 and G_2 as in Lemma 4.4.4. For $\mathcal{X} := \{K^5, K_{3,3}\}$, the lemma says that $G_1 \cap G_2$ is a K^2, with vertices x, y say, and that G_1 and G_2 too are edge-maximal without a TK^5 or $TK_{3,3}$. Hence, G_1 and G_2 are either a triangle or 3-connected by the induction hypothesis. Since they cannot contain K^5 or $K_{3,3}$ even as an ordinary minor (Lemma 4.4.2), they are thus planar by Lemma 4.4.3.

For each $i = 1, 2$ separately, choose a drawing of G_i, a face f_i with the edge xy on its boundary, and a vertex $z_i \neq x, y$ on the boundary of f_i. Let K be a TK^5 or $TK_{3,3}$ in the abstract graph $G + z_1 z_2$ (Fig. 4.4.4).

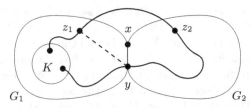

Fig. 4.4.4. A TK^5 or $TK_{3,3}$ in $G + z_1 z_2$

If all the branch vertices of K lie in the same G_i, then either $G_i + xz_i$ or $G_i + yz_i$ (or G_i itself, if z_i is already adjacent to x or y, respectively) contains a TK^5 or $TK_{3,3}$; this contradicts Corollary 4.2.11, since these graphs are planar by the choice of z_i. Since $G + z_1 z_2$ does not contain four independent paths between $(G_1 - G_2)$ and $(G_2 - G_1)$, these subgraphs cannot both contain a branch vertex of a TK^5, and cannot both contain two branch vertices of a $TK_{3,3}$. Hence K is a $TK_{3,3}$ with only one branch vertex v in, say, $G_2 - G_1$. But then also the graph $G_1 + v + \{vx, vy, vz_1\}$, which is planar by the choice of z_1, contains a $TK_{3,3}$. This contradicts Corollary 4.2.11. □

Theorem 4.4.6. (Kuratowski 1930; Wagner 1937)　　　　　　　　[4.5.1]
The following assertions are equivalent for graphs G:　　　　　　　[12.6.4]

  (i) *G is planar;*

 (ii) *G contains neither K^5 nor $K_{3,3}$ as a minor;*

(iii) *G contains neither K^5 nor $K_{3,3}$ as a topological minor.*

Proof. Combine Corollary 4.2.11 with Lemmas 4.4.2, 4.4.3 and 4.4.5.　　(4.2.11)
□

Corollary 4.4.7. *Every maximal planar graph with at least four vertices is 3-connected.*

Proof. Apply Lemma 4.4.5 and Theorem 4.4.6. □

4.5 Algebraic planarity criteria

One of the most conspicuous features of a plane graph G are its *facial cycles*, the cycles that bound a face. If G is 2-connected it is covered by its facial cycles, so in a sense these form a 'large' set. In fact, the set of facial cycles is large even in the sense that they generate the entire cycle space: every cycle in G is easily seen to be the sum of the facial cycles (see below). On the other hand, the facial cycles only cover G 'thinly', as every edge lies on at most two of them. Our first aim in this section is to show that the existence of such a large yet thinly spread family of cycles is not only a conspicuous feature of planarity but lies at its very heart: it characterizes it.

facial cycles

　　Let $G = (V, E)$ be any graph. We call a subset $\mathcal{F}$ of its edge space $\mathcal{E}(G)$ *sparse* if every edge of G lies in at most two sets of $\mathcal{F}$. For example, the cut space $\mathcal{B}(G)$ has a sparse basis: according to Proposition 1.9.2 it is generated by the cuts $E(v)$ formed by all the edges at a given vertex v, and an edge $xy \in G$ lies in $E(v)$ only for $v = x$ and for $v = y$.

sparse

Theorem 4.5.1. (MacLane 1937)　　　　　　　　　　　　　　[4.6.3]
A graph is planar if and only if its cycle space has a sparse basis.

(1.9.1)
(1.9.5)
(4.1.1)
(4.2.2)
(4.2.6)
(4.4.6)

Proof. The assertion being trivial for graphs of order at most 2, we consider a graph G of order at least 3. If $\kappa(G) \leqslant 1$, then G is the union of two proper induced subgraphs G_1, G_2 with $|G_1 \cap G_2| \leqslant 1$. Then $\mathcal{C}(G)$ is the direct sum of $\mathcal{C}(G_1)$ and $\mathcal{C}(G_2)$, and hence has a sparse basis if and only if both $\mathcal{C}(G_1)$ and $\mathcal{C}(G_2)$ do (proof?). Moreover, G is planar if and only if both G_1 and G_2 are: this follows at once from Kuratowski's theorem, but also from easy geometrical considerations. The assertion for G thus follows inductively from those for G_1 and G_2. For the rest of the proof, we now assume that G is 2-connected.

We first assume that G is planar and choose a drawing. By Proposition 4.2.6, the face boundaries of G are cycles, so they are elements of $\mathcal{C}(G)$. We shall show that the face boundaries generate all the cycles in G; then $\mathcal{C}(G)$ has a sparse basis by Lemma 4.2.2. Let $C \subseteq G$ be any cycle, and let f be its inner face. By Lemma 4.2.2, every edge e with $\mathring{e} \subseteq f$ lies on exactly two face boundaries $G[f']$ with $f' \subseteq f$, and every edge of C lies on exactly one such face boundary. Hence the sum in $\mathcal{C}(G)$ of all those face boundaries is exactly C.

Conversely, let $\{C_1, \ldots, C_k\}$ be a sparse basis of $\mathcal{C}(G)$. Then, for every edge $e \in G$, also $\mathcal{C}(G - e)$ has a sparse basis. Indeed, if e lies in just one of the sets C_i, say in C_1, then $\{C_2, \ldots, C_k\}$ is a sparse basis of $\mathcal{C}(G - e)$; if e lies in two of the C_i, say in C_1 and C_2, then $\{C_1 + C_2, C_3, \ldots, C_k\}$ is such a basis. (Note that the two bases are indeed subsets of $\mathcal{C}(G - e)$ by Proposition 1.9.1.) Thus every subgraph of G has a cycle space with a sparse basis. For our proof that G is planar, it thus suffices to show that the cycle spaces of K^5 and $K_{3,3}$ (and hence those of their subdivisions) do *not* have a sparse basis: then G cannot contain a TK^5 or $TK_{3,3}$, and so is planar by Kuratowski's theorem.

Let us consider K^5 first. By Theorem 1.9.5, $\dim \mathcal{C}(K^5) = 6$; let $\mathcal{B} = \{C_1, \ldots, C_6\}$ be a sparse basis, and put $C_0 := C_1 + \ldots + C_6$. As $\mathcal{B}$ is linearly independent, none of the sets $C_0, \ldots, C_6$ is empty, so each of them contains at least three edges (cf. Proposition 1.9.1). Moreover, as every edge from C_0 lies in just one of $C_1, \ldots, C_6$, the set $\{C_0, \ldots, C_6\}$ is still sparse. But this implies that K^5 should have more edges than it does, i.e. we obtain the contradiction of

$$21 = 7 \cdot 3 \leqslant |C_0| + \ldots + |C_6| \leqslant 2\,\|K^5\| = 20\,.$$

For $K_{3,3}$, Theorem 1.9.5 gives $\dim \mathcal{C}(K_{3,3}) = 4$; let $\mathcal{B} = \{C_1, \ldots, C_4\}$ be a sparse basis, and put $C_0 := C_1 + \ldots + C_4$. As $K_{3,3}$ has girth 4, each C_i contains at least four edges. We then obtain the contradiction of

$$20 = 5 \cdot 4 \leqslant |C_0| + \ldots + |C_4| \leqslant 2\,\|K_{3,3}\| = 18\,.$$

□

A constructive proof of the backward implication of MacLane's theorem is indicated in Exercise 30. That proof shows that the generating set we chose in our proof of the forward implication, the set of face boundaries, is canonical in the following sense: given any set $\mathcal{D}$ of cycles in a 2-connected planar graph G that generates $\mathcal{C}(G)$ and is such that every edge of G lies on exactly two of those cycles, there is a drawing of G in which the cycles in $\mathcal{D}$ are precisely the face boundaries.

It is one of the hidden beauties of planarity theory that two such abstract and seemingly unintuitive results about generating sets in cycle spaces as MacLane's theorem and Tutte's theorem 3.2.6 conspire to produce a very tangible planarity criterion for 3-connected graphs:

Theorem 4.5.2. (Kelmans 1978)
A 3-connected graph is planar if and only if every edge lies on at most (equivalently: exactly) two non-separating induced cycles.

Proof. The forward implication follows from Proposition 4.2.7 and Lemma 4.2.2 (and Proposition 4.2.6 for the 'exactly two' version); the backward implication follows from Theorems 3.2.6 and 4.5.1.   □

(3.2.6)
(4.2.2)
(4.2.6)
(4.2.7)

Let us conclude this section with another characterization of planarity, one with a very different flavour. A *linear extension* of a partial ordering $\leqslant$ of a set P is a total ordering $\leqslant'$ on P which includes $\leqslant$ as a subset of P^2. Thus, for any $p \leqslant q$ in P we still have $p \leqslant' q$, and for incomparable $p, q \in P$ we have either $p <' q$ or $p >' q$ in addition. The *dimension* of the partially ordered set $(P, \leqslant)$ is the least number of linear extensions of $\leqslant$ on P whose intersection is exactly $\leqslant$: for any incomparable $p, q \in P$ there must be a linear extension $\leqslant'$ with $p <' q$ and another linear extension $\leqslant''$ with $p >'' q$ in this collection.

With every graph $G = (V, E)$ one can associate its *incidence poset*, the partially ordered set $(V \cup E, \leqslant)$ in which $v < e$ if and only if v is a vertex and e is an edge at v. (Thus, as a relation, $<$ is the same as $\in$.)

linear extension

poset dimension

incidence poset

Theorem 4.5.3. (Schnyder 1989)
A graph is planar if and only if its incidence poset has dimension $\leqslant 3$.

4.6 Plane duality

In this section we shall use MacLane's theorem to uncover another connection between planarity and algebraic structure: a connection between
the duality of plane graphs, defined below, and the duality of the cycle
and cut space hinted at in Chapters 1.9 and 2.4.

plane
multigraph   A *plane multigraph* is a pair $G = (V, E)$ of finite sets (of *vertices*
and *edges*, respectively) satisfying the following conditions:

(i) $V \subseteq \mathbb{R}^2$;

(ii) every edge is either an arc between two vertices or a polygon
containing exactly one vertex (its *endpoint*);

(iii) apart from its own endpoint(s), an edge contains no vertex and
no point of any other edge.

We shall use terms defined for plane graphs freely for plane multigraphs.
Note that, as in abstract multigraphs, both loops and double edges count
as cycles.

Let us consider the plane multigraph G shown in Figure 4.6.1. Let
us place a new vertex inside each face of G and link these new vertices
up to form another plane multigraph G^*, as follows: for every edge e of
G we link the two new vertices in the faces incident with e by an edge e^*
crossing e; if e is incident with only one face, we attach a loop e^* to the
new vertex in that face, again crossing the edge e. The plane multigraph
G^* formed in this way is then dual to G in the following sense: if we
apply the same procedure as above to G^*, we obtain a plane multigraph
very similar to G; in fact, G itself may be reobtained from G^* in this way.

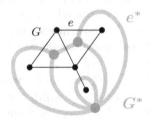

Fig. 4.6.1. A plane graph and its dual

To make this idea more precise, let $G = (V, E)$ and (V^*, E^*) be any
two plane multigraphs, and put $F(G) =: F$ and $F((V^*, E^*)) =: F^*$. We
plane dual   call (V^*, E^*) a *plane dual* of G, and write $(V^*, E^*) =: G^*$, if there are
G^* bijections

$$F \to V^* \qquad E \to E^* \qquad V \to F^*$$
$$f \mapsto v^*(f) \qquad e \mapsto e^* \qquad v \mapsto f^*(v)$$

satisfying the following conditions:

(i) $v^*(f) \in f$ for all $f \in F$;

(ii) $|e^* \cap G| = |\mathring{e}^* \cap \mathring{e}| = |e \cap G^*| = 1$ for all $e \in E$, and in each of e and e^* this point is an inner point of a straight line segment;

(iii) $v \in f^*(v)$ for all $v \in V$.

Every connected plane multigraph has a plane dual. Indeed, to satisfy condition (i) we start by picking from each face f of G a point $v^*(f)$ as a vertex for G^*. We can then link these vertices up by independent arcs as required by (ii), and using the connectedness of G show that there is indeed a bijection $V \to F^*$ satisfying (iii) (Exercise 35).

If G_1^* and G_2^* are two plane duals of G, then clearly $G_1^* \simeq G_2^*$; in fact, one can show that the natural bijection $v_1^*(f) \mapsto v_2^*(f)$ is a topological isomorphism between G_1^* and G_2^*. In this sense, we may speak of *the* plane dual G^* of G.

Finally, G is in turn a plane dual of G^*. Indeed, this is witnessed by the inverse maps of the bijections from the definition of G^*: setting $v^*(f^*(v)) := v$ and $f^*(v^*(f)) := f$ for $f^*(v) \in F^*$ and $v^*(f) \in V^*$, we see that conditions (i) and (iii) for G^* transform into (iii) and (i) for G, while condition (ii) is symmetrical in G and G^*. As duals are easily seen to be connected (Exercise 34), this symmetry implies that connectedness is also a necessary condition for G to have a dual.

Perhaps the most interesting aspect of plane duality is that it relates geometrically two types of edges sets—cycles and bonds—that we have previously seen to be algebraically related (Theorem 1.9.4):

Proposition 4.6.1. *For any connected plane multigraph G, an edge set* $E \subseteq E(G)$ *is the edge set of a cycle in G if and only if* $E^* := \{ e^* \mid e \in E \}$ *is a bond in G^*.*

[6.5.2]

Proof. By conditions (i) and (ii) in the definition of G^*, two vertices $v^*(f_1)$ and $v^*(f_2)$ of G^* lie in the same component of $G^* - E^*$ if and only if f_1 and f_2 lie in the same region of $\mathbb{R}^2 \setminus \bigcup E$: every $v^*(f_1)$–$v^*(f_2)$ path in $G^* - E^*$ is an arc between f_1 and f_2 in $\mathbb{R}^2 \setminus \bigcup E$, and conversely every such arc P (with $P \cap V(G) = \emptyset$) defines a walk in $G^* - E^*$ between $v^*(f_1)$ and $v^*(f_2)$.

(4.1.1)
(4.2.4)

Now if $C \subseteq G$ is a cycle and $E = E(C)$ then, by the Jordan curve theorem and the above correspondence, $G^* - E^*$ has exactly two components. So E^* is a bond of G^*, a minimal non-empty cut.

Conversely, if $E \subseteq E(G)$ is such that E^* is a cut in G^* then, by Proposition 4.2.4 and the above correspondence, E contains the edges of a cycle $C \subseteq G$. If E^* is a bond, E cannot contain any further edges (by the implication shown before). Hence $E = E(C)$. $\qquad\square$

abstract
dual
Proposition 4.6.1 suggests the following generalization of plane duality to abstract multigraphs.[5] Call a multigraph G^* an *abstract dual* of a multigraph G if $E(G^*) = E(G)$ and the bonds in G^* are precisely the edge sets of cycles in G. (Neither G nor G^* need be connected now.)

This correspondence between cycles and bonds extends to the spaces they generate:

Proposition 4.6.2. *If G^* is an abstract dual of G, then the cut space of G^* is the cycle space of G, i.e.,*

$$\mathcal{B}(G^*) = \mathcal{C}(G).$$

(1.9.3)
Proof. Since the cycles of G are precisely the bonds of G^*, the subspace $\mathcal{C}(G)$ they generate in $\mathcal{E}(G) = \mathcal{E}(G^*)$ is the same as the subspace generated by the bonds in G^*. By Lemma 1.9.3, this is the space $\mathcal{B}(G^*)$. $\square$

(1.9.4)
By Theorem 1.9.4, Proposition 4.6.2 implies at once that if G^* is an abstract dual of G then G is an abstract dual of G^*. One can show that if G is 3-connected, then G^* is unique (up to isomorphism and the (3.1.3)
(3.1.2) addition of isolated vertices). By Lemma 3.1.3, a non-empty subset of $E(G) = E(G^*)$ is the edge set of a block of G if and only if it is the edge set of a block of G^*. By Lemma 3.1.2, this implies that the blocks of G^* are duals of the blocks of G.

Although the notion of abstract duality arose as a generalization of plane duality, it could have been otherwise. We knew already from Theorem 1.9.4 that the cycles and the bonds of a graph form natural and related sets of edges. It would not have been unthinkable to ask whether, for some graphs, the orthogonality between these collections of edge sets might give them sufficiently similar intersection patterns that a collection forming the cycles in one graph could form the bonds in another, and vice versa. In other words, for which graphs can we move their entire edge set to a new set of vertices, redefining incidences, so that precisely those sets of edges that used to form cycles now become bonds (and vice versa)? Put in this way, it seems surprising that this could ever be achieved, let alone for such a large and natural class of graphs as all planar graphs.

As one of the highlights of classical planarity theory we now show that the planar graphs are *precisely* those for which this can be done. Admitting an abstract dual thus appears as a new planarity criterion. Conversely, the theorem can be read as a surprising topological characterization of the equally fundamental property of admitting an abstract dual:

---

[5] In what follows we shall use some lemmas from earlier chapters that were stated for graphs only. These lemmas extend to multigraphs with proofs unchanged.

Theorem 4.6.3. (Whitney 1932)
A graph is planar if and only if it has an abstract dual.

Proof. Let G be a planar graph, and consider any drawing. Every
component C of this drawing has a plane dual C^*. Consider these C^*
as abstract multigraphs, and let G^* be their disjoint union. Then the
bonds of G^* are precisely those of the C^*, which by Proposition 4.6.1
correspond to the cycles in G.

    Conversely, suppose that G has an abstract dual G^*. For a proof
that G is planar, it suffices by Theorem 4.5.1 and Proposition 4.6.2 to
show that $\mathcal{B}(G^*)$ has a sparse basis. By Proposition 1.9.2, it does. $\qquad\square$

(1.9.2)
(4.5.1)

    The duality theory for both abstract and plane graphs can be ex-
tended to infinite graphs. As these can have infinite bonds, their duals
must then have 'infinite cycles'. Such things do indeed exist, and are
fascinating: they arise as topological circles in a space formed by the
graph and its *ends*; see Chapter 8.6.

Exercises

1.   Show that every graph can be embedded in $\mathbb{R}^3$ with all edges straight.

2.⁻ Show directly by Lemma 4.1.2 that $K_{3,3}$ is not planar.

3.   Here is an inductive 'proof' that every maximal plane graph of order $\geqslant 4$
is a plane triangulation of minimum degree 3. The induction starts
with K^4. For the induction step, consider an arbitrary maximal plane
graph G of order $n \geqslant 4$, and consider all possible ways of extending it to
a maximal plane graph G' of order $n+1$ by adding a new vertex v. No
matter how this is done, v will come to sit in a face of G, which by the
inductive assumption is bounded by a triangle. Since G' is maximally
planar, v must be joined to all three vertices of that triangle. Clearly,
G' is another plane triangulation, and $\delta(G') = d(v) = 3$.

    (i)⁻ Find the flaw in this 'proof'.

    (ii) Find a counterexample, and explain why the 'proof' overlooks it.

4.   Show that every planar graph is a union of three forests.

5.   The ancient Greeks loved regular plane graphs whose faces were bounded
by cycles of the same length.

    (i) Show that such graphs exist for only finitely many pairs (d, ℓ) of
degree $d \geqslant 3$ and cycle length ℓ. Can you give an upper bound?

    (ii)⁺Show that there are only finitely many such plane graphs, up
to topological isomorphism.

6.   A *fullerene* is a molecule that is made up entirely of carbon atoms
forming a cubic plane graph all whose faces are pentagons or hexagons.
Show that, since carbon atoms can form double bonds, every such graph
can be realized in principle by (4-valent) carbon atoms.

7. A football is made of pentagons and hexagons, not necessarily of regular shape. They are sewn together so that their seams form a cubic planar graph. How many pentagons does the football have?

8.$^-$ (continued from Exercises 6 and 7)
    Fullerenes are less stable if they contain adjacent pentagons. Show that stable fullerenes have at least 60 carbon atoms.

9. Let G be a graph of order n that is embedded in a surface of Euler characteristic χ and cannot be embedded in a simpler surface (one of larger Euler characteristic). Show that G has at most $3n - 3\chi$ edges.

    (Hint. You may use that every face of such an embedded graph is a topological disc. Such embeddings satisfy the general Euler formula, $n - m + \ell = \chi$.)

10. Find a direct proof for planar graphs of Tutte's theorem on the cycle space of 3-connected graphs (Theorem 3.2.6).

11.$^-$ Show that the two plane graphs in Figure 4.3.1 are not combinatorially (and hence not topologically) isomorphic.

12. Show that the two graphs in Figure 4.3.2 are combinatorially but not topologically isomorphic.

13.$^-$ Show that our definition of equivalence for planar embeddings does indeed define an equivalence relation.

14. Find a 2-connected planar graph whose drawings are all topologically isomorphic but whose planar embeddings are not all equivalent.

15.$^+$ Show that every plane graph is combinatorially isomorphic to a plane graph whose edges are all straight.

    (Hint. Given a plane triangulation, construct inductively a graph-theoretically isomorphic plane graph whose edges are straight. Which additional property of the inner faces could help with the induction?)

Do not use Kuratowski's theorem in the following two exercises.

16. Show that any minor of a planar graph is planar. Deduce that a graph is planar if and only if it is the minor of a grid. (*Grids* are defined in Chapter 12.3.)

17. (i) Show that the planar graphs can in principle be characterized as in Kuratowski's theorem: that there exists a set $\mathcal{X}$ of graphs such that a graph G is planar if and only if G has no minor in $\mathcal{X}$.

    (ii) Can every graph property be characterized in this way? If not, which can?

18. Does every planar graph have a drawing with all inner faces convex?

19. Modify the proof of Lemma 4.4.3 so that all inner faces become convex.

20. Does every minimal non-planar graph G (i.e., every non-planar graph G whose proper subgraphs are all planar) contain an edge e such that $G - e$ is maximally planar? Does the answer change if we define 'minimal' with respect to minors rather than subgraphs?

21. Show that adding a new edge to a maximal planar graph of order at least 6 always produces both a TK^5 and a $TK_{3,3}$ subgraph.

22. Prove the general Kuratowski theorem from its 3-connected case by manipulating plane graphs, i.e. avoiding Lemma 4.4.5.

    (This is not intended as an exercise in elementary topology; for the topological parts of the proof, a rough sketch will do.)

23.⁻ A graph is called *outerplanar* if it has a drawing in which every vertex lies on the boundary of the outer face. Show that a graph is outerplanar if and only if it contains neither K^4 nor $K_{2,3}$ as a minor.

24. Show that a 2-connected plane graph is bipartite if and only if every face is bounded by an even cycle.

25. Let $G = G_1 \cup G_2$, where $|G_1 \cap G_2| \leqslant 1$. Show that $\mathcal{C}(G)$ has a sparse basis if both $\mathcal{C}(G_1)$ and $\mathcal{C}(G_2)$ have one.

26. Find a cycle space basis among the face boundaries of a 2-connected plane graph.

27. Show that a 3-connected graph of order n has at least $n/2$ peripheral cycles. Is this lower bound sharp?

28.⁺ Find an algebraic proof of Euler's formula for 2-connected plane graphs, along the following lines. Define the *face space* $\mathcal{F}$ (over $\mathbb{F}_2$) of such a graph in analogy to its vertex space $\mathcal{V}$ and edge space $\mathcal{E}$. Define *boundary maps* $\mathcal{F} \to \mathcal{E} \to \mathcal{V}$ in the obvious way, specifying them first on single faces or edges (i.e., on the standard bases of $\mathcal{F}$ and $\mathcal{E}$) and then extending these maps linearly to all of $\mathcal{F}$ and $\mathcal{E}$. Determine the kernels and images of these homomorphisms, and derive Euler's formula from the dimensions of those subspaces of $\mathcal{F}$, $\mathcal{E}$ and $\mathcal{V}$.

A family of subgraphs of G is said to form a *double cover* of G if every edge of G lies in exactly two of those subgraphs. A double cover by cycles is a *cycle double cover*.

29. Let G be a 2-connected graph whose cycle space is generated by a sparse set $\mathcal{B}$ of cycles. From MacLane's theorem we know that G even admits a double cover by cycles generating $\mathcal{C}(G)$: the face boundaries in any drawing of G. Show directly (without using MacLane's theorem) that $\mathcal{B}$ extends to a cycle double cover $\mathcal{D}$ of G.

30.⁺ (for topologists) Prove the non-trivial implication in MacLane's theorem constructively, as follows. Assume that the given graph G is 2-connected and, by the previous exercise, has a double cover $\mathcal{D}$ by cycles generating $\mathcal{C}(G)$. For each of these cycles C take a disc and identify its boundary with C.

    (i) Show that the space obtained is a surface, i.e., a compact 2-manifold without boundary.

    (ii) Use Theorem 1.9.5 to show that this surface has Euler characteristic at least 2. (This implies that it must be the sphere, a fact you may assume as known.)

31. Deduce from the last two exercises that, given any 2-connected planar
     graph and a sparse basis $\mathcal{B}$ of $\mathcal{C}(G)$ consisting of cycles, there is a
     drawing of G in which the cycles in $\mathcal{B}$ are precisely the boundaries of
     the inner faces.

32.$^+$ Let C be a closed curve in the plane that intersects itself at most once
     in any given point of the plane, and where every such self-intersection
     is a proper crossing. Call C *alternating* if we can turn these crossings
     into over- and underpasses in such a way that when we run along the
     curve the overpasses alternate with the underpasses.

     (i) Prove that every such curve is alternating, or find a counterex-
         ample.

     (ii) Does the solution to (i) change if the curves considered are not
          closed?

33.$^-$ What does the plane dual of a plane tree look like?

34.$^-$ Show that the plane dual of a plane multigraph is connected.

35.$^+$ Show that a connected plane multigraph has a plane dual.

36. Show that any two plane duals of a plane multigraph are combinatori-
     ally isomorphic.

37. Let G^* be an abstract dual of G, and let $e = e^*$ be an edge. Prove the
     following two assertions:

     (i) G^*/e^* is an abstract dual of $G - e$.

     (ii)$^+$ $G^* - e^*$ is an abstract dual of G/e.

38. Find a connected graph that has two non-isomorphic abstract duals.
     Can you find a 2-connected example?

39. Let G, G^* be dual plane graphs. Prove the following statements:

     (i) If G is 2-connected, then G^* is 2-connected.

     (ii) If G is 3-connected, then G^* is 3-connected.

     (iii) If G is 4-connected, then G^* need not be 4-connected.

40. Give detailed proofs for the statements made after Proposition 4.6.2,
     except for the uniqueness of G^* (which is proved in Exercise 41 (ii)).

41. Let $G^* = (V^*, E^*)$ be a connected abstract dual of a connected multi-
     graph $G = (V, E)$. Does G have a drawing whose plane dual is isomor-
     phic to G^*? (Perhaps even 'canonically' isomorphic? In which sense?)

     (i) For 2-connected G, prove this using the approach of Exercise 30.

     (ii) Deduce that abstract duals of 3-connected graphs are unique.
          (What exactly could this mean? Suggest a definition of unique-
          ness that is stronger than 'up to isomorphism'.)

     (iii) Find a counterexample to the general statement.

42. Show that the following statements are equivalent for connected multi-graphs $G = (V, E)$ and $G' = (V', E)$ with the same edge set:

(i) G and G' are abstract duals of each other;

(ii) given any set $F \subseteq E$, the multigraph (V, F) is a tree if and only if $(V', E \smallsetminus F)$ is a tree.

Notes

There is a very thorough monograph on the embedding of graphs in surfaces, including the plane: B. Mohar & C. Thomassen, *Graphs on Surfaces*, Johns Hopkins University Press 2001. Proofs of the results cited in Section 4.1, as well as all references for this chapter, can be found there. A good account of the Jordan curve theorem, both polygonal and general, is given also in J. Stillwell, *Classical topology and combinatorial group theory*, Springer 1980.

The short proof of Corollary 4.2.10 uses a trick that deserves special mention: the so-called *double counting* of pairs, illustrated in the text by a bipartite graph whose edges can be counted alternatively by summing its degrees on the left or on the right. Double counting is a technique widely used in combinatorics, and there will be more examples later in the book.

The material of Section 4.3 is not normally standard for an introductory graph theory course, and the rest of the chapter can be read independently of this section. However, the results of Section 4.3 are by no means unimportant. In a way, they have fallen victim to their own success: the shift from a topological to a combinatorial setting for planarity problems which they achieve has made the topological techniques developed there dispensable for most of planarity theory.

In its original version, Kuratowski's theorem was stated only for topological minors; the version for general minors was added by Wagner in 1937. Our proof of the 3-connected case (Lemma 4.4.3) is a weakening of a proof due to C. Thomassen, Planarity and duality of finite and infinite graphs, *J. Comb. Theory B* **29** (1980), 244–271, which yields a drawing in which all the inner faces are convex (Exercise 19). The existence of such 'convex' drawings for 3-connected planar graphs follows already from the theorem of Steinitz (1922) that these graphs are precisely the 1-skeletons of 3-dimensional convex polyhedra. Compare also W.T. Tutte, How to draw a graph, *Proc. Lond. Math. Soc.* **13** (1963), 743–767.

As one readily observes, adding an edge to a maximal planar graph (of order at least 6) produces not only a topological K^5 or $K_{3,3}$, but both. In Chapter 7.3 we shall see that, more generally, every graph with n vertices and more than $3n - 6$ edges contains a TK^5 and, with one easily described class of exceptions, also a $TK_{3,3}$ (Ex. 25, Ch. 7).

Theorem 4.5.2 is widely known as 'Tutte's planarity criterion', because it follows at once from Tutte's 1963 Theorem 3.2.6 and the even earlier planarity criterion of MacLane, Theorem 4.5.1. However, Tutte appears to have been unaware of this. Theorem 4.5.2 was first noticed in the late 1970s, and proved independently of both Theorems 3.2.6 and 4.5.1, by A.K. Kelmans, The concept of a vertex in a matroid, the non-separating cycles in a graph and a new criterion for graph planarity, in *Algebraic Methods in Graph Theory*, Vol. 1, Conf. Szeged 1978, *Colloq. Math. Soc. János Bolyai* **25** (1981) 345–388. Kelmans also reproved Theorem 3.2.6 (being unaware of Tutte's proof), and noted that it can be combined with MacLane's criterion to a proof of Theorem 4.5.2.

Theorem 4.5.3 is due to W. Schnyder, Planar graphs and poset dimension, *Order* **5** (1989), 323–343. For an alternative proof and further references see F. Barrera-Cruz and P. Haxell, A note on Schnyder's theorem, *Order* **28** (2011), 221–226, arXiv:1606.08943.

The proper setting for cycle-bond duality in abstract finite graphs (and beyond) is the theory of *matroids*; see J.G. Oxley, *Matroid Theory*, Oxford University Press 1992, and H. Bruhn & R. Diestel, Infinite matroids in graphs, *Discrete Math.* **311** (2011), 1461–1471.arXiv:1011.4749 The axioms of infinite matroids are given in H. Bruhn, R. Diestel, M. Kriesell, R. Pendavingh & P. Wollan, Axioms for infinite matroids, *Adv. Math.* **239** (2013), 18–46, arXiv:1003.3919 Duality in infinite graphs is treated without matroids in H. Bruhn & R. Diestel, Duality in infinite graphs, *Comb. Probab. Comput.* **15** (2006), 75–90, and in R. Diestel & J. Pott, Dual trees must share their ends, arXiv:1106.1324.

5   Colouring

How many colours do we need to colour the countries of a map in such a way that adjacent countries are coloured differently? How many days have to be scheduled for committee meetings of a parliament if every committee intends to meet for one day and some members of parliament serve on several committees? How can we find a school timetable of minimum total length, based on the information of how often each teacher has to teach each class?

A *vertex colouring* of a graph $G = (V, E)$ is a map $c: V \to S$ such that $c(v) \neq c(w)$ whenever v and w are adjacent. The elements of the set S are called the available *colours*. All that interests us about S is its size: typically, we shall be asking for the smallest integer k such that G has a k-colouring, a vertex colouring $c: V \to \{1, \ldots, k\}$. This k is the *(vertex-) chromatic number* of G; it is denoted by $\chi(G)$. A graph G with $\chi(G) = k$ is called k-*chromatic*; if $\chi(G) \leqslant k$, we call G k-*colourable*.

vertex
colouring

chromatic
number
$\chi(G)$

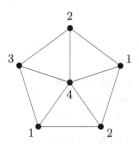

Fig. 5.0.1. A vertex colouring $V \to \{1, \ldots, 4\}$

Note that a k-colouring is nothing but a vertex partition into k independent sets, now called *colour classes*; the non-trivial 2-colourable graphs, for example, are precisely the bipartite graphs. Historically, the colouring terminology comes from the map colouring problem stated

colour
classes

© Reinhard Diestel 2017
R. Diestel, *Graph Theory*, Graduate Texts in Mathematics 173,
DOI 10.1007/978-3-662-53622-3_5

above, which leads to the problem of determining the maximum chromatic number of planar graphs. The committee scheduling problem, too, can be phrased as a vertex colouring problem—how?

edge colouring

An *edge colouring* of $G = (V, E)$ is a map $c \colon E \to S$ with $c(e) \neq c(f)$ for any adjacent edges e, f. The smallest integer k for which a k-*edge-colouring* exists, i.e. an edge colouring $c \colon E \to \{1, \dots, k\}$, is the *edge-chromatic number*, or *chromatic index*, of G; it is denoted by $\chi'(G)$. The third of our introductory questions can be modelled as an edge colouring problem in a bipartite multigraph (how?).

chromatic index $\chi'(G)$

Clearly, every edge colouring of G is a vertex colouring of its line graph $L(G)$, and vice versa; in particular, $\chi'(G) = \chi(L(G))$. The problem of finding good edge colourings may thus be viewed as a restriction of the more general vertex colouring problem to this special class of graphs. As we shall see, this relationship between the two types of colouring problem is reflected by a marked difference in our knowledge about their solutions: while there are only very rough estimates for χ, its sister χ' always takes one of two values, either Δ or $\Delta + 1$.

5.1 Colouring maps and planar graphs

If any result in graph theory has a claim to be known to the world outside, it is the following *four colour theorem* (which implies that every map can be coloured with at most four colours):

Theorem 5.1.1. (Four Colour Theorem)
Every planar graph is 4-colourable.

Some remarks about the proof of the four colour theorem and its history can be found in the notes at the end of this chapter. Here, we prove the following weakening:

Proposition 5.1.2. (Five Colour Theorem)
Every planar graph is 5-colourable.

(4.1.1)
(4.2.10)

Proof. Let G be a plane graph with $n \geq 6$ vertices and m edges. We assume inductively that every plane graph with fewer than n vertices can be 5-coloured. By Corollary 4.2.10,

n, m

$$d(G) = 2m/n \leq 2(3n - 6)/n < 6;$$

v

H

c

let $v \in G$ be a vertex of degree at most 5. By the induction hypothesis, the graph $H := G - v$ has a vertex colouring $c \colon V(H) \to \{1, \dots, 5\}$. If c uses at most 4 colours for the neighbours of v, we can extend it to a 5-colouring of G. Let us assume, therefore, that v has exactly 5 neighbours, and that these have distinct colours.

Let D be an open disc around v, so small that it meets only those
five straight edge segments of G that contain v. Let us enumerate these
segments according to their cyclic position in D as $s_1, \ldots, s_5$, and let
vv_i be the edge containing s_i $(i = 1, \ldots, 5;$ Fig. 5.1.1). Without loss of
generality we may assume that $c(v_i) = i$ for each i.

D

$s_1, \ldots, s_5$

$v_1, \ldots, v_5$

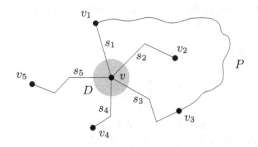

Fig. 5.1.1. The proof of the five colour theorem

Let us show first that every v_1–v_3 path $P \subseteq H - \{v_2, v_4\}$ separates
v_2 from v_4 in H. Clearly, this is the case if and only if the cycle $C :=$
$vv_1 P v_3 v$ separates v_2 from v_4 in G. We prove this by showing that v_2
and v_4 lie in different faces of C.

P

C

Let us pick an inner point x_2 of s_2 in D and an inner point x_4 of
s_4 in D. Then in $D \smallsetminus (s_1 \cup s_3) \subseteq \mathbb{R}^2 \smallsetminus C$ every point can be linked by
a polygonal arc to x_2 or to x_4. This implies that x_2 and x_4 (and hence
also v_2 and v_4) lie in different faces of C: otherwise D would meet only
one of the two faces of C, which would contradict the fact that v lies on
the frontier of both these faces (Theorem 4.1.1).

Given $i, j \in \{1, \ldots, 5\}$, let $H_{i,j}$ be the subgraph of H induced by
the vertices coloured i or j. We may assume that the component C_1 of
$H_{1,3}$ containing v_1 also contains v_3. Indeed, if we interchange the colours
1 and 3 at all the vertices of C_1, we obtain another 5-colouring of H;
if $v_3 \notin C_1$, then v_1 and v_3 are both coloured 3 in this new colouring,
and we may assign colour 1 to v. Thus, $H_{1,3}$ contains a v_1–v_3 path P.
As shown above, P separates v_2 from v_4 in H. Since $P \cap H_{2,4} = \emptyset$,
this means that v_2 and v_4 lie in different components of $H_{2,4}$. In the
component containing v_2, we now interchange the colours 2 and 4, thus
recolouring v_2 with colour 4. Now v no longer has a neighbour coloured 2,
and we may give it this colour. $\qquad\square$

$H_{i,j}$

As a backdrop to the two famous theorems above, let us cite another
well-known result:

Theorem 5.1.3. (Grötzsch 1959)
Every planar graph not containing a triangle is 3-colourable.

5.2 Colouring vertices

How do we determine the chromatic number of a given graph? How can we find a vertex-colouring with as few colours as possible? How does the chromatic number relate to other graph invariants, such as average degree, connectivity or girth?

Straight from the definition of the chromatic number we may derive the following upper bound:

Proposition 5.2.1. *Every graph G with m edges satisfies*

$$\chi(G) \leqslant \tfrac{1}{2} + \sqrt{2m + \tfrac{1}{4}}.$$

Proof. Let c be a vertex colouring of G with $k = \chi(G)$ colours. Then G has at least one edge between any two colour classes: if not, we could have used the same colour for both classes. Thus, $m \geqslant \tfrac{1}{2}k(k-1)$. Solving this inequality for k, we obtain the assertion claimed. $\qquad\square$

greedy algorithm

One obvious way to colour a graph G with not too many colours is the following *greedy algorithm*: starting from a fixed vertex enumeration $v_1, \dots, v_n$ of G, we consider the vertices in turn and colour each v_i with the first available colour—e.g., with the smallest positive integer not already used to colour any neighbour of v_i among $v_1, \dots, v_{i-1}$. In this way, we never use more than $\Delta(G) + 1$ colours, even for unfavourable choices of the enumeration $v_1, \dots, v_n$. If G is complete or an odd cycle, then this is even best possible.

In general, though, this upper bound of $\Delta + 1$ is rather generous, even for greedy colourings. Indeed, when we come to colour the vertex v_i in the above algorithm, we only need a supply of $d_{G[v_1,\dots,v_i]}(v_i) + 1$ rather than $d_G(v_i) + 1$ colours to proceed; recall that, at this stage, the algorithm ignores any neighbours v_j of v_i with $j > i$. Hence in most graphs, there will be scope for an improvement of the $\Delta + 1$ bound by choosing a particularly suitable vertex ordering to start with: one that picks vertices of large degree early (when most neighbours are ignored) and vertices of small degree last. Locally, the number $d_{G[v_1,\dots,v_i]}(v_i) + 1$ of colours required will be smallest if v_i has minimum degree in $G[v_1, \dots, v_i]$. But this is easily achieved: we just choose v_n first, with $d(v_n) = \delta(G)$, then choose as v_{n-1} a vertex of minimum degree in $G - v_n$, and so on.

colouring number col(G)

The least number k such that G has a vertex enumeration in which each vertex is preceded by fewer than k of its neighbours is called the *colouring number* $\mathrm{col}(G)$ of G. The enumeration we just discussed shows that $\mathrm{col}(G) \leqslant \max_{H \subseteq G} \delta(H) + 1$. But for $H \subseteq G$ clearly also $\mathrm{col}(G) \geqslant \mathrm{col}(H)$ and $\mathrm{col}(H) \geqslant \delta(H) + 1$, since the 'back-degree' of the last vertex in any enumeration of H is just its ordinary degree in H, which is at least $\delta(H)$. So we have proved the following:

Proposition 5.2.2. *Every graph G satisfies*

$$\chi(G) \leqslant \text{col}(G) = \max\{\,\delta(H) \mid H \subseteq G\,\}+1\,.$$

$\square$

The colouring number of a graph is closely related to its arboricity; see the remark following Theorem 2.4.3.

Proposition 5.2.2 shows that every k-chromatic graph has a subgraph of minimum degree at least $k-1$. In fact, it has a k-chromatic such subgraph:

Lemma 5.2.3. *Every k-chromatic graph has a k-chromatic subgraph of minimum degree at least $k-1$.*

[7.3]
[9.2.1]
[9.2.3]
[11.2.3]

Proof. Given G with $\chi(G) = k$, let $H \subseteq G$ be minimal with $\chi(H) = k$. If H had a vertex v of degree $d_H(v) \leqslant k-2$, we could extend a $(k-1)$-colouring of $H - v$ to one of H, contradicting the choice of H. $\square$

As we have seen, every graph G satisfies $\chi(G) \leqslant \Delta(G) + 1$, with equality for complete graphs and odd cycles. In all other cases, this general bound can be improved a little:

Theorem 5.2.4. (Brooks 1941)
Let G be a connected graph. If G is neither complete nor an odd cycle, then

$$\chi(G) \leqslant \Delta(G)\,.$$

Proof. We apply induction on $|G|$. If $\Delta(G) \leqslant 2$, then G is a path or a cycle, and the assertion is trivial. We therefore assume that $\Delta := \Delta(G) \geqslant 3$, and that the assertion holds for graphs of smaller order. Suppose that $\chi(G) > \Delta$.

Let $v \in G$ be a vertex and $H := G - v$. Then $\chi(H) \leqslant \Delta$: by induction, every component H' of H satisfies $\chi(H') \leqslant \Delta(H') \leqslant \Delta$ unless H' is complete or an odd cycle, in which case $\chi(H') = \Delta(H') + 1 \leqslant \Delta$ as every vertex of H' has maximum degree in H' and one such vertex is also adjacent to v in G.

Since H can be Δ-coloured but G cannot, we have the following:

Δ

v, H

> *Every Δ-colouring of H uses all the colours $1, \ldots, \Delta$ on the neighbours of v; in particular, $d(v) = \Delta$.* $\qquad (1)$

Given any Δ-colouring of H, let us denote the neighbour of v coloured i by v_i, $i = 1, \ldots, \Delta$. For all $i \neq j$, let $H_{i,j}$ denote the subgraph of H spanned by all the vertices coloured i or j.

$v_1, \ldots, v_\Delta$
$H_{i,j}$

$C_{i,j}$

> For all $i \neq j$, the vertices v_i and v_j lie in a common com-
> ponent $C_{i,j}$ of $H_{i,j}$. $\hspace{3em}$ (2)

Otherwise we could interchange the colours i and j in one of those components; then v_i and v_j would be coloured the same, contrary to (1).

$$C_{i,j} \text{ is always a } v_i\text{--}v_j \text{ path.} \hspace{3em} (3)$$

Indeed, let P be a v_i–v_j path in $C_{i,j}$. As $d_H(v_i) \leqslant \Delta - 1$, the neighbours of v_i have pairwise different colours: otherwise we could recolour v_i, contrary to (1). Hence the neighbour of v_i on P is its only neighbour in $C_{i,j}$, and similarly for v_j. Thus if $C_{i,j} \neq P$, then P has an inner vertex with three identically coloured neighbours in H; let u be the first such vertex on P (Fig. 5.2.1). Since at most $\Delta - 2$ colours are used on the neighbours of u, we may recolour u. But this makes $P\mathring{u}$ into a component of $H_{i,j}$, contradicting (2).

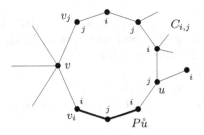

Fig. 5.2.1. The proof of (3) in Brooks's theorem

> For distinct i, j, k, the paths $C_{i,j}$ and $C_{i,k}$ meet only in v_i. $\hspace{2em}$ (4)

For if $v_i \neq u \in C_{i,j} \cap C_{i,k}$, then u has two neighbours coloured j and two coloured k, so we may recolour u. In the new colouring, v_i and v_j lie in different components of $H_{i,j}$, contrary to (2).

$v_1, \ldots, v_\Delta$

c

u

c'

$\hspace{2em}$ The proof of the theorem now follows easily. If the neighbours of v are pairwise adjacent, then each has Δ neighbours in $N(v) \cup \{v\}$ already, so $G = G[N(v) \cup \{v\}] = K^{\Delta+1}$. As G is complete, there is nothing to show. We may thus assume that $v_1 v_2 \notin G$, where $v_1, \ldots, v_\Delta$ derive their names from some fixed Δ-colouring c of H. Let $u \neq v_2$ be the neighbour of v_1 on the path $C_{1,2}$; then $c(u) = 2$. Interchanging the colours 1 and 3 in $C_{1,3}$, we obtain a new colouring c' of H; let v_i', $H_{i,j}'$, $C_{i,j}'$ etc. be defined with respect to c' in the obvious way. As a neighbour of $v_1 = v_3'$, our vertex u now lies in $C_{2,3}'$, since $c'(u) = c(u) = 2$. By (4) for c, however, the path $\mathring{v}_1 C_{1,2}$ retained its original colouring, so $u \in \mathring{v}_1 C_{1,2} \subseteq C_{1,2}'$. Hence $u \in C_{2,3}' \cap C_{1,2}'$, contradicting (4) for c'. $\hspace{2em}\square$

We have so far seen some necessary conditions for high chromaticity, in the form of upper bounds on χ. If $\chi(G) \geqslant k$, for example, then also $\Delta \geqslant k$ (unless G is complete or an odd cycle), and G has a subgraph of minimum degree at least $k - 1$. These conditions are far from sufficient, though: if $G = K_{n,n}$, say, they hold for all $k \leqslant n$ but $\chi(G) = 2$.

It would be nice also to have some sufficient conditions for $\chi \geqslant k$. If they are easy to check, they might provide useful certificates for why we are unable to colour a given graph with few colours. If they could even be shown to be necessary too, they would 'explain' why certain graphs are highly chromatic—just as the marriage condition in Hall's theorem 'explains' why certain matchings in bipartite graphs fail: its violation clearly prevents a graph from having the desired matching, and it is violated every time such a matching fails to exist.

For example, we might try to determine the class $\mathcal{X}_k$ of $\subseteq$-minimal graphs that cannot be coloured with fewer than k colours. As is easy to check (cf. Lemma 12.6.1.), a given graph G satisfies $\chi(G) \geqslant k$ if and only if it has a subgraph in $\mathcal{X}_k$, just as in Kuratowski's planarity theorem with minors or topological minors. So containing any graph from $\mathcal{X}_k$ is a certificate for $\chi \geqslant k$, and these certificates together 'explain' this phenomenon in the sense discussed.

(12.6.1)

But will these certificates be easy to find in an arbitrary k-chromatic graph, or at least easy to check? That is, will it be easy to verify that a given graph $X \in \mathcal{X}_k$ is indeed in $\mathcal{X}_k$, or even just that $\chi(X) \geqslant k$? We shall return to this question in a moment.

One obvious sufficient condition for $\chi(G) \geqslant k$ is that $K^k \subseteq G$. But this condition is not necessary: as Theorem 5.2.5 will show, k-chromatic graphs need not even contain a triangle. Hence while K^k certainly lies in $\mathcal{X}_k$, it is not its only element. Conversely, Lemma 5.2.3 implies that all the graphs in $\mathcal{X}_k$ have minimum degree at least $k - 1$; but not all graphs of minimum degree $k - 1$ are in $\mathcal{X}_k$, since they need not satisfy $\chi \geqslant k$.

The following theorem of Erdős implies that $\mathcal{X}_k$ cannot be finite. In fact, it implies that for no k is there a finite set $\mathcal{X}$ of graphs X with $\chi(X) \geqslant 3$ such that every k-chromatic graph has a subgraph in $\mathcal{X}$:

Theorem 5.2.5. (Erdős 1959)   [9.2.3]
For every integer k there exists a graph G with girth $g(G) > k$ and chromatic number $\chi(G) > k$.

Theorem 5.2.5 was first proved non-constructively using random graphs, and we shall give this proof in Chapter 11.2. Constructing graphs of large chromatic number and girth directly is not easy; cf. Exercise 24 for the simplest case.

The message of Erdős's theorem is that, contrary perhaps to what we had hoped, large chromatic number can occur as a purely global phenomenon: locally, around each vertex, a graph of large girth looks just like a tree, and in particular is 2-colourable there. But what exactly can cause high chromaticity as a global phenomenon remains a mystery.

Nevertheless, there exists a simple—though not always short— procedure to construct all the graphs of chromatic number at least k. For each $k \in \mathbb{N}$, let us define the class of k-*constructible* graphs recursively as follows:

k-con-
structible

(i) K^k is k-constructible.

(ii) If G is k-constructible and two vertices x, y of G are non-adjacent, then also $(G + xy)/xy$ is k-constructible.

(iii) If G_1, G_2 are k-constructible and there are vertices x, y_1, y_2 such that $G_1 \cap G_2 = \{x\}$ and $xy_1 \in E(G_1)$ and $xy_2 \in E(G_2)$, then also $(G_1 \cup G_2) - xy_1 - xy_2 + y_1 y_2$ is k-constructible (Fig. 5.2.2).

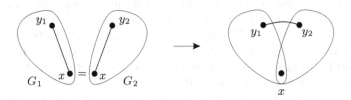

Fig. 5.2.2. The *Hajós construction* (iii)

One easily checks inductively that all k-constructible graphs—and hence their supergraphs—are at least k-chromatic. For example, any colouring of the graph $(G + xy)/xy$ in (ii) induces a colouring of G, and hence by inductive assumption uses at least k colours. Similarly, in any colouring of the graph constructed in (iii) the vertices y_1 and y_2 do not both have the same colour as x, so this colouring induces a colouring of either G_1 or G_2 and hence uses at least k colours.

It is remarkable, though, that the converse holds too:

Theorem 5.2.6. (Hajós 1961)
Let G be a graph and $k \in \mathbb{N}$. Then $\chi(G) \geqslant k$ if and only if G has a k-constructible subgraph.

Proof. Let G be a graph with $\chi(G) \geqslant k$; we show that G has a k-constructible subgraph. Suppose not; then $k \geqslant 3$. Adding some edges if necessary, let us make G edge-maximal with the property that none of its subgraphs is k-constructible. Now G is not a complete r-partite graph for any r: for then $\chi(G) \geqslant k$ would imply $r \geqslant k$, and G would contain the k-constructible graph K^k.

Since G is not a complete multipartite graph, non-adjacency is not an equivalence relation on $V(G)$. So there are vertices y_1, x, y_2 such that $y_1 x, x y_2 \notin E(G)$ but $y_1 y_2 \in E(G)$. Since G is edge-maximal without

x, y_1, y_2

a k-constructible subgraph, each edge xy_i lies in some k-constructible
subgraph H_i of $G + xy_i$ $(i = 1, 2)$. H_1, H_2

Let H_2' be an isomorphic copy of H_2 that contains x and $H_2 - H_1$ H_2'
but is otherwise disjoint from G, together with an isomorphism $v \mapsto v'$ v' etc.
from H_2 to H_2' that fixes $H_2 \cap H_2'$ pointwise. Then $H_1 \cap H_2' = \{x\}$, so

$$H := (H_1 \cup H_2') - xy_1 - xy_2' + y_1 y_2'$$

is k-constructible by (iii). One vertex at a time, let us identify in H each
vertex $v' \in H_2' - G$ with its partner v; since vv' is never an edge of H,
each of these identifications amounts to a construction step of type (ii).
Eventually, we obtain the graph

$$(H_1 \cup H_2) - xy_1 - xy_2 + y_1 y_2 \subseteq G ;$$

this is the desired k-constructible subgraph of G.   □

Does Hajós's theorem solve our Kuratowski-type problem for highly
chromatic graphs, which was to find a class of graphs of chromatic number at least k with the property that every such graph has a subgraph in
this class? Formally, it does—albeit with an infinite characterizing class
(the class of k-constructible graphs) that contains $\mathcal{X}_k$ properly. Unlike
Kuratowski's characterization of planar graphs, however, this does not—
at least not obviously—make Hajós's theorem a good characterization
of the graphs of chromatic number $< k$: as one can show, proving that
a given k-constructible graph is indeed k-constructible is just as hard as
proving that a graph of chromatic number $\geqslant k$ does indeed need at least
k colours. See the notes for details.

5.3 Colouring edges

Clearly, every graph G satisfies $\chi'(G) \geqslant \Delta(G)$. For bipartite graphs, we
have equality here:

Proposition 5.3.1. (König 1916)   [5.4.5]
Every bipartite graph G satisfies $\chi'(G) = \Delta(G)$.

Proof. We apply induction on $\|G\|$. For $\|G\| = 0$ the assertion holds.   (1.6.1)
Now assume that $\|G\| \geqslant 1$, and that the assertion holds for graphs with
fewer edges. Let $\Delta := \Delta(G)$, pick an edge $xy \in G$, and choose a Δ-
edge-colouring of $G - xy$ by the induction hypothesis. Let us refer to Δ, xy
the edges coloured α as α-*edges*, etc. α-edge

In $G - xy$, each of x and y is incident with at most $\Delta - 1$ edges.
Hence there are $\alpha, \beta \in \{1, \dots, \Delta\}$ such that x is not incident with an α, β

α-edge and y is not incident with a β-edge. If $\alpha = \beta$, we can colour the edge xy with this colour and are done; so we may assume that $\alpha \neq \beta$, and that x is incident with a β-edge.

Let us extend this edge to a maximal walk W from x whose edges are coloured β and α alternately. Since no such walk contains a vertex twice (why not?), W exists and is a path. Moreover, W does not contain y: if it did, it would end in y on an α-edge (by the choice of β) and thus have even length, so $W + xy$ would be an odd cycle in G (cf. Proposition 1.6.1). We now recolour all the edges on W, swapping α with β. By the choice of α and the maximality of W, adjacent edges of $G - xy$ are still coloured differently. We have thus found a Δ-edge-colouring of $G - xy$ in which neither x nor y is incident with a β-edge. Colouring xy with β, we extend this colouring to a Δ-edge-colouring of G. $\qquad\square$

Theorem 5.3.2. (Vizing 1964)
Every graph G satisfies

$$\Delta(G) \leqslant \chi'(G) \leqslant \Delta(G) + 1.$$

Proof. We prove the second inequality by induction on $\|G\|$. For $\|G\| = 0$ it is trivial. For the induction step let $G = (V, E)$ with $\Delta := \Delta(G) > 0$ be given, and assume that the assertion holds for graphs with fewer edges. Instead of '$(\Delta + 1)$-edge-colouring' let us just say 'colouring'.

For every edge $e \in G$ there exists a colouring of $G - e$, by the induction hypothesis. In such a colouring, the edges at a given vertex v use at most $d(v) \leqslant \Delta$ colours, so some colour $\beta \in \{1, \ldots, \Delta + 1\}$ is *missing at* v. For any other colour α, there is a unique maximal walk (possibly trivial) starting at v, whose edges are coloured alternately α and β. This walk is a path; we call it the α/β-*path from* v.

Suppose that G has no colouring. Then the following holds:

> Given $xy \in E$, and any colouring of $G - xy$ in which the
> colour α is missing at x and the colour β is missing at y, $\qquad(*)$
> the α/β-*path from y ends in x.*

Otherwise we could interchange the colours α and β along this path and colour xy with α, obtaining a colouring of G (contradiction).

Let $xy_0 \in G$ be an edge. By induction, $G_0 := G - xy_0$ has a colouring c_0. Let α be a colour missing at x in this colouring. Further, let $y_0, \ldots, y_k$ be a maximal sequence of distinct neighbours of x in G such that $c_0(xy_{i+1})$ is missing in c_0 at y_i for every $i < k$. For each of the graphs $G_i := G - xy_i$ we define a colouring c_i, setting

$$c_i(e) := \begin{cases} c_0(xy_{j+1}) & \text{for } e = xy_j \text{ with } j \in \{0, \ldots, i-1\} \\ c_0(e) & \text{otherwise;} \end{cases}$$

Margin notes:
V, E
Δ

colouring

missing

α/β-*path*

α

G_i

c_i

note that in each of these colourings the same colours are missing at x as in c_0.

Now let β be a colour missing at y_k in c_0. By (∗), the α/β-path P from y_k in G_k (with respect to c_k) ends in x, with an edge yx coloured β since α is missing at x. Since y cannot serve as y_{k+1}, by the maximality of the sequence $y_0, \dots, y_k$, we thus have $y = y_i$ for some $0 \leqslant i < k$ (Fig. 5.3.1). By definition of c_k, therefore, $\beta = c_k(xy_i) = c_0(xy_{i+1})$. By the choice of y_{i+1} this means that β was missing at y_i in c_0, and hence also in c_i. Now the α/β-path P' from y_i in G_i with respect to c_i starts with $y_i P y_k$, since the edges of $P\mathring{x}$ are coloured the same in c_i as in c_k. But in c_0, and hence in c_i, there is no edge at y_k coloured β. Therefore P' ends in y_k, contradicting (∗). $\square$

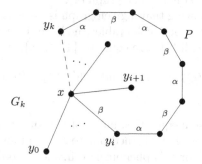

Fig. 5.3.1. The α/β-path P in $G_k = G - xy_k$

Vizing's theorem divides the finite graphs into two classes according to their chromatic index; graphs satisfying $\chi' = \Delta$ are called (imaginatively) *class 1*, those with $\chi' = \Delta + 1$ are *class 2*. There is no good characterization theorem that enables us to tell these classes apart, because no easily checkable 'certificate' is known for a graph to be class 2.

Regular graphs of large even order and large degree are class 1:

Theorem 5.3.3. (Csaba, Kühn, Lo, Osthus, Treglown 2016)
There exists an $n_0 \in \mathbb{N}$ such that, for all even $n \geqslant n_0$ and $d \geqslant n/2$, every d-regular graph G of order n satisfies $\chi'(G) = \Delta(G)$.

5.4 List colouring

In this section, we take a look at a relatively recent generalization of the concepts of colouring studied so far. This generalization may seem a little far-fetched at first glance, but it turns out to supply a fundamental link between the classical (vertex and edge) chromatic numbers of a graph and its other invariants.

Suppose we are given a graph $G = (V, E)$, and for each vertex of G a list of colours permitted at that particular vertex: when can we colour G (in the usual sense) so that each vertex receives a colour from its list? More formally, let $(S_v)_{v \in V}$ be a family of sets. We call a vertex colouring c of G with $c(v) \in S_v$ for all $v \in V$ a colouring *from the* *k-choosable* *lists* S_v. The graph G is called *k-list-colourable*, or *k-choosable*, if, for every family $(S_v)_{v \in V}$ with $|S_v| = k$ for all v, there is a vertex colouring *choice* of G from the lists S_v. The least integer k for which G is k-choosable is *number* the *list-chromatic number*, or *choice number* $\mathrm{ch}(G)$ of G.
$\mathrm{ch}(G)$

List-colourings of edges are defined analogously. The least integer k such that G has an edge colouring from any family of lists of size k $\mathrm{ch}'(G)$ is the *list-chromatic index* $\mathrm{ch}'(G)$ of G; formally, we just set $\mathrm{ch}'(G) := \mathrm{ch}(L(G))$, where $L(G)$ is the line graph of G.

In principle, showing that a given graph is k-choosable is more difficult than proving it to be k-colourable: the latter is just the special case of the former where all lists are equal to $\{1, \ldots, k\}$. Thus,

$$\mathrm{ch}(G) \geqslant \chi(G) \quad \text{and} \quad \mathrm{ch}'(G) \geqslant \chi'(G)$$

for all graphs G.

In spite of these inequalities, many of the known upper bounds for the chromatic number have turned out to be valid for the choice number, too. Examples for this phenomenon include Brooks's theorem and Proposition 5.2.2; in particular, graphs of large choice number still have subgraphs of large minimum degree. On the other hand, it is easy to construct graphs for which the two invariants are wide apart (Exercise 28). Taken together, these two facts indicate a little how far those general upper bounds on the chromatic number may be from the truth.

The following theorem shows that, in terms of its relationship to other graph invariants, the choice number differs fundamentally from the chromatic number. As mentioned before, there are 2-chromatic graphs of arbitrarily large minimum degree, e.g. the graphs $K_{n,n}$. The choice number, however, will be forced up by large values of invariants like δ, ε or κ:

Theorem 5.4.1. (Alon 1993)
There exists a function $f : \mathbb{N} \to \mathbb{N}$ such that, given any integer k, all graphs G with average degree $d(G) \geqslant f(k)$ satisfy $\mathrm{ch}(G) \geqslant k$.

The proof of Theorem 5.4.1 uses probabilistic methods as introduced in Chapter 11.

Although statements of the form $\mathrm{ch}(G) \leqslant k$ are formally stronger than the corresponding statement of $\chi(G) \leqslant k$, they can be easier to prove. A pretty example is the list version of the five colour theorem: every planar graph is 5-choosable. The proof of this does not use the five colour theorem (or even Euler's formula, on which the proof of the five colour theorem is based). We thus reobtain the five colour theorem as a corollary, with a very different proof.

Theorem 5.4.2. (Thomassen 1994)
Every planar graph is 5-choosable.

Proof. We shall prove the following assertion for all plane graphs G with at least 3 vertices:

(4.2.8)

> *Suppose that every inner face of G is bounded by a triangle and its outer face by a cycle $C = v_1 \ldots v_k v_1$. Suppose further that v_1 has already been coloured with the colour 1, and v_2 has been coloured 2. Suppose finally that with every other vertex of C a list of at least 3 colours is associated, and with every vertex of $G - C$ a list of at least 5 colours. Then the colouring of v_1 and v_2 can be extended to a colouring of G from the given lists.*

(∗)

Let us check first that (∗) implies the assertion of the theorem. Let any plane graph be given, together with a list of 5 colours for each vertex. Add edges to this graph until it is a maximal plane graph G. By Proposition 4.2.8, G is a plane triangulation; let $v_1 v_2 v_3 v_1$ be the boundary of its outer face. We now colour v_1 and v_2 (differently) from their lists, and extend this colouring by (∗) to a colouring of G from the lists given.

Let us now prove (∗), by induction on $|G|$. If $|G| = 3$, then $G = C$ and the assertion is trivial. Now let $|G| \geqslant 4$, and assume (∗) for smaller graphs. If C has a chord vw, then vw lies on two unique cycles $C_1, C_2 \subseteq C + vw$ with $v_1 v_2 \in C_1$ and $v_1 v_2 \notin C_2$. For $i = 1, 2$, let G_i denote the subgraph of G induced by the vertices lying on C_i or in its inner face (Fig. 5.4.1). Applying the induction hypothesis first to G_1 and then—with the colours now assigned to v and w—to G_2 yields the desired colouring of G.

vw

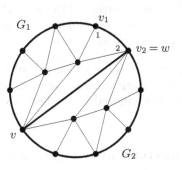

Fig. 5.4.1. The induction step with a chord vw; here the case of $w = v_2$

$u_1, \ldots, u_m$ If C has no chord, let $v_1, u_1, \ldots, u_m, v_{k-1}$ be the neighbours of v_k
in their natural cyclic order around v_k;[1] by definition of C, all those
neighbours u_i lie in the inner face of C (Fig. 5.4.2). As the inner faces
of C are bounded by triangles, $P := v_1 u_1 \ldots u_m v_{k-1}$ is a path in G, and
C' $C' := P \cup (C - v_k)$ a cycle.

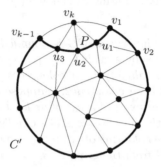

Fig. 5.4.2. The induction step without a chord

We now choose two different colours $j, \ell \neq 1$ from the list of v_k and
delete these colours from the lists of all the vertices u_i. Then every list of
a vertex on C' still has at least 3 colours, so by induction we may colour
C' and its interior, i.e. the graph $G - v_k$. At least one of the two colours
j, ℓ is not used for v_{k-1}, and we may assign that colour to v_k. □

As is often the case with induction proofs, the key to the proof above
lies in its delicately balanced strengthening of the assertion proved. Compared with ordinary colouring, the task of finding a suitable strengthening is helped greatly by the possibility to give different vertices lists of
different lengths, and thus to tailor the colouring problem more fittingly
to the structure of the graph. This suggests that maybe in other unsolved
colouring problems too it might be of advantage to aim straight for their
list version, i.e. to prove an assertion of the form $\mathrm{ch}(G) \leqslant k$ instead of
the formally weaker $\chi(G) \leqslant k$. Unfortunately, this approach fails for the
four colour theorem: planar graphs are *not* in general 4-choosable.

As mentioned before, the chromatic number of a graph and its choice
number may differ a lot. Surprisingly, however, no such examples are
known for edge colourings. Indeed it has been conjectured that none
exist:

List Colouring Conjecture. *Every graph G satisfies* $\mathrm{ch}'(G) = \chi'(G)$.

---
[1] as in the first proof of the five colour theorem

We shall prove the list colouring conjecture for bipartite graphs. As a tool we shall use orientations of graphs, defined in Chapter 1.10. If D is a directed graph and $v \in V(D)$, we denote by $N^+(v)$ the set, and by $d^+(v)$ the number, of vertices w such that D contains an edge directed from v to w.

$N^+(v)$
$d^+(v)$

To see how orientations come into play in the context of colouring, recall the greedy algorithm from Section 5.2. This colours the vertices of a graph G in turn, following a previously fixed ordering $(v_1, \ldots, v_n)$, with the smallest available colour. This ordering defines an orientation of G if we orient every edge $v_i v_j$ 'backwards', that is, from v_j to v_i if $i < j$. Then to determine a colour for v_j the algorithm only looks at previously coloured neighbours of v_j, those to which v_j sends a directed edge. In particular, if $d^+(v) < k$ for all vertices v, the algorithm will use at most k colours.

If we rewrite the proof of this fact (rather awkwardly) as a formal induction on k, the essential property of the set U of vertices coloured 1 is that every vertex in $G - U$ sends an edge to U: this ensures that $d_{G-U}^+(v) < d_G^+(v)$ for all $v \in G - U$, so we can colour $G - U$ with the remaining $k - 1$ colours by the induction hypothesis.

The following lemma generalizes these observations to list colouring, and to orientations D of G that do not necessarily come from a vertex enumeration but may contain some directed cycles. Let us call an independent set $U \subseteq V(D)$ a *kernel* of D if, for every vertex $v \in D - U$, there is an edge in D directed from v to a vertex in U. Note that kernels of non-empty directed graphs are themselves non-empty.

kernel

Lemma 5.4.3. *Let H be a graph and $(S_v)_{v \in V(H)}$ a family of lists. If H has an orientation D with $d^+(v) < |S_v|$ for every v, and such that every induced subgraph of D has a kernel, then H can be coloured from the lists S_v.*

Proof. We apply induction on $|H|$. For $|H| = 0$ we take the empty colouring. For the induction step, let $|H| > 0$. Let α be a colour occurring in one of the lists S_v, and let D be an orientation of H as stated. The vertices v with $\alpha \in S_v$ span a non-empty subgraph D' in D; by assumption, D' has a kernel $U \neq \emptyset$.

α
D'
U

Let us colour the vertices in U with α, and remove α from the lists of all the other vertices of D'. Since each of those vertices sends an edge to U, the modified lists S_v' for $v \in D - U$ again satisfy the condition $d^+(v) < |S_v'|$ in $D - U$. Since $D - U$ is an orientation of $H - U$, we can thus colour $H - U$ from those lists by the induction hypothesis. As none of these lists contains α, this extends our colouring $U \to \{\alpha\}$ to the desired list colouring of H. $\square$

In our proof of the list colouring conjecture for bipartite graphs we shall apply Lemma 5.4.3 only to colourings from lists of uniform length k.

However, note that keeping list lengths variable is essential for the proof of the lemma itself: its simple induction could not be performed with uniform list lengths.

Theorem 5.4.4. (Galvin 1995)
Every bipartite graph G satisfies ch$'(G) = \chi'(G)$.

Proof. Let $G =: (X \cup Y, E)$, where $\{X, Y\}$ is a vertex bipartition of G. We say that two edges of G *meet in X* if they share an end in X, and correspondingly for Y. Let $\chi'(G) =: k$, and let c be a k-edge-colouring of G.

Clearly, ch$'(G) \geqslant k$; we prove that ch$'(G) \leqslant k$. Our plan is to use Lemma 5.4.3 to show that the line graph H of G is k-choosable. To apply the lemma, it suffices to find an orientation D of H with $d^+(e) < k$ for every vertex e of H, and such that every induced subgraph of D has a kernel. To define D, consider adjacent $e, e' \in E$, say with $c(e) < c(e')$. If e and e' meet in X, we orient the edge $ee' \in H$ from e' towards e; if e and e' meet in Y, we orient it from e to e' (Fig 5.4.3).

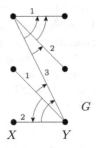

Fig. 5.4.3. Orienting the line graph of G

Let us compute $d^+(e)$ for given $e \in E = V(D)$. If $c(e) = i$, say, then every $e' \in N^+(e)$ meeting e in X has its colour in $\{1, \dots, i-1\}$, and every $e' \in N^+(e)$ meeting e in Y has its colour in $\{i+1, \dots, k\}$. As any two neighbours e' of e meeting e either both in X or both in Y are themselves adjacent and hence coloured differently, this implies $d^+(e) < k$ as desired.

It remains to show that every induced subgraph D' of D has a kernel. This, however, is immediate by the stable marriage theorem (2.1.4) for G, if we interpret the directions in D as expressing preference. Indeed, given a vertex $v \in X \cup Y$ and edges $e, e' \in V(D')$ at v, write $e <_v e'$ if the edge ee' of H is directed from e to e' in D. Then any stable matching in the graph $(X \cup Y, V(D'))$ for this set of preferences is a kernel in D'. $\square$

By Proposition 5.3.1, we now know the exact list-chromatic index of bipartite graphs:

Corollary 5.4.5. *Every bipartite graph G satisfies* ch$'(G) = \Delta(G)$. $\square$

5.5 Perfect graphs

As discussed in Section 5.2, a high chromatic number may occur as a purely global phenomenon: even when a graph has large girth, and thus locally looks like a tree, its chromatic number may be arbitrarily high. Since such 'global dependence' is obviously difficult to deal with, one may become interested in graphs where this phenomenon does not occur, i.e. whose chromatic number is high only when there is a local reason for it.

Before we make this precise, let us define two new invariants for a graph G. The greatest integer r such that $K^r \subseteq G$ is the *clique number* $\omega(G)$ of G, and the greatest integer r such that $\overline{K^r} \subseteq G$ (induced) is the *independence number* $\alpha(G)$ of G. Clearly, $\alpha(G) = \omega(\overline{G})$ and $\omega(G) = \alpha(\overline{G})$.

(margin: $\omega(G)$ $\alpha(G)$)

A graph is called *perfect* if every induced subgraph $H \subseteq G$ has chromatic number $\chi(H) = \omega(H)$, i.e. if the trivial lower bound of $\omega(H)$ colours always suffices to colour the vertices of H. Thus, while proving an assertion of the form $\chi(G) > k$ may in general be difficult, even in principle, for a given graph G, it can always be done for a perfect graph simply by exhibiting some K^{k+1} subgraph as a 'certificate' for non-colourability with k colours.

(margin: perfect)

At first glance, the structure of the class of perfect graphs appears somewhat contrived: although it is closed under induced subgraphs (if only by explicit definition), it is not closed under taking general subgraphs or supergraphs, let alone minors (examples?). However, perfection is an important notion in graph theory: the fact that several fundamental classes of graphs are perfect (as if by fluke) may serve as a superficial indication of this.[2]

What graphs, then, are perfect? Bipartite graphs are, for instance. Less trivially, the complements of bipartite graphs are perfect, too— a fact equivalent to König's duality theorem 2.1.1 (Exercise 39). The so-called *comparability graphs* are perfect, and so are the *interval graphs* (see the exercises); both these turn up in numerous applications.

In order to study at least one such example in some detail, we prove here that the chordal graphs are perfect: a graph is *chordal* (or *triangulated*) if each of its cycles of length at least 4 has a chord, i.e. if it contains no induced cycles other than triangles.

(margin: chordal)

To show that chordal graphs are perfect, we shall first characterize their structure. If G is a graph with induced subgraphs G_1, G_2 and S, such that $G = G_1 \cup G_2$ and $S = G_1 \cap G_2$, we say that G arises from G_1 and G_2 by *pasting* these graphs together along S.

(margin: pasting)

---

[2] The class of perfect graphs has duality properties with deep connections to optimization and complexity theory, which are far from understood. Theorem 5.5.6 shows the tip of an iceberg here; for more, the reader is referred to Lovász's survey cited in the notes.

[12.3.6]

Proposition 5.5.1. *A graph is chordal if and only if it can be constructed recursively by pasting along complete subgraphs, starting from complete graphs.*

Proof. If G is obtained from two chordal graphs G_1, G_2 by pasting them together along a complete subgraph, then G is clearly again chordal: any induced cycle in G lies in either G_1 or G_2, and is hence a triangle by assumption. Since complete graphs are chordal, this proves that all graphs constructible as stated are chordal.

Conversely, let G be a chordal graph. We show by induction on $|G|$ that G can be constructed as described. This is trivial if G is complete. We therefore assume that G is not complete, in particular that $|G| > 1$, and that all smaller chordal graphs are constructible as stated. Let $a, b \in G$ be two non-adjacent vertices, and let $X \subseteq V(G) \smallsetminus \{a, b\}$ be a minimal a–b separator. Let C denote the component of $G - X$ containing a, and put $G_1 := G[V(C) \cup X]$ and $G_2 := G - C$. Then G arises from G_1 and G_2 by pasting these graphs together along $S := G[X]$.

Since G_1 and G_2 are both chordal (being induced subgraphs of G) and hence constructible by induction, it suffices to show that S is complete. Suppose, then, that $s, t \in S$ are non-adjacent. By the minimality of $X = V(S)$ as an a–b separator, both s and t have a neighbour in C. Hence, there is an X-path from s to t in G_1; we let P_1 be a shortest such path. Analogously, G_2 contains a shortest X-path P_2 from s to t. But then $P_1 \cup P_2$ is a chordless cycle of length $\geqslant 4$ (Fig. 5.5.1), contradicting our assumption that G is chordal. $\qquad\square$

Margin notes: a, b X C G_1, G_2 S s, t

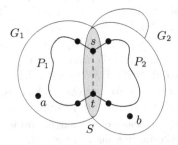

Fig. 5.5.1. If G_1 and G_2 are chordal, then so is G

Proposition 5.5.2. *Every chordal graph is perfect.*

Proof. Since complete graphs are perfect, it suffices by Proposition 5.5.1 to show that any graph G obtained from perfect graphs G_1, G_2 by pasting them together along a complete subgraph S is again perfect. So let $H \subseteq G$ be an induced subgraph; we show that $\chi(H) \leqslant \omega(H)$.

Let $H_i := H \cap G_i$ for $i = 1, 2$, and let $T := H \cap S$. Then T is again complete, and H arises from H_1 and H_2 by pasting along T. As an induced subgraph of G_i, each H_i can be coloured with $\omega(H_i)$ colours.

Since T is complete and hence coloured injectively, two such colourings, one of H_1 and one of H_2, may be combined into a colouring of H with $\max\{\omega(H_1), \omega(H_2)\} \leqslant \omega(H)$ colours—if necessary by permuting the colours in one of the H_i. □

By definition, every induced subgraph of a perfect graph is again perfect. The property of perfection can therefore be characterized by forbidden induced subgraphs: there exists a set $\mathcal{H}$ of imperfect graphs such that any graph is perfect if and only if it has no induced subgraph isomorphic to an element of $\mathcal{H}$. (For example, we may choose as $\mathcal{H}$ the set of all imperfect graphs with vertices in $\mathbb{N}$.)

Naturally, one would like to keep $\mathcal{H}$ as small as possible. It is one of the deepest results in graph theory that $\mathcal{H}$ need only contain two types of graph: the odd cycles of length $\geqslant 5$ and their complements. (Neither of these are perfect; cf. Theorem 5.5.4 below.) This fact, the famous *strong perfect graph conjecture* of Berge (1963), was proved only 40 years later:

Theorem 5.5.3. (Chudnovsky, Robertson, Seymour & Thomas 2006) *A graph G is perfect if and only if neither G nor $\overline{G}$ contains an odd cycle of length at least 5 as an induced subgraph.*

strong perfect graph theorem

In the context of perfect graphs, induced cycles of length at least 5 in G are usually referred to as *holes* in G, while holes in $\overline{G}$ are *antiholes* of G. In this jargon, the strong perfect graph theorem says that a graph is perfect if and only if it has neither holes nor antiholes.

The proof of the strong perfect graph theorem is long and technical, and it would not be too illuminating to attempt to sketch it. To shed more light on the notion of perfection, we instead give two direct proofs of its most important consequence: the *perfect graph theorem*, formerly Berge's *weak perfect graph conjecture*:

Theorem 5.5.4. (Lovász 1972) *A graph is perfect if and only if its complement is perfect.*

perfect graph theorem

The first proof we give for Theorem 5.5.4 is Lovász's original proof, which is still unsurpassed in its clarity and the amount of 'feel' for the problem it conveys. Our second proof, due to Gasparian (1996), is an elegant linear algebra proof of another theorem of Lovász's (Theorem 5.5.6), which easily implies Theorem 5.5.4.

Let us prepare our first proof of Theorem 5.5.4 by a lemma. Let G be a graph and $x \in G$ a vertex, and let G' be obtained from G by adding a vertex x' and joining it to x and all the neighbours of x. We say that G' is obtained from G by *expanding* the vertex x to an edge xx' (Fig. 5.5.2).

expanding a vertex

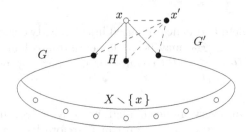

Fig. 5.5.2. Expanding the vertex x in the proof of Lemma 5.5.5

Lemma 5.5.5. *Any graph obtained from a perfect graph by expanding a vertex is again perfect.*

Proof. We use induction on the order of the perfect graph considered. Expanding the vertex of K^1 yields K^2, which is perfect. For the induction step, let G be a non-trivial perfect graph, and let G' be obtained from G by expanding a vertex $x \in G$ to an edge xx'. For our proof that G' is perfect it suffices to show $\chi(G') \leqslant \omega(G')$: every proper induced subgraph H of G' is either isomorphic to an induced subgraph of G or obtained from a proper induced subgraph of G by expanding x; in either case, H is perfect by assumption and the induction hypothesis, and can hence be coloured with $\omega(H)$ colours.

Let $\omega(G) =: \omega$; then $\omega(G') \in \{\omega, \omega+1\}$. If $\omega(G') = \omega+1$, then

$$\chi(G') \leqslant \chi(G) + 1 = \omega + 1 = \omega(G')$$

and we are done. So let us assume that $\omega(G') = \omega$. Then x lies in no $K^\omega \subseteq G$: together with x', this would yield a $K^{\omega+1}$ in G'. Let us colour G with ω colours. Since every $K^\omega \subseteq G$ meets the colour class X of x but not x itself, the graph $H := G - (X \smallsetminus \{x\})$ has clique number $\omega(H) < \omega$ (Fig. 5.5.2). Since G is perfect, we may thus colour H with $\omega - 1$ colours. Now X is independent, so the set $(X \smallsetminus \{x\}) \cup \{x'\} = V(G' - H)$ is also independent. We can therefore extend our $(\omega - 1)$-colouring of H to an ω-colouring of G', showing that $\chi(G') \leqslant \omega = \omega(G')$ as desired. □

Proof of Theorem 5.5.4. Applying induction on $|G|$, we show that the complement $\overline{G}$ of any perfect graph $G = (V, E)$ is again perfect. For $|G| = 1$ this is trivial, so let $|G| \geqslant 2$ for the induction step. Let $\mathcal{K}$ denote the set of all vertex sets of complete subgraphs of G. Put $\alpha(G) =: \alpha$, and let $\mathcal{A}$ be set of all independent vertex sets A in G with $|A| = \alpha$.

Every proper induced subgraph of $\overline{G}$ is the complement of a proper induced subgraph of G, and is hence perfect by induction. For the perfection of $\overline{G}$ it thus suffices to prove $\chi(\overline{G}) \leqslant \omega(\overline{G})$ $(= \alpha)$. To this end, we shall find a set $K \in \mathcal{K}$ such that $K \cap A \neq \emptyset$ for all $A \in \mathcal{A}$; then

$$\omega(\overline{G} - K) = \alpha(G - K) < \alpha = \omega(\overline{G}),$$

(margin notes)
x, x'
ω
X
H
$G = (V, E)$
$\mathcal{K}$
α
$\mathcal{A}$

so by the induction hypothesis

$$\chi(\overline{G}) \leqslant \chi(\overline{G} - K) + 1 = \omega(\overline{G} - K) + 1 \leqslant \omega(\overline{G})$$

as desired.

Suppose there is no such K; thus, for every $K \in \mathcal{K}$ there exists a set $A_K \in \mathcal{A}$ with $K \cap A_K = \emptyset$. Let us replace in G every vertex x by a complete graph G_x of order

$$k(x) := \left|\{ K \in \mathcal{K} \mid x \in A_K \}\right|,$$

joining all the vertices of G_x to all the vertices of G_y whenever x and y are adjacent in G. The graph G' thus obtained has vertex set $\bigcup_{x \in V} V(G_x)$, and two vertices $v \in G_x$ and $w \in G_y$ are adjacent in G' if and only if $x = y$ or $xy \in E$. Moreover, G' can be obtained by repeated vertex expansion from the graph $G[\{ x \in V \mid k(x) > 0 \}]$. Being an induced subgraph of G, this latter graph is perfect by assumption, so G' is perfect by Lemma 5.5.5. In particular,

$$\chi(G') \leqslant \omega(G'). \tag{1}$$

In order to obtain a contradiction to (1), we now compute in turn the actual values of $\omega(G')$ and $\chi(G')$. By construction of G', every maximal complete subgraph of G' has the form $G'[\bigcup_{x \in X} G_x]$ for some $X \in \mathcal{K}$. So there exists a set $X \in \mathcal{K}$ such that

$$\begin{aligned}
\omega(G') &= \sum_{x \in X} k(x) \\
&= \left|\{ (x, K) : x \in X,\ K \in \mathcal{K},\ x \in A_K \}\right| \\
&= \sum_{K \in \mathcal{K}} |X \cap A_K| \\
&\leqslant |\mathcal{K}| - 1; \tag{2}
\end{aligned}$$

the last inequality follows from the fact that $|X \cap A_K| \leqslant 1$ for all K (since A_K is independent but $G[X]$ is complete), and $|X \cap A_X| = 0$ (by the choice of A_X). On the other hand,

$$\begin{aligned}
|G'| &= \sum_{x \in V} k(x) \\
&= \left|\{ (x, K) : x \in V,\ K \in \mathcal{K},\ x \in A_K \}\right| \\
&= \sum_{K \in \mathcal{K}} |A_K| \\
&= |\mathcal{K}| \cdot \alpha.
\end{aligned}$$

As $\alpha(G') \leqslant \alpha$ by construction of G', this implies

$$\chi(G') \geqslant \frac{|G'|}{\alpha(G')} \geqslant \frac{|G'|}{\alpha} = |\mathcal{K}|. \tag{3}$$

Putting (2) and (3) together we obtain

$$\chi(G') \geqslant |\mathcal{K}| > |\mathcal{K}| - 1 \geqslant \omega(G'),$$

a contradiction to (1).   □

At first reading, the proof of Theorem 5.5.4 appears magical: it starts with an unmotivated lemma about expanding a vertex, shifts the problem to a strange graph G' obtained in this way, performs some double counting—and finished. With hindsight, however, we can understand it a little better.

The proof is completely natural up to the point where we assume that for every $K \in \mathcal{K}$ there is an $A_K \in \mathcal{A}$ such that $K \cap A_K = \emptyset$. To show that this contradicts our assumption that G is perfect, we would like to show next that its subgraph $\tilde{G}$ induced by all the A_K has a chromatic number that is too large, larger than its clique number. And, as always when we try to bound the chromatic number from below, our only hope is to bound $|\tilde{G}|/\alpha$ instead, i.e. to show that this is larger than $\omega(\tilde{G})$.

But is the bound of $|\tilde{G}|/\alpha$ likely to reflect the true value of $\chi(\tilde{G})$? In one special case it is: if the sets A_K happen to be disjoint, we have $|\tilde{G}| = |\mathcal{K}| \cdot \alpha$ and $\chi(\tilde{G}) = |\mathcal{K}|$, with the A_K as colour classes. Since every complete subgraph of $\tilde{G}$ meets each $A_{K'}$ at most once and misses its own A_K, we then have $\omega(\tilde{G}) < |\mathcal{K}| = \chi(\tilde{G})$ with the desired contradiction.

Of course, the sets A_K will not in general be disjoint. But we can make them so: by replacing every vertex x with $k(x)$ vertices, where $k(x)$ is the number of sets A_K it lives in! This is the idea behind G'. What remains is to endow G' with the right set of edges to make it perfect (assuming that G is perfect)—which leads straight to the definition of vertex expansion and Lemma 5.5.5.

Since the following characterization of perfection is symmetrical in G and $\overline{G}$, it clearly implies Theorem 5.5.4. As our proof of Theorem 5.5.6 will again be from first principles, we thus obtain a second and independent proof of Theorem 5.5.4.

Theorem 5.5.6. (Lovász 1972)
A graph G is perfect if and only if

$$|H| \leqslant \alpha(H) \cdot \omega(H) \tag{$*$}$$

for all induced subgraphs $H \subseteq G$.

Proof. Let us write $V(G) =: \{v_1, \ldots, v_n\}$, and put $\alpha := \alpha(G)$ and $\omega := \omega(G)$. The necessity of $(*)$ is immediate: if G is perfect, then every induced subgraph H of G can be partitioned into at most $\omega(H)$ colour classes each containing at most $\alpha(H)$ vertices, and $(*)$ follows.

v_i, n
α, ω

To prove sufficiency, we apply induction on $n = |G|$. Assume that every induced subgraph H of G satisfies $(*)$, and suppose that G is not perfect. By the induction hypothesis, every proper induced subgraph of G is perfect. Hence, every non-empty independent set $U \subseteq V(G)$ satisfies

$$\chi(G - U) = \omega(G - U) = \omega. \tag{1}$$

Indeed, while the first equality is immediate from the perfection of $G - U$, the second is easy: '$\leqslant$' is obvious, while $\chi(G - U) < \omega$ would imply $\chi(G) \leqslant \omega$, so G would be perfect contrary to our assumption.

Let us apply (1) to a singleton $U = \{u\}$ and consider an ω-colouring of $G - u$. Let K be the vertex set of any K^ω in G. Clearly,

$$\text{if } u \notin K \text{ then } K \text{ meets every colour class of } G - u; \tag{2}$$

$$\text{if } u \in K \text{ then } K \text{ meets all but exactly one colour class of } G - u. \tag{3}$$

Let $A_0 = \{u_1, \ldots, u_\alpha\}$ be an independent set in G of size α. Let $A_1, \ldots, A_\omega$ be the colour classes of an ω-colouring of $G - u_1$, let $A_{\omega+1}, \ldots, A_{2\omega}$ be the colour classes of an ω-colouring of $G - u_2$, and so on; altogether, this gives us $\alpha\omega + 1$ independent sets $A_0, A_1, \ldots, A_{\alpha\omega}$ in G. For each $i = 0, \ldots, \alpha\omega$, there exists by (1) a $K^\omega \subseteq G - A_i$; we denote its vertex set by K_i.

A_0

A_i

K_i

Note that if K is the vertex set of any K^ω in G, then

$$K \cap A_i = \emptyset \text{ for exactly one } i \in \{0, \ldots, \alpha\omega\}. \tag{4}$$

Indeed, if $K \cap A_0 = \emptyset$ then $K \cap A_i \neq \emptyset$ for all $i \neq 0$, by definition of A_i and (2). Similarly if $K \cap A_0 \neq \emptyset$, then $|K \cap A_0| = 1$, so $K \cap A_i = \emptyset$ for exactly one $i \neq 0$: apply (3) to the unique vertex $u \in K \cap A_0$, and (2) to all the other vertices $u \in A_0$.

Let J be the real $(\alpha\omega + 1) \times (\alpha\omega + 1)$ matrix with zero entries in the main diagonal and all other entries 1. Let $A = (a_{ij})$ be the real $(\alpha\omega + 1) \times n$ matrix whose rows are the incidence vectors of the sets A_i with $V(G)$: where $a_{ij} = 1$ if $v_j \in A_i$, and $a_{ij} = 0$ otherwise. Similarly, let B denote the real $n \times (\alpha\omega + 1)$ matrix whose columns are the incidence vectors of the sets K_j with $V(G)$. Now while $|A_i \cap K_i| = 0$ for all i by the choice of K_i, we have $A_i \cap K_j \neq \emptyset$ and hence $|A_i \cap K_j| = 1$ whenever $i \neq j$, by (4). Thus,

J

A

B

$$AB = J.$$

Since J is non-singular, this implies that A has rank $\alpha\omega + 1$. In particular, $n \geqslant \alpha\omega + 1$, which contradicts $(*)$ for $H := G$. $\qquad\square$

By Theorem 5.2.5, we cannot force a K^r subgraph, even for $r = 3$, by making the chromatic number of a graph large enough. However, this may be possible for graphs with certain specified properties, which can in turn lend some importance to these properties which they would not otherwise have. For example, the graphs with no odd hole or antihole seem like an odd class to study. But the fact that, by the strong perfect graph theorem, graphs of chromatic number k in this class have K^k subgraphs, makes them interesting: high chromatic number, for graphs in this class, will always have a local reason.

χ-bounded   Slightly more generally, a class $\mathcal{G}$ of graphs is called *χ-bounded* if there exists a function $f : \mathbb{N} \to \mathbb{N}$ such that $\chi(G) \leqslant f(r)$ for every graph $G \not\supseteq K^r$ in $\mathcal{G}$. In such graphs, then, we can force a K^r subgraph by making χ larger than $f(r)$.

Theorem 5.5.7. (Scott & Seymour 2014, (ii) with Chudnovsky 2015)

(i) *The graphs with no odd holes are χ-bounded with $f(r) = 2^{2^{r+1}}$.*

(ii) *For every integer ℓ, the graphs with no hole of length $> \ell$ are χ-bounded.*

The obvious question this raises is what we can say if both conditions are combined: given ℓ, are the graphs with no odd hole of length $> \ell$ still χ-bounded? This is indeed an old conjecture of Gýarfás, which motivated Theorem 5.5.7.

Exercises

1.⁻ Show that the four colour theorem does indeed solve the map colouring problem stated in the first sentence of the chapter. Conversely, does the 4-colourability of every map imply the four colour theorem?

2.⁻ Show that, for the map colouring problem above, it suffices to consider maps such that no point lies on the boundary of more than three countries. How does this affect the proof of the four colour theorem?

3.⁻ Try to turn the proof of the five colour theorem into one of the four colour theorem, as follows. Defining v and H as before, assume inductively that H has a 4-colouring; then proceed as before. Where does the proof fail?

4. Calculate the chromatic number of a graph in terms of the chromatic numbers of its blocks.

5. For every $n > 1$, find a bipartite graph on $2n$ vertices, ordered in such a way that the greedy algorithm uses n rather than 2 colours.

6. Consider the following approach to vertex colouring. First, find a maximal independent set of vertices and colour these with colour 1; then find a maximal independent set of vertices in the remaining graph and colour those 2, and so on. Compare this algorithm with the greedy algorithm: which is better?

7. Show that the bound of Proposition 5.2.2 is always at least as sharp as that of Proposition 5.2.1.

8. A k-chromatic graph G is called *critical* if $\chi(G - v) < k$ for every vertex $v \in G$. Determine the critical 3-chromatic graphs.

9.[+] Show that every critical k-chromatic graph is $(k - 1)$-edge-connected.

10. Formalize and prove the following statement: assuming large average degree drives the colouring number up but not the chromatic number.

11. Write $\mathrm{col}'(G)$ for the least number of colours used by the greedy algorithm for a suitable vertex ordering of a graph G. Does every G satisfy $\mathrm{col}'(G) = \mathrm{col}(G)$ or $\mathrm{col}'(G) = \chi(G)$? If so, which graphs satisfy which?

12. Find a function f such that every graph of arboricity at least $f(k)$ has colouring number at least k, and a function g such that every graph of colouring number at least $g(k)$ has arboricity at least k, for all $k \in \mathbb{N}$.

13. Given $k \in \mathbb{N}$, find a constant $c_k > 0$ such that every large enough graph G with $\alpha(G) \leqslant k$ contains a cycle of length at least $c_k |G|$.

14.[−] Find a graph G for which Brooks's theorem yields a significantly weaker bound on $\chi(G)$ than Proposition 5.2.2.

15.[+] Show that, in order to prove Brooks's theorem for a graph $G = (V, E)$, we may assume that $\kappa(G) \geqslant 2$ and $\delta(G) \geqslant 3$. Then prove the theorem under these assumptions, showing first the following two lemmas.

    (i) Let $v_1, \ldots, v_n$ be an enumeration of V. If every v_i $(i < n)$ has a neighbour v_j with $j > i$, and if $v_1 v_n, v_2 v_n \in E$ but $v_1 v_2 \notin E$, then the greedy algorithm uses at most $\Delta(G)$ colours.

    (ii) If G is not complete, it has a vertex v_n with non-adjacent neighbours v_1, v_2 that do not separate G.

16.[+] Show that the following statements are equivalent for a graph G:

    (i) $\chi(G) \leqslant k$;

    (ii) G has an orientation without directed paths of length k;

    (iii) G has an acyclic such orientation (one without directed cycles).

17. Given a graph G and $k \in \mathbb{N}$, let $P_G(k)$ denote the number of vertex colourings $V(G) \to \{1, \ldots, k\}$. Show that P_G is a polynomial in k of degree $n := |G|$, in which the coefficient of k^n is 1 and the coefficient of k^{n-1} is $-\|G\|$. (P_G is called the *chromatic polynomial* of G.)

    (Hint. Apply induction on $\|G\|$.)

18.[+] Determine the class of all graphs G for which $P_G(k) = k(k-1)^{n-1}$. (As in the previous exercise, let $n := |G|$, and let P_G denote the chromatic polynomial of G.)

19. Show that for every $k \in \mathbb{N}$ there is a unique $\subseteq$-minimal 'Kuratowski class' $\mathcal{X}_k$ of k-chromatic graphs such that every k-chromatic graph has a subgraph in $\mathcal{X}_k$, but that for $k \geqslant 3$ this class $\mathcal{X}_k$ is never finite.

20. In the definition of 'k-constructible', replace axioms (ii) and (iii) by

    (ii)′ *Every supergraph of a k-constructible graph is k-constructible.*

    (iii)′ *If x, y_1, y_2 are distinct vertices of a graph G and $y_1 y_2 \in E(G)$, and if both $G + x y_1$ and $G + x y_2$ are k-constructible, then G is k-constructible.*

    Show that a graph is k-constructible with respect to this new definition if and only if its chromatic number is at least k.

21.⁻ An $n \times n$-matrix with entries from $\{1, \ldots, n\}$ is called a *Latin square* if every element of $\{1, \ldots, n\}$ appears exactly once in each column and exactly once in each row. Recast the problem of constructing Latin squares as a colouring problem.

22. Without using Proposition 5.3.1, show that $\chi'(G) = k$ for every k-regular bipartite graph G.

23. Prove Proposition 5.3.1 from the statement of the previous exercise.

24.⁺ For every $k \in \mathbb{N}$, construct a triangle-free k-chromatic graph.

25.⁻ Let G be a graph, and let $k \in \mathbb{N}$.

    (i) Show that G has chromatic number at most k if and only if there exists a homomorphism from G to K^k.

    (ii) Show that G is bipartite if and only if there exists a homomorphism from G to K^2 or to an even cycle.

    (iii) Are there homomorphisms from C^{17} to C^7, from C^7 to C^{17}, from C^{16} to C^7, and from C^{17} to C^6?

26. Show that graphs of large girth and without a given minor are 'nearly bipartite' in the following sense. Let H be a fixed graph and C a fixed odd cycle. Use Theorem 7.2.6 to show that if G is a graph of sufficiently large girth (depending only on H and C) that does not contain H as a minor, then there is a homomorphism from G to C.

27.⁻ Without using Theorem 5.4.2, show that every plane graph is 6-list-colourable.

28. For every integer k, find a 2-chromatic graph whose choice number is at least k.

29.⁻ Find a general upper bound for $\mathrm{ch}'(G)$ in terms of $\chi'(G)$.

30. Compare the choice number of a graph with its colouring number: which is greater? Can you prove the analogue of Theorem 5.4.1 for the colouring number?

31.⁺ Prove that the choice number of K_2^r is r.

32. The *total chromatic number* $\chi''(G)$ of a graph $G = (V, E)$ is the least
     number of colours needed to colour the vertices and edges of G simulta-
     neously so that any adjacent or incident elements of $V \cup E$ are coloured
     differently. The *total colouring conjecture* says that $\chi''(G) \leqslant \Delta(G) + 2$.
     Bound the total chromatic number from above in terms of the list-
     chromatic index, and use this bound to deduce a weakening of the
     total colouring conjecture from the list colouring conjecture.

33.⁻ Does every oriented graph have a kernel? If not, does every graph have
      an orientation in which every induced subgraph has a kernel? If not,
      does every graph have an orientation that has a kernel?

34.⁺ Prove that every directed graph without odd directed cycles has a ker-
      nel.

35. Show that every bipartite planar graph is 3-list-colourable.

     (Hint. Apply the previous exercise and Lemma 5.4.3.)

36.⁻ Show that perfection is closed neither under edge deletion nor under
      edge contraction.

37.⁻ Deduce Theorem 5.5.6 from the strong perfect graph theorem.

38. Let $\mathcal{H}_1$ and $\mathcal{H}_2$ be two sets of imperfect graphs, each minimal with
     the property that a graph is perfect if and only if it has no induced
     subgraph in $\mathcal{H}_i$ $(i = 1, 2)$. Do $\mathcal{H}_1$ and $\mathcal{H}_2$ contain the same graphs, up
     to isomorphism?

39. Use König's Theorem 2.1.1 to show that the complement of any bipar-
     tite graph is perfect.

40. Using the results of this chapter, find a one-line proof of the following
     theorem of König, the dual of Theorem 2.1.1: in any bipartite graph
     without isolated vertices, the minimum number of edges meeting all
     vertices equals the maximum number of independent vertices.

41. A graph is called a *comparability graph* if there exists a partial ordering
     of its vertex set such that two vertices are adjacent if and only if they
     are comparable. Show that every comparability graph is perfect.

42. A graph G is called an *interval graph* if there exists a set $\{ I_v \mid v \in V(G) \}$
     of real intervals such that $I_u \cap I_v \neq \emptyset$ if and only if $uv \in E(G)$.

     (i) Show that every interval graph is chordal.

     (ii) Show that the complement of any interval graph is a compara-
          bility graph.

     (Conversely, a chordal graph is an interval graph if its complement is a
     comparability graph; this is a theorem of Gilmore and Hoffman (1964).)

43. Show that $\chi(H) \in \{\omega(H), \omega(H) + 1\}$ for every line graph H.

44.⁺ Characterize the graphs whose line graphs are perfect.

45. Show that a graph G is perfect if and only if every non-empty induced
     subgraph H of G contains an independent set $A \subseteq V(H)$ such that
     $\omega(H - A) < \omega(H)$.

46. Would the proof of Theorem 5.5.4 still go through if we let $\mathcal{K}$ consist of only the maximal sets of vertices spanning complete subgraphs in G? Or only of those that span a K^ω?

47.[+] Consider the graphs G for which every induced subgraph H has the property that every maximal complete subgraph of H meets every maximal independent vertex set in H.

    (i) Show that these graphs G are perfect.

    (ii) Show that these graphs G are precisely the graphs not containing an induced copy of P^3.

48.[+] Show that in every perfect graph G one can find a set $\mathcal{A}$ of independent vertex sets and a set $\mathcal{O}$ of vertex sets of complete subgraphs such that $\bigcup \mathcal{A} = V(G) = \bigcup \mathcal{O}$ and every set in $\mathcal{A}$ meets every set in $\mathcal{O}$.
(Hint. Lemma 5.5.5.)

49.[+] Let G be a perfect graph. As in the proof of Theorem 5.5.4, replace every vertex x of G with a perfect graph G_x (not necessarily complete). Show that the resulting graph G' is again perfect.

Notes

The authoritative reference work on all questions of graph colouring is T.R. Jensen & B. Toft, *Graph Coloring Problems*, Wiley 1995. Starting with a brief survey of the most important results and areas of research in the field, this monograph gives a detailed account of over 200 open colouring problems, complete with extensive background surveys and references. Most of the remarks below are discussed comprehensively in this book, and all the references for this chapter can be found there. A book specifically on edge colouring is L.M. Favrholdt, D. Scheide, M. Stiebitz & B. Toft, *Graph Edge Coloring: Vizing's Theorem and Goldberg's Conjecture*, Wiley 2012.

The *four colour problem*, whether every map can be coloured with four colours so that adjacent countries are shown in different colours, was raised by a certain Francis Guthrie in 1852. He put the question to his brother Frederick, who was then a mathematics undergraduate in Cambridge. The problem was first brought to the attention of a wider public when Cayley presented it to the London Mathematical Society in 1878. A year later, Kempe published an incorrect proof, which was in 1890 modified by Heawood into a proof of the five colour theorem. In 1880, Tait announced 'further proofs' of the four colour conjecture, which never materialized; see the notes for Chapter 10.

The first generally accepted proof of the four colour theorem was published by Appel and Haken in 1977. The proof builds on ideas that can be traced back as far as Kempe's paper, and were developed largely by Birkhoff and Heesch. Very roughly, the proof sets out first to show that every plane triangulation must contain at least one of 1482 certain 'unavoidable configurations'. In a second step, a computer is used to show that each of those configurations is 'reducible', i.e., that any plane triangulation containing such a configuration can be 4-coloured by piecing together 4-colourings of smaller plane triangulations. Taken together, these two steps amount to an inductive proof that all plane triangulations, and hence all planar graphs, can be 4-coloured.

Appel & Haken's proof has not been immune to criticism, not only because of their use of a computer. The authors responded with a 741 page long algorithmic version of their proof, which addresses the various criticisms and corrects a number of errors (e.g. by adding more configurations to the 'unavoidable' list): K. Appel & W. Haken, *Every Planar Map is Four Colorable*, American Mathematical Society 1989. A much shorter proof, which is based on the same ideas (and, in particular, uses a computer in the same way) but can be more readily verified both in its verbal and its computer part, has been given by N. Robertson, D. Sanders, P.D. Seymour & R. Thomas, The four-colour theorem, *J. Comb. Theory B* **70** (1997), 2–44.

A relatively short proof of Grötzsch's theorem was found by C. Thomassen, A short list color proof of Grötzsch's theorem, *J. Comb. Theory B* **88** (2003), 189–192. Although not touched upon in this chapter, colouring problems for graphs embedded in surfaces other than the plane form a substantial and interesting part of colouring theory; see B. Mohar & C. Thomassen, *Graphs on Surfaces*, Johns Hopkins University Press 2001.

The k-chromatic subgraph H with $\delta(H) \geqslant k - 1$ in Lemma 5.2.3 cannot in general be chosen with $\delta(H) = k - 1$. See Jensen & Toft, Chapter 5. In conjunction with Theorem 1.4.3, Lemma 5.2.3 implies that graphs of large chromatic number have highly connected subgraphs. Some of these also have large chromatic number themselves; this was proved by Alon, Kleitman, Saks, Seymour and Thomassen, Subgraphs of large connectivity and chromatic number in graphs of large chromatic number, *J. Graph Theory* **11** (1987), 367–371.

The proof of Brooks's theorem indicated in Exercise 15, where the greedy algorithm is applied to a carefully chosen vertex ordering, is due to Lovász (1973). Lovász (1968) was also the first to *construct* graphs of arbitrarily large girth and chromatic number, graphs whose existence Erdős had proved by probabilistic methods ten years earlier in Graph theory and probability, *Can. J. Math.* **11** (1959), 34–38. Another constructive proof can be found in J. Nešetřil & V. Rödl, Sparse Ramsey graphs, *Combinatorica* **4** (1984), 71–78.

A. Urquhart, The graph constructions of Hajós and Ore, *J. Graph Theory* **26** (1997), 211–215, showed that not only do the graphs of chromatic number at least k each *contain* a k-constructible graph (as by Hajós's theorem); they are in fact all themselves k-constructible. Note that, in the course of constructing a given graph, the order of the graphs constructed on the way can go both up and down, depending on which rule is applied at each step. This means that there is no obvious upper bound on the number of steps needed to construct a given graph, and indeed no such bound is known. In particular, Hajós's theorem does not provide bounded-length 'certificates' for the property of having chromatic number at least k. Unlike Kuratowski's theorem, it is therefore not a 'good characterization' in the sense of complexity theory. (See Chapter 12.7, the notes for Chapter 10, and the end of the notes for Chapter 12 for more details.)

Algebraic tools for showing that the chromatic number of a graph is large have been developed by Kleitman & Lovász (1982), by Alon & Tarsi (see Alon's paper cited below), and by Babson & Kozlov (2007).

Theorem 5.3.3 was proved by B. Csaba, D. Kühn, A. Lo, D. Osthus and

A. Treglown, Proof of the 1-factorization and Hamilton decomposition conjectures, *Memoirs of the AMS* (to appear).

List colourings were first introduced in 1976 by Vizing. Among other things, Vizing proved the list-colouring equivalent of Brooks's theorem. Voigt (1993) constructed a plane graph of order 238 that is not 4-choosable; thus, Thomassen's list version of the five colour theorem is best possible. A stimulating survey on the list-chromatic number and how it relates to the more classical graph invariants (including a proof of Theorem 5.4.1) is given by N. Alon, Restricted colorings of graphs, in (K. Walker, ed.) *Surveys in Combinatorics*, LMS Lecture Notes **187**, Cambridge University Press 1993. Both the list colouring conjecture and Galvin's proof of the bipartite case are originally stated for multigraphs. Kahn (1994) proved that the conjecture is asymptotically correct, as follows: given any $\epsilon > 0$, every graph G with large enough maximum degree satisfies $\mathrm{ch}'(G) \leqslant (1 + \epsilon)\Delta(G)$.

The total colouring conjecture (Exercise 32) was proposed around 1965 by Vizing and by Behzad; see Jensen & Toft for details.

A gentle introduction to the basic facts about perfect graphs and their applications is given by M. C. Golumbic, *Algorithmic Graph Theory and Perfect Graphs*, Academic Press 1980. A more comprehensive treatment is given in A. Schrijver, *Combinatorial optimization*, Springer 2003. Surveys on various aspects of perfect graphs are included in *Perfect Graphs* by J. Ramirez-Alfonsin & B. Reed (eds.), Wiley 2001. Our first proof of the perfect graph theorem, Theorem 5.5.4, follows Lovász's survey on perfect graphs in (L. W. Beineke and R. J. Wilson, eds.) *Selected Topics in Graph Theory 2*, Academic Press 1983. Our second proof, the proof of Theorem 5.5.6, is due to G. S. Gasparian, Minimal imperfect graphs: a simple approach, *Combinatorica* **16** (1996), 209–212.

Theorem 5.5.3 is proved in M. Chudnovsky, N. Robertson, P. D. Seymour and R. Thomas, The strong perfect graph theorem, *Ann. Math.* **164** (2006), 51–229, arXiv:math/0212070. This proof is elucidated by N. Trotignon in his 2013 survey on the arXiv:1301.5149, which also offers a short account of Lovász's proof of the (weak) perfect graph theorem. Chudnovsky, Cornuejols, Liu, Seymour and Vušković, Recognizing Berge graphs, *Combinatorica* **25** (2005), 143–186, constructed an $O(n^9)$ algorithm testing for odd holes and antiholes, and thus by the strong perfect graph theorem also for perfection.

Gyárfás's conjecture on χ-boundedness that prompted Theorem 5.5.7 is from A. Gyárfás, Problems from the world surrounding perfect graphs, Proceedings of the International Conference on Combinatorial Analysis and its Applications (Pokrzywna, 1985), *Zastos. Mat.* **19** (1987), 413–441. Part (i) of the theorem is due to A. Scott and P. D. Seymour, arXiv:1410.4118, while part (ii) was proved by M. Chudnovsky and these authors in arXiv:1506.02232.

The structure of graphs forced by forbidding some fixed induced subgraph or subgraphs, as in the strong perfect graph theorem, has been studied more widely since the proof of that theorem left a community of experts without an overriding goal. One of the central problems now studied is the *Erdős-Hajnal conjecture* that the graphs without some fixed induced subgraph have linearly large sets of vertices that are either independent or induce a complete subgraph. See Chapter 9.1 for a precise statement.

6

Flows

Let us view a graph $G = (V, E)$ as a network: its edges carry some kind of flow—of water, electricity, data or similar. How could we model this precisely?

For a start, we ought to know how much flow passes through each edge $e = xy$, and in which direction. In our model, we could assign a positive integer k to the pair (x, y) to express that a flow of k units passes through e from x to y, or assign $-k$ to (x, y) to express that k units of flow pass through e the other way, from y to x. For such an assignment $f: V^2 \to \mathbb{Z}$ we would thus have $f(x, y) = -f(y, x)$ whenever x and y are adjacent vertices.

Typically, a network will have only a few nodes where flow enters or leaves the network; at all other nodes, the total amount of flow into that node will equal the total amount of flow out of it. For our model this means that, at most nodes x, the function f will satisfy *Kirchhoff's law*

$$\sum_{y \in N(x)} f(x, y) = 0.$$

Kirchhoff's law

In this chapter, we call any map $f: V^2 \to \mathbb{Z}$ with the above two properties a 'flow' on G. Sometimes, we shall replace $\mathbb{Z}$ with another group, and as a rule we consider multigraphs rather than graphs.[1] As it turns out, the theory of those 'flows' is not only useful as a model for real flows: it blends so well with other parts of graph theory that some deep and surprising connections become visible, connections particularly with connectivity and colouring problems.

---

[1] For consistency, we shall phrase some of our proposition for graphs only: those whose proofs rely on assertions proved (for graphs) earlier in the book. However, all those results remain true for multigraphs.

© Reinhard Diestel 2017
R. Diestel, *Graph Theory*, Graduate Texts in Mathematics 173,
DOI 10.1007/978-3-662-53622-3_6

6.1 Circulations

G = (V, E)   Let $G = (V, E)$ be an (undirected) multigraph. Every edge $e = xy$ of G has two *directions*, (x, y) and (y, x). (These coincide if e is a loop, so loops have only one direction.) A triple (e, x, y) consisting of an edge together with one of its directions is an *oriented edge*. The oriented
edge   edges corresponding to e are its *orientations*, denoted by $\vec{e}$ and $\overleftarrow{e}$. Thus,
orientations $\{\vec{e}, \overleftarrow{e}\} = \{(e, x, y), (e, y, x)\}$, but we cannot generally say which is which. We write

$\vec{E}$
$$\vec{E} := \{ (e, x, y) \mid e \in E;\ x, y \in V;\ e = xy \}.$$

for the set of all oriented edges. We shall denote elements of $\vec{E}$ as $\vec{e}$, $\overleftarrow{e}$, etc. even if there is no previously defined edge e, and then use 'e' to refer to its underlying edge.

For an arbitrary set $\vec{F} \subseteq \vec{E}$ of oriented edges we put

$\overleftarrow{F}$
$$\overleftarrow{F} := \{ \overleftarrow{e} \mid \vec{e} \in \vec{F} \}.$$

Note that $\vec{E}$ itself is symmetrical: $\overleftarrow{E} = \vec{E}$. For two sets $X, Y \subseteq V$ of vertices, not necessarily disjoint, and $\vec{F} \subseteq \vec{E}$, we define

$\vec{F}(X, Y)$
$$\vec{F}(X, Y) := \{ (e, x, y) \in \vec{F} \mid x \in X;\ y \in Y;\ x \neq y \},$$

$\vec{F}(x, Y)$ abbreviate $\vec{F}(\{x\}, Y)$ to $\vec{F}(x, Y)$ etc., and write

$\vec{F}(x)$
$$\vec{F}(x) := \vec{F}(x, V) = \vec{F}(\{x\}, \overline{\{x\}}).$$

$\overline{X}$ Here, as below, $\overline{X}$ denotes the complement $V \smallsetminus X$ of a vertex set $X \subseteq V$. Note that any loops at vertices $x \in X \cap Y$ are disregarded in the definitions of $\vec{F}(X, Y)$ and $\vec{F}(x)$.

0   Let H be an abelian semigroup,[2] written additively with zero 0.
f   Given $X, Y \subseteq V$, not necessarily disjoint, and a function $f \colon \vec{E} \to H$, let

$f(X, Y)$
$$f(X, Y) := \sum_{\vec{e} \in \vec{E}(X, Y)} f(\vec{e}). \tag{1}$$

$f(x, Y)$ Instead of $f(\{x\}, Y)$ we again write $f(x, Y)$, etc.

From now on, we assume that H is an abelian group. We call f
circulation   a *circulation* on G (with values in H) if f satisfies the following two conditions:

(F1) $f(e, x, y) = -f(e, y, x)$ for all $(e, x, y) \in \vec{E}$ with $x \neq y$;

(F2) $f(v, V) = 0$ for all $v \in V$.

---

[2] This chapter contains no group theory. The only semigroups we ever consider for H are the natural numbers, the integers, the reals, the cyclic groups $\mathbb{Z}_k$, and their products $\mathbb{Z}_k \times \mathbb{Z}_m$.

If f satisfies (F1), then

$$f(X, X) = 0$$

for all $X \subseteq V$. If f satisfies (F2), then

$$f(X, V) = \sum_{x \in X} f(x, V) = 0.$$

Together, these two basic observations imply that, in a circulation, the net flow across any cut is zero:

Proposition 6.1.1. *If f is a circulation, then $f(X, \overline{X}) = 0$ for every set $X \subseteq V$.*

[6.3.1]
[6.5.2]
[6.6.1]

Proof. $f(X, \overline{X}) = f(X, V) - f(X, X) = 0 - 0 = 0.$ ☐

Since bridges form cuts by themselves, Proposition 6.1.1 implies that circulations are always zero on bridges:

Corollary 6.1.2. *If f is a circulation and $e = xy$ is a bridge in G, then $f(e, x, y) = 0$.* ☐

6.2 Flows in networks

In this section we give a brief introduction to the kind of network flow theory that is now a standard proof technique in areas such as matching and connectivity. By way of example, we shall prove a classic result of this theory, the so-called *max-flow min-cut* theorem of Ford and Fulkerson. This theorem alone implies Menger's theorem without much difficulty (Exercise 3), which indicates some of the natural power lying in this approach.

Consider the task of modelling a network with one source s and one sink t, in which the amount of flow through a given link between two nodes is subject to a certain capacity of that link. Our aim is to determine the maximum net amount of flow through the network from s to t. Somehow, this will depend both on the structure of the network and on the various capacities of its connections—how exactly, is what we wish to find out.

Let $G = (V, E)$ be a multigraph, $s, t \in V$ two fixed vertices, and $c: \vec{E} \to \mathbb{N}$ a map; we call c a *capacity function* on G, and the quadruple $N := (G, s, t, c)$ a *network*. Note that c is defined independently for the two orientations of an edge. A function $f: \vec{E} \to \mathbb{R}$ is a *flow* in N if it satisfies the following three conditions (Fig. 6.2.1):

$G = (V, E)$
s, t, c, N
network
flow

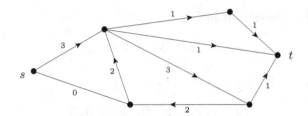

Fig. 6.2.1. A network flow in short notation: all values refer to
the direction indicated (capacities are not shown)

(F1) $f(e, x, y) = -f(e, y, x)$ for all $(e, x, y) \in \vec{E}$ with $x \neq y$;
(F2′) $f(v, V) = 0$ for all $v \in V \setminus \{s, t\}$;
(F3) $f(\vec{e}) \leqslant c(\vec{e})$ for all $\vec{e} \in \vec{E}$.

integral We call f *integral* if all its values are integers.
f Let f be a flow in N. If $S \subseteq V$ is such that $s \in S$ and $t \in \overline{S}$, we call
cut in N the pair $(S, \overline{S})$ a *cut in N*, and $c(S, \overline{S})$ the *capacity* of this cut.[3]
capacity Since f now has to satisfy only (F2′) rather than (F2), we no longer
have $f(X, \overline{X}) = 0$ for all $X \subseteq V$ (as in Proposition 6.1.1). However, the
value is the same for all cuts:

Proposition 6.2.1. *Every cut $(S, \overline{S})$ in N satisfies $f(S, \overline{S}) = f(s, V)$.*

Proof. As in the proof of Proposition 6.1.1, we have

$$
\begin{aligned}
f(S, \overline{S}) &= f(S, V) - f(S, S) \\
&\underset{\text{(F1)}}{=} f(s, V) + \sum_{v \in S \setminus \{s\}} f(v, V) - 0 \\
&\underset{\text{(F2′)}}{=} f(s, V).
\end{aligned}
$$

$\square$

total value The common value of $f(S, \overline{S})$ in Proposition 6.2.1 will be called the *total*
$|f|$ *value* of f and denoted by $|f|$;[4] the flow shown in Figure 6.2.1 has total
value 3.
By (F3), we have

$$
|f| = f(S, \overline{S}) \leqslant c(S, \overline{S})
$$

for every cut $(S, \overline{S})$ in N. Hence the total value of a flow in N is never
larger than the smallest capacity of a cut. The following *max-flow min-
cut* theorem states that this upper bound is always attained by some
flow:

---

[3] The number $c(S, \overline{S})$ is defined in (1) of Section 6.1.

[4] Thus, formally, $|f|$ may be negative. In practice, however, we can change the
sign of $|f|$ simply by swapping the roles of s and t.

Theorem 6.2.2. (Ford & Fulkerson 1956)

In every network, the maximum total value of a flow equals the minimum capacity of a cut.

Proof. Let $N = (G, s, t, c)$ be a network, and $G =: (V, E)$. We shall define a sequence $f_0, f_1, f_2, \ldots$ of integral flows in N of strictly increasing total value, i.e. with

$$|f_0| < |f_1| < |f_2| < \cdots$$

Clearly, the total value of an integral flow is again an integer, so in fact $|f_{n+1}| \geqslant |f_n| + 1$ for all n. Since all these numbers are bounded above by the capacity of any cut in N, our sequence will terminate with some flow f_n. Corresponding to this flow, we shall find a cut of capacity $c_n = |f_n|$. Since no flow can have a total value greater than c_n, and no cut can have a capacity less than $|f_n|$, this number is simultaneously the maximum and the minimum referred to in the theorem.

For f_0, we set $f_0(\vec{e}) := 0$ for all $\vec{e} \in \vec{E}$. Having defined an integral flow f_n in N for some $n \in \mathbb{N}$, we denote by S_n the set of all vertices v such that G contains an s–v walk $x_0 e_0 \ldots e_{\ell-1} x_\ell$ with

$$f_n(\vec{e_i}) < c(\vec{e_i})$$

for all $i < \ell$; here, $\vec{e_i} := (e_i, x_i, x_{i+1})$ (and, of course, $x_0 = s$ and $x_\ell = v$).

If $t \in S_n$, let $W = x_0 e_0 \ldots e_{\ell-1} x_\ell$ be the corresponding s–t walk; without loss of generality we may assume that W does not repeat any vertices. Let

$$\epsilon := \min \left\{ c(\vec{e_i}) - f_n(\vec{e_i}) \mid i < \ell \right\}.$$

Then $\epsilon > 0$, and since f_n (like c) is integral by assumption, ϵ is an integer. Let

$$f_{n+1} : \vec{e} \mapsto \begin{cases} f_n(\vec{e}) + \epsilon & \text{for } \vec{e} = \vec{e_i}, \ i = 0, \ldots, \ell - 1; \\ f_n(\vec{e}) - \epsilon & \text{for } \vec{e} = \overleftarrow{e_i}, \ i = 0, \ldots, \ell - 1; \\ f_n(\vec{e}) & \text{for } e \notin W. \end{cases}$$

Intuitively, f_{n+1} is obtained from f_n by sending additional flow of value ϵ along W from s to t (Fig. 6.2.2).

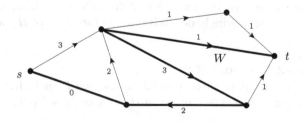

Fig. 6.2.2. An 'augmenting path' W with increment $\epsilon = 2$, for constant flow $f_n = 0$ and capacities $c = 3$

Clearly, f_{n+1} is again an integral flow in N. Let us compute its total value $|f_{n+1}| = f_{n+1}(s, V)$. Since W contains the vertex s only once, $\vec{e_0}$ is the only triple (e, x, y) with $x = s$ and $y \in V$ whose f-value was changed. This value, and hence that of $f_{n+1}(s, V)$ was raised. Therefore $|f_{n+1}| > |f_n|$ as desired.

If $t \notin S_n$, then $(S_n, \overline{S_n})$ is a cut in N. By (F3) for f_n, and the definition of S_n, we have

$$f_n(\vec{e}) = c(\vec{e})$$

for all $\vec{e} \in \vec{E}(S_n, \overline{S_n})$, so

$$|f_n| = f_n(S_n, \overline{S_n}) = c(S_n, \overline{S_n})$$

as desired. $\square$

Since the flow constructed in the proof of Theorem 6.2.2 is integral, we have also proved the following:

Corollary 6.2.3. *In every network (with integral capacity function) there exists an integral flow of maximum total value.* $\square$

6.3 Group-valued flows

Let $G = (V, E)$ be a multigraph. If f and g are two circulations on G

$f + g$ with values in the same abelian group H, then $(f + g) \colon \vec{e} \mapsto f(\vec{e}) + g(\vec{e})$

$-f$ and $-f \colon \vec{e} \mapsto -f(\vec{e})$ are again circulations. The circulations on G with values in H thus form a group in a natural way.

H-*flow*   An H-*flow* in our terminology[5] is a circulation $f \colon \vec{E} \to H$ that is *nowhere zero*, one that satisfies $f(\vec{e}) \neq 0$ for all $\vec{e} \in \vec{E}$. Note that the set of H-flows on G is not closed under addition: if two H-flows add up to zero on some oriented edge $\vec{e}$, then their sum is no longer an H-flow. By Corollary 6.1.2, a graph with an H-flow cannot have a bridge.

For finite groups H, the number of H-flows on G—and, in particular, their existence—surprisingly depends only on the order of H, not on H itself:

Theorem 6.3.1. (Tutte 1954)
For every multigraph G there exists a polynomial P such that, for any finite abelian group H, the number of H-flows on G is $P(|H| - 1)$.

---

[5] To avoid cumbersome repetitions of the phrase 'nowhere zero' before 'H-flow', we deviate slightly here from standard terminology. See the footnote in the notes.

Proof. Let $G =: (V, E)$; we use induction on $m := |E|$. Let us assume $\qquad$ (6.1.1)
first that all the edges of G are loops. Then, given any finite abelian
group H, every map $\vec{E} \to H \smallsetminus \{0\}$ is an H-flow on G. Since $|\vec{E}| = |E|$
when all edges are loops, there are $(|H| - 1)^m$ such maps, and $P := x^m$
is the polynomial sought.

Now assume there is an edge $e_0 = xy \in E$ that is not a loop; let $\qquad e_0 = xy$
$\vec{e_0} := (e_0, x, y)$ and $E' := E \smallsetminus \{e_0\}$. We consider the multigraphs $\qquad E'$

$$G_1 := G - e_0 \quad \text{and} \quad G_2 := G/e_0 \, .$$

By the induction hypothesis, there are polynomials P_i for $i = 1, 2$ such $\qquad P_1, P_2$
that, for any finite abelian group H and $k := |H| - 1$, the number of $\qquad k$
H-flows on G_i is $P_i(k)$. We shall prove that the number of H-flows on
G equals $P_2(k) - P_1(k)$; then $P := P_2 - P_1$ is the desired polynomial.

Let H be given, and denote the set of all H-flows on G by F. We $\qquad H$
are trying to show that $\qquad F$

$$|F| = P_2(k) - P_1(k) \, . \tag{1}$$

The H-flows on G_1 are precisely the restrictions to $\vec{E'}$ of those H-circu-
lations on G that are zero on e_0 but nowhere else. Let us denote the set
of these circulations on G by F_1; then $\qquad F_1$

$$|F_1| = P_1(k) \, .$$

Our aim is to show that, likewise, the H-flows on G_2 correspond bijectively to those circulations on G that are nowhere zero except possibly
on e_0. The set F_2 of those circulations on G then satisfies $\qquad F_2$

$$|F_2| = P_2(k) \, ,$$

and F_2 is the disjoint union of F_1 and F. This will prove (1), and hence
the theorem.

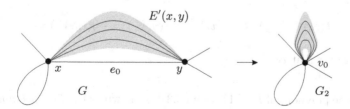

Fig. 6.3.1. Contracting the edge e_0

In G_2, let $v_0 := v_{e_0}$ be the vertex contracted from e_0 (Fig. 6.3.1; $\qquad v_0$
see Chapter 1.10). We are looking for a bijection $f \mapsto g$ between F_2

and the set of H-flows on G_2. Given f, let g be the restriction of f to $\vec{E}' \setminus \vec{E}'(y,x)$. (As the x–y edges $e \in E'$ become loops in G_2, they have only the one orientation (e, v_0, v_0) there; as its g-value, we choose $f(e,x,y)$.) Then g is indeed an H-flow on G_2; note that (F2) holds at v_0 by Proposition 6.1.1 for G, with $X := \{x,y\}$.

It remains to show that the map $f \mapsto g$ is a bijection. If we are given an H-flow g on G_2 and try to find an $f \in F_2$ with $f \mapsto g$, then $f(\vec{e})$ is already determined as $f(\vec{e}) = g(\vec{e})$ for all $\vec{e} \in \vec{E}' \setminus \vec{E}'(y,x)$; by (F1), we further have $f(\vec{e}) = -f(\overleftarrow{e})$ for all $\vec{e} \in \vec{E}'(y,x)$. Thus our map $f \mapsto g$ is bijective if and only if for given g there is always a unique way to define the remaining values of $f(\vec{e}_0)$ and $f(\overleftarrow{e}_0)$ so that f satisfies (F1) in e_0 and (F2) in x and y.

Now $f(\vec{e}_0)$ is already determined by (F2) for x and the known values of $f(\vec{e})$ for edges e at x, while $f(\overleftarrow{e}_0)$ is already determined by (F2) for y and the known values of $f(\vec{e})$ for edges e at y. Indeed, with

$$h := \sum_{\vec{e} \in \vec{E}'(x,y)} f(\vec{e}) \quad \left(= \sum_{e \in E'(x,y)} g(e, v_0, v_0) \right)$$

and $V' := V \setminus \{x,y\}$, (F2) will hold for f if and only if

$$0 = f(x,V) = f(\vec{e}_0) + h + f(x,V')$$

and

$$0 = f(y,V) = f(\overleftarrow{e}_0) - h + f(y,V'),$$

that is, if and only if we set

$$f(\vec{e}_0) := -f(x,V') - h \quad \text{and} \quad f(\overleftarrow{e}_0) := -f(y,V') + h.$$

Fortunately, defining $f(\vec{e}_0)$ and $f(\overleftarrow{e}_0)$ in this way also satisfies (F1) for f, as

$$f(\vec{e}_0) + f(\overleftarrow{e}_0) = -f(x,V') - f(y,V') = -g(v_0, V') = 0$$

by (F2) for g at v_0.   □

flow
polynomial   The polynomial P of Theorem 6.3.1 is known as the *flow polynomial* of G.

[6.4.5]   **Corollary 6.3.2.** *If H and H' are two finite abelian groups of equal order, then G has an H-flow if and only if G has an H'-flow.*   □

Corollary 6.3.2 has fundamental implications for the theory of algebraic flows: it indicates that crucial difficulties in existence proofs of H-flows are unlikely to be of a group-theoretic nature. On the other hand, being able to choose a convenient group can be quite helpful; we shall see a pretty example for this in Proposition 6.4.5.

Let $k \geqslant 1$ be an integer and $G = (V, E)$ a multigraph. A $\mathbb{Z}$-flow f on G such that $0 < |f(\vec{e})| < k$ for all $\vec{e} \in E$ is called a k-*flow*. Clearly, any k-flow is also an ℓ-flow for all $\ell > k$. Thus, we may ask which is the least integer k such that G admits a k-flow—assuming that such a k exists. We call this least k the *flow number* of G and denote it by $\varphi(G)$; if G has no k-flow for any k, we put $\varphi(G) := \infty$.

The task of determining flow numbers quickly leads to some of the deepest open problems in graph theory. We shall consider these later in the chapter. First, however, let us see how k-flows are related to the more general concept of H-flows.

There is an intimate connection between k-flows and $\mathbb{Z}_k$-flows. Let σ_k denote the natural homomorphism $i \mapsto \bar{i}$ from $\mathbb{Z}$ to $\mathbb{Z}_k$. By composition with σ_k, every k-flow defines a $\mathbb{Z}_k$-flow. As the following theorem shows, the converse holds too: from every $\mathbb{Z}_k$-flow on G we can construct a k-flow on G. In view of Corollary 6.3.2, this means that the general question about the existence of H-flows for arbitrary groups H reduces to the corresponding question for k-flows.

Theorem 6.3.3. (Tutte 1950)
A multigraph admits a k-flow if and only if it admits a $\mathbb{Z}_k$-flow.

Proof. Let g be a $\mathbb{Z}_k$-flow on a multigraph $G = (V, E)$; we construct a k-flow f on G. We may assume without loss of generality that G has no loops. Let F be the set of all functions $f \colon \vec{E} \to \mathbb{Z}$ that satisfy (F1), $|f(\vec{e})| < k$ for all $\vec{e} \in \vec{E}$, and $\sigma_k \circ f = g$; note that, like g, any $f \in F$ is nowhere zero.

Let us show first that $F \neq \emptyset$. Since we can express every value $g(\vec{e}) \in \mathbb{Z}_k$ as $\bar{i}$ with $|i| < k$ and then put $f(\vec{e}) := i$, there is clearly a map $f \colon \vec{E} \to \mathbb{Z}$ such that $|f(\vec{e})| < k$ for all $\vec{e} \in \vec{E}$ and $\sigma_k \circ f = g$. For each edge $e \in E$, let us choose one of its two orientations and denote this by $\vec{e}$. We may then define $f' \colon \vec{E} \to \mathbb{Z}$ by setting $f'(\vec{e}) := f(\vec{e})$ and $f'(\overleftarrow{e}) := -f(\vec{e})$ for every $e \in E$. Then f' is a function satisfying (F1) and with values in the desired range; it remains to show that $\sigma_k \circ f'$ and g agree not only on the chosen orientations $\vec{e}$ but also on their inverses $\overleftarrow{e}$. Since σ_k is a homomorphism, this is indeed so:

$$(\sigma_k \circ f')(\overleftarrow{e}) = \sigma_k(-f(\vec{e})) = -(\sigma_k \circ f)(\vec{e}) = -g(\vec{e}) = g(\overleftarrow{e}).$$

Hence $f' \in F$, so F is indeed non-empty.

f

K

x

X

X'

x'

W

f'

Our aim is to find an $f \in F$ that satisfies Kirchhoff's law (F2), and is thus a k-flow. As a candidate, let us consider an $f \in F$ for which the sum

$$K(f) := \sum_{x \in V} |f(x, V)|$$

of all deviations from Kirchhoff's law is least possible. We shall prove that $K(f) = 0$; then, clearly, $f(x, V) = 0$ for every x, as desired.

Suppose $K(f) \neq 0$. Since f satisfies (F1), and hence $\sum_{x \in V} f(x, V) = f(V, V) = 0$, there exists a vertex x with

$$f(x, V) > 0. \tag{1}$$

Let $X \subseteq V$ be the set of all vertices x' for which G contains a walk $x_0 e_0 \ldots e_{\ell-1} x_\ell$ from x to x' such that $f(e_i, x_i, x_{i+1}) > 0$ for all $i < \ell$; furthermore, let $X' := X \smallsetminus \{x\}$.

We first show that X' contains a vertex x' with $f(x', V) < 0$. By definition of X, we have $f(e, x', y) \leqslant 0$ for all edges $e = x'y$ such that $x' \in X$ and $y \in \overline{X}$. In particular, this holds for $x' = x$. Thus, (1) implies $f(x, X') > 0$. Then $f(X', x) < 0$ by (F1), as well as $f(X', X') = 0$. Therefore

$$\sum_{x' \in X'} f(x', V) = f(X', V) = f(X', \overline{X}) + f(X', x) + f(X', X') < 0,$$

so some $x' \in X'$ must indeed satisfy

$$f(x', V) < 0. \tag{2}$$

As $x' \in X$, there is an x–x' walk $W = x_0 e_0 \ldots e_{\ell-1} x_\ell$ such that $f(e_i, x_i, x_{i+1}) > 0$ for all $i < \ell$. We now modify f by sending some flow back along W, letting $f' : \vec{E} \to \mathbb{Z}$ be given by

$$f' : \vec{e} \mapsto \begin{cases} f(\vec{e}) - k & \text{for } \vec{e} = (e_i, x_i, x_{i+1}), \ i = 0, \ldots, \ell-1; \\ f(\vec{e}) + k & \text{for } \vec{e} = (e_i, x_{i+1}, x_i), \ i = 0, \ldots, \ell-1; \\ f(\vec{e}) & \text{for } e \notin W. \end{cases}$$

By definition of W, we have $|f'(\vec{e})| < k$ for all $\vec{e} \in \vec{E}$. Hence f', like f, lies in F.

How does the modification of f affect K? At all inner vertices v of W, as well as outside W, the deviation from Kirchhoff's law remains unchanged:

$$f'(v, V) = f(v, V) \qquad \text{for all } v \in V \smallsetminus \{x, x'\}. \tag{3}$$

For x and x', on the other hand, we have

$$f'(x, V) = f(x, V) - k \quad \text{and} \quad f'(x', V) = f(x', V) + k. \qquad (4)$$

Since g is a $\mathbb{Z}_k$-flow and hence

$$\sigma_k(f(x, V)) = g(x, V) = \overline{0} \in \mathbb{Z}_k$$

and

$$\sigma_k(f(x', V)) = g(x', V) = \overline{0} \in \mathbb{Z}_k,$$

$f(x, V)$ and $f(x', V)$ are both multiples of k. Thus $f(x, V) \geqslant k$ and $f(x', V) \leqslant -k$, by (1) and (2). But then (4) implies that

$$|f'(x, V)| < |f(x, V)| \quad \text{and} \quad |f'(x', V)| < |f(x', V)|.$$

Together with (3), this gives $K(f') < K(f)$, a contradiction to the choice of f.

   Therefore $K(f) = 0$ as claimed, and f is indeed a k-flow. $\square$

   Since the sum of two circulations with values in $\mathbb{Z}_k$ is another such circulation, $\mathbb{Z}_k$-flows are often easier to construct (by summing over suitable partial flows) than k-flows. In this way, Theorem 6.3.3 may be of considerable help in determining whether or not some given graph has a k-flow. In the following sections we shall meet a number of examples for this.
   Although Theorem 6.3.3 tells us whether a given multigraph admits a k-flow (assuming we know the value of its flow-polynomial for $k - 1$), it does not say anything about the number of such flows. By a recent result of Kochol, this number is also a polynomial in k, whose values can be bounded above and below by the corresponding values of the flow polynomial. See the notes for details.

6.4 k-Flows for small k

Trivially, a graph has a 1-flow (the empty set) if and only if it has no edges. In this section we collect a few simple examples of sufficient conditions under which a graph has a 2-, 3- or 4-flow. More examples can be found in the exercises.

Proposition 6.4.1. *A graph has a 2-flow if and only if all its degrees*   [6.6.1]
are even.

Proof. By Theorem 6.3.3, a graph $G = (V, E)$ has a 2-flow if and only if   (6.3.3)
it has a $\mathbb{Z}_2$-flow, i.e. if and only if the constant map $\vec{E} \to \mathbb{Z}_2$ with value $\overline{1}$
satisfies (F2). This is the case if and only if all degrees are even. $\square$

even
graph

For the remainder of this chapter, let us call a graph *even* if all its vertex degrees are even.

Proposition 6.4.2. *A cubic graph has a 3-flow if and only if it is bipartite.*

(1.6.1)
(6.3.3)

Proof. Let $G = (V, E)$ be a cubic graph. Let us assume first that G has a 3-flow, and hence also a $\mathbb{Z}_3$-flow f. We show that any cycle $C = x_0 \ldots x_\ell x_0$ in G has even length (cf. Proposition 1.6.1). Consider two consecutive edges on C, say $e_{i-1} := x_{i-1}x_i$ and $e_i := x_i x_{i+1}$. If f assigned the same value to these edges in the direction of the forward orientation of C, i.e. if $f(e_{i-1}, x_{i-1}, x_i) = f(e_i, x_i, x_{i+1})$, then f could not satisfy (F2) at x_i for any non-zero value of the third edge at x_i. Therefore f assigns the values $\bar{1}$ and $\bar{2}$ to the edges of C alternately, and in particular C has even length.

Conversely, let G be bipartite, with vertex bipartition $\{X, Y\}$. Since G is cubic, the map $\vec{E} \to \mathbb{Z}_3$ defined by $f(e, x, y) := \bar{1}$ and $f(e, y, x) := \bar{2}$ for all edges $e = xy$ with $x \in X$ and $y \in Y$ is a $\mathbb{Z}_3$-flow on G. By Theorem 6.3.3, then, G has a 3-flow. $\square$

What are the flow numbers of the complete graphs K^n? For odd $n > 1$, we have $\varphi(K^n) = 2$ by Proposition 6.4.1. Moreover, $\varphi(K^2) = \infty$, and $\varphi(K^4) = 4$; this is easy to see directly (and it follows from Propositions 6.4.2 and 6.4.5). Interestingly, K^4 is the only complete graph with flow number 4:

Proposition 6.4.3. *For all even $n > 4$, $\varphi(K^n) = 3$.*

(6.3.3)

Proof. Proposition 6.4.1 implies that $\varphi(K^n) \geqslant 3$ for even n. We show, by induction on n, that every $G = K^n$ with even $n > 4$ has a 3-flow.

For the induction start, let $n = 6$. Then G is the edge-disjoint union of three graphs G_1, G_2, G_3, with $G_1, G_2 = K^3$ and $G_3 = K_{3,3}$. Clearly G_1 and G_2 each have a 2-flow, while G_3 has a 3-flow by Proposition 6.4.2. The union of all these flows is a 3-flow on G.

Now let $n > 6$, and assume the assertion holds for $n-2$. Clearly, G is the edge-disjoint union of a K^{n-2} and a graph $G' = (V', E')$ with $G' = \overline{K^{n-2}} * K^2$. The K^{n-2} has a 3-flow by induction. By Theorem 6.3.3, it thus suffices to find a $\mathbb{Z}_3$-flow on G'. For every vertex z of the $\overline{K^{n-2}} \subseteq G'$, let f_z be a $\mathbb{Z}_3$-flow on the triangle $zxyz \subseteq G'$, where $e = xy$ is the edge of the K^2 in G'. Let $f : \vec{E'} \to \mathbb{Z}_3$ be the sum of these flows. Clearly, f is nowhere zero, except possibly in (e, x, y) and (e, y, x). If $f(e, x, y) \neq \bar{0}$, then f is the desired $\mathbb{Z}_3$-flow on G'. If $f(e, x, y) = \bar{0}$, then $f + f_z$ (for any z) is a $\mathbb{Z}_3$-flow on G'. $\square$

Proposition 6.4.4. *Every 4-edge-connected graph has a 4-flow.*

Proof. Let $G = (V, E)$ be a 4-edge-connected graph. By Corollary 2.4.2,   (2.4.2)
G has two edge-disjoint spanning trees T_i, $i = 1, 2$. For each edge $e \notin T_i$
let $C_{i,e}$ be the fundamental cycle with respect to T_i containing e, and
let $f_{i,e}$ be a $\mathbb{Z}_4$-flow of value $\bar{i}$ around $C_{i,e}$—more precisely: a circulation $f_{1,e}, f_{2,e}$
$\vec{E} \to \mathbb{Z}_4$ with values $\bar{i}$ and $-\bar{i}$ on the edges of $C_{i,e}$ and zero elsewhere.

Let $f_1 := \sum_{e \notin T_1} f_{1,e}$. Since each $e \notin T_1$ lies on only one cycle $C_{1,e'}$ f_1
(namely, for $e = e'$), f_1 takes only the values $\bar{1}$ and $-\bar{1} (= \bar{3})$ outside T_1.
Let

$$F := \{ e \in E(T_1) \mid f_1(e) = \bar{0} \}$$

and $f_2 := \sum_{e \in F} f_{2,e}$. As above, $f_2(e) = \bar{2} = -\bar{2}$ for all $e \in F$. Now f_2
$f := f_1 + f_2$ is the sum of circulations with values in $\mathbb{Z}_4$, and hence itself f
a circulation with values in $\mathbb{Z}_4$. Moreover, f is nowhere zero: on edges
in F it takes the value $\bar{2}$, on edges of $T_1 - F$ it agrees with f_1 (and is
hence non-zero by the choice of F), and on all edges outside T_1 it takes
one of the values $\bar{1}$ or $\bar{3}$. Hence, f is a $\mathbb{Z}_4$-flow on G, and the assertion
follows by Theorem 6.3.3. $\square$

Our next proposition describes the graphs with a 4-flow in terms
of those with a 2-flow. Given integers $m, n \geqslant 2$, write $\mathbb{Z}_m \times \mathbb{Z}_n$ for the $\mathbb{Z}_m \times \mathbb{Z}_n$
group whose elements are the pairs (a, b) with $a \in \mathbb{Z}_m$ and $b \in \mathbb{Z}_n$ and
where $(a, b) + (a', b') := (a + a', b + b')$.

Proposition 6.4.5.

    (i) *A graph has a 4-flow if and only if it is the union of two even
       subgraphs.*

   (ii) *A cubic graph has a 4-flow if and only if it is 3-edge-colourable.*

Proof. By Corollary 6.3.2 and Theorem 6.3.3, a graph has a 4-flow if and   (6.3.2)
only if it has a $\mathbb{Z}_2^2$-flow, where $\mathbb{Z}_2^2 := \mathbb{Z}_2 \times \mathbb{Z}_2$. Assertion (i) now follows   (6.3.3)
from Proposition 6.4.1.

    (ii) Let $G = (V, E)$ be a cubic graph. We assume first that G has a
$\mathbb{Z}_2^2$-flow f, and define an edge colouring $E \to \mathbb{Z}_2^2 \smallsetminus \{0\}$. As $a = -a$ for all
$a \in \mathbb{Z}_2^2$, we have $f(\vec{e}) = f(\overleftarrow{e})$ for every $\vec{e} \in \vec{E}$; let us colour the edge
e with this colour $f(\vec{e})$. Now if two edges with a common end v had
the same colour, then these two values of f would sum to zero; by (F2),
f would then assign zero to the third edge at v. As this contradicts the
definition of f, our edge colouring is correct.

    Conversely, since the three non-zero elements of $\mathbb{Z}_2^2$ sum to zero,
every 3-edge-colouring $c \colon E \to \mathbb{Z}_2^2 \smallsetminus \{0\}$ defines a $\mathbb{Z}_2^2$-flow on G by letting
$f(\vec{e}) = f(\overleftarrow{e}) = c(e)$ for all $\vec{e} \in \vec{E}$. $\square$

Corollary 6.4.6. *Every cubic 3-edge-colourable graph is bridgeless.*

                                                           $\square$

6.5 Flow-colouring duality

In this section we shall see a surprising connection between flows and colouring: every k-colouring of a plane multigraph gives rise to a k-flow on its dual, and vice versa. In this way, the investigation of k-flows on arbitrary graphs, not necessarily planar, appears as a natural generalization of the familiar map colouring problems in the plane.

$G = (V, E)$ Let $G = (V, E)$ and $G^* = (V^*, E^*)$ be dual plane multigraphs. (This implies that G and G^* are connected; see Chapter 4.6.) For simplicity, G^* let us assume that G and G^* have neither bridges nor loops and are non-trivial. For edge sets $F \subseteq E$, let us write

F^*
$$F^* := \{\, e^* \in E^* \mid e \in F \,\}.$$

Conversely, if a subset of E^* is given, we shall usually write it immediately in the form F^*, and thus let $F \subseteq E$ be defined implicitly via the bijection $e \mapsto e^*$.

Suppose we are given a circulation g on G^*: how can we employ the duality between G and G^* to derive from g some information about G? The most general property of all circulations is Proposition 6.1.1, which says that $g(X, \overline{X}) = 0$ for all $X \subseteq V^*$. By Proposition 4.6.1, the bonds $E^*(X, \overline{X})$ in G^* correspond precisely to the cycles in G. Thus if we take the composition f of the maps $e \mapsto e^*$ and g, and sum its values over the edges of a cycle in G, then this sum should again be zero. Our first aim is to formalize and prove this observation.

Of course, there is still a technical hitch: since g takes its arguments not in E^* but in $\overrightarrow{E^*}$, we cannot simply define f as above: we first have to refine the bijection $e \mapsto e^*$ into one from $\vec{E}$ to $\overrightarrow{E^*}$, i.e. assign to every $\vec{e} \in \vec{E}$ canonically one of the two orientations of e^*. This will be the purpose of our first lemma. After that, we shall show that f does indeed sum to zero along any cycle in G.

If $C = v_0 \ldots v_{\ell-1} v_0$ is a cycle with edges $e_i = v_i v_{i+1}$ (and $v_\ell := v_0$), we shall call

$\vec{C}$
$$\vec{C} := \{\, (e_i, v_i, v_{i+1}) \mid i < \ell \,\}$$

cycle with   a *cycle with orientation*. Note that this definition of $\vec{C}$ depends on the
orientation   vertex enumeration chosen to denote C: every cycle has two orientations. Conversely, of course, C can be reconstructed from the set $\vec{C}$. In practice, we shall therefore speak about C freely even when, formally, only $\vec{C}$ has been defined.

Lemma 6.5.1. *There exists a bijection* $*: \vec{e} \mapsto \vec{e}^*$ *from* $\vec{E}$ *to* $\overrightarrow{E^*}$ *with the following properties:*

    (i) *The underlying edge of* $\vec{e}^*$ *is always* e^*, *i.e.* $\vec{e}^*$ *is one of the two orientations* $\overrightarrow{e^*}, \overleftarrow{e^*}$ *of* e^*;

(ii) If $C \subseteq G$ is a cycle, $F := E(C)$, and if $X \subseteq V^*$ is such that $F^* = E^*(X, \overline{X})$, then there exists an orientation $\vec{C}$ of C with $\{ \vec{e}^* \mid \vec{e} \in \vec{C} \} = \vec{E}^*(X, \overline{X})$.

The proof of Lemma 6.5.1 is not entirely trivial: it is based on the so-called *orientability* of the plane, and we cannot give it here. Still, the assertion of the lemma is intuitively plausible. Indeed if we define for $e = vw$ and $e^* = xy$ the assignment $(e, v, w) \mapsto (e, v, w)^* \in \{(e^*, x, y), (e^*, y, x)\}$ simply by turning e and its ends clockwise onto e^* (Fig. 6.5.1), then the resulting map $\vec{e} \mapsto \vec{e}^*$ satisfies the two assertions of the lemma.

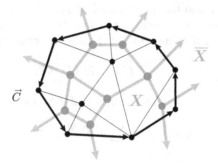

Fig. 6.5.1. Oriented cycle-cut duality

Consider a fixed bijection * as provided by Lemma 6.5.1. Given an abelian group H, let $f \colon \vec{E} \to H$ and $g \colon \vec{E}^* \to H$ be two maps such that f, g

$$f(\vec{e}) = g(\vec{e}^*)$$

for all $\vec{e} \in \vec{E}$. For $\vec{F} \subseteq \vec{E}$, we set

$$f(\vec{F}) := \sum_{\vec{e} \in \vec{F}} f(\vec{e}).$$

$f(\vec{C})$ etc.

Lemma 6.5.2.

(i) *The map g satisfies (F1) if and only if f does.*

(ii) *The map g is a circulation on G^* if and only if f satisfies (F1) and $f(\vec{C}) = 0$ for every cycle $\vec{C}$ with orientation.*

(1.9.3)
(4.6.1)
(6.1.1)

Proof. Assertion (i) follows from Lemma 6.5.1 (i) and the fact that $\vec{e} \mapsto \vec{e}^*$ is bijective.

For the forward implication of (ii), let us assume that g is a circulation on G^*, and consider a cycle $C \subseteq G$ with some given orientation. Let $F := E(C)$. By Proposition 4.6.1, F^* is a minimal cut in G^*, i.e.

$F^* = E^*(X, \overline{X})$ for some suitable $X \subseteq V^*$. By definition of f and g, Lemma 6.5.1 (ii) and Proposition 6.1.1 give

$$f(\vec{C}) = \sum_{\vec{e} \in \vec{C}} f(\vec{e}) = \sum_{\vec{d} \in \vec{E^*}(X, \overline{X})} g(\vec{d}) = g(X, \overline{X}) = 0$$

for one of the two orientations $\vec{C}$ of C. Then, by $f(\vec{C}) = -f(\overleftarrow{C})$, also the corresponding value for our given orientation of C must be zero.

For the backward implication it suffices by (i) to show that g satisfies (F2). Let $v \in V^*$ be given. By Lemma 1.9.3, the cut $E^*(v)$ is a disjoint union of bonds $D^* = E^*(X, \overline{X})$; let us name these so that always $v \in X$. Since every edge in these bonds is incident with v, we then have $\vec{E^*}(X, \overline{X}) \subseteq \vec{E^*}(v)$ also for the oriented edges.

By Proposition 4.6.1, each of the sets $D \subseteq E$ is the edge set of a cycle C in G, which by Lemma 6.5.1 (ii) has an orientation $\vec{C}$ such that

$$\{\, \vec{e}^* \mid \vec{e} \in \vec{C} \,\} = \vec{E^*}(X, \overline{X}).$$

Hence $g(X, \overline{X}) = f(\vec{C}) = 0$ by definition of f and g, giving

$$g(v, V^*) = \sum_X g(X, \overline{X}) = 0$$

as desired.   □

With the help of Lemma 6.5.2, we can now prove our colouring-flow duality theorem for plane multigraphs. If $P = v_0 \ldots v_\ell$ is a path with edges $e_i = v_i v_{i+1}$ ($i < \ell$), we set (depending on our vertex enumeration of P)

$$\vec{P} := \{\, (e_i, v_i, v_{i+1}) \mid i < \ell \,\}$$

$\vec{P}$

$v_0 \to v_\ell$
path

and call $\vec{P}$ a $v_0 \to v_\ell$ path. Again, P may be given implicitly by $\vec{P}$.

Theorem 6.5.3. (Tutte 1954)
For every dual pair G, G^* of plane multigraphs,

$$\chi(G) = \varphi(G^*).$$

(1.5.6)
V, E
V^*, E^*

Proof. Let $G =: (V, E)$ and $G^* =: (V^*, E^*)$. For $|G| \in \{1, 2\}$ the assertion is easily checked; we shall assume that $|G| \geqslant 3$, and apply induction on the number of bridges in G. If $e \in G$ is a bridge then e^* is a loop, and $G^* - e^*$ is a plane dual of G/e (why?). Hence, by the induction hypothesis,

$$\chi(G) = \chi(G/e) = \varphi(G^* - e^*) = \varphi(G^*);$$

for the first and the last equality we use that, by $|G| \geqslant 3$, e is not the only edge of G.

So all that remains to be checked is the induction start: let us assume that G has no bridge. If G has a loop, then G^* has a bridge, and $\chi(G) = \infty = \varphi(G^*)$ by convention. So we may also assume that G has no loop. Then $\chi(G)$ is finite; we shall prove for given $k \geqslant 2$ that G is k-colourable if and only if G^* has a k-flow. As G—and hence G^*—has neither loops nor bridges, we may apply Lemmas 6.5.1 and 6.5.2 to G and G^*. Let $\vec{e} \mapsto \vec{e}^*$ be a bijection between $\vec{E}$ and $\vec{E}^*$ as in Lemma 6.5.1.

We first assume that G^* has a k-flow. Then G^* also has a $\mathbb{Z}_k$-flow g. As before, let $f : \vec{E} \to \mathbb{Z}_k$ be defined by $f(\vec{e}) := g(\vec{e}^*)$. We shall use f to define a vertex colouring $c : V \to \mathbb{Z}_k$ of G.

Let T be a normal spanning tree of G, with root r, say. Put $c(r) := \overline{0}$. For every other vertex $v \in V$ let $c(v) := f(\vec{P})$, where $\vec{P}$ is the $r \to v$ path in T. To check that this is a proper colouring, consider an edge $e = vw \in E$. As T is normal, we may assume that $v < w$ in the tree-order of T. If e is an edge of T then $c(w) - c(v) = f(e, v, w)$ by definition of c, so $c(v) \neq c(w)$ since g (and hence f) is nowhere zero. If $e \notin T$, let $\vec{P}$ denote the $v \to w$ path in T. Then

$$c(w) - c(v) = f(\vec{P}) = -f(e, w, v) \neq \overline{0}$$

by Lemma 6.5.2 (ii).

Conversely, we now assume that G has a k-colouring c. Let us define $f : \vec{E} \to \mathbb{Z}$ by

$$f(e, v, w) := c(w) - c(v),$$

and $g : \vec{E}^* \to \mathbb{Z}$ by $g(\vec{e}^*) := f(\vec{e})$. Clearly, f satisfies (F1) and takes values in $\{\pm 1, \dots, \pm(k-1)\}$, so by Lemma 6.5.2 (i) the same holds for g. By definition of f, we further have $f(\vec{C}) = 0$ for every cycle $\vec{C}$ with orientation. By Lemma 6.5.2 (ii), therefore, g is a k-flow. $\qquad\square$

6.6 Tutte's flow conjectures

How can we determine the flow number of a graph? Indeed, does every (bridgeless) graph have a flow number, a k-flow for some k? Can flow numbers, like chromatic numbers, become arbitrarily large? Can we characterize the graphs admitting a k-flow, for given k?

Of these four questions, we shall answer the second and third in this section: we prove that every bridgeless graph has a 6-flow. In particular, a graph has a flow number if and only if it has no bridge. The question asking for a characterization of the graphs with a k-flow remains interesting for $k = 3, 4, 5$. Partial answers are suggested by the following three conjectures of Tutte, who initiated algebraic flow theory on graphs.

The oldest and best known of the Tutte conjectures is his *5-flow conjecture*:

Five-Flow Conjecture. (Tutte 1954)
Every bridgeless multigraph has a 5-flow.

Which graphs have a 4-flow? By Proposition 6.4.4, the 4-edge-connected graphs are among them. The Petersen graph (Fig. 6.6.1), on the other hand, is an example of a bridgeless graph without a 4-flow: since it is cubic but not 3-edge-colourable, it cannot have a 4-flow by Proposition 6.4.5 (ii).

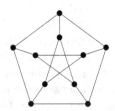

Fig. 6.6.1. The Petersen graph

Tutte's *4-flow conjecture* states that the Petersen graph must be present in every graph without a 4-flow:

Four-Flow Conjecture. (Tutte 1966)
Every bridgeless multigraph not containing the Petersen graph as a minor has a 4-flow.

By Proposition 1.7.3, we may replace the word 'minor' in the 4-flow conjecture by 'topological minor'.

Even if true, the 4-flow conjecture will not be best possible: a K^{11}, for example, contains the Petersen graph as a minor but has a 4-flow, even a 2-flow. The conjecture appears more natural for sparser graphs; a proof for cubic graphs was announced in 1998 by Robertson, Sanders, Seymour and Thomas.

snark   A cubic bridgeless graph or multigraph without a 4-flow (equivalently, without a 3-edge-colouring) is called a *snark*. The 4-flow conjecture for cubic graphs says that every snark contains the Petersen graph as a minor; in this sense, the Petersen graph has thus been shown to be the smallest snark. Snarks form the hard core both of the four colour theorem and of the 5-flow conjecture: the four colour theorem is equivalent to the assertion that no snark is planar (exercise), and it is not difficult to reduce the 5-flow conjecture to the case of snarks.[6] However, although the snarks form a very special class of graphs, none of the problems mentioned seems to become much easier by this reduction.[7]

---

[6] The same applies to another well-known conjecture, the *cycle double cover conjecture*; see Exercise 17.

[7] That snarks are elusive has been known to mathematicians for some time; cf. Lewis Carroll, *The Hunting of the Snark*, Macmillan 1876.

Three-Flow Conjecture. (Tutte 1972)
Every multigraph without a cut consisting of exactly one or exactly three edges has a 3-flow.

Again, the 3-flow conjecture will not be best possible: it is easy to construct graphs with three-edge cuts that have a 3-flow (exercise).

By our duality theorem (6.5.3), all three flow conjectures are true for planar graphs and thus motivated: the 3-flow conjecture translates to Grötzsch's theorem (5.1.3), the 4-flow conjecture to the four colour theorem (since the Petersen graph is not planar, it is not a minor of a planar graph), the 5-flow conjecture to the five colour theorem.

We finish this section with the main result of the chapter:

Theorem 6.6.1. (Seymour 1981)
Every bridgeless graph has a 6-flow.

Proof. Let $G = (V, E)$ be a bridgeless graph. Since 6-flows on the components of G will add up to a 6-flow on G, we may assume that G is connected; as G is bridgeless, it is then 2-edge-connected. Note that any two vertices in a 2-edge-connected graph lie in some common even connected subgraph—for example, in the union of two edge-disjoint paths linking these vertices by Menger's theorem (3.3.6 (ii)). We shall use this fact repeatedly. (3.3.6) (6.1.1) (6.4.1)

We shall construct a sequence $H_0, \dots, H_n$ of disjoint connected and even subgraphs of G, together with a sequence $F_1, \dots, F_n$ of non-empty sets of edges between them. The sets F_i will each contain only one or two edges, between H_i and $H_0 \cup \dots \cup H_{i-1}$. We write $H_i =: (V_i, E_i)$, $H_0, \dots, H_n$ $F_1, \dots, F_n$ V_i, E_i

$$H^i := (H_0 \cup \dots \cup H_i) + (F_1 \cup \dots \cup F_i)$$ H^i

and $H^i =: (V^i, E^i)$. Note that each $H^i = (H^{i-1} \cup H_i) + F_i$ is connected (induction on i). Our assumption that H_i is even implies by Proposition 6.4.1 (or directly by Proposition 1.2.1) that H_i has no bridge. V^i, E^i

As H_0 we choose any K^1 in G. Now assume that $H_0, \dots, H_{i-1}$ and $F_1, \dots, F_{i-1}$ have been defined for some $i > 0$. If $V^{i-1} = V$, we terminate the construction and set $i - 1 =: n$. Otherwise, we let $X_i \subseteq \overline{V^{i-1}}$ be minimal such that $X_i \neq \emptyset$ and n X_i

$$\left| E(X_i, \overline{V^{i-1}} \setminus X_i) \right| \leqslant 1 \tag{1}$$

(Fig. 6.6.2); such an X_i exists, because $\overline{V^{i-1}}$ is a candidate. Since G is 2-edge-connected, (1) implies that $E(X_i, V^{i-1}) \neq \emptyset$. By the minimality of X_i, the graph $G[X_i]$ is connected and bridgeless, i.e. 2-edge-connected or a K^1. As the elements of F_i we pick one or two edges F_i

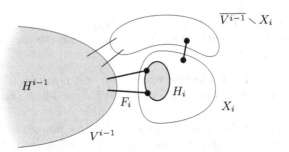

Fig. 6.6.2. Constructing the H_i and F_i

from $E(X_i, V^{i-1})$, if possible two. As H_i we choose any connected even subgraph of $G[X_i]$ containing the ends in X_i of the edges in F_i.

When our construction is complete, we set $H^n =: H$ and $E' := E \smallsetminus E(H)$. By definition of n, H is a spanning connected subgraph of G.

We now define, by 'reverse' induction, a sequence $f_n, \ldots, f_0$ of circulations on G with values in $\mathbb{Z}_3$. For every edge $e \in E'$, let $\vec{C_e}$ be a cycle (with orientation) in $H + e$ containing e, and f_e a positive flow around $\vec{C_e}$; formally, we let f_e be a circulation $\vec{E} \to \mathbb{Z}_3$ on G such that $f_e^{-1}(\vec{0}) = \vec{E} \smallsetminus (\vec{C_e} \cup \overleftarrow{C_e})$. Let f_n be the sum of all these f_e. Since each $e' \in E'$ lies on just one of the cycles C_e (namely, on $C_{e'}$), we have $f_n(\vec{e}) \neq \vec{0}$ for all $\vec{e} \in \vec{E'}$.

Assume now that circulations $f_n, \ldots, f_i$ have been defined for some $i \leqslant n$, and that

$$f_i(\vec{e}) \neq \vec{0} \quad \text{for all} \quad \vec{e} \in \vec{E'} \cup \bigcup_{j>i} \vec{F_j}, \tag{2}$$

where $\vec{F_j} := \{ \vec{e} \in \vec{E} \mid e \in F_j \}$. Our aim is to define f_{i-1} in such a way that (2) also holds for $i - 1$.

We first consider the case that $|F_i| = 1$, say $F_i = \{e\}$. We then let $f_{i-1} := f_i$, and thus have to show that f_i is non-zero on (the two orientations of) e. Our assumption of $|F_i| = 1$ implies by the choice of F_i that G contains no X_i-V^{i-1} edge other than e. Since G is 2-edge-connected, it therefore has at least—and thus, by (1), exactly—one edge e' between X_i and $\overline{V^{i-1}} \smallsetminus X_i$. We show that f_i is non-zero on e'; as $\{e, e'\}$ is a cut in G, this implies by Proposition 6.1.1 that f_i is also non-zero on e.

To show that f_i is non-zero on e', we use (2): we show that $e' \in E' \cup \bigcup_{j>i} F_j$, i.e. that e' lies in no H_k and in no F_j with $j \leqslant i$. Since e' has both ends in $\overline{V^{i-1}}$, it clearly lies in no F_j with $j \leqslant i$ and in no H_k with $k < i$. But every H_k with $k \geqslant i$ is a subgraph of $G[\overline{V^{i-1}}]$. Since e' is a bridge of $G[\overline{V^{i-1}}]$ but H_k has no bridge, this means that $e' \notin H_k$. Hence, f_{i-1} does indeed satisfy (2) for $i - 1$ in the case considered.

It remains to consider the case that $|F_i| = 2$, say $F_i = \{e_1, e_2\}$.
Since H_i and H^{i-1} are both connected, we can find a cycle C in $H^i = (H_i \cup H^{i-1}) + F_i$ that contains e_1 and e_2. If f_i is non-zero on both these edges, we again let $f_{i-1} := f_i$. Otherwise, there are orientations $\vec{e_1}$ and $\vec{e_2}$ of e_1 and e_2 such that, without loss of generality, $f_i(\vec{e_1}) = \bar{0}$ and $f_i(\vec{e_2}) \in \{\bar{0}, \bar{1}\}$. Let $\vec{C}$ be the orientation of C with $\vec{e_2} \in \vec{C}$, and let g be a flow of value $\bar{1}$ around $\vec{C}$ (formally: let $g\colon \vec{E} \to \mathbb{Z}_3$ be a circulation on G such that $g(\vec{e_2}) = \bar{1}$ and $g^{-1}(\bar{0}) = \vec{E} \setminus (\vec{C} \cup \overleftarrow{C})$). We then let $f_{i-1} := f_i + g$. By choice of the orientations $\vec{e_1}$ and $\vec{e_2}$, f_{i-1} is non-zero on both edges. Since f_{i-1} agrees with f_i on all of $\vec{E'} \cup \bigcup_{j>i} \vec{F_j}$ and (2) holds for i, we again have (2) also for $i-1$.

Eventually, f_0 will be a circulation on G with values in $\mathbb{Z}_3$ that is nowhere zero except possibly on edges of $H_0 \cup \ldots \cup H_n$. But there is a $\mathbb{Z}_2$-flow $H_0 \cup \ldots \cup H_n$ by Proposition 6.4.1. Together, these yield a $(\mathbb{Z}_3 \times \mathbb{Z}_2)$-flow on G. By Corollary 6.3.2 and Theorem 6.3.3, G has a 6-flow. $\square$

Exercises

1.⁻ Prove Proposition 6.2.1 by induction on $|S|$.

2. (i)⁻ Given $n \in \mathbb{N}$, find a capacity function for the network below such that the algorithm from the proof of the max-flow min-cut theorem will need more than n augmenting paths W if these are badly chosen.

(ii)⁺ Show that, if all augmenting paths are chosen as short as possible, their number is bounded by a function of the size of the network.

3.⁺ Derive Menger's Theorem 3.3.5 from the max-flow min-cut theorem.
(Hint. The edge version is easy. For the vertex version, apply the edge version to a suitable auxiliary graph.)

4.⁻ Let $f\colon \vec{E} \to H$ be a circulation on G and $g\colon H \to H'$ a group homomorphism. Show that $g \circ f$ is a circulation on G. Is $g \circ f$ an H'-flow if f is an H-flow?

5. View the group of circulations on a graph with values in $\mathbb{Z}_2$ as a vector space over $\mathbb{Z}_2$. Find a space in Chapter 1.9 to which it is isomorphic, and write down an explicit isomorphism.

6. Let H be an abelian group, $G = (V, E)$ a connected graph, T a spanning tree, and f a map from the orientations of the edges in $E \setminus E(T)$ to H that satisfies (F1). Show that f extends uniquely to a circulation on G with values in H.

e_1, e_2

C

7. (continued)

Let $\mathcal{V}_H = \mathcal{V}_H(G)$ be the group of all maps $V \to H$, and $\mathcal{E}_H = \mathcal{E}_H(G)$ the group of all maps $\vec{E} \to H$ satisfying (F1), both with pointwise addition. Every $\varphi \in \mathcal{V}_H$ defines a $\psi \in \mathcal{E}_H$ by $\psi(e, x, y) := \varphi(y) - \varphi(x)$.

(i) Show that these ψ form a subgroup $\mathcal{B}_H = \mathcal{B}_H(G)$ of $\mathcal{E}_H$ with
$$\mathcal{B}_H = \{ \psi \in \mathcal{E}_H \mid \psi(\vec{C}) = 0 \text{ for every oriented cycle } C \subseteq G \},$$
where $\psi(\vec{C}) := \sum_{\vec{e} \in \vec{C}} \psi(\vec{e})$.

(ii) Show that every map $\vec{E}(T) \to H$ satisfying (F1) extends uniquely to a map in $\mathcal{B}_H$.

8.$^+$ (continued)

Let $\mathcal{C}_H$ denote the group of all circulations on G with values in H.

(i) Show that $\mathcal{E}_H/\mathcal{B}_H$ is isomorphic to $\mathcal{C}_H$.

(ii) Show that $\mathcal{E}_H/\mathcal{C}_H$ is isomorphic to $\mathcal{B}_H$.

9.$^-$ Given $k \geqslant 1$, show that a graph has a k-flow if and only if each of its blocks has a k-flow.

10.$^-$ Show that $\varphi(G/e) \leqslant \varphi(G)$ whenever G is a multigraph and e an edge of G. Does this imply that, for every k, the class of all multigraphs admitting a k-flow is closed under taking minors?

11.$^-$ Work out the flow number of K^4 directly, without using any results from the text.

Do not use the 6-flow Theorem 6.6.1 for the following three exercises.

12. Show that $\varphi(G) < \infty$ for every bridgeless multigraph G.

13. Let G be a bridgeless connected graph with n vertices and m edges. By considering a normal spanning tree of G, show that $\varphi(G) \leqslant m - n + 2$.

14. Assume that a graph G has m spanning trees such that no edge of G lies in all of these trees. Show that $\varphi(G) \leqslant 2^m$.

15. Show that every 4-edge-connected graph has a 4-flow.

16. Show that every graph with a Hamilton cycle has a 4-flow. (A *Hamilton cycle* of G is a cycle in G that contains all the vertices of G.)

17. A family of (not necessarily distinct) subgraphs of a graph G is called a *double cover* of G if every edge of G lies on exactly two of these subgraphs. The *cycle double cover conjecture* asserts that every bridgeless multigraph admits a double cover by cycles. Prove the conjecture for graphs with a 4-flow.

18.$^-$ Determine the flow number of $C^5 * K^1$, the wheel with 5 spokes.

19. Find bridgeless graphs G and $H = G - e$ such that $2 < \varphi(G) < \varphi(H)$.

20. Prove Proposition 6.4.1 without using Theorem 6.3.3.

21.$^+$ Prove that a plane triangulation is 3-colourable if and only if all its vertices have even degree.

22. Show that the 3-flow conjecture for planar multigraphs is equivalent to Grötzsch's Theorem 5.1.3.

23. (i)⁻ Show that the four colour theorem is equivalent to the non-existence of a planar snark, i.e. to the statement that every cubic bridgeless planar multigraph has a 4-flow.

    (ii) Can 'bridgeless' in (i) be replaced by '3-connected'?

24.⁺ Show that a graph $G = (V, E)$ has a k-flow if and only if it has an orientation D that directs, for every $X \subseteq V$, at least $1/k$ of the edges in $E(X, \overline{X})$ from X towards $\overline{X}$.

25.⁻ Generalize the 6-flow Theorem 6.6.1 to multigraphs.

Notes

Network flow theory is an application of graph theory that has had a major and lasting impact on its development over decades. As is illustrated already by the fact that Menger's theorem can be deduced easily from the max-flow min-cut theorem (Exercise 3), the interaction between graphs and networks may go either way: while 'pure' results in areas such as connectivity, matching and random graphs have found applications in network flows, the intuitive power of the latter has boosted the development of proof techniques that have in turn brought about theoretic advances.

The classical reference for network flows is L.R. Ford & D.R. Fulkerson, *Flows in Networks*, Princeton University Press 1962. More recent and comprehensive accounts are given by R.K. Ahuja, T.L. Magnanti & J.B. Orlin, *Network flows*, Prentice-Hall 1993, by A. Frank in his chapter in the *Handbook of Combinatorics* (R.L. Graham, M. Grötschel & L. Lovász, eds.), North-Holland 1995, and by A. Schrijver, *Combinatorial optimization*, Springer 2003. An introduction to graph algorithms in general is given in A. Gibbons, *Algorithmic Graph Theory*, Cambridge University Press 1985.

If one recasts the maximum flow problem in linear programming terms, one can derive the max-flow min-cut theorem from the linear programming duality theorem; see A. Schrijver, *Theory of integer and linear programming*, Wiley 1986.

The more algebraic theory of group-valued flows and k-flows has been developed largely by Tutte; he gives a thorough account in his monograph W.T. Tutte, *Graph Theory*, Addison-Wesley 1984. The fact that the number of k-flows of a multigraph is a polynomial in k, whose values can be bounded in terms of the corresponding values of the flow polynomial, was proved by M. Kochol, Polynomials associated with nowhere-zero[8] flows, *J. Comb. Theory B* **84** (2002), 260–269.

Tutte's flow conjectures are covered also in F. Jaeger's survey, Nowhere-zero flow problems, in (L.W. Beineke & R.J. Wilson, eds.) *Selected Topics in Graph Theory 3*, Academic Press 1988. For the flow conjectures, see also

---

[8] In the literature, the term 'flow' is often used to mean what we have called 'circulation', i.e. flows are not required to be nowhere zero unless this is stated explicitly.

T.R. Jensen & B. Toft, *Graph Coloring Problems*, Wiley 1995. Seymour's 6-flow theorem is proved in P.D. Seymour, Nowhere-zero 6-flows, *J. Comb. Theory B* **30** (1981), 130–135. This paper also indicates how Tutte's 5-flow conjecture reduces to snarks. The proof of the 4-flow conjecture for cubic graphs announced in 1998 Robertson, Sanders, Seymour and Thomas has not yet been entirely written up. C. Thomassen, The weak 3-flow conjecture and the weak circular flow conjecture, *J. Comb. Theory B* **102** (2012), 521–529, proved that every graph of large enough edge-connectivity k has a 3-flow. Thomassen's proof yields this for $k = 8$, and was later improved to give $k = 6$.

Finally, Tutte discovered a 2-variable polynomial associated with a graph, which generalizes both its chromatic polynomial and its flow polynomial. What little is known about this *Tutte polynomial* can hardly be more than the tip of the iceberg: it has far-reaching, and largely unexplored, connections to areas as diverse as knot theory and statistical physics. See D.J.A. Welsh, *Complexity: knots, colourings and counting* (LMS Lecture Notes **186**), Cambridge University Press 1993.

7

Extremal
Graph Theory

In this chapter we study how global parameters of a graph, such as its edge density or chromatic number, can influence its local substructures. How many edges, for instance, do we have to give a graph on n vertices to be sure that, no matter how these edges are arranged, the graph will contain a K^r subgraph for some given r? Or at least a K^r minor? Will some sufficiently high average degree or chromatic number ensure that one of these substructures occurs?

Questions of this type are among the most natural ones in graph theory, and there is a host of deep and interesting results. Collectively, these are known as *extremal graph theory*.

Extremal graph problems in this sense fall neatly into two categories, as follows. If we are looking for ways to ensure by global assumptions that a graph G contains some given graph H as a *minor* (or topological minor), it will suffice to raise $\|G\|$ above the value of some linear function of $|G|$, i.e., to make $\varepsilon(G)$ large enough. The precise value of ε needed to force a desired minor or topological minor will be our topic in Section 7.2. Graphs whose number of edges is about[1] linear in their number of vertices are called *sparse*, so Section 7.2 is devoted to 'sparse extremal graph theory'. *sparse*

A particularly interesting way to force an H minor is to assume that $\chi(G)$ is large. Recall that if $\chi(G) \geqslant k+1$, say, then G has a subgraph G' with $2\varepsilon(G') \geqslant \delta(G') \geqslant k$ (Lemma 5.2.3). The question here is whether the effect of large χ is limited to this indirect influence via ε, or whether an assumption of $\chi \geqslant k+1$ can force bigger minors than

---

[1] Formally, the notions of sparse and dense (below) make sense only for classes of graphs whose order tends to infinity, not for individual graphs.

© Reinhard Diestel 2017

R. Diestel, *Graph Theory*, Graduate Texts in Mathematics 173,
DOI 10.1007/978-3-662-53622-3_7

the assumption of $2\varepsilon \geqslant k$ can. *Hadwiger's conjecture*, which we meet in Section 7.3, asserts that χ has this quality. The conjecture can be viewed as a generalization of the four colour theorem, and is regarded by many as the most challenging open problem in graph theory.

On the other hand, if we ask what global assumptions might imply the existence of some given graph H as a *subgraph*, it will not help to raise invariants such as ε or χ, let alone any of the other invariants discussed in Chapter 1. For as soon as H contains a cycle, there are graphs of arbitrarily large chromatic number not containing H as a subgraph (Theorem 5.2.5). In fact, unless H is bipartite, any function f such that $f(n)$ edges on n vertices force an H subgraph must grow quadratically with n (why?).

dense
edge density

Graphs with a number of edges about quadratic in their number of vertices are usually called *dense*; the number $\|G\|/\binom{|G|}{2}$, the proportion of its potential edges that G actually has, is the *edge density* of G. The question of exactly which edge density is needed to force a given subgraph is the archetypal extremal graph problem, and it is our first topic in this chapter (Section 7.1). Rather than attempting to survey the wide field of 'dense extremal graph theory', however, we shall concentrate on its two most important results: we first prove Turán's classical extremal graph theorem for $H = K^r$—a result that has served as a model for countless similar theorems for other graphs H—and then state the fundamental Erdős-Stone theorem, which gives precise asymptotic information for all H at once.

Although the Erdős-Stone theorem can be proved by elementary means, we shall use the opportunity of its proof to portray a powerful modern proof technique that has transformed much of extremal graph theory in recent years: Szemerédi *regularity lemma*. This lemma is presented and proved in Section 7.4. In Section 7.5, we outline a general method for applying it, and illustrate this in the proof of the Erdős-Stone theorem. Another application of the regularity lemma will be given in Chapter 9.2.

7.1 Subgraphs

Let H be a graph and $n \geqslant |H|$. How many edges will suffice to force an H subgraph in any graph on n vertices, no matter how these edges are arranged? Or, to rephrase the problem: which is the greatest possible number of edges that a graph on n vertices can have *without* containing a copy of H as a subgraph? What will such a graph look like? Will it be unique?

extremal

A graph $G \not\supseteq H$ on n vertices with the largest possible number of edges is called *extremal* for n and H; its number of edges is denoted by

ex(n, H). Clearly, any graph G that is extremal for some n and H will   ex(n, H)
also be edge-maximal with $H \not\subseteq G$. Conversely, though, edge-maximality
does not imply extremality: G may well be edge-maximal with $H \not\subseteq G$
while having fewer than ex(n, H) edges (Fig. 7.1.1).

Fig. 7.1.1. Two graphs that are edge-maximal with $P^3 \not\subseteq G$; is
the right one extremal?

As a case in point, we consider our problem for $H = K^r$ (with $r > 1$).
A moment's thought suggests some obvious candidates for extremality
here: all complete $(r-1)$-partite graphs are edge-maximal without containing K^r. But which among these have the greatest number of edges?
Clearly those whose partition sets are as equal as possible, i.e. differ in
size by at most 1: if V_1, V_2 are two partition sets with $|V_1| - |V_2| \geqslant 2$, we
may increase the number of edges in our complete $(r-1)$-partite graph
by moving a vertex from V_1 to V_2.

The unique complete $(r-1)$-partite graphs on $n \geqslant r-1$ vertices
whose partition sets differ in size by at most 1 are called *Turán graphs*;
we denote them by $T^{r-1}(n)$ and their number of edges by $t_{r-1}(n)$ $T^{r-1}(n)$
(Fig. 7.1.2). For $n < r - 1$ we shall formally continue to use these $t_{r-1}(n)$
definitions, with the proviso that—contrary to our usual terminology—
the partition sets may now be empty; then, clearly, $T^{r-1}(n) = K^n$ for
all $n \leqslant r-1$.

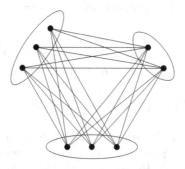

Fig. 7.1.2. The Turán graph $T^3(8)$

The following theorem tells us that $T^{r-1}(n)$ is indeed extremal for
n and K^r, and as such unique; in particular, ex(n, K^r) = $t_{r-1}(n)$.

Theorem 7.1.1. (Turán 1941)   [7.1.2]
For all integers r, n with $r > 1$, every graph $G \not\supseteq K^r$ with n vertices and   [9.2.2]
ex(n, K^r) edges is a $T^{r-1}(n)$.

We give two proofs: one using induction, the other by a short and direct local argument.

First proof. We apply induction on n. For $n \leqslant r - 1$ we have $G = K^n = T^{r-1}(n)$ as claimed. For the induction step, let now $n \geqslant r$.

K   Since G is edge-maximal without a K^r subgraph, G has a subgraph $K = K^{r-1}$. By the induction hypothesis, $G - K$ has at most $t_{r-1}(n - r + 1)$ edges, and each vertex of $G - K$ has at most $r - 2$ neighbours in K. Hence,

$$\|G\| \leqslant t_{r-1}(n - r + 1) + (n - r + 1)(r - 2) + \binom{r-1}{2} = t_{r-1}(n); \quad (1)$$

the equality on the right follows by inspection of the Turán graph $T^{r-1}(n)$ (Fig. 7.1.3).

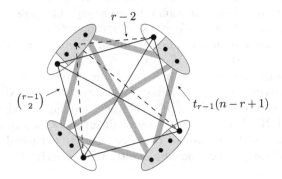

Fig. 7.1.3. The equation from (1) for $r = 5$ and $n = 14$

Since G is extremal for K^r (and $T^{r-1}(n) \not\supseteq K^r$), we have equality in (1). Thus, every vertex of $G - K$ has *exactly* $r - 2$ neighbours in K—
$x_1, \ldots, x_{r-1}$ just like the vertices $x_1, \ldots, x_{r-1}$ of K itself. For $i = 1, \ldots, r - 1$ let

$V_1, \ldots, V_{r-1}$
$$V_i := \{ v \in V(G) \mid vx_i \notin E(G) \}$$

be the set of all vertices of G whose $r - 2$ neighbours in K are precisely the vertices other than x_i. Since $K^r \not\subseteq G$, each of the sets V_i is independent, and they partition $V(G)$. Hence, G is $(r-1)$-partite. As $T^{r-1}(n)$ is the unique $(r-1)$-partite graph with n vertices and the maximum number of edges, our claim that $G = T^{r-1}(n)$ follows from the assumed extremality of G.   □

Written compactly as above, the proof of Turán's theorem may appear a little magical, perhaps even technical. When we look at it more closely, however, we can see how it evolves naturally from the initial idea of using a subgraph $K \simeq K^{r-1}$ as the seed for the $T^{r-1}(n)$ structure we are hoping to identify in G. Indeed, once we have fixed K and wonder how the rest of G might relate to it, we immediately observe that for every $v \in G - K$ there is a vertex $x \in K$ such that $vx \notin E(G)$. Turning then to the internal structure of $G - K$, we know from the induction hypothesis that the most edges it can possibly have is $t_{r-1}(n - r + 1)$, and only if $G - K \simeq T^{r-1}(n - r + 1)$. We do not know yet that $G - K$ can indeed have that many edges: all we know is that G, not necessarily $G - K$, has as many edges as possible without a K^r subgraph, and giving $G - K$ the structure of a $T^{r-1}(n - r + 1)$ might prevent us from having as many edges between $G - K$ and K as we might otherwise have. But this conflict does not in fact arise. Indeed, we can give $G - K$ this structure *and* have the theoretical maximum number of edges between $G - K$ and K (all except the necessary non-edges of type vx noted earlier) if we form G from K by expanding the $r - 1$ vertices of K to the $r - 1$ vertex classes of a $T^{r-1}(n)$. And since this is the only way in which both these aims can be achieved, $T^{r-1}(n)$ is once more the unique extremal graph on n vertices without a K^r.

In our second proof of Turán's theorem we shall use an operation called *vertex duplication*. By *duplicating* a vertex $v \in G$ we mean adding to G a new vertex v' and joining it to exactly the neighbours of v (but not to v itself).

vertex duplication

Second proof. We have already seen that among the complete k-partite graphs on n vertices the Turán graphs $T^k(n)$ have the most edges, and their degrees show that $T^{r-1}(n)$ has more edges than any $T^k(n)$ with $k < r - 1$. So it suffices to show that G is complete multipartite.

If not, then non-adjacency is not an equivalence relation on $V(G)$, and so there are vertices y_1, x, y_2 such that $y_1 x, xy_2 \notin E(G)$ but $y_1 y_2 \in E(G)$. If $d(y_1) > d(x)$, then deleting x and duplicating y_1 yields another K^r-free graph with more edges than G, contradicting the choice of G. So $d(y_1) \leqslant d(x)$, and similarly $d(y_2) \leqslant d(x)$. But then deleting both y_1 and y_2 and duplicating x twice yields a K^r-free graph with more edges than G, again contradicting the choice of G. $\qquad\square$

The Turán graphs $T^{r-1}(n)$ are dense: in order of magnitude, they have about n^2 edges. More exactly, for every n and r we have

$$t_{r-1}(n) \leqslant \tfrac{1}{2} n^2 \frac{r - 2}{r - 1},$$

with equality whenever $r - 1$ divides n (Exercise 7). It is therefore remarkable that just ϵn^2 more edges (for any fixed $\epsilon > 0$ and n large) give us not only a K^r subgraph (as does Turán's theorem) but a K_s^r for any given integer s—a graph itself teeming with K^r subgraphs:

Theorem 7.1.2. (Erdős & Stone 1946)
For all integers $r \geqslant 2$ and $s \geqslant 1$, and every $\epsilon > 0$, there exists an integer n_0 such that every graph with $n \geqslant n_0$ vertices and at least

$$t_{r-1}(n) + \epsilon n^2$$

edges contains K_s^r as a subgraph.

A proof of the Erdős-Stone theorem will be given in Section 7.5, as an illustration of how the regularity lemma may be applied. But the theorem can also be proved directly; see the notes for references.

The Erdős-Stone theorem is interesting not only in its own right: it also has a most interesting corollary. In fact, it was this entirely unexpected corollary that established the theorem as a kind of meta-theorem for the extremal theory of dense graphs, and thus made it famous.

Given a graph H and an integer n, consider the number $h_n :=$ $\mathrm{ex}(n, H)/\binom{n}{2}$: the maximum edge density that an n-vertex graph can have without containing a copy of H. Could it be that this critical density is essentially just a function of H, that h_n converges as $n \to \infty$? Theorem 7.1.2 implies this, and more: the limit of h_n is determined by a very simple function of a natural invariant of H—its chromatic number!

Corollary 7.1.3. *For every graph H with at least one edge,*

$$\lim_{n \to \infty} \mathrm{ex}(n, H) \binom{n}{2}^{-1} = \frac{\chi(H) - 2}{\chi(H) - 1}.$$

For the proof of Corollary 7.1.3 we need as a lemma that $t_{r-1}(n)$ never deviates much from the value it takes when $r - 1$ divides n (see above), and that $t_{r-1}(n)/\binom{n}{2}$ converges accordingly. The proof of the lemma is left as an easy exercise with hint (Exercise 8).

[7.1.2]   **Lemma 7.1.4.**

$$\lim_{n \to \infty} t_{r-1}(n) \binom{n}{2}^{-1} = \frac{r - 2}{r - 1}.$$

$\square$

r

Proof of Corollary 7.1.3. Let $r := \chi(H)$. Since H cannot be coloured with $r - 1$ colours, we have $H \not\subseteq T^{r-1}(n)$ for all $n \in \mathbb{N}$, and hence

$$t_{r-1}(n) \leqslant \mathrm{ex}(n, H).$$

On the other hand, $H \subseteq K_s^r$ for all sufficiently large s, so

$$\mathrm{ex}(n, H) \leqslant \mathrm{ex}(n, K_s^r)$$

for all those s. Let us fix such an s. For every $\epsilon > 0$, Theorem 7.1.2 implies that eventually (i.e. for large enough n)

$$\mathrm{ex}(n, K_s^r) < t_{r-1}(n) + \epsilon n^2.$$

Hence for n large,

$$
\begin{aligned}
t_{r-1}(n)/\binom{n}{2} &\leqslant \mathrm{ex}(n, H)/\binom{n}{2} \\
&\leqslant \mathrm{ex}(n, K_s^r)/\binom{n}{2} \\
&< t_{r-1}(n)/\binom{n}{2} + \epsilon n^2/\binom{n}{2} \\
&= t_{r-1}(n)/\binom{n}{2} + 2\epsilon/(1 - \tfrac{1}{n}) \\
&\leqslant t_{r-1}(n)/\binom{n}{2} + 4\epsilon \qquad \text{(assume } n \geqslant 2\text{)}.
\end{aligned}
$$

Therefore, since $t_{r-1}(n)/\binom{n}{2}$ converges to $\frac{r-2}{r-1}$ (Lemma 7.1.4), so does $\mathrm{ex}(n, H)/\binom{n}{2}$. $\qquad\square$

For bipartite graphs H, Corollary 7.1.3 says that substantially fewer than $\binom{n}{2}$ edges suffice to force an H subgraph. It turns out that

$$c_1 n^{2 - \frac{2}{r+1}} \leqslant \mathrm{ex}(n, K_{r,r}) \leqslant c_2 n^{2 - \frac{1}{r}}$$

for suitable constants c_1, c_2 depending on r; the lower bound is obtained by random graphs,[2] the upper bound is calculated in Exercise 12. If H is a forest, then $H \subseteq G$ as soon as $\varepsilon(G)$ is large enough, so $\mathrm{ex}(n, H)$ is at most linear in n (Exercise 14). Erdős and Sós conjectured in 1963 that $\mathrm{ex}(n, T) \leqslant \frac{1}{2}(k-1)n$ for all trees with $k \geqslant 2$ edges; as a general bound for all n, this is best possible for every T (Exercises 15–17).

Erdős-Sós conjecture

A related but rather different question is whether large values of ε or χ can force a graph G to contain a given tree T as an *induced* subgraph. Of course, we need some additional assumption for this to make sense— for example, to prevent G from just being a large complete graph. The weakest sensible such assumption is that G has bounded clique number, i.e., that $G \not\supseteq K^r$ for some fixed integer r. Then large average degree still does not force an induced copy of T—consider complete bipartite graphs—but large chromatic number might: according to a remarkable conjecture of Gyárfás (1975), there exists for every $r \in \mathbb{N}$ and every tree T an integer $k = k(T, r)$ such that every graph G with $\chi(G) \geqslant k$ and $\omega(G) < r$ contains T as an induced subgraph.

---

[2] see Chapter 11

7.2 Minors

In this section and the next we ask to what extent assumptions about invariants of a graph such as average degree, chromatic number, or girth can force it to contain another given graph as a minor or topological minor.

As a starting question, let us consider the analogue of Turán's theorem: how many edges on n vertices force a K^r minor or topological minor? The qualitative answer is that, unlike for K^r subgraphs where we might need as many as $\frac{1}{2}\frac{r-2}{r-1}n^2$ edges, a number of edges linear in n is enough: it suffices to assume that the graph has large enough average degree (depending on r).

Proposition 7.2.1. *Every graph of average degree at least 2^{r-2} has a K^r minor, for all $r \in \mathbb{N}$.*

Proof. We apply induction on r. For $r \leqslant 2$ the assertion is trivial. For the induction step let $r \geqslant 3$, and let G be any graph of average degree at least 2^{r-2}. Then $\varepsilon(G) \geqslant 2^{r-3}$; let H be a minimal minor of G with $\varepsilon(H) \geqslant 2^{r-3}$. Pick a vertex $x \in H$. By the minimality of H, x is not isolated. And each of its neighbours y has at least 2^{r-3} common neighbours with x: otherwise contracting the edge xy would lose us one vertex and at most 2^{r-3} edges, yielding a smaller minor H' with $\varepsilon(H') \geqslant 2^{r-3}$. The subgraph induced in H by the neighbours of x therefore has minimum degree at least 2^{r-3}, and hence has a K^{r-1} minor by the induction hypothesis. Together with x this yields the desired K^r minor of G. $\qquad\qquad\square$

In Proposition 7.2.1 we needed an average degree of 2^{r-2} to force a K^r minor by induction on r. Forcing a topological K^r minor is a little harder: we shall fix its branch vertices in advance and then construct its subdivided edges inductively, which requires an average degree of $2^{\binom{r}{2}}$ to start with. Apart from this difference, the proof follows the same idea:

Proposition 7.2.2. *Every graph of average degree at least $2^{\binom{r}{2}}$ has a topological K^r minor, for every integer $r \geqslant 2$.*

(1.2.2)
(1.3.1)

Proof. The assertion is clear for $r = 2$, so let us assume that $r \geqslant 3$. We show by induction on $m = r, \ldots, \binom{r}{2}$ that every graph G of average degree $d(G) \geqslant 2^m$ has a topological minor X with r vertices and m edges.

If $m = r$ then, by Propositions 1.2.2 and 1.3.1, G contains a cycle of length at least $\varepsilon(G) + 1 \geqslant 2^{r-1} + 1 \geqslant r + 1$, and the assertion follows with $X = C^r$.

Now let $r < m \leqslant \binom{r}{2}$, and assume the assertion holds for smaller m. Let G with $d(G) \geqslant 2^m$ be given; thus, $\varepsilon(G) \geqslant 2^{m-1}$. Since G has a component C with $\varepsilon(C) \geqslant \varepsilon(G)$, we may assume that G is connected.

Consider a maximal set $U \subseteq V(G)$ such that U is connected in G and $\varepsilon(G/U) \geqslant 2^{m-1}$; such a set U exists, because G itself has the form G/U with $|U| = 1$. Since G is connected, we have $N(U) \neq \emptyset$. $\qquad U$

Let $H := G[N(U)]$. If H has a vertex v of degree $d_H(v) < 2^{m-1}$, we $\qquad H$ may add it to U and obtain a contradiction to the maximality of U: when we contract the edge vv_U in G/U, we lose one vertex and $d_H(v) + 1 \leqslant 2^{m-1}$ edges, so ε will still be at least 2^{m-1}. Therefore $d(H) \geqslant \delta(H) \geqslant 2^{m-1}$. By the induction hypothesis, H contains a TY with $|Y| = r$ and $\|Y\| = m - 1$. Let x, y be two branch vertices of this TY that are non-adjacent in Y. Since x and y lie in $N(U)$ and U is connected in G, G contains an x–y path whose inner vertices lie in U. Adding this path to the TY, we obtain the desired TX. $\qquad\qquad\square$

In Chapter 3.5 we used the TK^r from Proposition 7.2.2 (stated there as Lemma 3.5.1) for a first proof that large enough connectivity $f(k)$ implies that a graph is k-linked. Later, in Theorem 3.5.3, we saw that connectivity as low as $2k$, coupled with an average degree of at least $16k$, is enough to imply this.

Conversely, we can use the more involved Theorem 3.5.3 to reduce the bound in Proposition 7.2.2 from exponential to quadratic, which is best possible up to a multiplicative constant (Exercise 24):

Theorem 7.2.3. *There is a constant $c \in \mathbb{R}$ such that, for every $r \in \mathbb{N}$, every graph G of average degree $d(G) \geqslant cr^2$ contains K^r as a topological minor.*

Proof. We prove the theorem with $c = 10$. Let G with $d(G) \geqslant 10r^2$ be given. By Theorem 1.4.3 for $k := r^2$, G has a subgraph H with $\kappa(H) \geqslant r^2$ and $\varepsilon(H) > \varepsilon(G) - r^2 \geqslant 4r^2$. For a TK^r in H, pick a set X of r vertices in H as branch vertices, and a set Y of $r(r-1)$ neighbours of X in H, $r - 1$ for each vertex in X, as initial subdividing vertices. These are r^2 vertices altogether; they can be chosen distinct, since $\delta(H) \geqslant \kappa(H) \geqslant r^2$. $\qquad$ (1.4.3) $\quad$ (3.5.3)

It remains to link up the vertices of Y in pairs, by disjoint paths in $H' := H - X$ corresponding to the edges of K^r. This can be done if Y is linked in H'. We show more generally that H' is $\frac{1}{2}r(r-1)$-linked, by checking that H' satisfies the premise of Theorem 3.5.3 for $k = \frac{1}{2}r(r-1)$. We have $\kappa(H') \geqslant \kappa(H) - r \geqslant r(r-1) = 2k$. And as H' was obtained from H by deleting at most $r|H|$ edges (as well as some vertices), we also have $\varepsilon(H') \geqslant \varepsilon(H) - r \geqslant 4r(r-1) = 8k$. $\qquad\qquad\square$

For small r one can try to determine the exact number of edges needed to force a TK^r subgraph on n vertices. For $r = 4$, this number is $2n - 2$; see Corollary 7.3.2. For $r = 5$, plane triangulations yield a lower bound of $3n - 5$ (Corollary 4.2.10). The converse, that $3n - 5$ edges do force a TK^5—not just either a TK^5 or a $TK_{3,3}$, as they do by Corollary 4.2.10 and Kuratowski's theorem—is already a difficult theorem (Mader 1998).

The average degree needed to force an arbitrary K^r minor is less than that for a TK^r, and it is known very precisely; see the notes for the value of c in the following result.

Theorem 7.2.4. (Kostochka 1982)
There exists a constant $c \in \mathbb{R}$ such that, for every $r \in \mathbb{N}$, every graph G of average degree $d(G) \geqslant cr\sqrt{\log r}$ contains K^r as a minor. Up to the value of c, this bound is best possible as a function of r.

The easier implication of the theorem, the fact that in general an average degree of $cr\sqrt{\log r}$ is needed to force a K^r minor, follows from considering random graphs as introduced in Chapter 11. The converse implication, that this average degree suffices, is proved by methods not dissimilar to the proof of Theorem 3.5.3.

Rather than proving Theorem 7.2.4, therefore, we devote the remainder of this section to another striking aspect of forcing minors: that we can force a K^r minor in a graph simply by raising its girth (as long as we do not merely subdivide edges). At first glance, this may seem almost paradoxical. But it looks more plausible if, rather than trying to force a K^r minor directly, we instead try to force a minor just of large minimum or average degree—which suffices by Theorem 7.2.4. For if the girth g of a graph is large then the ball $\{ v \mid d(x,v) < \lfloor g/2 \rfloor \}$ around a vertex x induces a tree with many leaves, each of which sends all but one of its incident edges away from the tree. Contracting enough disjoint such trees we can thus hope to obtain a minor of large average degree, which in turn will have a large complete minor.

The following lemma realizes this idea.

Lemma 7.2.5. *Let $d, k \in \mathbb{N}$ with $d \geqslant 3$, and let G be a graph of minimum degree $\delta(G) \geqslant d$ and girth $g(G) \geqslant 8k + 3$. Then G has a minor H of minimum degree $\delta(H) \geqslant d(d-1)^k$.*

Proof. Let $X \subseteq V(G)$ be maximal with $d(x,y) > 2k$ for all distinct $x, y \in X$. For each $x \in X$ put $T_x^0 := \{x\}$. Given $i < 2k$, assume that we have defined disjoint trees $T_x^i \subseteq G$ (one for each $x \in X$) whose vertices together are precisely the vertices at distance at most i from X in G. Joining each vertex at distance $i + 1$ from X to a neighbour at distance i, we obtain a similar set of disjoint trees T_x^{i+1}. As every vertex of G has distance at most $2k$ from X (by the maximality of X), the trees $T_x := T_x^{2k}$ obtained in this way partition the entire vertex set of G. Let H be the minor of G obtained by contracting every T_x.

To prove that $\delta(H) \geqslant d(d-1)^k$, note first that the T_x are induced subgraphs of G, because $\mathrm{diam}(T_x) \leqslant 4k$ and $g(G) > 4k + 1$. Similarly, there is at most one edge in G between any two trees T_x and T_y: two such edges, together with the paths joining their ends in T_x and T_y, would form a cycle of length at most $8k + 2 < g(G)$. So all the edges leaving T_x are preserved in the contraction.

How many such edges are there? Note that, for every vertex $u \in T_x^{k-1}$, all its $d_G(u) \geqslant d$ neighbours v also lie in T_x: since $d(v, x) \leqslant k$ and $d(x, y) > 2k$ for every other $y \in X$, we have $d(v, y) > k \geqslant d(v, x)$, so v was added to T_x rather than to T_y when those trees were defined. Therefore T_x^k, and hence also T_x, has at least $d(d-1)^{k-1}$ leaves. But every leaf of T_x sends at least $d-1$ edges away from T_x, so T_x sends at least $d(d-1)^k$ edges to (distinct) other trees T_y. □

Lemma 7.2.5 provides Theorem 7.2.4 with the following corollary:

Theorem 7.2.6. (Thomassen 1983)
There exists a function $f \colon \mathbb{N} \to \mathbb{N}$ such that every graph of minimum degree at least 3 and girth at least $f(r)$ has a K^r minor, for all $r \in \mathbb{N}$.

Proof. We prove the theorem with $f(r) := 8 \log r + 4 \log \log r + c$, for some constant $c \in \mathbb{R}$. Let $k = k(r) \in \mathbb{N}$ be minimal with $3 \cdot 2^k \geqslant c' r \sqrt{\log r}$, where $c' \in \mathbb{R}$ is the constant from Theorem 7.2.4. Then for a suitable constant $c \in \mathbb{R}$ we have $8k + 3 \leqslant 8 \log r + 4 \log \log r + c$, and the result follows by Lemma 7.2.5 and Theorem 7.2.4. □

Large girth can also be used to force a topological K^r minor. We now need some vertices of degree at least $r-1$ to serve as branch vertices, but if we assume a minimum degree of $r-1$ to secure these, we can even get by with a girth bound that is independent of r:

Theorem 7.2.7. (Kühn & Osthus 2002)   [7.3.9]
There exists a constant g such that $G \supseteq TK^r$ for every graph G satisfying $\delta(G) \geqslant r-1$ and $g(G) \geqslant g$.

7.3 Hadwiger's conjecture

As we saw in Section 7.2, an average degree of $c r \sqrt{\log r}$ suffices to force an arbitrary graph to have a K^r minor, and an average degree of cr^2 forces it to have a topological K^r minor. If we replace 'average degree' above with 'chromatic number' then, with almost the same constants c, the two assertions remain true: this is because every graph with chromatic number k has a subgraph of average degree at least $k-1$ (Lemma 5.2.3).

Although both functions above, $cr\sqrt{\log r}$ and cr^2, are best possible (up to the constant c) for the said implications with 'average degree', the question arises whether they are still best possible with 'chromatic number'—or whether some slower-growing function would do in that case. What lies hidden behind this problem about growth rates is a fundamental question about the nature of the invariant χ: can this invariant have some direct *structural* effect on a graph in terms of forcing concrete substructures, or is its effect no greater than that of the 'unstructural' property of having lots of edges somewhere, which it implies trivially?

Neither for general nor for topological minors is the answer to this question known. For general minors, however, the following conjecture of Hadwiger suggests a positive answer:

Conjecture. (Hadwiger 1943)
The following implication holds for every integer $r > 0$ and every graph G:

$$\chi(G) \geqslant r \implies G \succcurlyeq K^r.$$

Hadwiger's conjecture is trivial for $r \leqslant 2$, easy for $r = 3$ and $r = 4$ (exercises), and equivalent to the four colour theorem for $r = 5$ and $r = 6$. For $r \geqslant 7$ the conjecture is open, but it is true for line graphs (Exercise 34) and for graphs of large girth (Exercise 32; see also Corollary 7.3.9). Rephrased as $G \succcurlyeq K^{\chi(G)}$, it is true for almost all graphs.[3] In general, the conjecture for $r + 1$ implies it for r (exercise).

The Hadwiger conjecture for any fixed r is equivalent to the assertion that every graph without a K^r minor has an $(r-1)$-colouring. In this reformulation, the conjecture raises the question of what the graphs without a K^r minor look like: any sufficiently detailed structural description of those graphs should enable us to decide whether or not they can be $(r-1)$-coloured.

For $r = 3$, for example, the graphs without a K^r minor are precisely the forests (why?), and these are indeed 2-colourable. For $r = 4$, there is also a simple structural characterization of the graphs without a K^r minor:

[12.6.2]   **Proposition 7.3.1.** *A graph with at least three vertices is edge-maximal without a K^4 minor if and only if it can be constructed recursively from triangles by pasting[4] along K^2s.*

---

[3] See Chapter 11 for the notion of 'almost all'.

[4] This was defined formally in Chapter 5.5.

Proof. Recall first that every IK^4 contains a TK^4, because $\Delta(K^4) = 3$ (Proposition 1.7.3); the graphs without a K^4 minor thus coincide with those without a topological K^4 minor. The proof that any graph constructible as described is edge-maximal without a K^4 minor is left as an easy exercise; in order to deduce Hadwiger's conjecture for $r = 4$, we only need the converse implication anyhow. We prove this by induction on $|G|$.

(1.7.3)
(4.4.4)

Let G be given, edge-maximal without a K^4 minor. If $|G| = 3$ then G is itself a triangle, so let $|G| \geqslant 4$ for the induction step. Then G is not complete; let $S \subseteq V(G)$ be a separator of size $\kappa(G)$, and let C_1, C_2 be distinct components of $G - S$. Since S is a minimal separator, every vertex in S has a neighbour in C_1 and another in C_2. If $|S| \geqslant 3$, this implies that G contains three independent paths P_1, P_2, P_3 between a vertex $v_1 \in C_1$ and a vertex $v_2 \in C_2$. Since $\kappa(G) = |S| \geqslant 3$, the graph $G - \{v_1, v_2\}$ is connected and contains a (shortest) path P between two different P_i. Then $P \cup P_1 \cup P_2 \cup P_3$ is a TK^4, a contradiction.

Hence $\kappa(G) \leqslant 2$, and the assertion follows from Lemma 4.4.4[5] and the induction hypothesis. $\qquad\square$

One of the interesting consequences of Proposition 7.3.1 is that all the edge-maximal graphs without a K^4 minor have the same number of edges, and are thus all 'extremal':

Corollary 7.3.2. *Every edge-maximal graph G without a K^4 minor has $2|G| - 3$ edges.*

Proof. Induction on $|G|$. $\qquad\square$

Corollary 7.3.3. *Hadwiger's conjecture holds for $r = 4$.*

Proof. If G arises from G_1 and G_2 by pasting along a complete graph, then $\chi(G) = \max\{\chi(G_1), \chi(G_2)\}$ (see the proof of Proposition 5.5.2). Hence, Proposition 7.3.1 implies by induction on $|G|$ that all edge-maximal (and hence all) graphs without a K^4 minor can be 3-coloured. $\qquad\square$

It is also possible to prove Corollary 7.3.3 by a simple direct argument (Exercise 33).

By the four colour theorem, Hadwiger's conjecture for $r = 5$ follows from the following structure theorem for the graphs without a K^5 minor, just as it follows from Proposition 7.3.1 for $r = 4$. The proof of Theorem 7.3.4 is similar to that of Proposition 7.3.1, but considerably longer. We therefore state the theorem without proof:

---

[5] The proof of this lemma is elementary and can be read independently of the rest of Chapter 4.

Theorem 7.3.4. (Wagner 1937)
Let G be an edge-maximal graph without a K^5 minor. If $|G| \geqslant 4$ then G can be constructed recursively, by pasting along triangles and K^2s, from plane triangulations and copies of the graph W (Fig. 7.3.1).

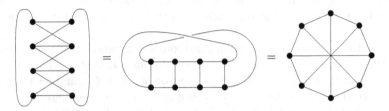

Fig. 7.3.1. Three representations of the *Wagner graph W*

(4.2.10)
       Using Corollary 4.2.10, one can easily compute which of the graphs constructed as in Theorem 7.3.4 have the most edges. It turns out that these *extremal* graphs without a K^5 minor have no more edges than those that are extremal with respect to $\{IK^5, IK_{3,3}\}$, i.e. the maximal planar graphs:

Corollary 7.3.5. *A graph with n vertices and no K^5 minor has at most $3n - 6$ edges.*   □

       Since $\chi(W) = 3$, Theorem 7.3.4 and the four colour theorem imply Hadwiger's conjecture for $r = 5$:

Corollary 7.3.6. *Hadwiger's conjecture holds for $r = 5$.*   □

       The Hadwiger conjecture for $r = 6$ is again substantially more difficult than the case $r = 5$, and again it relies on the four colour theorem. The proof shows (without using the four colour theorem) that any minimal-order counterexample arises from a planar graph by adding one vertex—so by the four colour theorem it is not a counterexample after all.

Theorem 7.3.7. (Robertson, Seymour & Thomas 1993)
Hadwiger's conjecture holds for $r = 6$.

       As mentioned earlier, the challenge posed by Hadwiger's conjecture is to devise a proof technique that makes better use of the assumption of $\chi \geqslant r$ than just using its consequence of $\delta \geqslant r - 1$ in a suitable subgraph, which we know cannot force a K^r minor (Theorem 7.2.4). So far, no such technique is known.

       If we resign ourselves to using just $\delta \geqslant r - 1$, we can still ask what additional assumptions might help in making this force a K^r minor. Theorem 7.2.7 says that an assumption of large girth has this effect;

see also Exercise 32. In fact, a much weaker assumption suffices: for
any fixed $s \in \mathbb{N}$ and all large enough d depending only on s, the graphs
$G \not\supseteq K_{s,s}$ of average degree at least d can be shown to have K^r minors
for r considerably larger than d. For Hadwiger's conjecture, this implies
the following:

Theorem 7.3.8. (Kühn & Osthus 2005)
*For every integer s there is an integer r_s such that Hadwiger's conjecture
holds for all graphs $G \not\supseteq K_{s,s}$ and $r \geqslant r_s$.*

The strengthening of Hadwiger's conjecture that graphs of chromatic number at least r contain K^r as a *topological* minor has become
known as *Hajós's conjecture*. It is false in general, but Theorem 7.2.7
implies it for graphs of large girth:

Corollary 7.3.9. *There is a constant g such that all graphs G of girth
at least g satisfy the implication $\chi(G) \geqslant r \Rightarrow G \supseteq TK^r$ for all r.*

Proof. Let g be the constant from Theorem 7.2.7. If $\chi(G) \geqslant r$ then, by
Lemma 5.2.3, G has a subgraph H of minimum degree $\delta(H) \geqslant r-1$. As
$g(H) \geqslant g(G) \geqslant g$, Theorem 7.2.7 implies that $G \supseteq H \supseteq TK^r$. $\square$

(5.2.3)
(7.2.7)

7.4 Szemerédi's regularity lemma

Some 40 years ago, in the course of the proof of a theorem about arithmetic progressions of integers, Szemerédi developed a graph-theoretical
tool that has since come to dominate methods in extremal graph theory
like none other: his *regularity lemma*. Very roughly, the lemma says
that all graphs can be approximated by random graphs in the following
sense: every graph can be partitioned, into a bounded number of equal
parts, so that most of its edges run between different parts and the edges
between any two parts are distributed fairly uniformly—just as we would
expect it if they had been generated at random.

In order to state the regularity lemma precisely, we need some definitions. Let $G = (V, E)$ be a graph, and let $X, Y \subseteq V$ be disjoint. Then
we denote by $\|X, Y\|$ the number of X–Y edges of G, and call

$$d(X, Y) := \frac{\|X, Y\|}{|X|\,|Y|}$$

the *density* of the pair (X, Y). (This is a real number between 0 and 1.)
Given some $\epsilon > 0$, we call a pair (A, B) of disjoint sets $A, B \subseteq V$ ϵ-*regular*
if all $X \subseteq A$ and $Y \subseteq B$ with

$$|X| \geqslant \epsilon |A| \quad \text{and} \quad |Y| \geqslant \epsilon |B|$$

$\|X, Y\|$

$d(X, Y)$

density

ϵ-*regular
pair*

satisfy

$$\left|d(X,Y) - d(A,B)\right| \leqslant \epsilon.$$

The edges in an ϵ-regular pair are thus distributed fairly uniformly, the more so the smaller the ϵ we started with.

exceptional set

Consider a partition $\{V_0, V_1, \ldots, V_k\}$ of V in which one set V_0 has been singled out as an *exceptional set*. (This exceptional set V_0 may be empty.[6]) We call such a partition an ϵ-*regular* partition of G if it satisfies the following three conditions:

ϵ-regular partition

(i) $|V_0| \leqslant \epsilon |V|$;

(ii) $|V_1| = \ldots = |V_k|$;

(iii) all but at most ϵk^2 of the pairs (V_i, V_j) with $1 \leqslant i < j \leqslant k$ are ϵ-regular.

The role of the exceptional set V_0 is one of pure convenience: it makes it possible to require that all the other partition sets have exactly the same size. Since condition (iii) affects only the sets $V_1, \ldots, V_k$, we may think of V_0 as a kind of bin: its vertices are disregarded when the regularity of the partition is assessed, but there are only few such vertices.

[7.1.2]
[9.2.2]

Theorem 7.4.1. (Regularity Lemma)
For every $\epsilon > 0$ and every integer $m \geqslant 1$ there exists an integer M such that every graph of order at least m admits an ϵ-regular partition $\{V_0, V_1, \ldots, V_k\}$ with $m \leqslant k \leqslant M$.

The regularity lemma thus says that, given any $\epsilon > 0$, every graph has an ϵ-regular partition into a bounded number of sets. The upper bound M on the number of partition sets ensures that for large graphs the partition sets are large too; note that ϵ-regularity is trivial when the partition sets are singletons, and a powerful property when they are large. The lemma also allows us to specify a lower bound m for the number of partition sets. This can be used to increase the proportion of edges running between different partition sets (i.e., of edges governed by the regularity assertion) over edges inside partition sets (about which we know nothing). See Exercise 38 for more details.

Note that the regularity lemma in this form is designed for use with dense graphs:[7] for sparse graphs it becomes trivial, because all densities of pairs—and hence their differences—tend to zero (Exercise 39).

The remainder of this section is devoted to the proof of the regularity lemma. Although the proof is not difficult, a reader meeting the

---

[6] So V_0 may be an exception also to our terminological rule that partition sets are not normally empty.

[7] Sparse versions were developed later; see the notes.

regularity lemma here for the first time is likely to draw more insight from seeing how the lemma is typically applied than from studying the technicalities of its proof. Any such reader is encouraged to skip to the start of Section 7.5 now and come back to the proof at his or her leisure.

We shall need the following inequality for reals $\mu_1, \ldots, \mu_k > 0$ and $e_1, \ldots, e_k \geqslant 0$:

$$\sum \frac{e_i^2}{\mu_i} \geqslant \frac{\left(\sum e_i\right)^2}{\sum \mu_i} . \tag{1}$$

This follows from the Cauchy-Schwarz inequality $\sum a_i^2 \sum b_i^2 \geqslant \left(\sum a_i b_i\right)^2$ by taking $a_i := \sqrt{\mu_i}$ and $b_i := e_i / \sqrt{\mu_i}$.

Let $G = (V, E)$ be a graph and $n := |V|$. For disjoint sets $A, B \subseteq V$ we define

$$q(A, B) := \frac{|A|\,|B|}{n^2}\, d^2(A, B) = \frac{\|A, B\|^2}{|A|\,|B|\, n^2} . \qquad q(A,B)$$

For partitions $\mathcal{A}$ of A and $\mathcal{B}$ of B we set

$$q(\mathcal{A}, \mathcal{B}) := \sum_{A' \in \mathcal{A};\ B' \in \mathcal{B}} q(A', B') , \qquad q(\mathcal{A}, \mathcal{B})$$

and for a partition $\mathcal{P} = \{C_1, \ldots, C_k\}$ of V we let

$$q(\mathcal{P}) := \sum_{i<j} q(C_i, C_j) . \qquad q(\mathcal{P})$$

However, if $\mathcal{P} = \{C_0, C_1, \ldots, C_k\}$ is a partition of V *with exceptional set* C_0, we treat C_0 as a set of singletons and define

$$q(\mathcal{P}) := q(\tilde{\mathcal{P}}) ,$$

where $\tilde{\mathcal{P}} := \{C_1, \ldots, C_k\} \cup \{\{v\} : v \in C_0\}$. $\qquad \tilde{\mathcal{P}}$

The function $q(\mathcal{P})$ plays a pivotal role in the proof of the regularity lemma. On the one hand, it measures the regularity of the partition $\mathcal{P}$: if $\mathcal{P}$ has too many irregular pairs (A, B), we may take the pairs (X, Y) of subsets violating the regularity of the pairs (A, B) and make those sets X and Y into partition sets of their own; as we shall prove, this refines $\mathcal{P}$ into a partition for which q is substantially greater than for $\mathcal{P}$. Here, 'substantial' means that the increase of $q(\mathcal{P})$ is bounded below by some constant depending only on ϵ. On the other hand,

$$q(\mathcal{P}) = \sum_{i<j} q(C_i, C_j)$$

$$= \sum_{i<j} \frac{|C_i|\,|C_j|}{n^2}\, d^2(C_i, C_j)$$

$$\leqslant \frac{1}{n^2} \sum_{i<j} |C_i|\,|C_j|$$

$$\leqslant 1.$$

The number of times that $q(\mathcal{P})$ can be increased by a constant is thus also bounded by a constant—in other words, after some bounded number of refinements our partition will be ϵ-regular! To complete the proof of the regularity lemma, all we have to do then is to note how many sets that last partition can possibly have if we start with a partition into m sets, and to choose this number as our promised bound M.

Let us make all this precise. We begin by showing that, when we refine a partition, the value of q will not decrease:

Lemma 7.4.2.

  (i) Let $C, D \subseteq V$ be disjoint. If $\mathcal{C}$ is a partition of C and $\mathcal{D}$ is a partition of D, then $q(\mathcal{C}, \mathcal{D}) \geqslant q(C, D)$.

  (ii) If $\mathcal{P}, \mathcal{P}'$ are partitions of V and $\mathcal{P}'$ refines $\mathcal{P}$, then $q(\mathcal{P}') \geqslant q(\mathcal{P})$.

Proof. (i) Let $\mathcal{C} =: \{C_1, \ldots, C_k\}$ and $\mathcal{D} =: \{D_1, \ldots, D_\ell\}$. Then

$$q(\mathcal{C}, \mathcal{D}) = \sum_{i,j} q(C_i, D_j)$$

$$= \frac{1}{n^2} \sum_{i,j} \frac{\|C_i, D_j\|^2}{|C_i|\,|D_j|}$$

$$\underset{(1)}{\geqslant} \frac{1}{n^2} \frac{\left(\sum_{i,j} \|C_i, D_j\|\right)^2}{\sum_{i,j} |C_i|\,|D_j|}$$

$$= \frac{1}{n^2} \frac{\|C, D\|^2}{\left(\sum_i |C_i|\right)\left(\sum_j |D_j|\right)}$$

$$= q(C, D).$$

  (ii) Let $\mathcal{P} =: \{C_1, \ldots, C_k\}$, and for $i = 1, \ldots, k$ let $\mathcal{C}_i$ be the partition of C_i induced by $\mathcal{P}'$. Then

$$q(\mathcal{P}) = \sum_{i<j} q(C_i, C_j)$$

$$\underset{(i)}{\leqslant} \sum_{i<j} q(\mathcal{C}_i, \mathcal{C}_j)$$

$$\leqslant q(\mathcal{P}'),$$

since $q(\mathcal{P}') = \sum_i q(\mathcal{C}_i) + \sum_{i<j} q(\mathcal{C}_i, \mathcal{C}_j)$. $\qquad\square$

Next, we show that refining a partition by subpartitioning an irregular pair of partition sets increases the value of q a little; since we are dealing here with a single pair only, the amount of this increase will still be less than any constant.

Lemma 7.4.3. *Let $\epsilon > 0$, and let $C, D \subseteq V$ be disjoint. If (C, D) is not ϵ-regular, then there are partitions $\mathcal{C} = \{C_1, C_2\}$ of C and $\mathcal{D} = \{D_1, D_2\}$ of D such that*

$$q(\mathcal{C}, \mathcal{D}) \geqslant q(C, D) + \epsilon^4 \frac{|C|\,|D|}{n^2}.$$

Proof. Suppose (C, D) is not ϵ-regular. Then there are sets $C_1 \subseteq C$ and $D_1 \subseteq D$ with $|C_1| \geqslant \epsilon|C|$ and $|D_1| \geqslant \epsilon|D|$ such that

$$|\eta| > \epsilon \qquad\qquad (2)$$

for $\eta := d(C_1, D_1) - d(C, D)$. Let $\mathcal{C} := \{C_1, C_2\}$ and $\mathcal{D} := \{D_1, D_2\}$, where $C_2 := C \smallsetminus C_1$ and $D_2 := D \smallsetminus D_1$. η

Let us show that $\mathcal{C}$ and $\mathcal{D}$ satisfy the conclusion of the lemma. We shall write $c_i := |C_i|$, $d_i := |D_i|$, $e_{ij} := \|C_i, D_j\|$, $c := |C|$, $d := |D|$ c_i, d_i, e_{ij}
and $e := \|C, D\|$. As in the proof of Lemma 7.4.2, c, d, e

$$q(\mathcal{C}, \mathcal{D}) = \frac{1}{n^2} \sum_{i,j} \frac{e_{ij}^2}{c_i d_j}$$

$$= \frac{1}{n^2}\left(\frac{e_{11}^2}{c_1 d_1} + \sum_{i+j>2} \frac{e_{ij}^2}{c_i d_j}\right)$$

$$\underset{(1)}{\geqslant} \frac{1}{n^2}\left(\frac{e_{11}^2}{c_1 d_1} + \frac{(e - e_{11})^2}{cd - c_1 d_1}\right).$$

By definition of η, we have $e_{11} = c_1 d_1 e / cd + \eta c_1 d_1$, so

$$
\begin{aligned}
n^2\, q(\mathcal{C}, \mathcal{D}) \;\geqslant\;& \frac{1}{c_1 d_1}\left(\frac{c_1 d_1 e}{cd} + \eta c_1 d_1\right)^2 \\
&+ \frac{1}{cd - c_1 d_1}\left(\frac{cd - c_1 d_1}{cd}e - \eta c_1 d_1\right)^2 \\
=\;& \frac{c_1 d_1 e^2}{c^2 d^2} + \frac{2e\eta c_1 d_1}{cd} + \eta^2 c_1 d_1 \\
&+ \frac{cd - c_1 d_1}{c^2 d^2}e^2 - \frac{2e\eta c_1 d_1}{cd} + \frac{\eta^2 c_1^2 d_1^2}{cd - c_1 d_1} \\
\geqslant\;& \frac{e^2}{cd} + \eta^2 c_1 d_1 \\
\underset{(2)}{\geqslant}\;& \frac{e^2}{cd} + \epsilon^4 cd
\end{aligned}
$$

since $c_1 \geqslant \epsilon c$ and $d_1 \geqslant \epsilon d$ by the choice of C_1 and D_1. $\qquad\square$

Finally, we show that if a partition has enough irregular pairs of partition sets to fall short of the definition of an ϵ-regular partition, then subpartitioning all those pairs at once results in an increase of q by a constant:

Lemma 7.4.4. *Let $0 < \epsilon \leqslant 1/4$, and let $\mathcal{P} = \{C_0, C_1, \ldots, C_k\}$ be a partition of V, with exceptional set C_0 of size $|C_0| \leqslant \epsilon n$ and $|C_1| = \ldots = |C_k| =: c$. If $\mathcal{P}$ is not ϵ-regular, then there is a partition $\mathcal{P}' = \{C_0', C_1', \ldots, C_\ell'\}$ of V with exceptional set C_0', where $k \leqslant \ell \leqslant k4^{k+1}$, such that $|C_0'| \leqslant |C_0| + n/2^k$, all other sets C_i' have equal size, and either $\mathcal{P}'$ is ϵ-regular or*

$$
q(\mathcal{P}') \geqslant q(\mathcal{P}) + \epsilon^5/2\,.
$$

Proof. For all $1 \leqslant i < j \leqslant k$, let us define a partition $\mathcal{C}_{ij}$ of C_i and a partition $\mathcal{C}_{ji}$ of C_j, as follows. If the pair (C_i, C_j) is ϵ-regular, we let $\mathcal{C}_{ij} := \{C_i\}$ and $\mathcal{C}_{ji} := \{C_j\}$. If not, then by Lemma 7.4.3 there are partitions $\mathcal{C}_{ij}$ of C_i and $\mathcal{C}_{ji}$ of C_j with $|\mathcal{C}_{ij}| = |\mathcal{C}_{ji}| = 2$ and

$$
q(\mathcal{C}_{ij}, \mathcal{C}_{ji}) \geqslant q(C_i, C_j) + \epsilon^4\frac{|C_i|\,|C_j|}{n^2} = q(C_i, C_j) + \frac{\epsilon^4 c^2}{n^2}\,. \tag{3}
$$

For each $i = 1, \ldots, k$, let $\mathcal{C}_i$ be the unique minimal partition of C_i that refines every partition $\mathcal{C}_{ij}$ with $j \neq i$. (In other words, if we consider two elements of C_i as equivalent whenever they lie in the same partition set of C_{ij} for every $j \neq i$, then C_i is the set of equivalence classes.) Thus, $|C_i| \leqslant 2^{k-1}$. Now consider the partition

$$C := \{C_0\} \cup \bigcup_{i=1}^{k} C_i \qquad\qquad\qquad C$$

of V, with C_0 as exceptional set. Then C refines P and $|C \smallsetminus \{C_0\}| \leqslant k2^{k-1}$, so

$$k \leqslant |C| \leqslant k2^k. \qquad (4)$$

Let $C_0 := \{\{v\} \mid v \in C_0\}$. Now if P is not ϵ-regular, then for more $\quad c_0$ than ϵk^2 of the pairs (C_i, C_j) with $1 \leqslant i < j \leqslant k$ the partition C_{ij} is non-trivial. Hence, by our definition of q for partitions with exceptional set, and Lemma 7.4.2 (i),

$$q(C) = \sum_{1 \leqslant i < j} q(C_i, C_j) + \sum_{1 \leqslant i} q(C_0, C_i) + \sum_{0 \leqslant i} q(C_i)$$

$$\geqslant \sum_{1 \leqslant i < j} q(C_{ij}, C_{ji}) + \sum_{1 \leqslant i} q(C_0, \{C_i\}) + q(C_0)$$

$$\underset{(3)}{\geqslant} \sum_{1 \leqslant i < j} q(C_i, C_j) + \epsilon k^2 \frac{\epsilon^4 c^2}{n^2} + \sum_{1 \leqslant i} q(C_0, \{C_i\}) + q(C_0)$$

$$= q(P) + \epsilon^5 \left(\frac{kc}{n} \right)^2$$

$$\geqslant q(P) + \epsilon^5/2 .$$

(For the last inequality, recall that $|C_0| \leqslant \epsilon n \leqslant \frac{1}{4} n$, so $kc \geqslant \frac{3}{4} n$.)

In order to turn C into our desired partition P', all that remains to do is to cut its sets up into pieces of some common size, small enough that all remaining vertices can be collected into the exceptional set without making this too large.

If $c < 4^k$, the ϵ-regular partition P' into $C_0' := C_0$ and the singletons $\{v\}$ with $v \in V \smallsetminus C_0$ is as desired, since there are ℓ such singletons for $k \leqslant \ell = kc < k4^k$.

Assume now that $c \geqslant 4^k$. Let $C_1', \ldots, C_\ell'$ be a maximal collection of disjoint sets of size $d := \lfloor c/4^k \rfloor \geqslant 1$ such that each C_i' is contained in some $\quad d$ $C \in C \smallsetminus \{C_0\}$, and put $C_0' := V \smallsetminus \bigcup C_i'$. Then $P' = \{C_0', C_1', \ldots, C_\ell'\}$ is $\quad P'$ indeed a partition of V. Moreover, $\tilde{P}'$ refines $\tilde{C}$, so

$$q(P') \geqslant q(C) \geqslant q(P) + \epsilon^5/2$$

by Lemma 7.4.2 (ii). Since each set $C_i' \neq C_0'$ is also contained in one of the sets $C_1, \ldots, C_k$, but no more than $c/d \leqslant 4^{k+1}$ sets C_i' can lie inside the same C_j (by the choice of d), we also have $k \leqslant \ell \leqslant k4^{k+1}$ as required. Finally, the sets $C'_1, \ldots, C'_\ell$ use all but at most d vertices from each set $C \neq C_0$ of $\mathcal{C}$. Hence,

$$|C'_0| \leqslant |C_0| + d\,|\mathcal{C}|$$

$$\underset{(4)}{\leqslant} |C_0| + \frac{c}{4^k}k2^k$$

$$= |C_0| + ck/2^k$$

$$\leqslant |C_0| + n/2^k.$$

$$\square$$

The proof of the regularity lemma now follows easily by repeated application of Lemma 7.4.4:

Proof of Theorem 7.4.1. Let $\epsilon > 0$ and $m \geqslant 1$ be given, assuming without loss of generality that $\epsilon \leqslant 1/4$. Let $s := 2/\epsilon^5$. This number s is an upper bound on the number of iterations of Lemma 7.4.4 that can be applied to a partition of a graph before it becomes ϵ-regular; recall that $q(\mathcal{P}) \leqslant 1$ for all partitions $\mathcal{P}$.

There is one formal requirement which a partition $\{C_0, C_1, \ldots, C_k\}$ with $|C_1| = \ldots = |C_k|$ has to satisfy before Lemma 7.4.4 can be (re-) applied: the size $|C_0|$ of its exceptional set must not exceed ϵn. With each iteration of the lemma, however, the size of the exceptional set can grow by up to $n/2^k$. (More precisely, by up to $n/2^\ell$, where ℓ is the number of other sets in the current partition; but $\ell \geqslant k$ by the lemma, so $n/2^k$ is certainly an upper bound for the increase.) We thus want to start with k large enough that even s increments of $n/2^k$ add up to at most $\frac{1}{2}\epsilon n$, and ensure that n large enough that, for any initial value of $|C_0| < k$, we have $|C_0| \leqslant \frac{1}{2}\epsilon n$. (If we give our starting partition k non-exceptional sets $C_1, \ldots, C_k$, we should allow an initial size of up to k for C_0, to be able to achieve $|C_1| = \ldots = |C_k|$.)

So let $k \geqslant m$ be large enough that $2^{k-1} \geqslant s/\epsilon$. Then $s/2^k \leqslant \epsilon/2$, and hence

$$k + \frac{s}{2^k}n \leqslant \epsilon n \tag{5}$$

whenever $k/n \leqslant \epsilon/2$, i.e. for all $n \geqslant 2k/\epsilon$.

Let us now choose M. This should be an upper bound on the number of (non-exceptional) sets in our partition after up to s iterations of Lemma 7.4.4, where in each iteration this number may grow from its current value r to at most $r4^{r+1}$. So let f be the function $x \mapsto x4^{x+1}$, and take $M := \max\{f^s(k), 2k/\epsilon\}$; the second term in the maximum ensures that any $n \geqslant M$ is large enough to satisfy (5).

We finally have to show that every graph $G = (V, E)$ of order at least m has an ϵ-regular partition $\{V_0, V_1, \ldots, V_{k'}\}$ with $m \leqslant k' \leqslant M$. So

let G be given, and let $n := |G|$. If $n \leqslant M$, we partition G into $k' := n$ n
singletons, choosing $V_0 := \emptyset$ and $|V_1| = \ldots = |V_{k'}| = 1$. This partition of
G is clearly ϵ-regular. Suppose now that $n > M$. Let $C_0 \subseteq V$ be minimal
such that our earlier $k =: k'$ divides $|V \smallsetminus C_0|$, and let $\{C_1, \ldots, C_k\}$ be
any partition of $V \smallsetminus C_0$ into sets of equal size. Then $|C_0| < k$, and
hence $|C_0| \leqslant \epsilon n$ by (5). Starting with $\{C_0, C_1, \ldots, C_k\}$ we apply Lemma
7.4.4 again and again, until the partition of G obtained is ϵ-regular;
this will happen after at most s iterations, since by (5) the size of the
exceptional set in the partitions stays below ϵn, so the lemma could
indeed be reapplied up to the theoretical maximum of s times. $\square$

7.5 Applying the regularity lemma

The purpose of this section is to illustrate how the regularity lemma
is typically applied in the context of (dense) extremal graph theory.
Suppose we are trying to prove that a certain edge density of a graph
G suffices to force the occurrence of some given subgraph H, and that
we have an ϵ-regular partition of G. For most of the pairs (V_i, V_j) of
partition sets, the edges between V_i and V_j are distributed fairly uniformly; their density, however, may depend on the pair. But since G
has many edges, this density cannot be too small for too many pairs:
some sizeable proportion of the pairs will have at least a certain positive
density. Moreover if G is large, then so are the pairs: recall that the
number of partition sets is bounded, and they have equal size. But any
large enough bipartite graph with equal partition sets, fixed positive edge
density (however small) and a uniform distribution of edges will contain
any given bipartite subgraph;[8] this will be made precise below. Writing
H as a union of bipartite subgraphs, say those induced by pairs of colour
classes of some vertex colouring of H, we shall obtain $H \subseteq G$ as desired.

These ideas will be formalized by Lemma 7.5.2 below. We shall then
use this and the regularity lemma to prove the Erdős-Stone theorem
from Section 7.1; another application will be given later, in the proof
of Theorem 9.2.2. We wind up the section with an informal review of
the application of the regularity lemma that we have seen, summarizing
what it can teach us for similar applications. In particular, we look at
how the various parameters involved depend on each other, and in which
order they have to be chosen to make the lemma work.

Let us begin by noting a simple consequence of the ϵ-regularity of a
pair (A, B). For any subset $Y \subseteq B$ that is not too small, most vertices
of A have about the expected number of neighbours in Y:

---

[8] Readers already acquainted with random graphs may find it instructive to compare this statement with Proposition 11.3.1.

Lemma 7.5.1. *Let (A, B) be an ϵ-regular pair, of density d say, and let $Y \subseteq B$ have size $|Y| \geqslant \epsilon|B|$. Then all but fewer than $\epsilon|A|$ of the vertices in A have (each) at least $(d - \epsilon)|Y|$ neighbours in Y.*

Proof. Let $X \subseteq A$ be the set of vertices with fewer than $(d - \epsilon)|Y|$ neighbours in Y. Then $\|X, Y\| < |X|(d - \epsilon)|Y|$, so

$$d(X, Y) = \frac{\|X, Y\|}{|X||Y|} < d - \epsilon = d(A, B) - \epsilon.$$

As (A, B) is ϵ-regular and $|Y| \geqslant \epsilon|B|$, this implies that $|X| < \epsilon|A|$. $\qquad\square$

Let G be a graph with an ϵ-regular partition $\{V_0, V_1, \ldots, V_k\}$, with exceptional set V_0 and $|V_1| = \ldots = |V_k| =: \ell$. Given $d \in [0, 1]$, let R be the graph on $\{V_1, \ldots, V_k\}$ in which two vertices V_i, V_j are adjacent if and only if they form an ϵ-regular pair in G of density $\geqslant d$. We shall call R a *regularity graph* of G with parameters ϵ, ℓ and d. Given $s \in \mathbb{N}$, let us now replace every vertex V_i of R by a set V_i^s of s vertices, and every edge by a complete bipartite graph between the corresponding s-sets. The resulting graph will be denoted by R_s. (For $R = K^r$, for example, we have $R_s = K_s^r$.)

The following lemma says that subgraphs of R_s can also be found in G, provided that $d > 0$, that ϵ is small enough, and that the V_i are large enough. In fact, the values of ϵ and ℓ required depend only on (d and) the maximum degree of the subgraph:

Lemma 7.5.2. *For all $d \in (0, 1]$ and $\Delta \geqslant 1$ there exists an $\epsilon_0 > 0$ with the following property: if G is any graph, H is a graph with $\Delta(H) \leqslant \Delta$, $s \in \mathbb{N}$, and R is any regularity graph of G with parameters $\epsilon \leqslant \epsilon_0$, $\ell \geqslant 2s/d^\Delta$ and d, then*

$$H \subseteq R_s \implies H \subseteq G.$$

Proof. Given d and Δ, choose $\epsilon_0 > 0$ small enough that $\epsilon_0 < d$ and

$$(d - \epsilon_0)^\Delta - \Delta\epsilon_0 \geqslant \tfrac{1}{2}d^\Delta; \tag{1}$$

such a choice is possible, since $(d - \epsilon)^\Delta - \Delta\epsilon \to d^\Delta$ as $\epsilon \to 0$. Now let G, H, s and R be given as stated. Let $\{V_0, V_1, \ldots, V_k\}$ be the ϵ-regular partition of G that gave rise to R; thus, $\epsilon \leqslant \epsilon_0$, $V(R) = \{V_1, \ldots, V_k\}$ and $|V_1| = \ldots = |V_k| = \ell \geqslant 2s/d^\Delta$. Let us assume that H is actually a subgraph of R_s (not just isomorphic to one), with vertices $u_1, \ldots, u_h$ say. Each vertex u_i lies in one of the s-sets V_j^s of R_s, which defines a map $\sigma: i \mapsto j$. Our aim is to define an embedding $u_i \mapsto v_i \in V_{\sigma(i)}$ of H in G as a subgraph; thus, $v_1, \ldots, v_h$ will be distinct, and $v_i v_j$ will be an edge of G whenever $u_i u_j$ is an edge of H.

(Margin notes:)
R

regularity graph
V_i^s

R_s

[9.2.2]

d, Δ, ϵ_0

G, H, R, R_s
V_i
ϵ, k, ℓ

u_i, h

σ
v_i

Our plan is to choose the vertices $v_1, \ldots, v_h$ inductively. Throughout the induction, we shall have a 'target set' $Y_i \subseteq V_{\sigma(i)}$ assigned to each u_i; this contains the vertices that are still candidates for the choice of v_i. Initially, Y_i is the entire set $V_{\sigma(i)}$. As the embedding proceeds, Y_i will get smaller and smaller (until it collapses to $\{v_i\}$ when v_i is chosen): whenever we choose a vertex v_j with $j < i$ and $u_j u_i \in E(H)$, we delete all those vertices from Y_i that are not adjacent to v_j. The set Y_i thus evolves as

$$V_{\sigma(i)} = Y_i^0 \supseteq \ldots \supseteq Y_i^i = \{v_i\},$$

where Y_i^j denotes the version of Y_i current after the definition of v_j and the resulting deletion of vertices from Y_i^{j-1}.

In order to make this approach work, we have to ensure that the target sets Y_i do not get too small. When we come to embed a vertex u_j, we consider all the indices $i > j$ with $u_j u_i \in E(H)$; there are at most Δ such i. For each of these i, we wish to select v_j so that

$$Y_i^j = N(v_j) \cap Y_i^{j-1} \tag{2}$$

is still relatively large: smaller than Y_i^{j-1} by no more than a constant factor such as $(d - \epsilon)$. Now this can be done by Lemma 7.5.1 (with $A = V_{\sigma(j)}$, $B = V_{\sigma(i)}$ and $Y = Y_i^{j-1}$): provided that Y_i^{j-1} still has size at least $\epsilon \ell$ (which induction will ensure), all but at most $\epsilon \ell$ choices of v_j will be such that the new set Y_i^j as in (2) satisfies

$$|Y_i^j| \geq (d - \epsilon)|Y_i^{j-1}|. \tag{3}$$

Excluding the bad choices for v_j for all the relevant values of i simultaneously, we find that all but at most $\Delta \epsilon \ell$ choices of v_j from $V_{\sigma(j)}$, and in particular from $Y_j^{j-1} \subseteq V_{\sigma(j)}$, satisfy (3) for all i.

It remains to show that the sets Y_i^{j-1} considered above as Y for Lemma 7.5.1 never fall below the size of $\epsilon \ell$, and that when we come to select $v_j \in Y_j^{j-1}$ we have a choice of at least s suitable candidates: since before u_j at most $s - 1$ vertices u were given an image in $V_{\sigma(j)}$, we can then choose v_j distinct from these.

But all this follows from our choice of ϵ_0. Indeed, the initial target sets Y_i^0 have size ℓ, and each Y_i shrinks at most Δ times by a factor of $(d - \epsilon)$ when some v_j with $j < i$ and $u_j u_i \in E(H)$ is defined. Thus,

$$|Y_i^{j-1}| - \Delta \epsilon \ell \underset{(3)}{\geq} (d - \epsilon)^\Delta \ell - \Delta \epsilon \ell \geq (d - \epsilon_0)^\Delta \ell - \Delta \epsilon_0 \ell \underset{(1)}{\geq} \tfrac{1}{2} d^\Delta \ell \geq s$$

for all $j \leq i$; in particular, we have $|Y_i^{j-1}| \geq \epsilon \ell$ and $|Y_j^{j-1}| - \Delta \epsilon \ell \geq s$ as desired. $\qquad \square$

We are now ready to prove the Erdős-Stone theorem.

(7.1.1)
(7.1.4)
(7.4.1)
r, s
γ

Proof of Theorem 7.1.2. Let $r \geqslant 2$ and $s \geqslant 1$ be given as in the statement of the theorem. For $s = 1$ the assertion follows from Turán's theorem, so we assume that $s \geqslant 2$. Let $\gamma > 0$ be given; this γ will play the role of the ϵ of the theorem. If any graph G with $|G| =: n$ has

$\|G\|$

$$\|G\| \geqslant t_{r-1}(n) + \gamma n^2$$

edges, then $\gamma < 1$. We want to show that $K_s^r \subseteq G$ if n is large enough.

Our plan is to use the regularity lemma to show that G has a regularity graph R dense enough to contain a K^r by Turán's theorem. Then R_s contains a K_s^r, so we may hope to use Lemma 7.5.2 to deduce that $K_s^r \subseteq G$.

d, Δ, ϵ_0
m, ϵ

On input $d := \gamma$ and $\Delta := \Delta(K_s^r)$ Lemma 7.5.2 returns an $\epsilon_0 > 0$. To apply the regularity lemma, let $m > 1/\gamma$ and choose $\epsilon > 0$ small enough that $\epsilon \leqslant \epsilon_0$,

$$\epsilon < \gamma/2 < 1, \tag{1}$$

and

δ

$$\delta := 2\gamma - \epsilon^2 - 4\epsilon - d - \frac{1}{m} > 0;$$

this is possible, since $2\gamma - d - \frac{1}{m} > 0$. On input ϵ and m, the regularity

M

lemma returns an integer M. Let us assume that

n

$$n \geqslant \frac{2Ms}{d^\Delta(1-\epsilon)}.$$

Since this number is at least m, the regularity lemma provides us with

k

an ϵ-regular partition $\{V_0, V_1, \ldots, V_k\}$ of G, where $m \leqslant k \leqslant M$; let

ℓ

$|V_1| = \ldots = |V_k| =: \ell$. Then

$$n \geqslant k\ell, \tag{2}$$

and

$$\ell = \frac{n - |V_0|}{k} \geqslant \frac{n - \epsilon n}{M} = n\frac{1-\epsilon}{M} \geqslant \frac{2s}{d^\Delta}$$

R

by the choice of n. Let R be the regularity graph of G with parameters ϵ, ℓ, d corresponding to the above partition. Then Lemma 7.5.2 will imply $K_s^r \subseteq G$ as desired if $K^r \subseteq R$ (and hence $K_s^r \subseteq R_s$).

Our plan was to show $K^r \subseteq R$ by Turán's theorem. We thus have to check that R has enough edges, i.e. that enough ϵ-regular pairs (V_i, V_j) have density at least d. This should follow from our assumption that G has at least $t_{r-1}(n) + \gamma n^2$ edges, i.e. an edge density of about $\frac{r-2}{r-1} + 2\gamma$: this lies substantially above the approximate density of $\frac{r-2}{r-1}$ of the Turán graph $T^{r-1}(k)$, and hence substantially above any density that G could derive from $t_{r-1}(k)$ dense pairs alone, even if all these had density 1.

Let us then estimate $\|R\|$ more precisely. How many edges of G lie outside ϵ-regular pairs? At most $\binom{|V_0|}{2}$ edges lie inside V_0, and by condition (i) in the definition of ϵ-regularity these are at most $\frac{1}{2}(\epsilon n)^2$ edges. At most $|V_0|k\ell \leqslant \epsilon n^2$ edges join V_0 to other partition sets. The at most ϵk^2 other pairs (V_i, V_j) that are not ϵ-regular contain at most ℓ^2 edges each, together at most $\epsilon k^2 \ell^2$. The ϵ-regular pairs of insufficient density $(< d)$ each contain no more than $d\ell^2$ edges, altogether at most $\frac{1}{2}k^2 d\ell^2$ edges. Finally, there are at most $\binom{\ell}{2}$ edges inside each of the partition sets $V_1, \ldots, V_k$, together at most $\frac{1}{2}\ell^2 k$ edges. All *other* edges of G lie in ϵ-regular pairs of density at least d, and thus contribute to edges of R. Since each edge of R corresponds to at most ℓ^2 edges of G, we thus have in total

$$\|G\| \leqslant \tfrac{1}{2}\epsilon^2 n^2 + \epsilon n^2 + \epsilon k^2 \ell^2 + \tfrac{1}{2}k^2 d\ell^2 + \tfrac{1}{2}\ell^2 k + \|R\|\,\ell^2.$$

Hence, for all sufficiently large n,

$$\|R\| \geqslant \tfrac{1}{2}k^2\, \frac{\|G\| - \tfrac{1}{2}\epsilon^2 n^2 - \epsilon n^2 - \epsilon k^2 \ell^2 - \tfrac{1}{2}dk^2 \ell^2 - \tfrac{1}{2}k\ell^2}{\tfrac{1}{2}k^2 \ell^2}$$

$$\underset{(1,2)}{\geqslant} \tfrac{1}{2}k^2 \left(\frac{t_{r-1}(n) + \gamma n^2 - \tfrac{1}{2}\epsilon^2 n^2 - \epsilon n^2}{n^2/2} - 2\epsilon - d - \frac{1}{k} \right)$$

$$\geqslant \tfrac{1}{2}k^2 \left(\frac{t_{r-1}(n)}{n^2/2} + 2\gamma - \epsilon^2 - 4\epsilon - d - \frac{1}{m} \right)$$

$$= \tfrac{1}{2}k^2 \left(t_{r-1}(n) \binom{n}{2}^{-1} \left(1 - \frac{1}{n}\right) + \delta \right)$$

$$> \tfrac{1}{2}k^2\, \frac{r-2}{r-1}$$

$$\geqslant t_{r-1}(k).$$

(The strict inequality follows from Lemma 7.1.4.) Therefore $K^r \subseteq R$ by Theorem 7.1.1, as desired. $\square$

Having seen a typical application of the regularity lemma in full detail, let us now step back and try to separate the wheat from the chaff: what were the main ideas, how do the various parameters depend on each other, and in which order were they chosen?

The task was to show that γn^2 more edges than can be accommodated on n vertices without creating a K^r force a K_s^r subgraph, provided that G is large enough. The plan was to do this using Lemma 7.5.2, which asks for the input of two parameters: d and Δ. As we wish to find a copy of $H = K_s^r$ in G, it is clear that we must choose $\Delta := \Delta(K_s^r)$. We shall return to the question of how to choose d in a moment.

Given d and Δ, Lemma 7.5.2 tells us how small we must choose ϵ to make the regularity lemma provide us with a suitable partition. The regularity lemma also requires the input of a lower bound m for the number of partition classes; we shall discuss this below, together with d.

All that remains now is to choose G large enough that the partition classes have size at least $2s/d^\Delta$, as required by Lemma 7.5.2. (The s here depends on the graph H we wish to embed, and $s := |H|$ would certainly be big enough. In our case, we can use the s from our $H = K_s^r$.) How large is 'large enough' for $|G|$ follows straight from the upper bound M on the number of partition classes returned by the regularity lemma: roughly, i.e. disregarding V_0, an assumption of $|G| \geqslant 2Ms/d^\Delta$ suffices.

So far, everything was entirely straightforward, and standard for any application of the regularity lemma of this kind. But now comes the interesting bit, the part specific to this proof: the observation that, if only d is small enough, our γn^2 'additional edges' force an 'additional dense ϵ-regular pair' of partition sets, giving us more than $t_{r-1}(k)$ dense ϵ-regular pairs in total (where 'dense' means 'of density at least d'), thus forcing R to contain a K^r and hence R_s to contain a K_s^r.

Let us examine why this is so. Suppose we have at most $t_{r-1}(k)$ dense ϵ-regular pairs . Inside these, G has at most

$$\tfrac{1}{2}k^2\frac{r-2}{r-1}\,\ell^2 \leqslant \tfrac{1}{2}n^2\frac{r-2}{r-1}$$

edges, even if we use those pairs to their full capacity of ℓ^2 edges each (where ℓ is again the common size of the partition sets other than V_0, so that $k\ell$ is nearly n). Thus, we have almost exactly our γn^2 additional edges left to accommodate elsewhere in the graph: either in ϵ-regular pairs of density less than d, or in some exceptional way, i.e. in irregular pairs, inside a partition set, or with an end in V_0. Now the number of edges in low-density ϵ-regular pairs is less than

$$\tfrac{1}{2}k^2d\ell^2 \leqslant \tfrac{1}{2}dn^2,$$

and hence less than half of our extra edges if $d \leqslant \gamma$. The other half, the remaining $\frac{1}{2}\gamma n^2$ edges, are more than can be accommodated in exceptional ways, provided we choose m large enough and ϵ small enough (giving an additional upper bound for ϵ). It is now a routine matter to compute the values of m and ϵ that will work.

Exercises

1.⁻ Show that $K_{1,3}$ is extremal without a P^3.

2.⁻ Given $k > 0$, determine the extremal graphs of chromatic number at most k.

3.⁻ Is there a graph that is edge-maximal without a K^3 minor but not extremal?

4. Determine the value of $\operatorname{ex}(n, K_{1,r})$ for all $r, n \in \mathbb{N}$.

5.⁺ Given $k > 0$, determine the extremal graphs without a matching of size k.

(Hint. Theorem 2.2.3 and Ex. 20, Ch. 2.)

6. Without using Turán's theorem, show that the maximum number of edges in a triangle-free graph of order $n > 1$ is $\lfloor n^2/4 \rfloor$.

7. Show that
$$t_{r-1}(n) \leqslant \tfrac{1}{2} n^2 \frac{r-2}{r-1},$$
with equality whenever $r - 1$ divides n.

8. Show that $t_{r-1}(n)/\binom{n}{2}$ converges to $(r-2)/(r-1)$ as $n \to \infty$.

(Hint. $t_{r-1}((r-1)\lfloor \frac{n}{r-1} \rfloor) \leqslant t_{r-1}(n) \leqslant t_{r-1}((r-1)\lceil \frac{n}{r-1} \rceil)$.)

9. Does every large enough graph G with at most $c\,|G|$ edges, where c is any constant, contain a set of 100 independent vertices?

10. Show that deleting at most $(m - s)(n - t)/s$ edges from a $K_{m,n}$ will never destroy all its $K_{s,t}$ subgraphs.

11. For $0 < s \leqslant t \leqslant n$ let $z(n, s, t)$ denote the maximum number of edges in a bipartite graph whose partition sets both have size n, and which does not contain a $K_{s,t}$. Show that $2\operatorname{ex}(n, K_{s,t}) \leqslant z(n, s, t) \leqslant \operatorname{ex}(2n, K_{s,t})$.

12.⁺ Let $1 \leqslant r \leqslant n$ be integers. Let G be a bipartite graph with bipartition $\{A, B\}$, where $|A| = |B| = n$, and assume that $K_{r,r} \not\subseteq G$. Show that
$$\sum_{x \in A} \binom{d(x)}{r} \leqslant (r-1)\binom{n}{r}.$$

Using the previous exercise, deduce that $\operatorname{ex}(n, K_{r,r}) \leqslant cn^{2-1/r}$ for some constant c depending only on r.

13. The *upper density* of an infinite graph G is the lim sup of the maximum edge densities of its (finite) n-vertex subgraphs as $n \to \infty$.

    (i) Show that, for every $r \in \mathbb{N}$, every infinite graph of upper density $> \frac{r-2}{r-1}$ has a K_s^r subgraph for every $s \in \mathbb{N}$.

    (ii) Deduce that the upper density of infinite graphs can only take the countably many values of $0, 1, \frac{1}{2}, \frac{2}{3}, \frac{3}{4}, \ldots$.

14. Given a tree T, find an upper bound for $\mathrm{ex}(n, T)$ that is linear in n and independent of the structure of T, i.e. depends only on $|T|$.

15. Show that the Erdős-Sós conjecture is best possible in the sense that, for every k and infinitely many n, there is a graph on n vertices and with $\frac{1}{2}(k-1)n$ edges that contains no tree with k edges.

16.⁻ Prove the Erdős-Sós conjecture for the case when the tree considered is a star.

17. Prove the Erdős-Sós conjecture for the case when the tree considered is a path.

    (Hint. Use Exercise 9 of Chapter 1.)

18. Can large average degree force the chromatic number up if we exclude some tree as an induced subgraph? More precisely: For which trees T is there a function $f: \mathbb{N} \to \mathbb{N}$ such that, for every $k \in \mathbb{N}$, every graph of average degree at least $f(k)$ either has chromatic number at least k or contains an induced copy of T?

19. Given two numerical graph invariants i_1 and i_2, write $i_1 \leqslant i_2$ if we can force i_2 to be arbitrarily high on some subgraph of G by assuming that $i_1(G)$ is large enough. (Formally: write $i_1 \leqslant i_2$ if there exists a function $f: \mathbb{N} \to \mathbb{N}$ such that, given any $k \in \mathbb{N}$, every graph G with $i_1(G) \geqslant f(k)$ has a subgraph H with $i_2(H) \geqslant k$.) If $i_1 \leqslant i_2$ as well as $i_1 \geqslant i_2$, write $i_1 \sim i_2$. Show that this is an equivalence relation for graph invariants, and sort the following invariants into equivalence classes ordered by $<$: minimum degree; average degree; connectivity; arboricity; chromatic number; colouring number; choice number; $\max\{r \mid K^r \subseteq G\}$; $\max\{r \mid TK^r \subseteq G\}$; $\max\{r \mid K^r \preccurlyeq G\}$; $\min \max d^+(v)$, where the maximum is taken over all vertices v of the graph, and the minimum over all its orientations.

20.⁺ Prove, from first principles and without using average or minimum degree arguments, the existence of a function $f: \mathbb{N} \to \mathbb{N}$ such that every graph of chromatic number at least $f(r)$ has a K^r minor.

    (Hint. Use induction on r. For the induction step $(r-1) \to r$ try to find a connected set U of vertices whose neighbours induce a subgraph that needs enough colours to contract to K^{r-1}. If no such set U exists, show that the given graph can be coloured with fewer colours than assumed.)

21. Given a graph G with $\varepsilon(G) \geqslant k \in \mathbb{N}$, find a minor $H \preccurlyeq G$ such that $\delta(H) \geqslant k \geqslant |H|/2$.

22.⁺ Find a constant c such that every graph with n vertices and at least $n + 2k(\log k + \log\log k + c)$ edges contains k edge-disjoint cycles (for all $k \in \mathbb{N}$). Deduce an edge-analogue of the Erdős-Pósa theorem (2.3.2).

    (Hint. Assuming $\delta \geqslant 3$, delete the edges of a short cycle and apply induction. The calculations are similar to the proof of Lemma 2.3.1.)

23. Simplify the proof of Theorem 7.2.3 by using Exercise 32 of Chapter 3.

24.$^{+}$ Show that any function h as in Lemma 3.5.1 satisfies the inequality $h(r) > \frac{1}{8}r^2$ for all even r, and hence that Theorem 7.2.3 is best possible up to the value of the constant c.

25. Characterize the graphs with n vertices and more than $3n - 6$ edges that contain no $TK_{3,3}$. In particular, determine $\mathrm{ex}(n, TK_{3,3})$.

   (Hint. You may use the theorem of Wagner that every edge-maximal graph without a $K_{3,3}$ minor can be constructed recursively from maximal planar graphs and copies of K^5 by pasting along K^2s.)

26.$^{-}$ Derive the four colour theorem from Hadwiger's conjecture for $r = 5$.

27.$^{-}$ Show that Hadwiger's conjecture for $r + 1$ implies the conjecture for r.

28.$^{-}$ Deduce the following weakening of Hadwiger's conjecture from known results: given any $\epsilon > 0$, every graph of chromatic number at least $r^{1+\epsilon}$ has a K^r minor, provided that r is large enough.

29.$^{-}$ Show that any graph constructed as in Proposition 7.3.1 is edge-maximal without a K^4 minor.

30. Prove the implication $\delta(G) \geqslant 3 \Rightarrow G \supseteq TK^4$.

   (Hint. You may use any result from Section 7.3.)

31. A multigraph is called *series-parallel* if it can be constructed recursively from a K^2 by the operations of subdividing and of doubling edges. Show that a 2-connected multigraph is series-parallel if and only if it has no (topological) K^4 minor.

32. Without using Theorem 7.3.8, prove Hadwiger's conjecture for all graphs of girth at least 11 and r large enough. Without using Corollary 7.3.9, show that there is a constant $g \in \mathbb{N}$ such that all graphs of girth at least g satisfy Hadwiger's conjecture, irrespective of r.

33.$^{+}$ Prove Hadwiger's conjecture for $r = 4$ from first principles.

34.$^{+}$ Prove Hadwiger's conjecture for line graphs.

35. Prove Corollary 7.3.5.

36.$^{-}$ In the definition of an ϵ-regular pair, what is the purpose of the requirement that $|X| \geqslant \epsilon |A|$ and $|Y| \geqslant \epsilon |B|$?

37.$^{-}$ Show that any ϵ-regular pair in G is also ϵ-regular in $\overline{G}$.

38. Consider a partition of a finite set V into k equally sized subsets. Show that the complete graph on V has about $k - 1$ as many edges between different partition sets as edges inside partition sets. Explain how this leads to the choice of $m := 1/\gamma$ in the proof of the Erdős-Stone theorem.

39. (i) Deduce the regularity lemma from the assumption that it holds, given $\epsilon > 0$ and $m \geqslant 1$, for all graphs of order at least some $n = n(\epsilon, m)$.

   (ii) Prove the regularity lemma for sparse graphs—more precisely, for every sequence $(G_n)_{n \in \mathbb{N}}$ of graphs G_n of order n such that $\|G_n\|/n^2 \to 0$ as $n \to \infty$.

Notes

The standard reference work for results and open problems in extremal graph theory (in a very broad sense) is still B. Bollobás, *Extremal Graph Theory*, Academic Press 1978. A kind of update on the book is given by its author in his chapter of the *Handbook of Combinatorics* (R.L. Graham, M. Grötschel & L. Lovász, eds.), North-Holland 1995. An instructive survey of extremal graph theory in the narrower sense of Section 7.1 is given by M. Simonovits in (L.W. Beineke & R.J. Wilson, eds.) *Selected Topics in Graph Theory 2*, Academic Press 1983. This paper focuses among other things on the particular role played by the Turán graphs. A more recent survey by the same author can be found in (R.L. Graham & J. Nešetřil, eds.) *The Mathematics of Paul Erdős*, Vol. 2, Springer 1996.

Turán's theorem is not merely one extremal result among others: it is the result that sparked off the entire line of research. Our first proof of Turán's theorem is essentially the original one; the second is a version of a proof of Zykov due to Brandt.

Túran's theorem has been generalized as follows. Suppose that, for some fixed $r \geqslant 3$, we wish to construct a graph on n vertices with at least γn^2 edges, where now $\frac{1}{2}\frac{r-2}{r-1} < \gamma < \frac{1}{2}$, in such a way as to create as few K^r subgraphs as possible. The *clique density theorem* says that, for fixed γ, the asymptotically best way to do this is to form a complete multipartite graph in which all classes have the same size except for one, which may be smaller. How many such classes there are depends on γ, but not on n: as in Turán's theorem, s classes will always give about γn^2 edges for $\gamma = \frac{1}{2}\frac{s-1}{s}$. The clique density theorem had been conjectured by Lovász and Simonovits in 1983, and was finally proved for all r by C. Reiher, The clique density theorem, *Ann. Math.* **184** (2016), 683–707, arXiv:1212.2454.

Our version of the Erdős-Stone theorem is a slight simplification of the original. A direct proof, not using the regularity lemma, is given in L. Lovász, *Combinatorial Problems and Exercises* (2nd edn.), North-Holland 1993. Its most fundamental application, Corollary 7.1.3, was only found 20 years after the theorem, by Erdős and Simonovits (1966).

Of our two bounds on $\mathrm{ex}(n, K_{r,r})$ the upper one is thought to give the correct order of magnitude. For vastly off-diagonal complete bipartite graphs this was verified by J. Kollár, L. Rónyai & T. Szabó, Norm-graphs and bipartite Turán numbers, *Combinatorica* **16** (1996), 399–406, who proved that $\mathrm{ex}(n, K_{r,s}) \geqslant c_r n^{2-\frac{1}{r}}$ when $s > r!$.

Details about the Erdős-Sós conjecture, including an approximate solution for large k, can be found in the survey by Komlós and Simonovits cited below. The case where the tree T is a path (Exercise 17) was proved by Erdős & Gallai in 1959. It was this result, together with the easy case of stars (Exercise 16) at the other extreme, that inspired the conjecture as a possible unifying result. A proof of the precise conjecture for large graphs was announced in 2009 by Ajtai, Komlós, Simonovits and Szemerédi, but has not been made publicly available.

The Erdős-Sós conjecture says that graphs of average degree greater than $k-1$ contain every tree with k edges. Loebl, Komlós and Sós have conjectured a 'median' version, which appears to be easier: that if at least half the vertices of a graph have degree greater than $k - 1$ it contains every tree with k edges. An approximate version of this conjecture has been proved by Hladký, Komlós, Piguet, Simonovis, Stein and Szemerédi in arXiv:1408.3870.

Theorem 7.2.3 was first proved by B. Bollobás & A. G. Thomason, Proof of a conjecture of Mader, Erdős and Hajnal on topological complete subgraphs, *Eur. J. Comb.* **19** (1998), 883–887, and independently by J. Komlós & E. Szemerédi, Topological cliques in graphs II, *Comb. Probab. Comput.* **5** (1996), 79–90. For large G, the latter authors show that the constant c in the theorem can be brought down to about $\frac{1}{2}$, which is not far from the lower bound of $\frac{1}{8}$ given in Exercise 24.

Theorem 7.2.4 was first proved in 1982 by Kostochka, and in 1984 with a better constant by Thomason. For references and more insight, also in these early proofs, see A. G. Thomason, The extremal function for complete minors, *J. Comb. Theory B* **81** (2001), 318–338. There, Thomason determines the smallest possible value of the constant c in Theorem 7.2.4 asymptotically for large r. It can be written as $c = \alpha + o(1)$, where $\alpha = 0.53131\ldots$ is an explicit constant and $o(1)$ stands for a function of r tending to zero as $r \to \infty$.

Surprisingly, the average degree needed to force an *incomplete* minor H of order r remains at $cr\sqrt{\log r}$, with $c = \alpha\gamma(H) + o(1)$, where γ is a graph invariant $H \mapsto [0,1]$ that is bounded away from 0 for dense H, and $o(1)$ is a function of $|H|$ tending to 0 as $|H| \to \infty$. See J. S. Myers & A. G. Thomason, The extremal function for noncomplete minors, *Combinatorica* **25** (2005), 725–753.

As Theorem 7.2.4 is best possible, there is no constant c such that all graphs of average degree at least cr have a K^r minor. Strengthening this assumption to $\kappa \geqslant cr$, however, can force a K^r minor in all large enough graphs; this was proved by T. Böhme, K. Kawarabayashi, J. Maharry and B. Mohar, Linear connectivity forces large complete bipartite minors, *J. Comb. Theory B* **99** (2009), 557–582. Their proof rests on a structure theorem for graphs of large tree-width not containing a given minor, which was proved only later by R. Diestel, K. Kawarabayashi, Th. Müller & P. Wollan, On the excluded minor structure theorem for graphs of large tree-width, *J. Comb. Theory B* **102** (2012), 1189–1210, arXiv:0910.0946. A simple direct argument that bypasses the use of this structure theorem was found by J.-O. Fröhlich and Th. Müller, Linear connectivity forces large complete bipartite minors: an alternative approach, *J. Comb. Theory B* **101** (2011), 502–508, arXiv:0906.2568.

The fact that large enough girth can force minors of arbitrarily high minimum degree, and hence large complete minors, was discovered by Thomassen in 1983. The reference can be found in W. Mader, Topological subgraphs in graphs of large girth, *Combinatorica* **18** (1998), 405–412, from which our Lemma 7.2.5 is extracted. Our girth assumption of $8k + 3$ has been reduced to about $4k$ by D. Kühn and D. Osthus, Minors in graphs of large girth, *Random Struct. Alg.* **22** (2003), 213–225, which is conjectured to be best possible.

The original reference for Theorem 7.2.7 can be found in D. Kühn and D. Osthus, Improved bounds for topological cliques in graphs of large girth, *SIAM J. Discrete Math.* **20** (2006), 62–78, where they re-prove their theorem with $g \leqslant 27$. See also D. Kühn & D. Osthus, Subdivisions of K_{r+2} in graphs of average degree at least $r + \varepsilon$ and large but constant girth, *Comb. Probab. Comput.* **13** (2004), 361–371.

The proof of Hadwiger's conjecture for $r = 4$ hinted at in Exercise 33 was given by Hadwiger himself, in the 1943 paper containing his conjecture. Like Hadwiger's conjecture, Hajós's conjecture has (later) been proved for graphs of large girth (Corollary 7.3.9) and for line graphs; see C. Thomassen, Hajós' conjecture for line graphs, *J. Comb. Theory B* **97** (2007), 156–157. A counterexample to the general Hajós conjecture was found as early as 1979 by Catlin. A little later, Erdős and Fajtlowicz proved that Hajós's conjecture is false for 'almost all' graphs, while Bollobás, Catlin and Erdős showed that Hadwiger's conjecture is true for 'almost all graphs' (see Chapter 11). Proofs of Wagner's Theorem 7.3.4 (with Hadwiger's conjecture for $r = 5$ as a corollary) can be found in Bollobás's *Extremal Graph Theory* (see above) and in Halin's *Graphentheorie* (2nd ed.), Wissenschaftliche Buchgesellschaft 1989. Hadwiger's conjecture for $r = 6$ was proved by N. Robertson, P.D. Seymour and R. Thomas, Hadwiger's conjecture for K_6-free graphs, *Combinatorica* **13** (1993), 279–361.

For infinite graphs, the following weakening of the assertion of Hadwiger's conjecture is true: every graph of chromatic number $\alpha \geqslant \aleph_0$ contains every K_β with $\beta < \alpha$ as a minor, even as a topological minor. This was proved by R. Halin, Unterteilungen vollständiger Graphen in Graphen mit unendlicher chromatischer Zahl, *Abh. Math. Sem. Univ. Hamburg* **31** (1967), 156–165. The case of $\alpha = \aleph_0$ is Exercise 14 in Chapter 8; the proof for $\alpha > \aleph_0$ is included in R. Diestel, *Graph Decompositions*, Oxford University Press 1990.

The investigation of graphs not containing a given graph as a minor, or topological minor, has a long history. It probably started with Wagner's 1935 PhD thesis, in which he sought to 'detopologize' the four colour problem by classifying the graphs without a K^5 minor. His hope was to be able to show abstractly that all those graphs were 4-colourable; since the graphs without a K^5 minor include the planar graphs, this would amount to a proof of the four colour conjecture involving no topology whatsoever. The result of Wagner's efforts, Theorem 7.3.4, falls tantalizingly short of this goal: although it succeeds in classifying the graphs without a K^5 minor in structural terms, planarity re-emerges as one of the criteria used in the classification. From this point of view, it is instructive to compare Wagner's K^5 theorem with similar classification theorems, such as his analogue for K^4 (Proposition 7.3.1), where the graphs are decomposed into parts from a *finite* set of irreducible graphs. See R. Diestel, *Graph Decompositions*, Oxford University Press 1990, for more such classification theorems.

Despite its failure to resolve the four colour problem, Wagner's K^5 structure theorem had consequences for the development of graph theory like few others. To mention just two: it prompted Hadwiger to make his famous conjecture; and it inspired much of the work of Robertson and Seymour on minors (Chapter 12), in particular the notion of a tree-decomposition and the structure theorem for graphs without a K^n minor (Theorem 12.6.6). Wagner himself responded to Hadwiger's conjecture with a proof in 1964 that, to force a K^r minor, it does suffice to raise the chromatic number of a graph to *some* value depending only on r (Exercise 20). This theorem, along with its analogue for topological minors proved independently by Dirac and by Jung, prompted the question which average degree suffices to force the desired minor. This was first addressed by Mader, whose seminal proofs of Propositions 7.2.1 and 7.2.2 were part of his PhD thesis in 1967.

Theorem 7.3.8 is a consequence of the more fundamental result of D. Kühn and D. Osthus, Complete minors in $K_{s,s}$-free graphs, *Combinatorica* **25** (2005) 49–64, that every graph without a $K_{s,s}$ subgraph that has average degree $r \geqslant r_s$ has a K^p minor for $p = \lfloor r^{1+\frac{1}{2(s-1)}}/(\log r)^3 \rfloor$. This was improved further by M. Krivelevich and B. Sudakov, Minors in expanding graphs, *Geom. Funct. Anal.* **19** (2009), 294–331, arXiv:0707.0133.

As in Gyárfás's conjecture, one may ask under what additional assumptions large average degree forces an *induced* subdivision of a given graph H. This was answered for arbitrary H by D. Kühn and D. Osthus, Induced subdivisions in $K_{s,s}$-free graphs of large average degree, *Combinatorica* **24** (2004) 287–304, who proved that for all $r, s \in \mathbb{N}$ there exists $d \in \mathbb{N}$ such that every graph $G \not\supseteq K_{s,s}$ with $d(G) \geqslant d$ contains a TK^r as an induced subgraph.

Gyárfás's conjecture itself, that excluding a fixed tree as an induced subgraph bounds the chromatic number of a graph in terms of its clique number, is still open. Excluding all induced subdivisions of a fixed tree, however, does achieve this: this was proved by A.D. Scott, Induced trees in graphs of large chromatic number, *J. Graph Theory* **24** (1997), 297–311. On the other hand, excluding all induced subdivisions of an arbitrary fixed graph H need not bound the chromatic number of a graph G in terms of its clique number; see J. Kozik et al, Triangle-free intersection graphs of line segments with large chromatic number, *J. Comb. Theory B* **105** (2014), 6–10, arXiv:1209.1595.

The regularity lemma is proved in E. Szemerédi, Regular partitions of graphs, *Colloques Internationaux CNRS* **260**—*Problèmes Combinatoires et Théorie des Graphes*, Orsay (1976), 399–401. Our rendering follows an account by Scott (personal communication). A broad survey on the regularity lemma and its applications is given by J. Komlós & M. Simonovits in (D. Miklós, V.T. Sós & T. Szőnyi, eds.) *Paul Erdős is 80*, Vol. 2, Proc. Colloq. Math. Soc. János Bolyai (1996); the concept of a regularity graph and Lemma 7.5.2 are taken from this paper. The regularity lemma was adapted to sparse graphs by A.D. Scott, Szemerédi's regularity lemma for matrices and sparse graphs, *Comb. Probab. Comput.* **20** (2011), 455–466. The statement of the lemma remains the same, only the definition of an ϵ-regular pair is adapted in the obvious way depending on the graph G considered: a pair (A, B) of disjoint sets of vertices of G is now called ϵ-*regular* if all subsets $X \subseteq A$ and $Y \subseteq B$ with $|X| \geqslant \epsilon |A|$ and $|Y| \geqslant \epsilon |B|$ satisfy $|d(X, Y) - d(A, B)| \leqslant \epsilon p$, where $p := \|G\|/\binom{|G|}{2}$.

8 Infinite Graphs

The study of infinite graphs is an attractive, but often neglected, part of graph theory. This chapter aims to give an introduction that starts gently, but then moves on in several directions to display both the breadth and some of the depth that this field has to offer. Our overall theme will be to highlight the typical kinds of phenomena that will always appear when graphs are infinite, and to show how they can lead to deep and fascinating problems.

Perhaps the most typical such phenomena occur already when the graphs are 'only just' infinite, when they have only countably many vertices and perhaps only finitely many edges at each vertex. This is not surprising: after all, some of the most basic structural features of graphs, such as paths, are intrinsically countable. Problems that become really interesting only for uncountable graphs tend to be interesting for reasons that have more to do with sets than with graphs, and are studied in *combinatorial set theory*. This, too, is a fascinating field, but not our topic in this chapter. The problems we shall consider will all be interesting for countable graphs, and set-theoretic problems will not arise.

The terminology we need is exactly the same as for finite graphs, except when we wish to describe an aspect of infinite graphs that has no finite counterpart. One important such aspect is the eventual behaviour of the infinite paths in a graph, which is captured by the notion of *ends*. The ends of a graph can be thought of as additional limit points at infinity to which its infinite paths converge. This convergence is described formally in terms of a natural topology placed on the graph together with its ends. In Sections 6–8 we shall therefore assume familiarity with the basic concepts of point-set topology; reminders of the relevant definitions will be included as they arise.

© Reinhard Diestel 2017
R. Diestel, *Graph Theory*, Graduate Texts in Mathematics 173,
DOI 10.1007/978-3-662-53622-3_8

8.1 Basic notions, facts and techniques

This section gives a gentle introduction to the aspects of infinity most commonly encountered in graph theory.[1]

After just a couple of definitions, we begin by looking at a few obvious properties of infinite sets, and how they can be employed in the context of graphs. We then illustrate how to use the three most basic common tools in infinite graph theory: Zorn's lemma, transfinite induction, and something called 'compactness'. We complete the section with the combinatorial definition of an end; topological aspects will be treated in Section 8.6.

locally finite   A graph is *locally finite* if all its vertices have finite degrees. An infinite graph (V, E) of the form

$$V = \{x_0, x_1, x_2, \ldots\} \qquad E = \{x_0 x_1, x_1 x_2, x_2 x_3, \ldots\}$$

rays   is called a *ray*, and a *double ray* is an infinite graph (V, E) of the form

$$V = \{\ldots, x_{-1}, x_0, x_1, \ldots\} \qquad E = \{\ldots, x_{-1} x_0, x_0 x_1, x_1 x_2, \ldots\};$$

in both cases the x_n are assumed to be distinct. Thus, up to isomorphism, there is only one ray and one double ray, the latter being the unique infinite 2-regular connected graph. In the context of infinite *path*   graphs, finite paths, rays and double rays are all called *paths*.

tail   The subrays of a ray or double ray are its *tails*. Formally, every ray has infinitely many tails, but any two of them differ only by a finite initial segment. The union of a ray R with infinitely many disjoint finite *comb*   paths having precisely their first vertex on R is a *comb*; the last vertices *teeth, spine* of those paths are the *teeth* of this comb, and R is its *spine*. (If such a path is trivial, which we allow, then its unique vertex lies on R and also counts as a tooth; see Figure 8.1.1.)

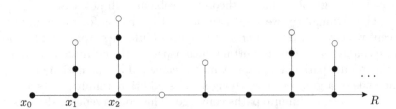

Fig. 8.1.1. A comb with white teeth and spine $R = x_0 x_1 \ldots$

---

[1] This introductory section is deliberately kept informal, with the emphasis on ideas rather than definitions that do not belong in a graph theory book. A more formal reminder of those basic definitions about infinite sets and numbers that we shall need is given in Appendix A at the end of the book.

Let us now look at a few very basic properties of infinite sets, and see how they appear in some typical arguments about graphs.

$$\textit{An infinite set minus a finite subset is still infinite.} \qquad (1)$$

This trivial property is eminently useful when the infinite set in question plays the role of 'supplies' that keep an iterated process going. For example, let us show that if a graph G is infinitely connected (that is, if G is k-connected for every $k \in \mathbb{N}$), then G contains a subdivision of $K^{\aleph_0}$, the complete graph of order $|\mathbb{N}|$. We embed $K^{\aleph_0}$ in G (as a topological minor) in one infinite sequence[2] of steps, as follows. We begin by enumerating its vertices. Then at each step we embed the next vertex in G, connecting it to the images of its earlier neighbours by paths in G that avoid any other vertices used so far. The point here is that each new path has to avoid only finitely many previously used vertices, which is not a problem since deleting any finite set of vertices keeps G infinitely connected.

$K^{\aleph_0}$

If G, too, is countable, can we then also find a $TK^{\aleph_0}$ as a spanning subgraph of G? Although embedding $K^{\aleph_0}$ in G topologically as above takes infinitely many steps, it is by no means guaranteed that the $TK^{\aleph_0}$ constructed uses all the vertices of G. However, it is not difficult to ensure this: since we are free to choose the image of each new vertex of $K^{\aleph_0}$, we can choose this as the next unused vertex from some fixed enumeration of $V(G)$. In this way, every vertex of G gets chosen eventually, unless it becomes part of the $TK^{\aleph_0}$ before its time, as a subdividing vertex on one of the paths.

$$\textit{Unions of countably many countable sets are countable.} \qquad (2)$$

This fact can be applied in two ways: to show that sets that come to us as countable unions are 'small', but also to rewrite a countable set deliberately as a disjoint union of infinitely many infinite subsets. For an example of the latter type of application, let us show that an infinitely edge-connected countable graph has infinitely many edge-disjoint spanning trees. (Note that the converse implication is trivial.) The trick is to construct the trees simultaneously, in one infinite sequence of steps. We first use (2) to partition $\mathbb{N}$ into infinitely many infinite subsets N_i ($i \in \mathbb{N}$). Then at step n we look which N_i contains n, and add a further vertex v to the ith tree T_i. As before, we choose v minimal in some fixed enumeration of $V(G)$ among the vertices not yet in T_i, and join v to T_i by a path avoiding the finitely many edges used so far.

Clearly, a countable set cannot have uncountably many disjoint subsets. However,

---

[2] We reserve the term 'infinite sequence' for sequences indexed by the set of natural numbers. (In the language of well-orderings: for sequences of order type ω.)

> *A countable set can have uncountably many subsets whose* (3)
> *pairwise intersections are all finite.*

This is a remarkable property of countable sets, and a good source of counterexamples to rash conjectures. Can you prove it without looking at Figure 8.1.4?

Another common pitfall in dealing with infinite sets is to assume that the intersection of an infinite nested sequence $A_0 \supseteq A_1 \supseteq \ldots$ of infinite (or uncountable) sets must still be infinite (or uncountable). It need not be; in fact it may be empty. (Examples?)

Before we move on to our discussion of common infinite proof techniques, let us look at one more type of construction. One often wants to construct a graph G with a property that is in some sense local, a property that has more to do with the finite subgraphs of G than with G itself. Rather than formalize what exactly this should mean, let us consider an example: given two large integers k and g, let us construct a graph G that is k-connected and has girth at least g.[3]

We start with a cycle of length g; call it G_0. This graph has the right girth, but it is not k-connected. To cure this defect for the vertices of G_0, join every pair of them by k new independent paths, keeping all these paths internally disjoint. If we choose the paths long enough, the resulting graph G_1 will again have girth g, and no two vertices of G_0 can be separated in it by fewer than k other vertices. Of course, G_1 is not k-connected either. But we can repeat the construction step for the pairs of vertices of G_1, extending G_1 to G_2, and so on. The limit graph $G = \bigcup_{n \in \mathbb{N}} G_n$ will again have girth g, since any short cycle would have appeared in some G_n on the way. And, unlike all the G_n, it will be k-connected: since every two vertices are contained in some common G_n, they cannot be separated by fewer than k other vertices in G_{n+1}, let alone in G.

There are a few basic proof techniques that are found frequently in infinite combinatorics. The two most common of these are the use of Zorn's lemma and transfinite induction. Rather than describing these formally,[4] we illustrate their use by a simple example.

Proposition 8.1.1. *Every connected graph contains a spanning tree.*

First proof (by Zorn's lemma).
Given a connected graph G, consider the set of all trees $T \subseteq G$, ordered by the subgraph relation. Since G is connected, any maximal such tree contains every vertex of G, i.e. is a spanning tree of G.

---

[3] There are finite such graphs, but they are much harder to construct; we shall prove their existence by random methods in Chapter 11.2.

[4] Appendix A offers brief introductions to both, enough to enable the reader to use these tools with confidence in practice.

To prove that a maximal tree exists, we have to show that for any chain $\mathcal{C}$ of such trees there is an upper bound: a tree $T^* \subseteq G$ containing every tree in $\mathcal{C}$ as a subgraph. We claim that $T^* := \bigcup \mathcal{C}$ is such a tree.

To show that T^* is connected, let $u, v \in T^*$ be two vertices. Then in $\mathcal{C}$ there is a tree T_u containing u and a tree T_v containing v. One of these is a subgraph of the other, say $T_u \subseteq T_v$. Then T_v contains a path from u to v, and this path is also contained in T^*.

To show that T^* is acyclic, suppose it contains a cycle C. Each of the edges of C lies in some tree in $\mathcal{C}$. These trees form a finite subchain of $\mathcal{C}$, which has a maximal element T. Then $C \subseteq T$, a contradiction. $\square$

Transfinite induction and recursion are very similar to finite inductive proofs and constructions, respectively. Basically, one proceeds step by step, and may at each step assume as known what was shown or constructed before. The only difference is that one may 'start again' after performing any infinite number of steps. This is formalized by the use of ordinals rather than natural numbers for counting the steps; see Appendix A.

Just as with finite graphs, it is usually more intuitive to construct a desired object (such as a spanning tree) step by step, rather than starting with some unknown 'maximal' object and then proving that it has the desired properties. More importantly, a step-by-step construction is almost always the best way to *find* the desired object: only later, when one understands the construction well, can one devise an inductive ordering (one whose chains have upper bounds) in which the desired objects appear as the maximal elements. Thus, although Zorn's lemma may at times provide an elegant way to wrap up a constructive proof, it cannot in general replace a good understanding of transfinite induction—just as a preference for elegant direct definitions of finite objects cannot, for a thorough understanding, replace the more pedestrian algorithmic approach.

Our second proof of Proposition 8.1.1 illustrates both the constructive and the proof aspect of transfinite induction in the typical intertwined way. We define larger and larger subgraphs $T_\alpha \subseteq G$ inductively. At each step α we prove that T_α is a tree. The definition of T_α will assume and use that subgraphs T_β for all $\beta < \alpha$ have been previously defined, but not only this: it needs to assume that they are nested trees. This fact, therefore, has to be proved along with the recursive definition, always 'just before' it is needed.

Second proof (by transfinite induction).
Let G be a connected graph. We define trees $T_\alpha \subseteq G$ recursively so that

$$T_\beta \subseteq T_\alpha \text{ for all } \beta < \alpha. \qquad (*_\alpha)$$

Let T_0 consist of a single vertex. Given a limit ordinal $\alpha > 0$, let $T_\alpha := \bigcup_{\beta < \alpha} T_\beta$. Since the T_β are trees satisfying $(*_\beta)$, our new T_α is also a tree (as in the first proof), and it clearly satisfies $(*_\alpha)$.

Given a successor $\alpha = \beta + 1$, we first check whether $G - T_\beta = \emptyset$. If so, then T_β is a spanning tree and we terminate the recursion. If not, then $G - T_\beta$ has a vertex v_α that sends an edge e_α to a vertex in T_β. Then T_α, obtained from T_β by adding v_α and e_α, is a tree satisfying $(*_\alpha)$.

It remains to check that our recursion does indeed terminate. But if $v_{\beta+1}$ gets defined for all $\beta < \gamma$ then $\beta \mapsto v_{\beta+1}$ is an injective map showing that $|\gamma| \leqslant |G|$. This cannot hold for all ordinals γ; it fails, for example, when γ represents a well-ordering of the power set of $V(G)$.   □

Why did these proofs work so smoothly? The reason is that the forbidden substructures, cycles, were finite and therefore could not arise unexpectedly at limit steps. If we wanted to construct a *rayless* spanning tree, on the other hand, one that contains no ray, then the edges of partial finite trees T_β might combine to form a ray in $T_\alpha = \bigcup_{\beta < \alpha} T_\beta$ when α is a limit. And indeed, here lies the challenge in most transfinite constructions: to make the right choices at successor steps to ensure that the structure will also be as desired at limits.

compactness
proofs

Our third basic proof technique, somewhat mysteriously referred to as *compactness* (see below for why), offers a formalized way of making the right choices in certain standard cases. These are cases where nothing unexpected happens at limits, but a choice that looks good at the time it is made may lead to a dead end after another *finite* number of steps—unlike the creation of a cycle, which is visible at once.

For example, let G be a graph whose finite subgraphs are all k-colourable. It is natural then to try to construct a k-colouring of G as a limit of k-colourings of its finite subgraphs. Now each finite subgraph will have several k-colourings; will it matter which we choose? Clearly, it will. When $G' \subseteq G''$ are two finite subgraphs and u, v are vertices of G' that receive the same colour in every k-colouring of G'' (and hence also in any k-colouring of G), we must not give them different colours in the colouring we choose for G', even if such a colouring exists. However if we do manage, somehow, to colour the finite subgraphs of G compatibly, we shall automatically have a colouring of all of G.

All compactness proofs deal with situations similar to this. We wish to solve a problem about an infinite structure, and we know how to solve it for all the finite substructures. 'Compactness' enables us to combine these partial solutions to an overall solution if the partial solutions are compatible in the right way.

For countable structures, all this—the choices, the dead ends, the compatibility requirement on the finite solutions, and how they combine to an overall solution—can be made visible in a particularly intuitive way, by a graph:

[8.2.1]
[8.2.6]
[8.6.1]
[8.6.10]
[8.7.3]
[9.1.3]

Lemma 8.1.2. (König's Infinity Lemma)
Let $V_0, V_1, \ldots$ be an infinite sequence of disjoint non-empty finite sets, and let G be a graph on their union. Assume that every vertex v in a set V_n with $n \geqslant 1$ has a neighbour $f(v)$ in V_{n-1}. Then G contains a ray $v_0 v_1 \ldots$ with $v_n \in V_n$ for all n.

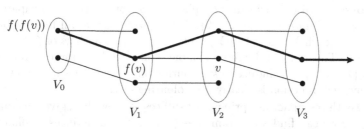

Fig. 8.1.2. König's infinity lemma

Proof. Let $\mathcal{P}$ be the set of all finite paths of the form $v\, f(v)\, f(f(v)) \ldots$ ending in V_0. Since V_0 is finite but $\mathcal{P}$ is infinite, infinitely many of the paths in $\mathcal{P}$ end at the same vertex $v_0 \in V_0$. Of these paths, infinitely many also agree on their penultimate vertex $v_1 \in V_1$, because V_1 is finite. Of those paths, infinitely many agree even on their vertex v_2 in V_2—and so on. Although the set of paths considered decreases from step to step, it is still infinite after any finite number of steps, so v_n gets defined for every $n \in \mathbb{N}$. By definition, each vertex v_n is adjacent to v_{n-1} on one of those paths, so $v_0 v_1 \ldots$ is indeed a ray. $\qquad\square$

The following 'compactness theorem', the first of its kind in graph theory, answers our question about colourings:

Theorem 8.1.3. (de Bruijn & Erdős, 1951)
Let $G = (V, E)$ be a graph and $k \in \mathbb{N}$. If every finite subgraph of G has chromatic number at most k, then so does G.

First proof (for G countable, by the infinity lemma).
Let $v_0, v_1, \ldots$ be an enumeration of V and put $G_n := G[v_0, \ldots, v_n]$. Write V_n for the set of all k-colourings of G_n with colours in $\{1, \ldots, k\}$. Define a graph on $\bigcup_{n \in \mathbb{N}} V_n$ by inserting all edges cc' such that $c \in V_n$ and $c' \in V_{n-1}$ is the restriction of c to $\{v_0, \ldots, v_{n-1}\}$. Let $c_0 c_1 \ldots$ be a ray in this graph with $c_n \in V_n$ for all n. Then $c := \bigcup_{n \in \mathbb{N}} c_n$ is a colouring of G with colours in $\{1, \ldots, k\}$. $\qquad\square$

Applications of the infinity lemma such as this one rely on the fact that a countable graph can be exhausted by a nested sequence of finite subgraphs. Appendix A offers a version of the infinity lemma that works for arbitrary graphs, in which these finite subgraphs need not be sequentially ordered. This general version is still very intuitive and can be used conveniently in many settings, including a proof of Theorem 8.1.3.

The essence of compactness proofs is often encoded more directly, if less graphically, in terms of just sets and functions. The *compactness principle* from Appendix A is such a version that is particularly easy to apply. We illustrate its application by another proof of Theorem 8.1.3:

Second proof (for arbitrary graphs, by the compactness principle).
Let $X := V$ and $S := \{1, \ldots, k\}$. Let $\mathcal{F}$ be the set of all finite subsets of V. For each $Y \in \mathcal{F}$ let $\mathcal{A}(Y)$ be the set of k-colourings of $G[Y]$. Our proof of Theorem 8.1.3 will be complete once we have found a function $V \to \{1, \ldots, k\}$ that induces a k-colouring on every finite subgraph $G[Y]$, as any such function is clearly a k-colouring of G.

By the compactness principle it suffices to show that, given any finite $\mathcal{Y} \subseteq \mathcal{F}$, we can find a function $V \to \{1, \ldots, k\}$ that induces a colouring on every $G[Y]$ with $Y \in \mathcal{Y}$. But this is easy: just take a k-colouring of the finite graph $G[\bigcup \mathcal{Y}]$, and extend it arbitrarily to the rest of V. $\square$

Our last proof of Theorem 8.1.3 appeals directly to compactness as defined in topology. Recall that a topological space is *compact* if its closed sets have the 'finite intersection property', which means that the overall intersection $\bigcap \mathcal{A}$ of a set $\mathcal{A}$ of closed sets is non-empty whenever every finite subset of $\mathcal{A}$ has a non-empty intersection. By Tychonoff's theorem of general topology, any product of compact spaces is compact in the usual product topology.

Third proof (for arbitrary graphs, by Tychonoff's theorem).
Consider the product space

$$X := \prod_V \{1, \ldots, k\} = \{1, \ldots, k\}^V$$

of $|V|$ copies of the finite set $\{1, \ldots, k\}$ endowed with the discrete topology. By Tychonoff's theorem, this is a compact space. Its basic open sets have the form

$$O_h := \{ f \in X : f|_U = h \},$$

where h is some map from a finite set $U \subseteq V$ to $\{1, \ldots, k\}$.

For every finite set $U \subseteq V$, let A_U be the set of all $f \in X$ whose restriction to U is a k-colouring of $G[U]$. These sets A_U are closed (as well as open—why?), and for any finite set $\mathcal{U}$ of finite subsets of V we have $\bigcap_{U \in \mathcal{U}} A_U \neq \emptyset$, because $G[\bigcup \mathcal{U}]$ has a k-colouring. By the finite intersection property of the sets A_U, their overall intersection is non-empty, and every element of this intersection is a k-colouring of G. $\square$

Although our three compactness proofs look formally different, it is instructive to compare them in detail, checking how the requirements in one are reflected in the other (cf. Exercise 16).

As mentioned before, the standard use for compactness proofs is to transfer theorems from finite to infinite graphs, or conversely. This is not always quite as straightforward as above; often, the statement has to be modified a little to make it susceptible to a compactness argument.

As an example—see Exercises 17–28 for more—let us prove the locally finite version of the following famous conjecture. Call a bipartition of the vertex set of a graph *unfriendly* if every vertex has at least as many neighbours in the other class as in its own. Clearly, every finite graph has an unfriendly partition: just take any partition that maximizes the number of edges between the partition classes. At the other extreme, it can be shown by set-theoretic methods that uncountable graphs need not have such partitions. Thus, intriguingly, it is the countable case that has remained unsolved:

Unfriendly Partition Conjecture. *Every countable graph admits an unfriendly partition of its vertex set.*

Proof for locally finite graphs. Let $G = (V, E)$ be an infinite but locally finite graph, and enumerate its vertices as $v_0, v_1, \ldots$. For every $n \in \mathbb{N}$, let $\mathcal{V}_n$ be the set of partitions of $V_n := \{v_0, \ldots, v_n\}$ into two sets U_n and W_n such that every vertex $v \in V_n$ with $N_G(v) \subseteq V_n$ has at least as many neighbours in the other class as in its own. Since the conjecture holds for finite graphs, the sets $\mathcal{V}_n$ are non-empty. For all $n \geqslant 1$, every $(U_n, W_n) \in \mathcal{V}_n$ induces a partition (U_{n-1}, W_{n-1}) of V_{n-1}, which lies in $\mathcal{V}_{n-1}$. By the infinity lemma, there is an infinite sequence of partitions $(U_n, W_n) \in \mathcal{V}_n$, one for every $n \in \mathbb{N}$, such that each is induced by the next. Then $(\bigcup_{n \in \mathbb{N}} U_n, \bigcup_{n \in \mathbb{N}} W_n)$ is an unfriendly partition of G. $\square$

The trick that made this proof possible was to require, for the partitions of V_n, correct positions only of vertices that send no edge out of V_n: this weakening is necessary to ensure that partitions from $\mathcal{V}_n$ induce partitions in $\mathcal{V}_{n-1}$; but since, by local finiteness, every vertex has this property eventually (for large enough n), the weaker assumption suffices to ensure that the limit partition is unfriendly.

Let us complete this section with an introduction to the one important concept of infinite graph theory that has no finite counterpart, the notion of an end. An *end*[5] of a graph G is an equivalence class of rays in G, where two rays are considered equivalent if, for every finite set $S \subseteq V(G)$, both have a tail in the same component of $G - S$. This

end

---

[5] Not to be confused with the ends, or endvertices, of an edge. In the context of infinite graphs, we use the term 'endvertices' to avoid confusion.

is indeed an equivalence relation: note that, since S is finite, there is exactly one such component for each ray. If two rays are equivalent— and only then—they can be linked by infinitely many disjoint paths: just choose these inductively, taking as S the union of the vertex sets of the first finitely many paths to find the next. The set of ends of G is denoted $\Omega(G)$ by $\Omega(G)$, and we write $G = (V, E, \Omega)$ to express that G has vertex, edge and end sets V, E, Ω.

For example, let us determine the ends of the 2-way infinite ladder shown in Figure 8.1.3. Every ray in this graph contains vertices arbitrarily far to the left or vertices arbitrarily far to the right, but not both. These two types of rays are clearly equivalence classes, so the ladder has exactly two ends. (In Figure 8.1.3 these are shown as two isolated dots—one on the left, the other on the right.)

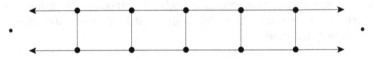

Fig. 8.1.3. The 2-way ladder has two ends

The ends of a tree are particularly simple: two rays in a tree are equivalent if and only if they share a tail, and for every fixed vertex v each end contains exactly one ray starting at v. Even a locally finite tree can have uncountably many ends. The prototype example (see Exercise 37) *binary* is the *binary tree* T_2, the rooted tree in which every vertex has exactly *tree* T_2 two upper neighbours. Often, the vertex set of T_2 is taken to be the set of finite 0–1 sequences (with the empty sequence as the root), as indicated in Figure 8.1.4. The ends of T_2 then correspond bijectively to

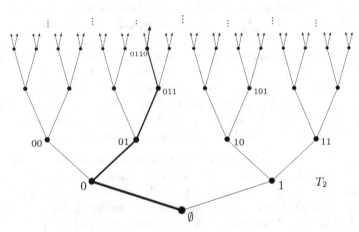

Fig. 8.1.4. The binary tree T_2 has continuum many ends, one for every infinite 0–1 sequence

its rays starting at $\emptyset$, and hence to the infinite 0–1 sequences.

These examples suggest that the ends of a graph can be thought of as 'points at infinity' to which its rays converge. We shall formalize this in Section 8.6, where we define a natural topology on a graph and its ends in which rays will indeed converge to their respective ends.

The maximum number of disjoint rays in an end is the *(combinatorial) vertex-degree* of that end, the maximum number of edge-disjoint rays in it is its *(combinatorial) edge-degree*. These maxima are indeed attained: if an end contains a set of k (edge-) disjoint rays for every integer k, it also contains an infinite set of (edge-) disjoint rays (Exercise 43). Thus, every end has a vertex-degree and an edge-degree in $\mathbb{N} \cup \{\infty\}$.

end degrees

8.2 Paths, trees, and ends

There are two fundamentally different aspects to the infinity of an infinite connected graph: one of 'length', expressed in the presence of rays, and one of 'width', expressed locally by infinite degrees. The infinity lemma tells us that at least one of these must occur:

Proposition 8.2.1. *Every infinite connected graph has a vertex of infinite degree or contains a ray.*

Proof. Let G be an infinite connected graph with all degrees finite. Let v_0 be a vertex, and for every $n \in \mathbb{N}$ let V_n be the set of vertices at distance n from v_0. Induction on n shows that the sets V_n are finite, and hence that $V_{n+1} \neq \emptyset$ (because G is infinite and connected). Furthermore, the neighbour of a vertex $v \in V_{n+1}$ on any shortest v–v_0 path lies in V_n. By Lemma 8.1.2, G contains a ray. $\qquad\square$

(8.1.2)

Often it is useful to have more detailed information on how this ray or vertex of infinite degree lies in G. The following lemma enables us to find it 'close to' any given infinite set of vertices.

Lemma 8.2.2. (Star-Comb Lemma)
Let U be an infinite set of vertices in a connected graph G. Then G contains either a comb with all teeth in U or a subdivision of an infinite star with all leaves in U.

[8.6.3]

Proof. As G is connected, it contains a path between two vertices in U. This path is a tree $T \subseteq G$ every edge of which lies on a path in T between two vertices in U. By Zorn's lemma there is a maximal such tree T^*. Since U is infinite and G is connected, T^* is infinite. If T^* has a vertex of infinite degree, it contains the desired subdivided star.

Suppose now that T^* is locally finite. Then T^* contains a ray R (Proposition 8.2.1). Let us construct a sequence $P_1, P_2, \ldots$ of disjoint R–U paths in T^*. Having chosen P_i for every $i < n$ for some n, pick $v \in R$ so that vR meets none of those paths P_i. The first edge of vR lies on a path P in T^* between two vertices in U; let us think of P as traversing this edge in the same direction as R, and choose P minimal. Then vP has the form $vRwP$, where $P_n := wP$ is an R–U path. And $P_n \cap P_i = \emptyset$ for all $i < n$, because $P_i \cup Rw \cup P_n$ contains no cycle. $\qquad\square$

We shall often apply Lemma 8.2.2 in locally finite graphs, in which case it always yields a comb.

Recall that a rooted tree $T \subseteq G$ is *normal* in G if the endvertices of every T-path in G are comparable in the tree-order of T. If T is a spanning tree, the only T-paths are edges of G that are not edges of T.

Normal spanning trees are perhaps the single most important structural tool in infinite graph theory. As in finite graphs, they exhibit the separation properties of the graph they span.[6] Moreover, their *normal* *normal ray* *rays*, those that start at the root, reflect its end structure:

[8.6.8]   **Lemma 8.2.3.** *If T is a normal spanning tree of G, then every end of* G *contains exactly one normal ray of T.*

(1.5.5)   *Proof.* Let $\omega \in \Omega(G)$ be given. Apply the star-comb lemma in T with U the vertex set of any ray $R \in \omega$. If the lemma gives a subdivided star with leaves in U and centre z, say, then the finite down-closure $\lceil z \rceil$ of z in T separates infinitely many vertices $u > z$ of U pairwise in G (Lemma 1.5.5). This contradicts our choice of U.

So T contains a comb with teeth on R. Let $R' \subseteq T$ be its spine. Since every ray in T has an increasing tail (Exercise 4), we may assume that R' is a normal ray. Since R' is equivalent to R, it lies in ω.

Conversely, distinct normal rays of T are separated in G by the (finite) down-closure of their greatest common vertex (Lemma 1.5.5), so they cannot belong to the same end of G. $\qquad\square$

Not all connected graphs have a normal spanning tree; complete uncountable graphs, for example, have none. (Why not?) The quest to characterize the graphs that have a normal spanning tree is not entirely over, and it has held some surprises.[7] One of the most useful sufficient conditions is that the graph contains no $TK^{\aleph_0}$; see Theorem 12.6.9. For our purposes, the following result suffices:

[8.7.2]   **Theorem 8.2.4.** (Jung 1967)
Every countable connected graph has a normal spanning tree.

---

[6] Lemma 1.5.5 continues to hold for infinite graphs, with the same proof.

[7] One of these is Theorem 8.6.2; for more see the notes.

Proof. The proof follows that of Proposition 1.5.6; we only sketch the   (1.5.6)
differences. Starting with a single vertex, we construct an infinite se-
quence $T_0 \subseteq T_1 \subseteq \ldots$ of finite normal trees in G, all with the same root,
whose union T will be a normal spanning tree.

To ensure that T spans G, we fix an enumeration $v_0, v_1, \ldots$ of $V(G)$
and see to it that T_n contains v_n. It is clear that T will be a tree (since
any cycle in T would lie in some T_n, and every two vertices of T lie in
a common T_n and can be linked there), and clearly the tree order of T
induces that of the T_n. Finally, T will be normal, because the endvertices
of any edge of G that is not an edge of T lie in some T_n: since that T_n
is normal, they must be comparable there, and hence in T.

It remains to specify how to construct T_{n+1} from T_n. If $v_{n+1} \in T_n$,
put $T_{n+1} := T_n$. If not, let C be the component of $G - T_n$ containing v_{n+1}. Let x be the greatest element of the chain $N(C)$ in T_n, and
let T_{n+1} be the union of T_n and an x–v_{n+1} path P with $\mathring{P} \subseteq C$. Then
the neighbourhood in T_{n+1} of any new component $C' \subseteq C$ of $G - T_{n+1}$
is a chain in T_{n+1}, so T_{n+1} is again normal.   □

One of the most basic problems in an infinite setting that has no
finite equivalent is whether or not 'arbitrarily many', in some context,
implies 'infinitely many'. Suppose we can find k disjoint rays in some
given graph G, for every $k \in \mathbb{N}$; does G also contain an infinite set of
disjoint rays?

The answer to the corresponding question for finite paths (of any
fixed length) is clearly 'yes', since a finite path P can never get in the way
of more than $|P|$ disjoint other paths. A badly chosen ray, however, can
meet infinitely many other rays, preventing them from being selected for
the same disjoint set. Rather than collecting our disjoint rays greedily,
we therefore have to construct them carefully and all simultaneously.

The proof of the following theorem is a nice example of a construction in an infinite sequence of steps, where the final object emerges only
at the limit step. Each of the steps in the sequence will involve a non-
trivial application of Menger's theorem (3.3.1).

Theorem 8.2.5. (Halin 1965)

(i) *If an infinite graph G contains k disjoint rays for every $k \in \mathbb{N}$,
then G contains infinitely many disjoint rays.*

(ii) *If an infinite graph G contains k edge-disjoint rays for every $k \in \mathbb{N}$,
then G contains infinitely many edge-disjoint rays.*

Proof. (i) We construct our infinite system of disjoint rays inductively   (3.3.1)
in ω steps. After step n, we shall have found n disjoint rays $R_1^n, \ldots, R_n^n$
and chosen initial segments $R_i^n x_i^n$ of these rays. In step $n+1$ we choose
the rays $R_1^{n+1}, \ldots, R_{n+1}^{n+1}$ so as to extend these initial segments, i.e. so
that $R_i^n x_i^n$ is a proper initial segment of $R_i^{n+1} x_i^{n+1}$, for $i = 1, \ldots, n$.

Then, clearly, the graphs $R_i^* := \bigcup_{n \in \mathbb{N}} R_i^n x_i^n$ will form an infinite family $(R_i^*)_{i \in \mathbb{N}}$ of disjoint rays in G.

For $n = 0$ the empty set of rays is as required. So let us assume that $R_1^n, \ldots, R_n^n$ have been chosen, and describe step $n + 1$. For simplicity, let us abbreviate $R_i^n =: R_i$ and $x_i^n =: x_i$. Let $\mathcal{R}$ be any set of $|R_1 x_1 \cup \ldots \cup R_n x_n| + n^2 + 1$ disjoint rays (which exists by assumption), and immediately delete those rays from $\mathcal{R}$ that meet any of the paths $R_1 x_1, \ldots, R_n x_n$; then $\mathcal{R}$ still contains at least $n^2 + 1$ rays.

We begin by repeating the following step as often as possible. If there exists an $i \in \{1, \ldots, n\}$ such that R_i^{n+1} has not yet been defined and $\mathring{x}_i R_i$ meets at most n of the rays currently in $\mathcal{R}$, we delete those rays from $\mathcal{R}$, put $R_i^{n+1} := R_i$, and choose as x_i^{n+1} the successor of x_i on R_i. Having performed this step as often as possible, we let I denote the set of those $i \in \{1, \ldots, n\}$ for which R_i^{n+1} is still undefined, and put $|I| =: m$. Then $\mathcal{R}$ still contains at least $n^2 + 1 - (n - m)n \geqslant m^2 + 1$ rays. Every R_i with $i \in I$ meets more than $n \geqslant m$ of the rays in $\mathcal{R}$; let z_i be its first vertex on the mth ray it meets. Then $Z := \bigcup_{i \in I} x_i R_i z_i$ meets at most m^2 of the rays in $\mathcal{R}$; we delete all the other rays from $\mathcal{R}$, choosing one of them as R_{n+1}^{n+1} (with x_{n+1}^{n+1} arbitrary).

On each remaining ray $R \in \mathcal{R}$ we now pick a vertex $y = y(R)$ after its last vertex in Z, and put $Y := \{ y(R) \mid R \in \mathcal{R} \}$. Let H be the union of Z and all the paths Ry $(R \in \mathcal{R})$. Then $X := \{ x_i \mid i \in I \}$ cannot be separated from Y in H by fewer than m vertices, because these would miss both one of the m rays R_i with $i \in I$ and one of the m rays in $\mathcal{R}$ that meet $x_i R_i z_i$ for this i. So by Menger's theorem (3.3.1) there are m disjoint X–Y paths $P_i = x_i \ldots y_i$ $(i \in I)$ in H. For each $i \in I$ let R_i' denote the ray from $\mathcal{R}$ that contains y_i, choose as R_i^{n+1} the ray $R_i x_i P_i y_i R_i'$, and put $x_i^{n+1} := y_i$.

(ii) The proof is similar; see Exercise 42 and its hint.   □

Does Theorem 8.2.5 generalize to other graphs than rays? Let us call a graph H *ubiquitous* with respect to a relation $\leqslant$ between graphs (such as the subgraph relation $\subseteq$, or the minor relation $\preccurlyeq$) if $nH \leqslant G$ for all $n \in \mathbb{N}$ implies $\aleph_0 H \leqslant G$, where nH denotes the disjoint union of n copies of H. Ubiquity appears to be closely related to questions of well-quasi-ordering as discussed in Chapter 12. Non-ubiquitous graphs exist for all the standard graph orderings; see Exercise 46 for an example of a locally finite graph that is not ubiquitous under the subgraph relation.

Ubiquity conjecture. (Andreae 2002)
Every locally finite connected graph is ubiquitous with respect to the minor relation.

Just as in Theorem 8.2.5 one can show that an end contains infinitely many disjoint rays as soon as the number of disjoint rays in it is not finitely bounded, and similarly for edge-disjoint rays (Exercise 43). Hence, the maxima in our earlier definitions of the vertex- and edge-degrees of an end exist as claimed. Ends of infinite vertex-degree are called *thick*; ends of finite vertex-degree are *thin*.   *thick/thin*

The $\mathbb{N} \times \mathbb{N}$ *grid*, for example, the graph on $\mathbb{N}^2$ in which two vertices   *grid*
(n, m) and (n', m') are adjacent if and only if $|n - n'| + |m - m'| = 1$, has only one end, which is thick. In fact, the $\mathbb{N} \times \mathbb{N}$ grid is a kind of prototype for thick ends: every graph with a thick end contains it as a minor. This is another classical result of Halin, which we prove in the remainder of this section.

For technical reasons, we shall prove Halin's theorem for hexagonal rather than square grids. These may seem a little unwieldy at first, but have the advantage that they can be found as topological rather than ordinary minors (Proposition 1.7.3), which makes them much easier to handle. We shall define the hexagonal grid H^∞ so that it is a subgraph of the $\mathbb{N} \times \mathbb{N}$ grid, and it will be easy to see that, conversely, the $\mathbb{N} \times \mathbb{N}$ grid is a minor of H^∞. (See also Exercise 57, Ch. 12.)

To define our standard copy of the *hexagonal quarter grid* H^∞, we H^∞
delete from the $\mathbb{N} \times \mathbb{N}$ grid H the vertex $(0, 0)$, the vertices (n, m) with $n > m$, and all edges $(n, m)(n + 1, m)$ such that n and m have equal parity (Fig. 8.2.1). Thus, H^∞ consists of the *vertical rays*

$$U_0 := H\left[\{ (0, m) \mid 1 \leqslant m \}\right]$$
$$U_n := H\left[\{ (n, m) \mid n \leqslant m \}\right] \quad (n \geqslant 1)$$ U_n

and between these a set of *horizontal edges*,

$$E := \{ (n, m)(n + 1, m) \mid n \not\equiv m \pmod{2} \}.$$

To enumerate these edges, as $e_1, e_2, \ldots$ say, we order them colexicograph- $e_1, e_2, \ldots$
ically: the edge $(n, m)(n + 1, m)$ precedes the edge $(n', m')(n' + 1, m')$ if $m < m'$, or if $m = m'$ and $n < n'$ (Fig. 8.2.1).

Theorem 8.2.6. (Halin 1965)
Whenever a graph contains a thick end, it has a TH^∞ subgraph whose rays belong to that end.

Proof. Given two infinite sets $\mathcal{P}, \mathcal{P}'$ of finite or infinite paths, let us write   (8.1.2)
$\mathcal{P} \geqslant \mathcal{P}'$ if $\mathcal{P}'$ consists of final segments of paths in $\mathcal{P}$. (Thus, if $\mathcal{P}$ is a $\leqslant$
set of rays, then so is $\mathcal{P}'$.)

Let G be any graph with a thick end ω. Our task is to find disjoint ω
rays in ω that can serve as 'vertical' (subdivided) rays U_n for our desired grid, and to link these up by suitable disjoint 'horizontal' paths. We begin by constructing a sequence $R_0, R_1, \ldots$ of rays (of which we shall later choose some tails R'_n as 'vertical rays'), together with path systems

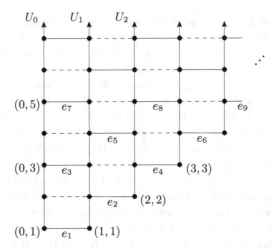

Fig. 8.2.1. The hexagonal quarter grid H^∞.

$\mathcal{P}_n$ between the R_n and suitable $R_{p(n)}$ with $p(n) < n$ (from which we shall later choose the 'horizontal paths'). We shall aim to find the R_n in 'supply sets' $\mathcal{R}_0 \geqslant \mathcal{R}_1 \geqslant \ldots$ of unused rays. After the nth construction step we shall have constructed the subgraph $H_n := \bigcup_{i=0}^n \left(R_i \cup \bigcup \mathcal{P}_i\right)$.

H_n

We start with any infinite set $\mathcal{R}$ of disjoint rays in ω; this exists by our assumption that ω is a thick end. Pick $R_0 \in \mathcal{R}$, put $\mathcal{P}_0 := \emptyset$, and let $\mathcal{R}_0 := \mathcal{R} \smallsetminus \{R_0\}$. At step $n \geqslant 1$ of the construction we shall choose the following:

R_0
$\mathcal{R}_0$

(1) a ray $R_n \in \omega$ disjoint from H_{n-1};

(2) an integer $p(n) < n$;

(3) an infinite set $\mathcal{P}_n$ of disjoint R_n–$R_{p(n)}$ paths avoiding every other R_i;

(4) an infinite set $\mathcal{R}_n \leqslant \mathcal{R}_{n-1}$ of disjoint rays in $G - H_n$.

Let $n \geqslant 1$ be given. As a first candidate for R_n consider any ray $R \in \mathcal{R}_{n-1}$. By (4) for smaller values of n, we have both $R \in \omega$ (since $\mathcal{R}_{n-1} \leqslant \ldots \leqslant \mathcal{R}_0 \subseteq \mathcal{R}$) and $R \cap H_{n-1} = \emptyset$, as required for R_n in (1).

R

Next, let us try to find a set $\mathcal{P}_n$ and $p(n)$ to go with R as R_n. Since H_{n-1} contains $R_0 \in \omega$, there exists an infinite set $\mathcal{P}$ of disjoint R–H_{n-1} paths in G. If $\mathcal{P}$ has an infinite subset of paths all ending on the same R_i $(i < n)$, we delete all other paths from $\mathcal{P}$. If not, then $\mathcal{P}$ has an infinite subset of paths all ending at an inner vertex of a path in $\mathcal{P}_i$, for the same $i < n$. We extend them back along this path of $\mathcal{P}_i$ until they hit R_i, and delete all other paths from $\mathcal{P}$. In both cases we put $p(n) := i$, satisfying (2), and have found an infinite set $\mathcal{P}$ of disjoint R–$R_{p(n)}$ paths that avoid all the other R_i $(i < n)$.

$p(n)$, $\mathcal{P}$

For later use we record:

> *Every path in $\mathcal{P}$ consists of an R–H_{n-1} path followed by a (possibly trivial) path in H_{n-1}.* $\qquad(*)$

What can still prevent us from choosing R as R_n and $\mathcal{P}$ as $\mathcal{P}_n$ is condition (4): if $\mathcal{P}$ meets all but finitely many rays in $\mathcal{R}_{n-1}$ infinitely, we cannot find an infinite set $\mathcal{R}_n \leqslant \mathcal{R}_{n-1}$ of rays avoiding $\mathcal{P}$.

If this happens, the only option we have is to make a virtue of necessity and use the rich intersection of $\mathcal{P}$ and $\mathcal{R}_{n-1}$ for an altogether different construction of R_n, $\mathcal{P}_n$ and $\mathcal{R}_n$. This will require some work. But we may assume the following, as the result of our effort so far:

> *Whenever $R' \in \mathcal{R}_{n-1}$ and $\mathcal{P}' \leqslant \mathcal{P}$ is an infinite set of R'–$R_{p(n)}$ paths, there is a ray $R'' \neq R'$ in $\mathcal{R}_{n-1}$ that meets $\mathcal{P}'$ infinitely.* $\qquad(**)$

For if $(**)$ failed, we could choose R' as R_n and $\mathcal{P}'$ as $\mathcal{P}_n$, and for $\mathcal{R}_n$ select from every ray $R'' \neq R'$ in $\mathcal{R}_{n-1}$ a tail avoiding $\mathcal{P}'$. This would satisfy conditions (1)–(4) for n.

Consider the paths in $\mathcal{P}$ as linearly ordered by the natural order of their starting vertices on R. This induces an ordering on every $\mathcal{P}' \leqslant \mathcal{P}$. If $\mathcal{P}'$ is a set of R'–$R_{p(n)}$ paths for some ray R', we shall call this ordering of $\mathcal{P}'$ *compatible* with R' if the ordering it induces on the first vertices of its paths coincides with the natural ordering of those vertices on R'.

Starting with $R =: R_{n-1}^0$ and $\mathcal{P} =: \mathcal{P}^0$, let us construct sequences $R_{n-1}^0, R_{n-1}^1, \dots$ and $\mathcal{P}^0 \geqslant \mathcal{P}^1 \geqslant \dots$ such that every R_{n-1}^k is a tail of a ray in $\mathcal{R}_{n-1}$ and each $\mathcal{P}^k$ is an infinite set of R_{n-1}^k–$R_{p(n)}$ paths whose ordering is compatible with R_{n-1}^k. The first path of $\mathcal{P}^k$ in this ordering will be denoted by P_k, its starting vertex on R_{n-1}^k by v_k, and the path in $\mathcal{P}^{k-1}$ containing P_k (if $k \geqslant 1$) by P_k^- (Fig. 8.2.2). To define R_{n-1}^k

$R_{n-1}^0, \mathcal{P}^0$

P_k, v_k
P_k^-

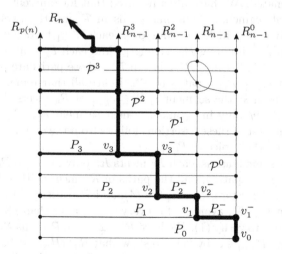

Fig. 8.2.2. Constructing R_n from condition $(**)$

and $\mathcal{P}^k$ for $k \geqslant 1$, we use $(**)$ with $R' \supseteq R_{n-1}^{k-1}$ and $\mathcal{P}' = \mathcal{P}^{k-1}$ to find in $\mathcal{R}_{n-1}$ a ray $R'' \not\supseteq R_{n-1}^{k-1}$ that meets $\mathcal{P}^{k-1}$ infinitely; let R_{n-1}^k be a tail of R'' that avoids the finitely many paths in $\mathcal{P}$ containing $P_0, \ldots, P_{k-1}$. Let P_k^- be a path in $\mathcal{P}^{k-1}$ that meets R_{n-1}^k and let v be its 'highest' vertex on R_{n-1}^k, that is, the last vertex of R_{n-1}^k in $V(P_k^-)$. Replacing R_{n-1}^k with its tail vR_{n-1}^k, we can arrange that P_k^- has only the vertex v on R_{n-1}^k. Then $P_k := vP_k^-$ is an R_{n-1}^k–$R_{p(n)}$ path starting at $v_k = v$. We may now select an infinite set $\mathcal{P}^k \leqslant \mathcal{P}^{k-1}$ of R_{n-1}^k–$R_{p(n)}$ paths compatible with R_{n-1}^k and containing P_k as its first path.

Note that P_k cannot be a subpath of any P_i with $i < k$, since P_k contains $v_k \in R_{n-1}^k$ but $R_{n-1}^k \cap P_i = \emptyset$. As $\mathcal{P}^k \leqslant \mathcal{P}^i$, this means that P_k and P_i are subpaths of disjoint paths in $\mathcal{P}$. Similarly, for $i < k$ the rays R_{n-1}^k and R_{n-1}^i cannot be tails of the same ray in $\mathcal{R}_{n-1}$, and are therefore tails of disjoint rays in $\mathcal{R}_{n-1}$, because $\bigcup \mathcal{P}^k$ meets R_{n-1}^k infinitely but avoids R_{n-1}^i. (Indeed, as R_{n-1}^k is disjoint from R_{n-1}^{k-1} by definition, the paths in $\mathcal{P}^k$ are proper final segments of paths in $\mathcal{P}^{k-1} \leqslant \mathcal{P}^i$, and the paths in $\mathcal{P}_i$ meet R_{n-1}^i only in their first vertex.)

For each k, let v_{k+1}^- denote the starting vertex of P_{k+1}^- on R_{n-1}^k, and put $R_n^k := \mathring{v}_{k+1}^- R_{n-1}^k$. Then let

$$R_n := v_0 R_{n-1}^0 v_1^- P_1^- v_1 R_{n-1}^1 v_2^- P_2^- v_2 R_{n-1}^2 \cdots$$

$$\mathcal{P}_n := \{P_0, P_1, P_2, \ldots\}$$

$$\mathcal{R}_n := \{\, R_n^k \mid k \in \mathbb{N} \,\}.$$

To verify that R_n is indeed a ray, we have to check that the various path segments it is composed of meet only at the vertices at which they are concatenated. We have already noted that for different k the paths P_k^- are final segments of disjoint paths in $\mathcal{P}$ and the rays R_{n-1}^k are tails of disjoint rays in $\mathcal{R}_{n-1}$. Moreover, each R_{n-1}^k by definition avoids the paths in $\mathcal{P}$ containing $P_0, \ldots, P_{k-1}$. And each P_{k+1}^- avoids the rays R_{n-1}^i with $i < k$, because P_{k+1}^- lies in $\mathcal{P}^k$, whose paths are proper final segments of R_{n-1}^i–$R_{p(n)}$ paths in $\mathcal{P}^i$. Hence all that remains to check is that, for each k, the segment $v_k R_{n-1}^k v_{k+1}^-$ of R_n meets the previous segment $v_k^- P_k^- v_k$ only in v_k and the next segment $v_{k+1}^- P_{k+1}^- v_{k+1}$ only in v_{k+1}^-. The first of these assertions follows from the definition of R_{n-1}^k, the second by the choice of P_{k+1}^-.

For the same reasons, each P_k meets R_n only in v_k (so $\mathcal{P}_n$ is indeed a set of R_n–$R_{p(n)}$ paths), and the paths in $\mathcal{R}_n$ meet neither R_n nor the paths in $\mathcal{P}_n$. Therefore $\mathcal{R}_n$ satisfies (4), and $\mathcal{P}_n$ satisfies (3); recall that the paths in $\mathcal{P} \geqslant \mathcal{P}_n$ avoid R_i for every $i < n$ other than $p(n)$.

It remains to verify (1). We have $R_n \in \omega$, since $\mathcal{P}_n$ joins R_n disjointly to $R_{p(n)}$, and $R_{p(n)} \in \omega$ by (1). To show that $R_n \cap H_{n-1} = \emptyset$, recall that the 'vertical' segments of R_n lie in rays from $\mathcal{R}_{n-1}$, so by (4) they do not meet H_{n-1}. And a 'horizontal' segment of R_n can meet H_{n-1} only if it does so also in its last vertex v_k, by $(*)$. But v_k also lies on a vertical segment, and hence not in H_{n-1}. Thus, $R_n \cap H_{n-1} = \emptyset$ as desired.

Let us now use our rays R_n and path systems $\mathcal{P}_n$ to construct the desired grid. In the tree on $\mathbb{N}$ defined by joining each n to $p(n)$, apply the infinity lemma (8.1.2) to the distance classes from 0 to find a ray $n_0 n_1 \ldots$ with vertices $n_0 < n_1 < \ldots$ or, if one of these classes is infinite, a vertex n_0 with infinitely many neighbours $n_1, n_2, \ldots$ greater than n_0. We treat these two cases in turn, assuming for notational simplicity that $n_i = i$ for all i. (In other words, we discard any R_n with $n \notin \{n_0, n_1, \ldots\}$.)

In the first case, each $\mathcal{P}_n$ is an infinite set of disjoint R_n–R_{n-1} paths. Our aim is to choose tails R'_n of our rays R_n that will correspond to the vertical rays $U_n \subseteq H^\infty$, and paths $S_1, S_2, \ldots$ between the R'_n that will correspond to the horizontal edges $e_1, e_2, \ldots$ of H^∞. We shall find the paths $S_1, S_2, \ldots$ inductively, choosing the R'_n as needed as we go along (but also in the order of increasing n, starting with $R'_0 := R_0$). At every step of the construction, we shall have selected only finitely many S_k and only finitely many R'_n.

Let k and n be minimal such that S_k and R'_n are still undefined. We describe how to choose S_k, and R'_n if the definition of S_k requires it. Let i be such that e_k joins U_{i-1} to U_i in H^∞. If $i = n$, let R'_n be a tail of R_n that avoids the finitely many paths $S_1, \ldots, S_{k-1}$; otherwise, R'_i has already been defined, and so has R'_{i-1}. Now choose $S_k \in \mathcal{P}_i$ 'high enough' between R'_{i-1} and R'_i to mirror the position of e_k in H^∞, and to avoid $S_1 \cup \ldots \cup S_{k-1}$. Then S_k will also avoid every other R'_j already defined: by (3) for i if $j < i$, and by (1) for j if $j > i$. Since every R'_n is chosen so as to avoid all previously defined S_k, and every S_k avoids all previously defined R'_j (except R'_{i-1} and R'_i), the R'_n and S_k are pairwise disjoint for all $n, k \in \mathbb{N}$, except for the required incidences. Our construction thus yields the desired subdivision of H^∞.

In the second case, every $\mathcal{P}_n$ is a set of disjoint R_n–R_0 paths. We now use only the R_n with $n \geqslant 1$ for vertical rays of H^∞, because R_0 will be needed for the horizontal paths. More precisely, we choose rays $R'_n \subseteq R_n$ for $n \geqslant 1$, and paths S_k between them, inductively as before, except that S_k now consists of three parts: an initial segment from $\mathcal{P}_{i-1}$, followed by a middle segment on R_0, and a final segment from $\mathcal{P}_i$. Such S_k can again be found, since at every stage of the construction only a finite part of R_0 has been used. $\qquad\square$

8.3 Homogeneous and universal graphs

Unlike finite graphs, infinite graphs offer the possibility to represent an entire graph property $\mathcal{P}$ by just one specimen, a single graph that contains all the graphs in $\mathcal{P}$ up to some fixed cardinality. Such graphs are called 'universal' for this property.

More precisely, if $\leqslant$ is a graph relation (such as the minor, topological minor, subgraph, or induced subgraph relation up to isomorphism), *universal* we call a countable graph G^* *universal* in $\mathcal{P}$ (for $\leqslant$) if $G^* \in \mathcal{P}$ and $G \leqslant G^*$ for every countable graph $G \in \mathcal{P}$.

Is there a graph that is universal in the class of *all* countable graphs? Suppose a graph R has the following property:

> *Whenever U and W are disjoint finite sets of vertices in R,*
> *there exists a vertex $v \in R - U - W$ that is adjacent in R*   (*)
> *to all the vertices in U but to none in W.*

Then R is universal even for the strongest of all graph relations, the induced subgraph relation. Indeed, in order to embed a given countable graph G in R we just map its vertices $v_1, v_2, \ldots$ to R inductively, making sure that v_n gets mapped to a vertex $v \in R$ adjacent to the images of all the neighbours of v_n in $G[v_1, \ldots, v_n]$ but not adjacent to the image of any non-neighbour of v_n in $G[v_1, \ldots, v_n]$. Clearly, this map is an isomorphism between G and the subgraph of R induced by its image.

[11.3.5]
R
Theorem 8.3.1. (Erdős and Rényi 1963)
There exists a unique countable graph R with property (*).

Proof. To prove existence, we construct a graph R with property (*) inductively. Let $R_0 := K^1$. For all $n \in \mathbb{N}$, let R_{n+1} be obtained from R_n by adding for every set $U \subseteq V(R_n)$ a new vertex v joined to all the vertices in U but to none outside U. (In particular, the new vertices form an independent set in R_{n+1}.) Clearly $R := \bigcup_{n \in \mathbb{N}} R_n$ has property (*).

To prove uniqueness, let $R = (V, E)$ and $R' = (V', E')$ be two graphs with property (*), each given with a fixed vertex enumeration. We construct a bijection $\varphi \colon V \to V'$ in an infinite sequence of steps, defining $\varphi(v)$ for one new vertex $v \in V$ at each step.

At every odd step we look at the first vertex v in the enumeration of V for which $\varphi(v)$ has not yet been defined. Let U be the set of those of its neighbours u in R for which $\varphi(u)$ has already been defined. This is a finite set. Using (*) for R', find a vertex $v' \in V'$ outside the image of φ (which is a finite set), so that v' is adjacent in R' to all the vertices in $\varphi(U)$ but to no other vertex in the image of φ. Put $\varphi(v) := v'$.

At even steps in the definition process we do the same thing with the roles of R and R' interchanged: we look at the first vertex v' in the enumeration of V' that does not yet lie in the image of φ, and set

$\varphi(v) = v'$ for a new vertex v that matches the adjacencies and non-adjacencies of v' among the vertices for which φ (resp. φ^{-1}) has already been defined.

By our minimum choices of v and v', the bijection gets defined on all of V and all of V', and it is clearly an isomorphism. $\square$

The graph R in Theorem 8.3.1 is usually called the *Rado graph*, named after Richard Rado who gave one of its earliest explicit definitions. The method of constructing a bijection in alternating steps, as in the uniqueness part of the proof, is known as the *back-and-forth* technique.

Rado graph

The Rado graph R is unique in another rather fascinating respect. We shall hear more about this in Chapter 11.3, but in a nutshell it is the following. If we generate a countably infinite *random* graph by admitting its pairs of vertices as edges independently with some fixed positive probability $p \in (0,1)$, then with probability 1 the resulting graph has property $(*)$, and is hence isomorphic to R! In the context of infinite graphs, the Rado graph is therefore also called *the* (countably infinite) *random graph*.

'the' random graph

As one would expect of a random graph, the Rado graph shows a high degree of uniformity. One aspect of this is its resilience against small changes: the deletion of finitely many vertices or edges, and similar local changes, leave it 'unchanged' and result in just another copy of R (Exercise 50).

The following rather extreme aspect of uniformity, however, is still surprising: no matter how we partition the vertex set of R into two parts, at least one of the parts will induce another isomorphic copy of R. Trivial examples aside, the Rado graph is the only countable graph with this property, and hence unique in yet another respect:

Proposition 8.3.2. *The Rado graph is the unique countable graph G other than $K^{\aleph_0}$ and $\overline{K^{\aleph_0}}$ such that, no matter how $V(G)$ is partitioned into two parts, one of the parts induces an isomorphic copy of G.*

Proof. We first show that the Rado graph R has the partition property. Let $\{V_1, V_2\}$ be a partition of $V(R)$. If $(*)$ fails in both $R[V_1]$ and $R[V_2]$, say for sets U_1, W_1 and U_2, W_2, respectively, then $(*)$ fails for $U = U_1 \cup U_2$ and $W = W_1 \cup W_2$ in R, a contradiction.

To show uniqueness, let $G = (V, E)$ be a countable graph with the partition property. Let V_1 be its set of isolated vertices, and V_2 the rest. If $V_1 \neq \emptyset$ then $G \not\simeq G[V_2]$, since G has isolated vertices but $G[V_2]$ does not. Hence $G = G[V_1] \simeq \overline{K^{\aleph_0}}$. Similarly, if G has a vertex adjacent to all other vertices, then $G = K^{\aleph_0}$.

Assume now that G has no isolated vertex and no vertex joined to all other vertices. If G is not the Rado graph then there are sets U, W for which $(*)$ fails in G; choose these with $|U \cup W|$ minimum. Assume first that $U \neq \emptyset$, and pick $u \in U$. Let V_1 consist of u and all vertices outside $U \cup W$ that are not adjacent to u, and let V_2 contain the remaining vertices. As u is isolated in $G[V_1]$, we have $G \not\simeq G[V_1]$ and hence $G \simeq G[V_2]$. By the minimality of $|U \cup W|$, there is a vertex $v \in G[V_2] - U - W$ that is adjacent to every vertex in $U \smallsetminus \{u\}$ and to none in W. But v is also adjacent to u, because it lies in V_2. So U, W and v satisfy $(*)$ for G, contrary to assumption.

Finally, assume that $U = \emptyset$. Then $W \neq \emptyset$. Pick $w \in W$, and consider the partition $\{V_1, V_2\}$ of V where V_1 consists of w and all its neighbours outside W. As before, $G \not\simeq G[V_1]$ and hence $G \simeq G[V_2]$. Therefore U and $W \smallsetminus \{w\}$ satisfy $(*)$ in $G[V_2]$, with $v \in V_2 \smallsetminus W$ say, and then U, W, v satisfy $(*)$ in G. $\square$

Another indication of the high degree of uniformity in the structure of the Rado graph is its large automorphism group. For example, R is easily seen to be *vertex-transitive*: given any two vertices x and y, there is an automorphism of R mapping x to y.

homoge-
neous

In fact, much more is true: using the back-and-forth technique, one can easily show that the Rado graph is *homogeneous*: every isomorphism between two finite induced subgraphs can be extended to an automorphism of the entire graph (Exercise 51).

Which other countable graphs are homogeneous? The complete graph $K^{\aleph_0}$ and its complement are again obvious examples. Moreover, for every integer $r \geqslant 3$ there is a homogeneous K^r-free graph R^r, constructed as follows. Let $R_0^r := K^1$, and let R_{n+1}^r be obtained from R_n^r by joining, for every subgraph $H \not\supseteq K^{r-1}$ of R_n^r, a new vertex v_H to every vertex in H. Then let $R^r := \bigcup_{n \in \mathbb{N}} R_n^r$. Clearly, as the new vertices v_H of R_{n+1}^r are independent, there is no K^r in R_{n+1}^r if there was none in R_n^r, so $R^r \not\supseteq K^r$ by induction on n. Just like the Rado graph, R^r is clearly universal among the K^r-free countable graphs, and by the back-and-forth argument from the proof of Theorem 8.3.1 it is easily seen to be homogeneous.

R^r

By the following deep theorem of Lachlan and Woodrow, the countable homogeneous graphs we have seen so far are essentially all:

Theorem 8.3.3. (Lachlan & Woodrow 1980)
Every countably infinite homogeneous graph is one of the following:

- *a disjoint union of complete graphs of the same order, or the complement of such a graph;*
- *the graph R^r or its complement, for some $r \geqslant 3$;*
- *the Rado graph R.*

To conclude this section, let us return to our original problem: for which graph properties is there a graph that is universal with this property? Most investigations into this problem have addressed it from a more general model-theoretic point of view, and have therefore been based on the strongest of all graph relations, the induced subgraph relation. As a consequence, most of these results are negative; see the notes.

From a graph-theoretic point of view, it seems more promising to look instead for universal graphs for the weaker subgraph relation, or even the topological minor or minor relation. For example, while there is no universal planar graph for subgraphs or induced subgraphs, there is one for minors:

Theorem 8.3.4. *There exists a universal planar graph for the minor relation.*

So far, this theorem is the only one of its kind. But it should be possible to find more. For instance: for which graphs X is there a minor-universal graph in the class $\mathrm{Forb}_{\preccurlyeq}(X) = \{\, G \mid X \npreccurlyeq G \,\}$?

8.4 Connectivity and matching

In this section we look at infinite versions of Menger's theorem and of the matching theorems from Chapter 2. This area of infinite graph theory is one of its best developed fields, with several deep results. One of these, however, stands out among the rest: a version of Menger's theorem that had been conjectured by Erdős decades ago, and was proved only fairly recently by Aharoni and Berger. The techniques developed for its proof inspired, over the years, much of the theory in this area.

We shall prove this theorem for countable graphs, which will take up most of this section. Although the countable case is much easier, the techniques it requires already give a good impression of the general proof. We then wind up with an overview of infinite matching theorems and a conjecture conceived in the same spirit.

Recall that Menger's theorem, in its simplest form, says that if A and B are sets of vertices in a finite graph G, not necessarily disjoint, and if $k = k(G, A, B)$ is the minimum number of vertices separating A from B in G, then G contains k disjoint A–B paths. (Clearly, it cannot contain more.) The same holds, and is easily deduced from the finite case, when G is infinite but k is still finite:

Proposition 8.4.1. *Let G be any graph, $k \in \mathbb{N}$, and let A, B be two sets of vertices in G that can be separated by k but no fewer than k vertices. Then G contains k disjoint A–B paths.*

(3.3.1)   *Proof.* By assumption, every set of disjoint A–B paths has cardinality at
            most k. Choose one, $\mathcal{P}$ say, of maximum cardinality. Suppose $|\mathcal{P}| < k$.
            Then no set X consisting of one vertex from each path in $\mathcal{P}$ separates A
            from B. For each X, let P_X be an A–B path avoiding X. Let H be the
            union of $\bigcup \mathcal{P}$ with all these paths P_X. This is a finite graph in which no
            set of $|\mathcal{P}|$ vertices separates A from B. So $H \subseteq G$ contains more than
            $|\mathcal{P}|$ paths from A to B by Menger's theorem (3.3.1), which contradicts
            the choice of $\mathcal{P}$. $\square$

            When k is infinite, however, the result suddenly becomes trivial.
            Indeed, let $\mathcal{P}$ be any maximal set of disjoint A–B paths in G. Then the
            union of all these paths separates A from B, so $\mathcal{P}$ must be infinite. But
            then the cardinality of this union is no bigger than $|\mathcal{P}|$. Thus, $\mathcal{P}$ contains
            $|\mathcal{P}| = |\bigcup \mathcal{P}| \geqslant k$ disjoint A–B paths, as desired.

            Of course, this is no more than a trick played on us by infinite car-
            dinal arithmetic: although, numerically, the A–B separator consisting of
            all the inner vertices of paths in $\mathcal{P}$ is no bigger than $|\mathcal{P}|$, it uses far more
            vertices to separate A from B than should be necessary. Or put another
            way: when our path systems and separators are infinite, their cardinal-
            ities alone are no longer a sufficiently fine tool to distinguish carefully
            chosen 'small' separators from unnecessarily large and wasteful ones.

            To overcome this problem, Erdős suggested an alternative form of
            Menger's theorem, which for finite graphs is clearly equivalent to the
            standard version. Recall that an A–B separator X is said to lie *on* a set $\mathcal{P}$
Erdős-   of disjoint A–B paths if X consists of a choice of exactly one vertex from
Menger   each path in $\mathcal{P}$. The following so-called *Erdős-Menger conjecture*, now
conjecture a theorem, influenced much of the development of infinite connectivity
            and matching theory:

            Theorem 8.4.2. (Aharoni & Berger 2009)
            *Let G be any graph, and let $A, B \subseteq V(G)$. Then G contains a set $\mathcal{P}$ of
            disjoint A–B paths and an A–B separator on $\mathcal{P}$.*

            The next few pages give a proof of Theorem 8.4.2 for countable G.

            Of the three proofs we gave for the finite case of Menger's theorem,
            only the last has any chance of being adaptable to the infinite case: the
            others were by induction on $|G|$ or $|\bigcup \mathcal{P}|$, and both these parameters
            may now be infinite. The third proof, however, looks more promising:
            recall that, by Lemmas 3.3.2 and 3.3.3, it provided us with a tool to
            either find a separator on a given system of A–B paths, or to construct
            another system of A–B paths that covers more vertices in A and in B.

            Lemmas 3.3.2 and 3.3.3 (whose proofs work for infinite graphs too)
            will indeed form a cornerstone of our proof for Theorem 8.4.2. However,
            it will not do just to apply these lemmas infinitely often. Indeed, al-
            though any finite number of applications of Lemma 3.3.2 leaves us with

another system of disjoint A–B paths, an infinite number of iterations may leave nothing at all: each edge may be toggled on and off infinitely often by successive alternating paths, so that no 'limit system' of A–B paths will be defined. We shall therefore take another tack: starting at A, we grow simultaneously as many disjoint paths towards B as possible.

To make this precise, we need some terminology. Given a set $X \subseteq V(G)$, let us write $G_{X \to B}$ for the subgraph of G induced by X and all the components of $G - X$ that meet B.

$G_{X \to B}$

Let $W = (W_a \mid a \in A)$ be a family of disjoint paths such that every W_a starts in a. We call W an $A \to B$ *wave* in G if the set Z of final vertices of paths in W separates A from B in G. (Note that W may contain infinite paths, which have no final vertex.) Sometimes, we shall wish to consider $A \to B$ waves in subgraphs of G that contain A but not all of B. For this reason we do not formally require that $B \subseteq V(G)$.

wave

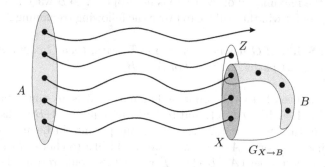

Fig. 8.4.1. A small $A \to B$ wave W with boundary X

When W is a wave, then the set $X \subseteq Z$ of those vertices in Z that either lie in B or have a neighbour in $G_{Z \to B} - Z$ is a minimal A–B separator in G; note that $z \in Z$ lies in X if and only if it can be linked to B by a path that has no vertex other than z on W. We call X the *boundary* of W, and often use (W, X) as shorthand for the wave W together with its boundary X. If all the paths in W are finite and $X = Z$, we call the wave W *large*; otherwise it is *small*. We shall call W *proper* if at least one of the paths in W is non-trivial, or if all its paths are trivial but its boundary is a proper subset of A. Every small wave, for example, is proper. Note that while some $A \to B$ wave always exists, e.g. the family $(\{a\} \mid a \in A)$ of singleton paths, G need not have a proper $A \to B$ wave. (For example, if A consists of two vertices of $G = K^{10}$ and B of three other vertices, there is no proper $A \to B$ wave.)

boundary (W, X)

large/small *proper*

If $(\mathcal{U}, X)$ is an $A \to B$ wave in G and $(\mathcal{V}, Y)$ is an $X \to B$ wave in $G_{X \to B}$, then the family $W = \mathcal{U} + \mathcal{V}$ obtained from $\mathcal{U}$ by appending the paths of $\mathcal{V}$ (to those paths of $\mathcal{U}$ that end in X) is clearly an $A \to B$ wave in G, with boundary Y. Note that W is large if and only if both $\mathcal{V}$ and $\mathcal{U}$ are large. W is greater than $\mathcal{U}$ in the following sense.

$\mathcal{U} + \mathcal{V}$

Given two path systems $\mathcal{U} = (U_a \mid a \in A)$ and $\mathcal{W} = (W_a \mid a \in A)$,
$\leqslant$ write $\mathcal{U} \leqslant \mathcal{W}$ if $U_a \subseteq W_a$ for every $a \in A$. Given a chain $(\mathcal{W}^i, X^i)_{i \in I}$
of waves in this ordering, with $\mathcal{W}^i = (W_a^i \mid a \in A)$ say, let $\mathcal{W}^* = (W_a^* \mid a \in A)$ be defined by $W_a^* := \bigcup_{i \in I} W_a^i$. Then $\mathcal{W}^*$ is an $A \to B$
wave: any A–B path is finite but meets every X^i, so at least one of its
vertices lies in X^i for arbitrarily large $(\mathcal{W}^i, X^i)$ and hence is the final
vertex of a path in $\mathcal{W}^*$. Clearly $\mathcal{W}^i \leqslant \mathcal{W}^*$ for all $i \in I$; we call $\mathcal{W}^*$ the
limit wave *limit* of the waves $\mathcal{W}^i$.

maximal As every chain of $A \to B$ waves is bounded above by its limit wave,
wave Zorn's lemma implies that G has a maximal $A \to B$ wave $\mathcal{W}$; let X be
its boundary. This wave $(\mathcal{W}, X)$ forms the first step in our proof for
Theorem 8.4.2: if we can now find disjoint paths in $G_{X \to B}$ linking all
the vertices of X to B, then X will be an A–B separator on these paths
preceded by the paths of $\mathcal{W}$ that end in X.

By the maximality of $\mathcal{W}$, there is no proper $X \to B$ wave in $G_{X \to B}$.
For our proof it will thus suffice to prove the following (renaming X as A):

Lemma 8.4.3. *If G has no proper $A \to B$ wave, then G contains a set
of disjoint A–B paths linking all of A to B.*

Our approach to the proof of Lemma 8.4.3 is to enumerate the
$a_1, a_2, \ldots$ vertices in $A =: \{a_1, a_2, \ldots\}$, and to find the required A–B paths $P_n = $
P_n $a_n \ldots b_n$ in turn for $n = 1, 2, \ldots$. Since our premise in Lemma 8.4.3 is
that G has no proper $A \to B$ wave, we would like to choose P_1 so that
$G - P_1$ has no proper $(A \smallsetminus \{a_1\}) \to B$ wave: this would restore the same
premise to $G - P_1$, and we could proceed to find P_2 in $G - P_1$ in the
same way.

We shall not be able to choose P_1 quite like this, but we shall be
able to do something almost as good. We shall construct P_1 so that
deleting it (as well as a few more vertices outside A) leaves a graph that
has a large maximal $(A \smallsetminus \{a_1\}) \to B$ wave $(\mathcal{W}, A')$. We then earmark
the paths $W_n = a_n \ldots a_n'$ $(n \geqslant 2)$ of this wave as initial segments for the
paths P_n. By the maximality of $\mathcal{W}$, there is no proper $A' \to B$ wave in
$G_{A' \to B}$. In other words, we have restored our original premise to $G_{A' \to B}$,
and can find there an A'–B path $P_2' = a_2' \ldots b_2$. Then $P_2 := a_2 W_2 a_2' P_2'$
is our second path for Lemma 8.4.3, and we continue inductively inside
$G_{A' \to B}$.

linkable Given a set $\hat{A}$ of vertices in G, let us call a vertex $a \notin \hat{A}$ *linkable*
for $(G, \hat{A}, B)$ if $G - \hat{A}$ contains an a–B path P and a set $X \supseteq V(P)$
of vertices such that $G - X$ has a large maximal $\hat{A} \to B$ wave. (The
first such a we shall be considering will be a_1, and $\hat{A}$ will be the set
$\{a_2, a_3, \ldots\}$.)

Lemma 8.4.4. *Let $a^* \in A$ and $\hat{A} := A \smallsetminus \{a^*\}$, and assume that G has
no proper $A \to B$ wave. Then a^* is linkable for $(G, \hat{A}, B)$.*

Proof of Lemma 8.4.3 (assuming Lemma 8.4.4). Let G be as in Lemma 8.4.3, i.e. assume that G has no proper $A \to B$ wave. We construct subgraphs $G_1, G_2, \ldots$ of G satisfying the following statement (Fig. 8.4.2):

> G_n contains a set $A^n = \{a_n^n, a_{n+1}^n, a_{n+2}^n, \ldots\}$ of distinct vertices such that G_n has no proper $A^n \to B$ wave. In G there are disjoint paths P_i $(i < n)$ and W_i^n $(i \geqslant n)$ (∗) starting at a_i. The P_i are disjoint from G_n and end in B. The W_i^n end in a_i^n and are otherwise disjoint from G_n.

Clearly, the paths $P_1, P_2, \ldots$ will satisfy Lemma 8.4.3.

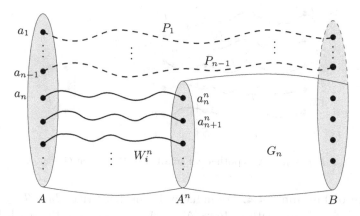

Fig. 8.4.2. G_n has no proper $A^n \to B$ wave

Let $G_1 := G$, and put $a_i^1 := a_i$ and $W_i^1 := \{a_i\}$ for all $i \geqslant 1$. Since by assumption G has no proper $A \to B$ wave, these definitions satisfy (∗) for $n = 1$. Suppose now that (∗) has been satisfied for n. Put $\hat{A}^n := A^n \smallsetminus \{a_n^n\}$. By Lemma 8.4.4 applied to G_n, we can find in $G_n - \hat{A}^n$ an a_n^n–B path P and a set $X_n \supseteq V(P)$ such that $G_n - X_n$ has a large maximal $\hat{A}^n \to B$ wave $(\mathcal{W}, A^{n+1})$. Let P_n be the path $W_n^n \cup P$. For $i \geqslant n + 1$, let W_i^{n+1} be W_i^n followed by the path of $\mathcal{W}$ starting at a_i^n, and call its last vertex a_i^{n+1}. By the maximality of $\mathcal{W}$ there is no proper $A^{n+1} \to B$ wave in $G_{n+1} := (G_n - X_n)_{A^{n+1} \to B}$, so (∗) is satisfied for $n + 1$. ∎

To complete our proof of Theorem 8.4.2, it remains to prove Lemma 8.4.4. For this, we need another lemma:

Lemma 8.4.5. *Let x be a vertex in $G - A$. If G has no proper $A \to B$ wave but $G - x$ does, then every $A \to B$ wave in $G - x$ is large.*

(3.3.2)
(3.3.3)

Proof. Suppose $G - x$ has a small $A \to B$ wave $(\mathcal{W}, X)$. Put $B' :=$ $X \cup \{x\}$, and let $\mathcal{P}$ denote the set of A–X paths in $\mathcal{W}$ (Fig. 8.4.3). If G contains an A–B' separator S on $\mathcal{P}$, then replacing in $\mathcal{W}$ every $P \in \mathcal{P}$ with its initial segment ending in S we obtain a small (and hence proper) $A \to B$ wave in G, which by assumption does not exist. By Lemmas 3.3.3 and 3.3.2, therefore, G contains a set $\mathcal{P}'$ of disjoint A–B' paths exceeding $\mathcal{P}$. The set of last vertices of these paths contains X properly, and hence must be all of $B' = X \cup \{x\}$. But B' separates A from B in G, so we can turn $\mathcal{P}'$ into an $A \to B$ wave in G by adding as singleton paths any vertices of A it does not cover. As x lies on $\mathcal{P}'$ but not in A, this is a proper wave, which by assumption does not exist. $\qquad\square$

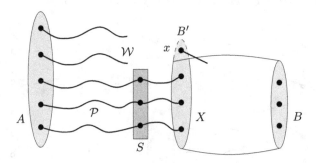

Fig. 8.4.3. A hypothetical small $A \to B$ wave in $G - x$

Proof of Lemma 8.4.4. We inductively construct trees $T_0 \subseteq T_1 \subseteq \dots$

$\mathcal{W}_n$

in $G - (\hat{A} \cup B)$ and path systems $\mathcal{W}_0 \leqslant \mathcal{W}_1 \leqslant \dots$ in G so that each $\mathcal{W}_n$ is a large maximal $\hat{A} \to B$ wave in $G - T_n$.

Let $\mathcal{W}_0 := (\{a\} \mid a \in \hat{A})$. Clearly, $\mathcal{W}_0$ is an $\hat{A} \to B$ wave in $G - a^*$, and it is large and maximal: if not, then $G - a^*$ has a proper $\hat{A} \to B$ wave, and adding the trivial path $\{a^*\}$ to this wave turns it into a proper $A \to B$ wave (which by assumption does not exist). If $a^* \in B$, the existence of $\mathcal{W}_0$ makes a^* linkable for $(G, \hat{A}, B)$. So we assume that $a^* \notin B$. Now $T_0 := \{a^*\}$ and $\mathcal{W}_0$ are as desired.

A_n

Suppose now that T_n and $\mathcal{W}_n$ have been defined, and let A_n denote the set of last vertices of the paths in $\mathcal{W}_n$. Since $\mathcal{W}_n$ is large, A_n is its

G_n

boundary, and since $\mathcal{W}_n$ is maximal, $G_n := (G - T_n)_{A_n \to B}$ has no proper $A_n \to B$ wave (Fig. 8.4.4).

Note that A_n does not separate A from B in G: if it did, then $\mathcal{W}_n \cup \{a^*\}$ would be a small $A \to B$ wave in G, which does not exist. Hence, $G - A_n$ contains an A–B path P, which meets T_n because

P_n
p_n, t_n

$(\mathcal{W}_n, A_n)$ is a wave in $G - T_n$. Let P_n be such a path P, chosen so that its vertex p_n following its last vertex t_n in T_n is chosen minimal in some fixed enumeration of $V(G)$. Note that $p_n P_n \subseteq G_n - A_n$, by definition of G_n.

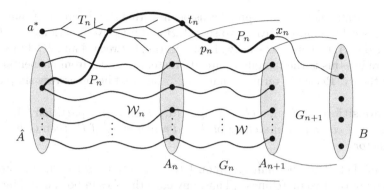

Fig. 8.4.4. As $\mathcal{W}_n$ is maximal, G_n has no proper $A_n \to B$ wave

Now $P'_n = a^*T_nt_nP_n$ is an a^*–B path in $G - \hat{A} - A_n$. If $G_n - p_nP_n$ has no proper $A_n \to B$ wave, then $\mathcal{W}_n$ is large and maximal not only in $G - T_n$ but also in $G - T_n - p_nP_n$, and a^* is linkable for $(G, \hat{A}, B)$ with a^*–B path P'_n and $X = V(T_n \cup p_nP_n)$. We may therefore assume that $G_n - p_nP_n$ has a proper $A_n \to B$ wave.

Let x_n be the first vertex on p_nP_n such that $G_n - p_nP_nx_n$ has a proper $A_n \to B$ wave. Then $G'_n := G_n - p_nP_n\mathring{x}_n$ has no proper $A_n \to B$ wave but $G'_n - x_n$ does, so by Lemma 8.4.5 every $A_n \to B$ wave in $G'_n - x_n = G_n - p_nP_nx_n$ is large. Let W be a maximal such wave, put $\mathcal{W}_{n+1} := \mathcal{W}_n + W$, and let $T_{n+1} := T_n \cup t_nP_nx_n$. Then $\mathcal{W}_{n+1}$ is a large maximal $\hat{A} \to B$ wave in $G - T_{n+1}$. If $x_n \in B$, then T_{n+1} contains a path linking a^* to B, which satisfies the lemma with $\mathcal{W}_{n+1}$ and $X = V(T_{n+1})$. We may therefore assume that $x_n \notin B$, giving $T_{n+1} \subseteq G - (\hat{A} \cup B)$ as required.

T_{n+1}

Put $T^* := \bigcup_{n\in\mathbb{N}} T_n$. Then the $\mathcal{W}_n$ are $\hat{A} \to B$ waves in $G - T^*$; let $(\mathcal{W}^*, A^*)$ be their limit. Our aim is to show that A^* separates A from B not only in $G - T^*$ but even in G: then $(\mathcal{W}^* \cup \{a^*\}, A^*)$ is a small $A \to B$ wave in G, a contradiction.

Suppose there exists an A–B path Q in $G - A^*$. Let t be its last vertex in T^*. Since T^* does not meet B, there is a vertex p following t on Q. Since T^* contains every p_n but not p, the path $P = a^*T^*tQ$ was never chosen as P_n. Now let n be large enough that $t \in T_n$, and that p precedes p_n in our fixed enumeration of $V(G)$. (As $p_n \in T_{n+1} - T_n$, the p_n are pairwise distinct.) The fact that P was not chosen as P_n then means that its portion pQ outside T_n meets A_n, say in a vertex q. Now $q \notin A^*$ by the choice of Q. Let W be the path in $\mathcal{W}_n$ that joins $\hat{A}$ to q; this path too avoids A^*. But then WqQ contains an $\hat{A}$–B path in $G - T^*$ avoiding A^*, which contradicts the definition of A^*. □

The proof of Theorem 8.4.2 for countable G is now complete.

Turning now to matching, let us begin with a simple problem that is intrinsically infinite. Given two sets A, B and injective functions $A \to B$ and $B \to A$, is there necessarily also a bijection between A and B? Indeed there is—this is the famous Cantor-Bernstein theorem from elementary set theory. Recast in terms of matchings, the proof becomes very simple:

Proposition 8.4.6. *Let G be a bipartite graph, with bipartition $\{A, B\}$ say. If G contains a matching of A and a matching of B, then G has a 1-factor.*

Proof. Let H be the multigraph on $V(G)$ whose edge set is the disjoint union of the two matchings. (Thus, any edge that lies in both matchings becomes a double edge in H.) Every vertex in H has degree 1 or 2. In fact, it is easy to check that every component of H is an even cycle or an infinite path. Picking every other edge from each component, we obtain a 1-factor of G.   □

The corresponding path problem in non-bipartite graphs, with sets of disjoint A–B paths instead of matchings, is less trivial. Let us say *covers* that a set $\mathcal{P}$ of paths in G *covers* a set U of vertices if every vertex in U is an endvertex of a path in $\mathcal{P}$.

Theorem 8.4.7. (Pym 1969)
Let G be a graph, and let $A, B \subseteq V(G)$. Suppose that G contains two sets of disjoint A–B paths, one covering A and one covering B. Then G contains a set of disjoint A–B paths covering $A \cup B$.

Some hints for a proof of Theorem 8.4.7 are included with Exercise 64.

Next, let us see how the standard matching theorems for finite graphs—König, Hall, Tutte, Gallai-Edmonds—extend to infinite graphs. For locally finite graphs, they all have straightforward extensions by compactness; see Exercises 24–27. But there are also very satisfactory extensions to graphs of arbitrary cardinality. Their proofs form a coherent body of theory and are much deeper, so we shall only be able to state those results and point out how some of them are related. But, as with Menger's theorem, the statements themselves are interesting too: finding the 'right' restatement of a given finite result to make a substantial infinite theorem is by no means easy, and most of them were found only as the theory itself developed over the years.

Let us start with bipartite graphs. The following Erdős-Menger-type extension of König's theorem (2.1.1) is now a corollary of Theorem 8.4.2:

Theorem 8.4.8. (Aharoni 1984)
Every bipartite graph has a matching, M say, and a vertex cover of its edge set that consists of exactly one vertex from every edge in M.

What about an infinite version of the marriage theorem (2.1.2)? The finite theorem says that a matching exists as soon as every subset S of the first partition class has enough neighbours in the second. But how do we measure 'enough' in an infinite graph? Just as in Menger's theorem, comparing cardinalities is not enough (Exercise 25).

However, there is a neat way of rephrasing the marriage condition for a finite graph without appealing to cardinalities. Call a subset X of one partition class *matchable* to a subset Y of the other if the subgraph *matchable* spanned by X and Y contains a matching of X. Now if S is *minimal* with $|S| > |N(S)|$ then, by the marriage theorem, S is 'larger' than $N(S)$ also in the sense that S is not matchable to $N(S)$ but $N(S)$ is matchable to S. (Indeed, by the minimality of S and the marriage theorem, any $S' \subseteq S$ with $|S'| = |S| - 1$ can be matched to $N(S)$. As $|S'| = |S| - 1 \geqslant |N(S)|$, this matching covers $N(S)$.) Thus, if there is any obstruction S of the type $|S| > |N(S)|$ to a perfect matching, there is also one where S is larger than $N(S)$ in this other sense: that S is not matchable to $N(S)$ but $N(S)$ is matchable to S.

Rewriting the marriage condition in this way does indeed yield an infinite version of Hall's theorem, which follows from Theorem 8.4.8 just as the marriage theorem follows from König's theorem:

Corollary 8.4.9. *A bipartite graph with bipartition $\{A, B\}$ contains a matching of A unless there is a set $S \subseteq A$ such that S is not matchable to $N(S)$ but $N(S)$ is matchable to S.*

Proof. Consider a matching M and a cover U as in Theorem 8.4.8. Then $U \cap B \supseteq N(A \smallsetminus U)$ is matchable to $A \smallsetminus U$, by the edges of M. And if $A \smallsetminus U$ is matchable to $N(A \smallsetminus U)$, then adding this matching to the edges of M incident with $A \cap U$ yields a matching of A.   □

Applied to a finite graph, Corollary 8.4.9 implies the marriage theorem: if $N(S)$ is matchable to S but not conversely, then clearly $|S| > |N(S)|$. Similarly, the finite version of Corollary 8.4.9 implies the finite case of the following sufficient condition for the existence of a matching of A:

Theorem 8.4.10. (Milner & Shelah 1974)
A bipartite graph with bipartition $\{A, B\}$ contains a matching of A if $d(a) \geqslant 1$ for every $a \in A$ and $d(a) \geqslant d(b)$ for every edge ab with $a \in A$.

Let us now turn to non-bipartite graphs. If a finite graph has a 1-factor, then the set of vertices covered by any *partial matching*—one that *partial* leaves some vertices unmatched—can be increased by an augmenting *matching* path, an alternating path whose first and last vertex are unmatched (Ex. 1, Ch. 2). In an infinite graph we no longer insist that augmenting

augmenting path paths be finite, as long as they have a first vertex. Then, starting at any unmatched vertex with an edge of the 1-factor that we are assuming to exist, we can likewise find a unique maximal alternating path that will either be a ray or end at another unmatched vertex. Switching edges along this path we can then improve our current matching to increase the set of matched vertices, just as in a finite graph.

The existence of an inaugmentable partial matching, therefore, is an obvious obstruction to the existence of a 1-factor. The following theorem asserts that this obstruction is the only one:

Theorem 8.4.11. (Steffens 1977)
A countable graph has a 1-factor if and only if for every partial matching there exists an augmenting path.

Unlike its finite counterpart, Theorem 8.4.11 is far from trivial: augmenting a given matching 'blindly' need not lead to a well-defined matching at limit steps, since a given edge may get toggled on and off infinitely often (in which case its status will be undefined at the limit—example?). We therefore cannot simply find the desired 1-factor inductively.

In fact, Theorem 8.4.11 does not extend to uncountable graphs (Exercise 67). However, from the obstruction of inaugmentable partial matchings one can derive a Tutte-type condition that does extend. $\mathcal{C}'_{G-S}$ Given a set S of vertices in a graph G, let us write $\mathcal{C}'_{G-S}$ for the set of factor-critical components of $G - S$, and G'_S for the bipartite graph with G'_S vertex set $S \cup \mathcal{C}'_{G-S}$ and edge set $\{\, sC \mid \exists\, c \in C : sc \in E(G) \,\}$.

Theorem 8.4.12. (Aharoni 1988)
A graph G has a 1-factor if and only if, for every set $S \subseteq V(G)$, the set $\mathcal{C}'_{G-S}$ is matchable to S in G'_S.

Applied to a finite graph, Theorem 8.4.12 implies Tutte's 1-factor theorem (2.2.1): if $\mathcal{C}'_{G-S}$ is not matchable to S in G'_S, then by the marriage theorem there is a subset S' of S that sends edges to more than $|S'|$ components in $\mathcal{C}'_{G-S}$ that are also components of $G - S'$, and these components are odd because they are factor-critical.

Theorems 8.4.8 and 8.4.12 also imply an infinite version of the Gallai-Edmonds theorem (2.2.3):

Corollary 8.4.13. *Every graph $G = (V, E)$ has a set S of vertices that is matchable to $\mathcal{C}'_{G-S}$ in G'_S and such that every component of $G - S$ not in $\mathcal{C}'_{G-S}$ has a 1-factor. Given any such set S, the graph G has a 1-factor if and only if $\mathcal{C}'_{G-S}$ is matchable to S in G'_S.*

Proof. Given a pair (S, M) where $S \subseteq V$ and M is a matching of S in G'_S, and given another such pair (S', M'), write $(S, M) \leqslant (S', M')$ if

$$S \subseteq S' \subseteq V \smallsetminus \bigcup \{ V(C) \mid C \in \mathcal{C}'_{G-S} \}$$

and $M \subseteq M'$. Since $\mathcal{C}'_{G-S} \subseteq \mathcal{C}'_{G-S'}$ for any such S and S', Zorn's lemma implies that there is a maximal such pair (S, M). S, M

For the first statement, we have to show that every component C of $G - S$ that is not in $\mathcal{C}'_{G-S}$ has a 1-factor. If it does not, then by Theorem 8.4.12 there is a set $T \subseteq V(C)$ such that $\mathcal{C}'_{C-T}$ is not matchable to T in C'_T. By Corollary 8.4.9, this means that $\mathcal{C}'_{C-T}$ has a subset $\mathcal{C}$ that is not matchable in C'_T to the set $T' \subseteq T$ of its neighbours, while T' is matchable to $\mathcal{C}$; let M' be such a matching. Then $(S, M) < (S \cup T', M \cup M')$, contradicting the maximality of (S, M).

Of the second statement, only the backward implication is non-trivial. Our assumptions now are that $\mathcal{C}'_{G-S}$ is matchable to S in G'_S and vice versa (by the choice of S), so Proposition 8.4.6 yields that G'_S has a 1-factor. This defines a matching of S in G that picks one vertex x_C from every component $C \in \mathcal{C}'_{G-S}$ and leaves the other components of $G - S$ untouched. Adding to this matching a 1-factor of $C - x_C$ for every $C \in \mathcal{C}'_{G-S}$ and a 1-factor of every other component of $G - S$, we obtain the desired 1-factor of G. $\square$

Infinite matching theory may seem rather mature and complete as it stands, but there are still fascinating unsolved problems in the Erdős-Menger spirit concerning related discrete structures, such as partially ordered sets or hypergraphs. We conclude with one about graphs.

Call an infinite graph G *perfect* if every induced subgraph $H \subseteq G$ has a complete subgraph K of order $\chi(H)$, and *strongly perfect* if K can always be chosen so that it meets every colour class of some $\chi(H)$-colouring of H. (Exercise 69 gives an example of a perfect graph that is not strongly perfect.) Call G *weakly perfect* if the chromatic number of every induced subgraph $H \subseteq G$ is at most the supremum of the orders of its complete subgraphs.   *strongly perfect*   *weakly perfect*

Conjecture. (Aharoni & Korman 1993)
Every weakly perfect graph without infinite independent sets of vertices is strongly perfect.

8.5 Recursive structures

In this section we introduce another tool that is commonly used in infinite graph theory: to define a class of graphs recursively, so as to be able later to prove assertions about these graphs by (transfinite) induction. Rather than attempting a systematic treatment of this technique we give two examples; more can be found in the exercises.

Our first example is very simple: it describes the structure of a tree by recursively pruning away leaves and isolated ends. Let T be any tree, equipped with a root and the corresponding tree-order on its vertices. We recursively label the vertices of T by ordinals, as follows. Given an ordinal α, assume that we have decided for every $\beta < \alpha$ which of the vertices of T to label β, and let T_α be the subgraph of T induced by the vertices that are still unlabelled. Assign label α to every vertex t of T_α whose up-closure $\lfloor t \rfloor_{T_\alpha} = \lfloor t \rfloor_T \cap T_\alpha$ in T_α is a chain. The recursion T^* terminates at the first α not used to label any vertex; for this α we put $T_\alpha =: T^*$.

For each α, the vertices labelled α form an up-set in T_α: if $\lfloor t \rfloor_{T_\alpha}$ is a chain, then so is $\lfloor t' \rfloor_{T_\alpha}$ for every $t' \in \lfloor t \rfloor_{T_\alpha}$. Every T_α, therefore, is a down-set in T (induction on α) and hence connected. Thus, T_α is a tree, and the set of vertices labelled α induces in T_α a disjoint union of paths.

recursively prunable Let us call T *recursively prunable* if every vertex of T gets labelled in this way, i.e., if $T^* = \emptyset$. We may then be able to prove assertions about T, or about graphs containing T as a normal spanning tree, by dealing in turn with those chains as they get deleted. The following proposition shows that the recursively prunable trees form a natural class also in structural terms:

Proposition 8.5.1. *A rooted tree is recursively prunable if and only if it contains no subdivision of the infinite binary tree T_2 as a subgraph.*

Proof. Let T be any rooted tree. Suppose first that T is not recursively prunable, i.e. that $T^* \neq \emptyset$. Since no vertex of T^* gets labelled when the recursion terminates, every $t \in T^*$ has two incomparable vertices of T^* above it. As T^* is connected, it is now easy to find a subdivision of T_2 in T^* inductively, along the levels of T_2.

Conversely, suppose that T contains a subdivision T' of T_2. We shall see in a moment that T' can be chosen 'upwards' in T, that is, in such a way that the tree-order which T induces on its vertices agrees with its own tree-order as induced by T_2. If this is the case, then every vertex of T' has two incomparable vertices of T' above it (in both orders). Hence there can be no minimal ordinal α such that a vertex of T' is labelled α. Thus all of T' remains unlabelled, and $\emptyset \neq T' \subseteq T^*$ as desired.

It remains to show that T' can indeed be chosen in this way. Let T' be any subdivision of T_2 in T, and let u be minimal in the tree-order of T among the vertices of T'. Induction on the levels of the tree $\lfloor u\rfloor_{T'}$ shows that $\leqslant_{T'}$ and $\leqslant_T$ agree on $\lfloor u\rfloor_{T'}$: any upper neighbour in T' of a vertex $t \in \lfloor u\rfloor_{T'}$ must lie above t also in T, since the unique lower neighbour of t in T is either not in T' (if $t = u$), or by induction it is the unique lower neighbour of t also in T'. Pick any branch vertex v of T' in $\lfloor u\rfloor_{T'}$. Then $\lfloor v\rfloor_{T'}$ is the desired subdivision of T_2 in T. $\qquad\square$

The charm of the recursive pruning discussed above lies in the fact that it removes the 'messy bits' of a given tree in an automated sort of way: we do not have to know where they are, but if our given tree contains a 'clean' ever-branching subtree, then the recursion will reveal it.

And there is another way of viewing it. We might think of rooted paths (paths with a first vertex, which we take to be the root) as particularly basic objects, and call them *rooted trees of rank* 0. We could then define rooted trees of higher ordinal rank inductively, taking as the rooted trees of rank α those that do not have any rank $\beta < \alpha$ but in which it is possible to delete a path starting at the root so as to leave components that each have some rank $< \alpha$ when taken with the induced tree-order. Then the rooted trees that are assigned a rank in this way are precisely the recursively prunable ones, and those of rank $\leqslant \alpha$ are precisely those whose labels do not exceed α (Exercise 71).

We now apply the same idea to graphs that are not trees. Let us assign *rank* 0 to all the finite graphs. Given an ordinal $\alpha > 0$, we assign *rank* α to every graph G that does not already have a rank $\beta < \alpha$ and which has a finite set U of vertices such that every component of $G - U$ has some rank $< \alpha$. *rank*

When disjoint graphs G_i have ranks $\alpha_i < \alpha$, their union clearly has a rank of at most α; if the union is finite, it has rank $\max_i \alpha_i$. Induction on α shows that subgraphs of graphs of rank α also have a rank of at most α. Conversely, joining finitely many new vertices to a graph, no matter how, will not change its rank.

Not every graph has a rank. Indeed, the ray cannot have a rank, since deleting finitely many of its vertices always leaves a component that is also a ray. As subgraphs of graphs with a rank also have a rank, this means that only rayless graphs can have a rank. But all these do:

Lemma 8.5.2. *A graph has a rank if and only if it is rayless.*

Proof. Consider a graph G that has no rank. Then one of its components, C_0 say, has no rank; let v_0 be a vertex in C_0. Now $C_0 - v_0$ has a component C_1 that has no rank; let v_1 be a neighbour of v_0 in C_1. Continuing inductively, we find a ray $v_0 v_1 \ldots$ in G. $\qquad\square$

Because of Lemma 8.5.2, we call the ranking defined above the *ranking of rayless graphs*. As an application of this ranking, we now prove the unfriendly partition conjecture from Section 8.1 for rayless graphs.

Theorem 8.5.3. *Every countable rayless graph G has an unfriendly partition.*

Proof. To help with our formal notation, we shall think of a partition of a set V as a map $\pi\colon V \to \{0,1\}$. We apply induction on the rank of G. When this is zero then G is finite, and an unfriendly partition can be obtained by maximizing the number of edges across the partition.
Suppose now that G has rank $\alpha > 0$, and assume the theorem as true for graphs of smaller rank.

α

Let U be a finite set of vertices in G such that each of the components $C_0, C_1, \ldots$ of $G - U$ has rank $< \alpha$. Partition U into the set U_0 of vertices that have finite degree in G, the set U_1 of vertices that have infinitely many neighbours in some C_n, and the set U_2 of vertices that have infinite degree but only finitely many neighbours in each C_n.

U
$C_0, C_1, \ldots$
U_0, U_1, U_2

For every $n \in \mathbb{N}$ let $G_n := G[U \cup V(C_0) \cup \ldots \cup V(C_n)]$. This is a graph of some rank $\alpha_n < \alpha$, so by induction it has an unfriendly partition π_n. Each of these π_n induces a partition of U. Let π_U be a partition of U induced by π_n for infinitely many n, say for $n_0 < n_1 < \ldots$. Choose n_0 large enough that G_{n_0} contains all the neighbours of vertices in U_0, and the other n_i large enough that every vertex in U_2 has more neighbours in $G_{n_i} - G_{n_{i-1}}$ than in $G_{n_{i-1}}$, for all $i > 0$. Let π be the partition of G defined by letting $\pi(v) := \pi_{n_i}(v)$ for all $v \in G_{n_i} - G_{n_{i-1}}$ and all i, where $G_{n_{-1}} := \emptyset$. Note that $\pi|_U = \pi_{n_0}|_U = \pi_U$.

$G_0, G_1, \ldots$

$n_0, n_1, \ldots$

π

Let us show that π is unfriendly. We have to check that every vertex is *happy with* π, i.e., that it has at least as many neighbours in the opposite class under π as in its own.[8] To see that a vertex $v \in G - U$ is happy with π, let i be minimal such that $v \in G_{n_i}$ and recall that v was happy with π_{n_i}. As both v and its neighbours in G lie in $U \cup V(G_{n_i} - G_{n_{i-1}})$, and π agrees with π_{n_i} on this set, v is happy also with π. Vertices in U_0 are happy with π, because they were happy with π_{n_0}, and π agrees with π_{n_0} on U_0 and all its neighbours. Vertices in U_1 are also happy. Indeed, every $u \in U_1$ has infinitely many neighbours in some C_n, and hence in some $G_{n_i} - G_{n_{i-1}}$. Then u has infinitely many opposite neighbours in $G_{n_i} - G_{n_{i-1}}$ under π_{n_i}. Since π_{n_i} agrees with π on both U and $G_{n_i} - G_{n_{i-1}}$, our vertex u has infinitely many opposite neighbours also under π. Vertices in U_2, finally, are happy with every π_{n_i}. By our choice of n_i, at least one of their opposite neighbours under π_{n_i} must lie in $G_{n_i} - G_{n_{i-1}}$. Since π_{n_i} agrees with π on both U_2 and $G_{n_i} - G_{n_{i-1}}$, this gives every $u \in U_2$ at least one opposite neighbour under π in every $G_{n_i} - G_{n_{i-1}}$. Hence u has infinitely many opposite neighbours under π, which clearly makes it happy. $\square$

---

[8] It is only by tradition that such partitions are called 'unfriendly'; our vertices love them.

8.6 Graphs with ends: the complete picture

In this section we shall develop a deeper understanding of the global structure of infinite graphs, especially locally finite ones, that can be attained only by studying their ends. This structure is intrinsically topological, because topology best captures our intuition about convergence.[9]

Our first goal will be to make precise our intuitive idea that the ends of a graph are the 'points at infinity' to which its rays converge. To do so, we shall define a topological space $|G|$ associated with a graph $G = (V, E, \Omega)$ and its ends.[10] By considering topological versions of paths, cycles and spanning trees in this space, we shall then be able to extend to infinite graphs some parts of finite graph theory that would not otherwise have infinite counterparts; see the notes for more examples. Thus, the ends of an infinite graph turn out to be more than a curious phenomenon: they form an integral part of the picture, without which it cannot be properly understood.

V, E, Ω

To build the space $|G|$ formally, we start with the set $V \cup \Omega$. For every edge $e = uv$ we add a set $\mathring{e} = (u, v)$ of continuum many points, making these sets $\mathring{e}$ disjoint from each other and from $V \cup \Omega$. We then choose for each e some fixed bijection between $\mathring{e}$ and the real interval $(0, 1)$, and extend this bijection to one between $[u, v] := \{u\} \cup \mathring{e} \cup \{v\}$ and $[0, 1]$. This bijection defines a metric on $[u, v]$; we call $[u, v]$ a *topological edge* with *inner points* $x \in \mathring{e}$. Given any $F \subseteq E$ we write $\mathring{F} := \bigcup \{ \mathring{e} \mid e \in F \}$. When we speak of a 'graph' $H \subseteq G$, we shall often also mean its corresponding point set $V(H) \cup \mathring{E}(H)$.

(u, v)

$[u, v]$

$\mathring{F}$

Having thus defined the point set of $|G|$, let us choose a basis of open sets to define its topology. For every edge uv, declare as open all subsets of (u, v) that correspond, by our fixed bijection between (u, v) and $(0, 1)$, to an open set in $(0, 1)$. For every vertex u and $\epsilon > 0$, declare as open the 'open star around u of radius ϵ', that is, the set of all points on edges $[u, v]$ at distance less than ϵ from u, measured individually for each edge in its metric inherited from $[0, 1]$. Finally, for every end ω and every finite set $S \subseteq V$, there is a unique component $C(S, \omega)$ of $G - S$ that contains rays from ω. Let $\Omega(S, \omega) := \{ \omega' \in \Omega \mid C(S, \omega') = C(S, \omega) \}$. For every $\epsilon > 0$, write $\mathring{E}_\epsilon(S, \omega)$ for the set of all inner points of S–$C(S, \omega)$ edges at distance less than ϵ from their endpoint in $C(S, \omega)$. Then declare as open all sets of the form

$C(S, \omega)$

$$\hat{C}_\epsilon(S, \omega) := C(S, \omega) \cup \Omega(S, \omega) \cup \mathring{E}_\epsilon(S, \omega).$$

$\hat{C}_\epsilon(S, \omega)$

This completes the definition of $|G|$, whose open sets are the unions of the sets we explicitly chose as open above.

$|G|$

---

[9] Only point-set topology is needed for the text. See the exercises for more.

[10] The notation of $|G|$ comes from topology and clashes with our notation for the order of G. But there is little danger of confusion, so we keep both.

closure $\overline{X}$ The *closure* of a set $X \subseteq |G|$ will be denoted by $\overline{X}$. For example, $\overline{V} = V \cup \Omega$ (because every neighbourhood of an end contains a vertex), and the closure of a ray is obtained by adding its end. More generally, the closure of the set of teeth of a comb contains a unique end, the end of its spine. Conversely, if $U \subseteq V$ and $R \in \omega \in \Omega \cap \overline{U}$, there is a comb with spine R and teeth in U (Exercise 77). In particular, the closure of the subgraph $C(S, \omega)$ considered above is the set $C(S, \omega) \cup \Omega(S, \omega)$.

standard
subspace   The subspaces X of $|G|$ we shall be interested in are usually the closure of a subgraph H of G, i.e., of the form $X = \overline{U} \cup \mathring{D}$ for $H = (U, D)$.

$V(X), E(X)$ We write $V(X)$ for U and $E(X)$ for D, and call such subspaces *standard*. We also refer to such X as $\overline{H}$, or even as $\overline{D}$ if H has no isolated vertices, and then say that X is *spanned by* D. Note that the ends in X are always ends of G, not of H; in particular, they need not have a ray in H.

By definition, $|G|$ is always Hausdorff; indeed one can show that it is normal. When G is connected and locally finite, then $|G|$ is compact:[11]

Proposition 8.6.1. *If G is connected and locally finite, then $|G|$ is a compact Hausdorff space.*

(8.1.2)   *Proof.* Let $\mathcal{O}$ be an open cover of $|G|$; we show that $\mathcal{O}$ has a finite subcover. Pick a vertex $v_0 \in G$, write D_n for the (finite) set of vertices at distance n from v_0, and put $S_n := D_0 \cup \ldots \cup D_{n-1}$. For every $v \in D_n$, let $C(v)$ denote the component of $G - S_n$ containing v, and let $\hat{C}(v)$ be

$\hat{C}(v)$ its closure together with all inner points of $C(v)$–S_n edges. Then $G[S_n]$ and these $\hat{C}(v)$ together partition $|G|$.

We wish to prove that, for some n, each of the sets $\hat{C}(v)$ with $v \in D_n$ is contained in some $O(v) \in \mathcal{O}$. For then we can take a finite subcover of $\mathcal{O}$ for $G[S_n]$ (which is compact, being a finite union of edges and vertices), and add to it these finitely many sets $O(v)$ to obtain the desired finite subcover for $|G|$.

Suppose there is no such n. Then for each n the set V_n of vertices $v \in D_n$ such that no set from $\mathcal{O}$ contains $\hat{C}(v)$ is non-empty. Moreover, for every neighbour $u \in D_{n-1}$ of $v \in V_n$ we have $C(v) \subseteq C(u)$ because $S_{n-1} \subseteq S_n$, and hence $u \in V_{n-1}$; let $f(v)$ be such a vertex u. By the infinity lemma (8.1.2) there is a ray $R = v_0 v_1 \ldots$ with $v_n \in V_n$ for all n. Let ω be its end, and let $O \in \mathcal{O}$ contain ω. Since O is open, it contains a basic open neighbourhood of ω: there exist a finite set $S \subseteq V$ and $\epsilon > 0$ such that $\hat{C}_\epsilon(S, \omega) \subseteq O$. Now choose n large enough that S_n contains S and all its neighbours. Then $C(v_n)$ lies inside a component of $G - S$. As $C(v_n)$ contains the ray $v_n R \in \omega$, this component must be $C(S, \omega)$. Thus

$$\hat{C}(v_n) \subseteq \hat{C}_\epsilon(S, \omega) \subseteq O \in \mathcal{O},$$

contradicting the fact that $v_n \in V_n$.   □

---

[11] Topologists call $|G|$ the *Freudenthal compactification* of G.

If G has a vertex of infinite degree then $|G|$ cannot be compact. (Why not?) But $\Omega \subseteq |G|$ can be compact; see Exercise 85 for when it is.

What else can we say about the space $|G|$ in general? For example, is it metrizable? Using a normal spanning tree T of G, it is indeed not difficult to define a metric on $|G|$ that induces its topology. But not every connected graph has a normal spanning tree, and it is not easy to determine in graph-theoretical terms which graphs do. Surprisingly, though, it is possible to deduce the existence of a normal spanning tree from that of a defining metric on $|G|$. Thus whenever $|G|$ is metrizable, a metric can be made visible in a natural and structural way.

Theorem 8.6.2. *For a connected graph G, the following assertions are equivalent:*

(i) *The space $|G|$ is metrizable.*

(ii) *G has a normal spanning tree.*

(iii) *All minors of G have countable colouring number.*

The proof of the equivalence of (i) and (ii) in Theorem 8.6.2 is indicated in Exercises 41 and 86. More on (iii) can be found in the notes.

Our next aim is to review, or newly define, some topological notions of paths and connectedness, of cycles, and of spanning trees. By substituting these topological notions with respect to $|G|$ for the corresponding graph-theoretical notions with respect to G one can extend to locally finite infinite graphs a number of theorems about paths, cycles and spanning trees in finite graphs whose ordinary infinite versions are false. We shall do this, as a case in point, for the tree packing theorem of Nash-Williams and Tutte, Theorem 2.4.1; see the notes for more.

Let X be an arbitrary Hausdorff space. (Later, this will be a subspace of $|G|$.) X is *(topologically) connected* if it is not a union of two disjoint non-empty open subsets.[12] Note that continuous images of connected spaces are connected. For example, since the real interval $[0,1]$ is connected,[13] so are its continuous images in X. *X* *connected*

A *homeomorphic* image of $[0,1]$ in X is an *arc* in X; it *links* the images of 0 and 1, which are its *endpoints*. Every finite path in G defines an arc in $|G|$ in an obvious way. Similarly, every ray defines an arc linking its starting vertex to its end, and a double ray in G forms an arc with the two ends of its tails if these ends are distinct. *arc*

The *(topological) degree* of an end ω of G in a standard subspace X of $|G|$ is the supremum, in fact maximum, of all integers k such that X contains k arcs that end in ω and are otherwise disjoint. *end degrees in subspaces*

---

[12] These subsets would be complements of each other, and hence also be closed. Note that 'open' and 'closed' means open and closed *in* X: when X is a subspace of $|G|$ with the subspace topology, the two sets need not be open or closed in $|G|$.

[13] This takes a few lines to prove—can you prove it?

$G = (V, E, \Omega)$ For the remainder of this section let, unless otherwise mentioned, $G = (V, E, \Omega)$ be a fixed connected locally finite graph.

Unlike ordinary paths, arcs in $|G|$ can jump across a cut without containing an edge from it—but only if the cut is infinite:

[8.7.1]   **Lemma 8.6.3.** (Jumping Arc Lemma)
Let $F \subseteq E$ be a cut of G with sides V_1, V_2.

(i) *If F is finite, then $\overline{V_1} \cap \overline{V_2} = \emptyset$, and there is no arc in $|G| \setminus \mathring{F}$ with one endpoint in V_1 and the other in V_2.*

(ii) *If F is infinite, then $\overline{V_1} \cap \overline{V_2} \neq \emptyset$, and there will be such an arc if both V_1 and V_2 are connected in G.*

(8.2.2)   *Proof.* (i) Suppose that F is finite. Let S be the set of vertices incident with edges in F. Then S is finite and separates V_1 from V_2, so for every $\omega \in \Omega$ the connected graph $C(S, \omega)$ misses either V_1 or V_2. But then so does every basic open set of the form $\hat{C}_\epsilon(S, \omega)$. Therefore no end ω lies in the closure of both V_1 and V_2.

As $|G| \setminus \mathring{F} = \overline{G[V_1]} \cup \overline{G[V_2]}$ and this union is disjoint, no connected subset of $|G| \setminus \mathring{F}$ can meet both V_1 and V_2. Since arcs are continuous images of $[0, 1]$ and hence connected, there is no V_1–V_2 arc in $|G| \setminus \mathring{F}$.

(ii) Suppose now that F is infinite. Since G is locally finite, the set U of endvertices of F in V_1 is also infinite. By the star-comb lemma (8.2.2), there is a comb in G with teeth in U; let ω be the end of its spine. Then every basic open neighbourhood $\hat{C}_\epsilon(S, \omega)$ of ω meets $U \subseteq V_1$ infinitely and hence also meets V_2, giving $\omega \in \overline{V_1} \cap \overline{V_2}$.

To obtain a V_1–V_2 arc in $|G| \setminus \mathring{F}$, all we need now is an arc in $\overline{G[V_1]}$ and another in $\overline{G[V_2]}$, both ending in ω. Such arcs exist if the graphs $G[V_i]$ are connected: we can then pick a sequence of vertices in V_i converging to ω, and apply the star-comb lemma in $G[V_i]$ to obtain a comb whose spine is a ray in $G[V_i]$ converging to ω. Concatenating these two rays yields the desired jumping arc. $\square$

To some extent, arcs in $|G|$ assume the role that paths play in finite graphs. So arcs are important—but how do we find them? It is not always possible to construct arcs as explicitly as in the proof of Lemma 8.6.3 (ii). Figure 8.6.1, for example, shows an arc that goes through continuum many ends; such arcs cannot be constructed greedily by following a ray into its end and emerging from that end on another ray, etc.

There are two basic methods to obtain an arc between two given points, say two vertices x and y. One is to use compactness to obtain, as a limit of finite x–y paths, a *topologial x–y path*, a continuous map $\pi \colon [0, 1] \to |G|$ sending 0 to x and 1 to y. A lemma from general topology then tells us that this path can be made injective:

[8.7.3]   **Lemma 8.6.4.** *The image of a topological x–y path in a Hausdorff space contains an x–y arc.*

To illustrate this method, we will use it in the proof of Theorem 8.7.3.

Another method is to prove that the subspace in which we wish to find our x–y arc is topologically connected, and use this to deduce that it contains the desired arc. Our next three lemmas provide the tools needed to implement this approach in practice; we shall then illustrate its use in the proof of Theorem 8.6.9.

Being linked by an arc is an equivalence relation on the points of our Hausdorff space X: every x–y arc A has a first point p on any y–z arc A' (because A' is closed), and the obvious segments Ap and pA' together form an x–z arc in X. The corresponding equivalence classes are the *arc-components* of X. If X has only one arc-component, then X is *arc-connected*.

arc-
component

arc-
connected

Since $[0,1]$ is connected, arc-connectedness implies connectedness. The converse implication is false in general, even for spaces $X \subseteq |G|$ with G locally finite. But it holds in all the cases that matter:

Lemma 8.6.5. *Connected standard subspaces of $|G|$ are arc-connected.* [8.7]

Our proof of Theorem 8.7.3 will show how one can prove Lemma 8.6.5. Two further proofs are indicated in Exercises 88 and 129.

Lemma 8.6.6. *Arc-components of standard subspaces of $|G|$ are closed.* [8.7]

Proof. Let A be an arc-component of a standard subspace of $|G|$. Since A is connected, so is its closure $\overline{A}$. If $\overline{A} \smallsetminus A \neq \emptyset$ then its points are limits of vertices in A (why?), so $\overline{A}$ is again standard. Hence $\overline{A}$ is arc-connected, either because $\overline{A} = A$ or by Lemma 8.6.5. But then $\overline{A} = A$, by definition of $\overline{A}$. Hence A is closed, as claimed. $\square$

Connected standard subspaces of $|G|$ containing two given points are much easier to construct than an arc between two points. This has to do with the fact that they can be described in purely graph-theoretical terms, with reference only to finite subgraphs of G rather than to $|G|$. The description can be viewed as a topological analogue of the fact that a subgraph H of G is connected if and only if it contains an edge from every cut of G that separates two of its vertices:

Lemma 8.6.7. *A standard subspace of $|G|$ is connected if and only if it* [8.7.1]
contains an edge from every finite cut of G of which it meets both sides.

Proof. Let $X \subseteq |G|$ be a standard subspace. For the forward implication, suppose that G has a finite cut $F = E(V_1, V_2)$ such that X meets both V_1 and V_2 but has no edge in F. Then

$$X \subseteq |G| \smallsetminus \mathring{F} = \overline{G[V_1]} \cup \overline{G[V_2]},$$

and this union is disjoint by Lemma 8.6.3 (i). The induced partition of X into non-empty closed subsets of X shows that X is not connected.

The backward implication holds vacuously if X meets more than one component of G; we may therefore assume that G is connected. If X is not connected, we can partition it into disjoint non-empty open subsets O_1 and O_2. As X is standard, $U_i := O_i \cap V(X) \neq \emptyset$ for both i. Let $\mathcal{P}$ be a maximal set of edge-disjoint U_1–U_2 paths in G, and put

$$F := \bigcup \{ E(P) \mid P \in \mathcal{P} \}.$$

Then $E(X) \cap F = \emptyset$, and no component of $G - F$ meets both U_1 and U_2. Extending $\{U_1, U_2\}$ to a partition of V in such a way that each component of $G - F$ has all its vertices in one class, we obtain a cut $F' \subseteq F$ of G of which X meets both sides. As $E(X) \cap F = \emptyset$, it thus suffices to show that F is finite.

If F is infinite, then so is $\mathcal{P}$. As G is locally finite, the vertices of each $P \in \mathcal{P}$ are incident with only finitely many edges of G. We can thus inductively find an infinite subset of $\mathcal{P}$ consisting of paths that are not only edge-disjoint but disjoint. As G is connected, the endvertices in U_1 of these paths have a limit point ω in $|G|$ (Proposition 8.6.1), which is also a limit point of their endvertices in U_2. Since both O_1 and O_2 are closed in $|G|$, we thus have $\omega \in O_1 \cap O_2$, contradicting the choice of the O_i. $\qquad\square$

circle　　A *circle* in a topological space is a homeomorphic image of the unit circle $S^1 \subseteq \mathbb{R}^2$. For example, if G is the 2-way infinite ladder shown in Figure 8.1.3, and we delete all its rungs (the vertical edges), what remains is a disjoint union of two double rays; its closure in $|G|$, obtained by adding the two ends of G, is a circle. Similarly, the double ray 'round the outside' of the 1-way ladder forms a circle together with the unique end of that ladder.

It is not hard to show that no arc in $|G|$ can consist entirely of ends. This implies that every circle in $|G|$ is a standard subspace; the set of
circuit　　edges spanning it will be called its *circuit*.

A more adventurous example of a circle is shown in Figure 8.6.1. Suppose G is the graph obtained from the binary tree T_2 by joining for every finite 0–1 sequence ℓ the vertices $\ell 01$ and $\ell 10$ by a new edge e_ℓ. Together with all the (uncountably many) ends of G, the double rays $D_\ell \ni e_\ell$ shown in the figure form an arc A in $|G|$, whose union with the bottom double ray D is a circle in $|G|$ (Exercise 94). Note that no two of the double rays in A are consecutive: between any two there lies a third (cf. Exercise 95).

topological
spanning
tree　　A *topological spanning tree* of G is a connected standard subspace T of $|G|$ that contains every vertex but contains no circle. Since standard subspaces are closed, T also contains every end, and by Lemma 8.6.5 it is even arc-connected. With respect to the deletion or addition of edges, it is both minimally connected and maximally 'acirclic' (Exercise 99).

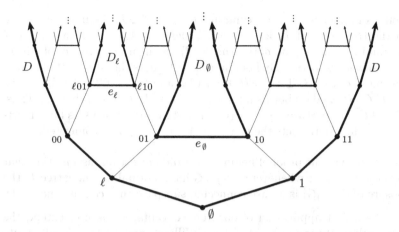

Fig. 8.6.1. The Wild Circle

One might expect that the closure $\overline{T}$ of an ordinary spanning tree T of G is always a topological spanning tree of $|G|$, but this is not the case: $\overline{T}$ may well contain a circle (Figure 8.6.2). Conversely, a subgraph whose closure is a topological spanning tree may well be disconnected: the 'vertical' rays in the $\mathbb{N} \times \mathbb{N}$ grid, for example, form a topological spanning tree together with the unique end.

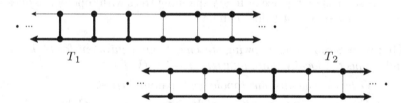

Fig. 8.6.2. T_1 is a topological spanning tree, but T_2 contains three circles

Topological spanning trees can be constructed much as spanning trees of finite graphs: Lemma 8.6.11 will find one by iteratively deleting edges from $|G|$, but they can also be built up 'from below' (Exercise 102). Their mere existence even comes as a corollary of Theorem 8.2.4:   (8.2.4)

Lemma 8.6.8. *The closure in $|G|$ of any normal spanning tree of G is a topological spanning tree of G.*   [8.7]

Proof. Let T be a normal spanning tree of G. By Lemma 8.2.3, every   (1.5.5) (8.2.3)
end ω of G contains a normal ray R of T. Then $R \cup \{\omega\}$ is an arc linking ω to the root of T, so $\overline{T}$ is arc-connected.

It remains to check that $\overline{T}$ contains no circle. Suppose it does, and let A be the u–v arc obtained from that circle by deleting the inner

f

points of an edge $f = uv$ it contains. Clearly, $f \in T$. Assume that $u < v$ in the tree-order of T, let T_u and T_v denote the components of $T - f$ containing u and v, and notice that $V(T_v)$ is the up-closure $\lfloor v \rfloor$ of v in T.

Now let $S := \lceil u \rceil$. By Lemma 1.5.5 (ii), $\lfloor v \rfloor$ is the vertex set of a component C of $G - S$. Thus, $V(C) = V(T_v)$ and $V(G - C) = V(T_u)$, so the set $E(C, S)$ of edges between these sets meets $E(T)$ precisely in f. Thus, $\overline{C}$ and $\overline{G - C}$ partition $|G| \setminus \mathring{E}(C, S) \supseteq A$ into two open sets both meeting A. This contradicts the fact that A is topologically connected. $\qquad \square$

Note that the proof of Lemma 8.6.8 did not use our assumption that G is locally finite: whenever a graph G has a normal spanning tree T, the closure of T in $|G|$ is an arc-connected subspace that contains no circle.

As a first application of our new concepts, let us now extend the tree packing theorem (2.4.1) of Nash-Williams and Tutte to locally finite graphs. Its naive extension, with ordinary spanning trees, fails. Indeed, for every $k \in \mathbb{N}$ one can construct a $2k$-edge-connected locally finite graph that is left disconnected by the deletion of the edges in any one finite circuit (Exercise 19). Such a graph will have at least $k(\ell - 1)$ edges across any vertex partition into ℓ sets, but it cannot have more than two edge-disjoint spanning trees: adding an edge of one of these to another creates a (finite) fundamental circuit there, whose deletion would disconnect any third spanning tree.

As soon as we replace ordinary spanning trees with topological ones, however, Theorem 2.4.1 does extend:

$G = (V, E)$

Theorem 8.6.9. *The following statements are equivalent for all $k \in \mathbb{N}$ and connected locally finite multigraphs $G = (V, E)$:*

    (i) *G has k edge-disjoint topological spanning trees.*

    (ii) *For every finite partition of V, into ℓ sets say, G has at least $k(\ell - 1)$ cross-edges.*

We begin our proof of Theorem 8.6.9 with a compactness extension of the finite theorem. This yields a weaker, 'finitary', statement at the limit (cf. Lemma 8.6.7):

Lemma 8.6.10. *If for every finite partition of V, into ℓ sets say, G has at least $k(\ell - 1)$ cross-edges, then G has k edge-disjoint spanning submultigraphs whose closures in $|G|$ are topologically connected.*

(2.4.1)
(8.1.2)

Proof. Pick an enumeration $v_0, v_1, \ldots$ of V. For every $n \in \mathbb{N}$ let G_n be the finite multigraph obtained from G by contracting every component of $G - \{v_0, \ldots, v_n\}$ to a vertex, deleting any loops but no parallel edges that arise in the contraction. Then $G[v_0, \ldots, v_n]$ is an induced submultigraph of G_n. Let $\mathcal{V}_n$ denote the set of all k-tuples $(H_n^1, \ldots, H_n^k)$ of edge-disjoint connected spanning submultigraphs of G_n.

Since every partition P of $V(G_n)$ induces a partition of V, since G has enough cross-edges for that partition, and since all these cross-edges are also cross-edges of P, Theorem 2.4.1 implies that $\mathcal{V}_n \neq \emptyset$. As every $(H_n^1, \ldots, H_n^k) \in \mathcal{V}_n$ induces an element $(H_{n-1}^1, \ldots, H_{n-1}^k)$ of $\mathcal{V}_{n-1}$, the infinity lemma (8.1.2), yields a sequence $(H_n^1, \ldots, H_n^k)_{n \in \mathbb{N}}$ of k-tuples, one from each $\mathcal{V}_n$, with a limit $(H^1, \ldots, H^k)$ defined by the nested unions

$$H^i := \bigcup_{n \in \mathbb{N}} H_n^i[v_0, \ldots, v_n].$$

These H^i are edge-disjoint for distinct i (because the H_n^i are), but they need not be connected. To show that they have connected closures, it suffices by Lemma 8.6.7 to show that each of them has an edge in every finite cut F of G. Given F, choose n large enough that all the edges of F lie in $G[v_0, \ldots, v_n]$. Then F is also a cut of G_n. Now consider the k-tuple $(H_n^1, \ldots, H_n^k)$ which the infinity lemma picked from $\mathcal{V}_n$. Each of these H_n^i is a connected spanning submultigraph of G_n, so it contains an edge from F. But H_n^i agrees with H^i on $\{v_0, \ldots, v_n\}$, so H^i too contains this edge from F. □

Lemma 8.6.11. *Every connected standard subspace of $|G|$ that contains V also contains a topological spanning tree of G.*

Proof. Let X be a connected standard subspace of $|G|$ containing V. Then G too must be connected, so it is countable. Let $e_0, e_1, \ldots$ be an enumeration of $E(X)$, and consider these edges in turn. Starting with $X_0 := X$, define $X_{n+1} := X_n \smallsetminus \mathring{e}_n$ if this keeps X_{n+1} connected; if not, put $X_{n+1} := X_n$. Finally, let $T := \bigcap_{n \in \mathbb{N}} X_n$.

Since T is closed and contains V, it is still a standard subspace. And T has an edge in every finite cut of G, because X does and its last edge in that cut will never be deleted. So T is connected, by Lemma 8.6.7. But T contains no circle: that would contain an edge, which should have got deleted since deleting an edge from a circle cannot destroy connectedness. □

Proof of Theorem 8.6.9. The implication (ii)→(i) follows from our two lemmas. For (i)→(ii), let G have edge-disjoint topological spanning trees $T_1, \ldots, T_k$, and consider a partition P of V into ℓ sets. If there are infinitely many cross-edges, there is nothing to show; so we assume there are only finitely many. For each $i \in \{1, \ldots, k\}$, let T_i' be the multigraph of order ℓ which the edges of T_i induce on P.

To establish that G has at least $k(\ell - 1)$ cross-edges, we show that the multigraphs T_i' are connected. If not, then some T_i' has a vertex partition crossed by no edge of T_i. This partition induces a cut of G that contains no edge of T_i. By our assumption that G has only finitely many cross-edges, this cut is finite. By Lemma 8.6.7, this contradicts the connectedness of T_i. □

8.7 The topological cycle space

As a more comprehensive application of our new theory, let us now look
at how the cycle space theory of finite graphs extends to locally finite
$G = (V, E)$ graphs $G = (V, E)$ with infinite circuits and topological spanning trees.

Every two points of a topological spanning tree T are joined by a
unique arc in T: existence follows from Lemma 8.6.5, while uniqueness
is proved as for finite graphs. Adding a new edge e to T therefore creates
fundamental   a unique circle in $T \cup e$; its edges form the *fundamental circuit* C_e of e
circuit C_e   with respect to T. Note that C_e can be infinite.

Similarly, for every edge $f \in E(T)$ the space $T \smallsetminus \mathring{f}$ has exactly two
fundamental   arc-components; the set of edges between these is the *fundamental cut* D_f
cut D_f   of T. Since the two arc-components of $T \smallsetminus \mathring{f}$ are closed (Lemma 8.6.6)
but disjoint, Lemma 8.6.3 (ii) implies that D_f is finite.

As in finite graphs, we have $e \in D_f$ if and only if $f \in C_e$, for all
$f \in E(T)$ and $e \in E \smallsetminus E(T)$. Topological spanning trees that are the
(8.6.8)   closure of a normal spanning tree, as in Lemma 8.6.8, are particularly
useful in this context: their fundamental circuits and cuts are both finite.

For locally finite graphs there will be two cycle spaces: the usual
'finitary' one from Chapter 1.9, and a new 'topological' one based on
topological circuits. The former will be a subspace of the latter, much
as the space of all finite cuts is a subspace of the space of all cuts. These
four spaces are cross-related by matroid duality in a surprising way; see
the notes and Exercise 118.

thin sum   Call a family $(D_i)_{i \in I}$ of subsets of E *thin* if no edge lies in D_i for
infinitely many i. Let the *thin sum* $\sum_{i \in I} D_i$ of this family be the set of
topological   all edges that lie in D_i for an odd number of indices i. The *topological
cycle space*   *cycle space* $\mathcal{C}(G)$ of G is the subspace of its edge space $\mathcal{E}(G)$ consisting
$\mathcal{C}(G)$ of all thin sums of circuits.

We say that a given set $\mathcal{Z}$ of circuits *generates* $\mathcal{C}(G)$ if every element
generates   of $\mathcal{C}(G)$ is a thin sum of elements of $\mathcal{Z}$. For example, the topological
cycle space of the ladder in Figure 8.1.3 can be generated by all its
squares (the 4-element circuits), or by the infinite circuit consisting of
all horizontal edges and all squares but one. Similarly, the 'wild circuit'
of Figure 8.6.1 is the thin sum of all the finite face boundaries of that
graph, which thus generate it.

finitary   Let us use $\mathcal{C}_{\mathrm{fin}}(G)$ to denote the *finitary cycle space* of G as defined
cycle space   in Chapter 1.9: the (finite) sums of its finite circuits. Clearly $\mathcal{C}_{\mathrm{fin}}(G) \subseteq$
$\mathcal{C}_{\mathrm{fin}}(G)$ $\mathcal{C}(G)$. We shall see later that $\mathcal{C}_{\mathrm{fin}}(G)$ contains all the finite elements
of $\mathcal{C}(G)$, but this is not obvious from the definitions; see Exercise 115.
When G is finite, however, clearly $\mathcal{C}_{\mathrm{fin}}(G) = \mathcal{C}(G)$.

As shown in Chapter 1.9, a finite set of edges of G lies in $\mathcal{C}_{\mathrm{fin}}(G)$ if
and only if it meets every cut of G evenly, and the fundamental circuits of
any ordinary spanning tree generate $\mathcal{C}_{\mathrm{fin}}(G)$ by finite sums: just copy the
proofs given there. For $\mathcal{C}(G)$ we have the following topological analogue:

Theorem 8.7.1. *The following statements are equivalent for every set D of edges of a locally finite connected graph G:*

(i) $D \in \mathcal{C}(G)$;

(ii) D *meets every finite cut F of G in an even number of edges;*

(iii) D *is a thin sum of fundamental circuits of any topological spanning tree of G.*

Proof. The implication (iii)→(i) holds by definition of $\mathcal{C}(G)$ and the fact that G has a topological spanning tree (Lemma 8.6.11).

Let us prove (i)→(ii). By assumption, D is a thin sum of circuits. Only finitely many of these can meet F, so it suffices to show that every circuit meets F evenly. This follows from Lemma 8.6.3 (i): given a circle C in $|G|$, the segments of C between its edges in F (if any) are arcs whose vertices all lie on the same side of the cut F. These sides alternate as we follow C round. Therefore, there is an even number of such arcs, and hence also of edges that C has in F.

It remains to prove (ii)→(iii). Write C_e for the fundamental circuit of an edge $e \notin E(T)$, and D_f for the fundamental cut of an edge $f \in E(T)$. Recall that, by Lemma 8.6.3 (ii), these D_f are finite cuts. We show that

$$D = \sum_{e \in D \smallsetminus E(T)} C_e. \tag{$*$}$$

This sum is well defined: since $f \in C_e \Leftrightarrow e \in D_f$ and fundamental cuts are finite, the C_e in this sum form a thin family. To prove $(*)$ we show that $D' := D + \sum_{e \in D \smallsetminus E(T)} C_e = \emptyset$.

Note first that $D' \subseteq E(T)$: any chord of T that lies in D also lies in exactly one of the C_e in the sum. Hence any $f \in D'$ is the unique edge of T, and hence of D', in the finite cut D_f, giving $|D' \cap D_f| = 1$. This is a contradiction, since D meets D_f evenly by (ii), and every C_e does by Lemma 8.6.3. $\qquad\square$

Corollary 8.7.2. *$\mathcal{C}(G)$ is generated by finite circuits.*

Proof. Apply Theorem 8.7.1 with the closure of a normal spanning tree, which is a topological spanning tree by Lemma 8.6.8. $\qquad\square$

Our second aim in this section is to prove the analogue of Proposition 1.9.1 (ii) for the topological cycle space: that its elements D are not only thin sums but even disjoint unions of circuits. For finite graphs, it was easy to find these circuits greedily: we would 'follow the edges of D round' until a circuit was found, delete it, and repeat.

This will still be our overall strategy when G is infinite. But it is no longer straightforward now to isolate a single circuit from D. For example, without using our knowledge that the edge set D of the wild circle in the graph G of Figure 8.6.1 is a circuit, we can see at once that

(8.2.4)
(8.6.3)
(8.6.7)

(8.2.4)

it must lie in $\mathcal{C}(G)$: it is the thin sum of all the finite circuits bounding a face. Our proof must therefore be able to 'decompose' D into disjoint circuits. Since D itself is the only circuit contained in D, the proof thus has to reconstruct the complicated wild circle just from the information that $D \in \mathcal{C}(G)$. And it has to do so generically, without appealing to the special structure of this particular graph.

Theorem 8.7.3. *For every locally finite graph G, every element of $\mathcal{C}(G)$ is a disjoint union of circuits.*

(1.9.1)
(8.1.2)
(8.6.4)

Proof. We may assume that G is connected, and hence countable. Let $D \in \mathcal{C}(G)$ be given, and enumerate its edges. We inductively construct a sequence of disjoint circuits $C \subseteq D$ each containing the smallest edge in our enumeration of D that is not yet contained in the circuits constructed before. Then all these circuits will form the desired partition of D.

D', e — Suppose we have already constructed finitely many disjoint circuits all contained in D. Deleting these edges from D leaves a set D' of edges that is again in $\mathcal{C}(G)$; let e be its smallest edge in our enumeration of D. We shall find a topological path π between the endvertices of e in the standard subspace that $D' \smallsetminus \{e\}$ spans in $|G|$. By Lemma 8.6.4, the image of π will contain an arc A between these vertices, and $A \cup e$ will be the circle defining our next circuit.

$e = v_0 v_1$ — Enumerate the vertices of G as $v_0, v_1, \ldots$, with $e = v_0 v_1$. Let $S_n := \{v_0, \ldots, v_n\}$. For each $n \geqslant 1$, let G_n be the finite multigraph obtained from G by contracting every component of $G - S_n$ to a vertex, deleting

G_n — any loops but keeping parallel edges that arise in the contraction. Note that both $V(G_n)$ and $E(G_n)$ are finite, and that $G[S_n] \subseteq G_n$. Let v'_n

v'_n, V_n — denote the vertex of $G_{n-1} - S_{n-1}$ whose branch set V_n contains v_n.

We may think of $E(G_n)$ as a subset of $E(G)$. Then the cuts of G_n are also cuts of G. By Theorem 8.7.1, D' meets these evenly; in particular, every vertex of G_n is incident with an even number of edges in D'. Hence $D' \cap E(G_n) \in \mathcal{C}(G_n)$, by Proposition 1.9.1, so G_n contains a cycle through e that has all its edges in D'. Let P_n be the unique v_0–v_1 walk in this cycle that does not contain e and does not repeat any vertices.

Let V_n be the set of all v_0–v_1 walks in $G_n - e$ in which none of the vertices $v_0, \ldots, v_n$, and hence no edge, occurs more than once. Then $P_n \in V_n \neq \emptyset$, and V_n is finite. Every walk $W \in V_n$ with $n \geqslant 2$ induces a walk $W' \in V_{n-1}$ consisting of the edges that W has in G_{n-1}, traversed in the same order and direction.[14] Thus, W' arises from W by replacing any subwalk of vertices and edges not in G_{n-1} with v'_n. The vertices of any such subwalk of W will be v_n or vertices of $G_n - S_n$ whose branch set is contained in V_n. By the infinity lemma, there exists a choice of

---

[14] These are well defined: every edge $e \in W$ that is an edge of G_{n-1} has at least one endvertex in S_{n-1}, which either precedes it in W or follows it. In W', this vertex will likewise precede or follow e, respectively.

walks $W_n \in \mathcal{V}_n$ such that $W_n' = W_{n-1}$ for all $n \geq 2$.

Our next aim is to turn these walks W_n into topological paths $\pi_n \colon [0,1] \to |G_n|$ that traverse them from v_0 to v_1 and reflect their compatibility. We shall define these π_n for $n = 1, 2, \ldots$ in turn, as follows.

For $n = 1$, note that W_1 has exactly two edges: at least two, because e has no parallel edge, and at most two, because every edge of G_1 is adjacent to either v_0 or v_1. Let π_1 map $[0, \frac{1}{3}]$ onto the first edge, $[\frac{1}{3}, \frac{2}{3}]$ to the unique inner vertex of W_1, and $[\frac{2}{3}, 1]$ onto the second edge.

For $n \geq 2$, assume inductively that π_{n-1} traverses the edges of W_{n-1} in their given order and direction, and that π_{n-1} 'pauses' at each vertex $v \in G_{n-1} - S_{n-1}$ on W_{n-1} for a non-singleton closed interval $I \subseteq [0,1]$, mapping I constantly to that vertex. (Thus, if W_{n-1} visits v five times, then $\pi_{n-1}^{-1}(v)$ is a disjoint union of five such intervals.) We start our definition of π_n by letting $\pi_n(\lambda) := \pi_{n-1}(\lambda)$ for all λ with $\pi_{n-1}(\lambda) \in |G_n|$.

Every other $\lambda \in [0,1]$ satisfies $\pi_{n-1}(\lambda) = v_n'$. These λ form a disjoint union of closed intervals, one for every occurrence of v_n' on W_{n-1}. Recall that W_n arises from W_{n-1} by replacing each occurrence of v_n' by a subwalk of W_n whose vertices are either v_n or vertices of $G_n - S_n$ whose branch set is contained in V_n. For every occurrence of v_n' on W_{n-1}, let π_n on the corresponding interval I with $\pi_{n-1}(I) = \{v_n'\}$ traverse this subwalk of W_n, once more pausing for a non-singleton interval at any vertex that this subwalk has in $G_n - S_n$.

These maps π_n tend to a limit $\pi \colon [0,1] \to |G|$, defined as follows. Let $\lambda \in [0,1]$ be given. If $\pi_n(\lambda) \in |G|$ for some n, then $\pi_m(\lambda) = \pi_n(\lambda)$ for all $m > n$, and we let $\pi(\lambda) := \pi_n(\lambda)$. Otherwise $\pi_n(\lambda) \in V(G_n) \smallsetminus S_n$ for all n; let U_n be the branch set of this vertex $u_n := \pi_n(\lambda)$ of G_n in G. By our inductive construction of the maps π_n, we have $U_1 \supseteq U_2 \supseteq \ldots$. Since U_n spans a component $C_n = C_n(\lambda)$ of $G - S_n$, we can find a ray in G that has a tail in each C_n; let $\pi(\lambda)$ be the end ω of this ray. Note that ω, and hence $\pi(\lambda)$, is well defined: every end $\omega' \neq \omega$ is separated from ω by some S_n, and then fails to have a ray in C_n.

For a proof that π is our desired topological v_0–v_1 path in $|G|$, we need to check continuity at every λ. If $\pi(\lambda) = \pi_n(\lambda)$ for some n, then π agrees with π_n also in a small neighbourhood of λ, so this follows from the continuity of π_n. Otherwise $\pi(\lambda)$ is an end, ω say. Then ω has a neighbourhood basis in $|G|$ consisting of open sets $\hat{C}_\epsilon(S_n, \omega)$. Here $C(S_n, \omega)$ is the component $C_n(\lambda)$ defined earlier, since ω has a ray in it.

Now λ is an inner point of an interval $I \subseteq [0,1]$ which π_n maps to the vertex $u_n = \pi_n(\lambda)$. By construction, $\pi(I) \subseteq \overline{C_n(\lambda)} \subseteq \hat{C}_\epsilon(S_n, \omega)$, completing our continuity proof for π. $\square$

Corollary 8.7.4. *$\mathcal{C}(G)$ is closed under infinite thin sums.*

Proof. Consider a thin sum $\sum_{i \in I} D_i$ of elements of $\mathcal{C}(G)$. By Theorem 8.7.3, each D_i is a disjoint union of circuits. Together, these form a thin family, whose sum lies in $\mathcal{C}(G)$ and equals $\sum_{i \in I} D_i$. $\square$

8.8 Infinite graphs as limits of finite ones

In the last section we saw how the space $|G|$, for a locally finite graph G, seems to appear as a 'limit' of the finite minors G_n of G obtained by contracting the components left on deleting its first n vertices. We now make this relationship between $|G|$ and the G_n more formal. Clarifying this can help a lot with transferring theorems for finite graphs to infinite ones—which, after all, is the idea behind considering $|G|$ in the first place.

directed   Let $(P, \leqslant)$ be a *directed* partially ordered set, one such that for all p, q there exists an r such that $p \leqslant r$ and $q \leqslant r$. A subset $Q \subseteq P$ is *cofinal*   *cofinal* in P if for every $p \in P$ there exists some $q \in Q$ with $p \leqslant q$.

For every $p \in P$ let X_p be a compact Hausdorff topological space; later, these will represent finite graphs. Assume that we have continuous *inverse* maps $f_{qp} \colon X_q \to X_p$ for all $q > p$, which are *compatible* in that, whenever *system* $r > q > p$, we have $f_{qp} \circ f_{rq} = f_{rp}$. The family $\mathcal{X} = (X_p \mid p \in P)$, *bonding* together with these *bonding maps* f_{qp}, is called an *inverse system*. *maps*

The set X of all $x = (x_p \mid p \in P)$ with $x_p \in X_p$ and $f_{qp}(x_q) = x_p$ for all $p < q$ in P is the *inverse limit* $X = \varprojlim \mathcal{X}$ of $\mathcal{X}$. We give it the *inverse limit* subspace topology from the product space $\prod_{p \in P} X_p$ which, like the X_p, is Hausdorff and compact by Tychonoff's theorem.

The space $X = \varprojlim \mathcal{X}$ is the intersection, over all $q \in P$, of the sets $X_{<q}$ of all $(x_p \mid p \in P) \in \prod_p X_p$ that satisfy $f_{qp}(x_q) = x_p$ for all $p < q$. Using the fact that the X_p are Hausdorff and the maps f_{qp} are continuous, one can show that these subsets $X_{<q}$ of $\prod_p X_p$ are closed. Thus, $X = \bigcap_{q \in P} X_{<q}$ is closed in the compact space $\prod_p X_p$, and therefore compact.

As P is directed, the sets $X_{<q}$ have the finite intersection property, as long as the X_p are non-empty. Then $X = \bigcap_q X_{<q}$ is also non-empty:

Lemma 8.8.1. $X = \varprojlim (X_p \mid p \in P)$ *is a compact Hausdorff space. It is non-empty if* $X_p \neq \emptyset$ *for all* $p \in P$.   □

G, V, E, Ω Given a graph $G = (V, E, \Omega)$, consider as $P = P(G)$ the set of all $P(G), P$ finite partitions of V with only finitely many cross-edges. Letting $p \leqslant q$ whenever q refines p makes P into a directed partially ordered set. For each p, let G/p be the finite multigraph on p whose edges are the cross-G/p edges of p.[15] The vertices of G/p that are non-singleton partition classes *dummy* are its *dummy vertices*. The other vertices of G/p, those of the form $\{v\}$, *vertices* we consider to be vertices of G and refer to them as v.

On the compact spaces $X_p := |G/p|$ we have compatible quotient X_p, f_{qp} maps $f_{qp} \colon X_q \to X_p$ for $q > p$ which send the vertices of G/q to the vertices of G/p that contain them as subsets; which are the identity on the edges of G/q that are also edges of G/p; and which send any other edge of G/q

---

[15] If the partition classes $U \in p$ are connected in G, then G/p is the minor of G obtained by contracting them. But we do not require them to be connected.

to the dummy vertex of G/p that contains both its endvertices in G/q. Let

$$\|G\| := \varprojlim \, (\, X_p \mid p \in P \,),$$

$\|G\|$

with these f_{qp} as bonding maps.

Theorem 8.8.2. *If G is locally finite and connected, then $\|G\|$ is homeomorphic to $|G|$.*

Proof. As $\|G\|$ is compact and $|G|$ is Hausdorff, it suffices to construct a continuous bijection $\sigma \colon \|G\| \to |G|$. Let $x = (\, x_p \mid p \in P \,) \in \|G\|$ be given.

If there exists $p \in P$ such that x_p is not a dummy vertex of G/p, then $x_p \in |G| \setminus \Omega$ and we let $\sigma(x) := x_p$. To see that this is well defined, consider two such points x_p and $x_{p'}$ and pick $q > p, p'$. Then x_q is not a dummy vertex either, and $x_p = x_q = x_{p'}$ by the definition of f_{qp} and $f_{qp'}$.

Suppose now that x_p is a dummy vertex for every p. For every $n \in \mathbb{N}$ let S_n be the set of the first n vertices of G in some fixed enumeration, and let $p_n \in P$ consist of the vertices in S_n as singleton partition classes and the vertex sets of the components of $G - S_n$ as the remaining partition classes. This sequence $p_0, p_1, \ldots$ is cofinal in P, since every $p \in P$ is refined by every p_n with n large enough that all the cross-edges of p have their endvertices in S_n.

As $f_{qp}(x_q) = x_p$ whenever $p = p_m < p_n = q$, the connected vertex sets $U_n = x_{p_n}$ form a descending sequence $U_0 \supseteq U_1 \supseteq \ldots$. It is straightforward to construct a ray R in G that has a tail in $G[U_n]$ for every n. Let ω be the end of R.

For every $p \in P$ the set $U = x_p$ contains every U_n with $p < p_n$ as a subset. As the p_n are cofinal in P, every $G[x_p]$ thus contains a tail of R. Conversely, for every end $\omega' \neq \omega$ there is an n such that $G[U_n]$ contains no ray from ω'. Thus, ω is the unique end of G that has a ray in $G[x_p]$ for every $p \in P$. Let $\sigma(x) := \omega$. This completes the definition of σ.

To see that σ is injective, consider distinct points $x, x' \in \|G\|$, differing in their components $x_p \neq x'_p$ say. If p can be chosen so that one of these is not a dummy vertex of G/p, then clearly $\sigma(x) \neq \sigma(x')$. Otherwise $U = x_p$ and $U' = x'_p$ are disjoint vertex sets in G separated by finitely many edges, and $\sigma(x)$ is an end with a ray in $G[U]$ while $\sigma(x')$ is an end with a ray in $G[U']$. Thus again, $\sigma(x) \neq \sigma(x')$.

To see that σ is surjective, let $x \in |G|$ be given. If x is not an end, choose $p(x) \in P$ so as to contain the vertex x, or the endvertices of the edge containing x, as singleton partition classes. For every $q \geqslant p(x)$ in P let $x_q := x$, and for every $p' < q$ for some such q let $x_{p'} := f_{qp'}(x)$. Then $(\, x_p \mid p \in P \,)$ is a well-defined point in $\|G\|$ which σ maps to x.

If x is an end, it has a ray in $G[x_p]$ for exactly one dummy vertex x_p of G/p for every $p \in P$. These satisfy $f_{qp}(x_q) = x_p$ whenever $p < q$, so $(\, x_p \mid p \in P \,)$ is a point in $\|G\|$ which σ maps to x.

Let us show that σ is continuous at every point $x = (x_p \mid p \in P)$
of $\|G\|$. If $\sigma(x)$ is not an end, there exists some $p(x) \in P$ such that $\sigma(x) =$
$x_{p(x)}$, which is a point in $X_{p(x)}$ but not a dummy vertex. Then every
basic open neighbourhood O of $\sigma(x)$ in $|G|$ is also a basic neighbourhood
of this same point $x_{p(x)}$ in $X_{p(x)}$. Then the set $\prod_{p \in P} O_p$ with $O_{p(x)} = O$
and $O_p = X_p$ for all $p \neq p(x)$ is a basic open neighbourhood of x
in $\prod_p X_p$. Its intersection with $\|G\|$ is an open neighbourhood of x
in $\|G\|$ which σ maps to O.

If $\sigma(x)$ is an end, ω say, consider any basic open neighbourhood
$O = \hat{C}_\epsilon(S, \omega)$ of ω in $|G|$. Let $p(\omega) \in P$ be the partition of V into the
vertex sets of the components of $G - S$ and the singletons in S. Then
$V(C)$ is a dummy vertex of $G/p(\omega)$; call it $x_{p(\omega)}$. Let $O_{p(\omega)} \subseteq X_{p(\omega)}$
consist of $x_{p(\omega)}$ and the inner points in O of any C–S edges; these are
also points of $X_{p(\omega)}$. As earlier, x has a basic open neighbourhood $\prod_p O_p$
in $\prod_p X_p$ with $O_p = X_p$ for all $p \neq p(\omega)$, whose intersection with $\|G\|$
maps to O under σ. $\square$

Note that our proof did not use that $|G|$ is compact: we reobtain
Proposition 8.6.1 as a corollary.

In the proof of Theorem 8.8.2 we found it convenient to work with
a cofinal sequence in P instead of the entire set P. This is justified more
generally by the following easy lemma:

Lemma 8.8.3. *Let $(X_p \mid p \in P)$ be an inverse system of compact spaces,
let $Q \subseteq P$ be cofinal in P, and consider $(X_p \mid p \in Q)$ with the same
bonding maps. Mapping every point $(x_p \mid p \in P)$ to its restriction
$(x_p \mid p \in Q)$ then defines a homeomorphism from $\varprojlim (X_p \mid p \in P)$ to
$\varprojlim (X_p \mid p \in Q)$.* $\square$

By Theorem 8.8.2 and this lemma, our familiar $|G|$ for locally finite
G is the inverse limit of the finite contraction minors G_n of G defined as
in Section 8.6. Indeed, for the cofinal sequence $p_0, p_1, \ldots$ in P defined in
the proof of the theorem, we have $G_n = G/p_n$, and by the lemma $|G|$ is
the inverse limit of the corresponding compact spaces X_{p_n}.

Just like $|G|$ itself, every standard subspace X' of $X = |G|$ can be
obtained as an inverse limit of finite multigraphs. Indeed, the projections
$f_p \colon X \to X_p$ defined by $(x_p \mid p \in P) \mapsto x_p$ are continuous, so their images
$X'_p \subseteq X_p$ of X' are compact since X' is, and the f_{qp} send X'_q to X'_p. Thus,
$(X'_p \mid p \in P)$ is an inverse system with bonding maps $f'_{qp} := f_{qp} \restriction X'_q$,
and $X' = \varprojlim (X'_p \mid p \in P)$.

More typically, we would like to find a standard subspace X' with
certain desired properties—for example, a topological spanning tree. We
can then try to construct some X'_p whose inverse limit is X'. It may not
be straightforward, however, to find such compatible X'_p for all $p \in P$.
Here, Lemma 8.8.3 can help: it is only necessary to find them for all

p in some cofinal $Q \subseteq P$. For example, we can construct spanning trees inductively in all the G_n by expanding a dummy vertex in the tree $T_n \subseteq G_n$ to a star in $T_{n+1} \subseteq G_{n+1}$. Then our given bonding maps $X_{p_n} \to X_{p_m}$ will map the subspace X'_{p_n} induced by T_n to that induced by T_m, and these X'_{p_n} will have a topological spanning tree in $X = |G|$ as their inverse limit. This construction is possible only because the partition classes of the p_n are connected in G; we could not perform it on all of $P(G)$.

Arcs and circles in $|G|$, or in a standard subspace, can be obtained easily by applying the following *lifting lemma* with $Y = [0,1]$ or $Y = S^1$. Let $(X_p \mid p \in P)$ be any inverse system of compact spaces, with bonding maps f_{qp} say, and let X be its inverse limit. Let Y be a topological space with continuous *compatible* maps $g_p: Y \to X_p$: maps that commute with the f_{qp} in that $g_p = f_{qp} \circ g_q$ whenever $p < q$. Let us call the family $(g_p \mid p \in P)$ *eventually injective* if for all distinct $y, y' \in Y$ there exists some $p \in P$ with $g_p(y) \neq g_p(y')$.

compatible maps

Lemma 8.8.4. *There is a unique continuous map $g: Y \to X$ that commutes with the projections $f_p: X \to X_p$ in that $g_p = f_p \circ g$ for all $p \in P$. If the g_p are eventually injective, then g is injective.* □

lifting lemma

For example, suppose we wish to find an arc in X between some points x and y. We can find a topological x–y path $g: [0,1] \to X$ by finding topological $f_p(x)$–$f_p(y)$ paths $g_p: [0,1] \to X_p$ that commute with the f_{qp}. If we can make these g_p eventually injective, then g will be injective, and its image will be the desired arc.

Similarly, if we can find compatible circles $g_p: S^1 \to X_p$ that are eventually injective, whose images contain all the vertices of G/p, and which commute with the f_{qp}, then g will define a *Hamilton circle* of G, a circle in $|G|$ that traverses every vertex.

Hamilton circle

Exercises

1.⁻ Show that a connected graph is countable if all its vertices have countable degrees.

2.⁻ Given countably many sequences $\sigma^i = s^i_1, s^i_2, \ldots$ ($i \in \mathbb{N}$) of natural numbers, find one sequence $\sigma = s_1, s_2, \ldots$ that beats every σ^i eventually, i.e. such that for every i there exists an $n(i)$ such that $s_n > s^i_n$ for all $n \geqslant n(i)$.

3.⁻ Can a countable set have uncountably many subsets whose intersections have finitely bounded size?

4.⁻ Let T be an infinite rooted tree. Show that every ray in T has an increasing tail, that is, a tail whose sequence of vertices increases in the tree-order associated with T and its root.

5.⁻ Let G be an infinite graph and $A, B \subseteq V(G)$. Show that if no finite set of vertices separates A from B in G, then G contains an infinite set of disjoint A–B paths.

6.⁻ In Proposition 8.1.1, the existence of a spanning tree was proved using Zorn's lemma 'from below', to find a maximal acyclic subgraph. For finite graphs, one can also use induction 'from above', to find a minimal spanning connected subgraph. What happens if we apply Zorn's lemma 'from above' to find such a subgraph?

For the next two exercises it may help to consider the cycle space of the given graph, defined as for finite graphs in Chapter 1.9.

7.⁻ Show that if a graph has a spanning tree with infinitely many chords then all its spanning trees have infinitely many chords.

8. Show that if a graph contains infinitely many distinct cycles then it contains infinitely many edge-disjoint cycles.

9. Let G be a countable infinitely connected graph. Show that G has, for every $k \in \mathbb{N}$, an infinitely connected spanning subgraph of girth at least k.

10. Construct, for any given $k \in \mathbb{N}$, a planar k-connected graph. Can you construct one whose girth is also at least k? Can you construct an infinitely connected planar graph?

11. Theorem 8.1.3 implies that there exists an $\mathbb{N} \to \mathbb{N}$ function f_χ such that, for every $k \in \mathbb{N}$, every infinite graph of chromatic number at least $f_\chi(k)$ has a finite subgraph of chromatic number at least k. (E.g., let f_χ be the identity on $\mathbb{N}$.) Find similar functions f_δ and f_κ for the minimum degree and connectivity, or show that no such functions exist.

12. Let $k \in \mathbb{N}$, and let a, b be two vertices in a graph G.

   (i) Show that there are only finitely many minimal a–b separators of order k, and only finitely many minimal a–b cuts of order k. (A cut is an a–b cut if it separates a from b.)

   (ii) Deduce that every edge of G lies in only finitely many bonds of k edges.

13. Use the infinity lemma to show that a rayless connected graph of minimum degree d has a finite subgraph of minimum degree d.

14. A theorem of Halin says that every graph of chromatic number $\alpha \geqslant \aleph_0$ contains a TK^β for every cardinal $\beta < \alpha$. Prove this for $\alpha = \aleph_0$.

15.⁻ Prove Theorem 8.1.3 for arbitrary graphs using the generalized infinity lemma from Appendix A.

16. Give a proof of Theorem 8.1.3 for countable graphs that is based on the fact that, in this case, the topological space X defined in the third proof of the theorem is sequentially compact. (Thus, every infinite sequence of points in X has a convergent subsequence: there is an $x \in X$ such that every neighbourhood of x contains a tail of the subsequence.)

17.$^{-}$ Extend Nash-Williams's tree covering theorem (2.4.3) to infinite graphs.

18.$^{+}$ Extend the packing-covering theorem (2.4.4) to infinite graphs.

19.$^{+}$ For every $k \in \mathbb{N}$, construct a k-connected locally finite graph such that the deletion of the edge set of any cycle disconnects that graph. Deduce that the tree packing theorem (2.4.1) of Nash-Williams and Tutte fails for infinite graphs.

(Hint. Start with a k-connected finite graph G_0. If G_0 has a cycle C such that deleting $E(C)$ does not disconnect G_0, graft some more copies of G_0 on to $E(C)$ to give C that property. Continue inductively.)

20. Derive the generalized infinity lemma and the compactness principle in Appendix A from each other.

21. In the text, the unfriendly partition conjecture is proved for locally finite graphs, using the infinity lemma.

    (i) Give an alternative proof using the compactness principle from Appendix A.

    (ii) The proof in the text, by the infinity lemma, required a modification of the statement. Is this still necessary? Which step in the proof using the compactness principle reflects the requirement in the infinity lemma that every admissible partial solution must induce an admissible solution on a smaller substructure? Where is the local finiteness used?

22. (i) Prove the unfriendly partition conjecture for countable graphs with all degrees infinite.

    (ii) Can you adapt the proof to cover also those countable graphs that have finitely many vertices of finite degree?

23. Rephrase Gallai's partition theorem of Exercise 46, Chapter 1, in terms of degrees, and extend the equivalent version to locally finite graphs.

24. Prove Theorem 8.4.8 for locally finite graphs. Does your proof extend to arbitrary countable graphs?

25. Extend the marriage theorem to locally finite graphs, but show that it fails for countable graphs with infinite degrees.

26. Show that every locally finite factor-critical graph is finite.

27.$^{+}$ Show that a locally finite graph G has a 1-factor if and only if, for every finite set $S \subseteq V(G)$, the graph $G - S$ has at most $|S|$ odd (finite) components. Find a counterexample that is not locally finite.

28.$^{+}$ Extend Kuratowski's theorem to countable graphs.

29.$^{-}$ A vertex $v \in G$ is said to $dominate$ an end ω of G if any of the following three assertions holds; show that they are equivalent.

    (i) For some ray $R \in \omega$ there is an infinite v–$(R - v)$ fan in G.

    (ii) For every ray $R \in \omega$ there is an infinite v–$(R - v)$ fan in G.

    (iii) No finite subset of $V(G - v)$ separates v from a ray in ω.

30. Show that a graph G contains a $TK^{\aleph_0}$ if and only if some end of G is dominated by infinitely many vertices.

31.[+] Let G be a *finitely separable* graph, one in which any two vertices can be separated by finitely many edges.

   (i) Show that any two rays in G that cannot be separated by finitely many edges are dominated by a common vertex.

   (ii) Is the assumption of finite separability necessary for (i) to hold?

32. Construct a countable graph with uncountably many thick ends. Can you find a locally finite such graph?

33. Show that a locally finite connected vertex-transitive graph has exactly 0, 1, 2 or infinitely many ends.

34.[+] Show that the automorphisms of a graph $G = (V, E)$ act naturally on its ends, i.e., that every automorphism $\sigma: V \to V$ can be extended to a map $\sigma: \Omega(G) \to \Omega(G)$ such that $\sigma(R) \in \sigma(\omega)$ whenever R is a ray in an end ω. Prove that, if G is connected, every automorphism σ of G fixes a finite set of vertices or an end. If σ fixes no finite set of vertices, can it fix more than one end? More than two?

35.[−] Show that a locally finite spanning tree of a graph G contains a ray from every end of G.

36. A ray in a graph *follows* another ray if the two have infinitely many vertices in common. Show that if T is a normal spanning tree of G then every ray of G follows a unique normal ray of T.

37. Use normal spanning trees to show that a countable connected graph has either countably many or continuum many ends.

38. Show that the following assertions are equivalent for connected countable graphs G.

   (i) G has a locally finite spanning tree.

   (ii) For no finite separator $X \subseteq V(G)$ does $G - X$ have infinitely many components.

39. Show that every (countable) planar 3-connected graph has a locally finite spanning tree.

40. Prove the following infinite version of the Erdős-Pósa theorem: an infinite graph G either contains infinitely many disjoint cycles or it has a finite set Z of vertices such that $G - Z$ is a forest.

41. Let G be a connected graph. Call a set $U \subseteq V(G)$ *dispersed* if every ray in G can be separated from U by a finite set of vertices. (In the topology of Section 8.6, these are precisely the closed subsets of $V(G)$.)

   (i) Prove that G has a normal spanning tree if and only if $V(G)$ is a countable union of dispersed sets.

   (ii) Deduce that if G has a normal spanning tree then so does every connected minor of G.

42.[+] Prove Theorem 8.2.5 (ii).

43. (i) Prove that if a given end of a graph contains k disjoint rays for every $k \in \mathbb{N}$ then it contains infinitely many disjoint rays.

(ii)[+] Prove that if a given end of a graph contains k edge-disjoint rays for every $k \in \mathbb{N}$ then it contains infinitely many edge-disjoint rays.

44. Prove that if a graph contains k disjoint double rays for every $k \in \mathbb{N}$ then it contains infinitely many disjoint double rays.

45. Show that, in the ubiquity conjecture, the host graphs G considered can be assumed to be locally finite too.

46. Show that the modified comb below is not ubiquitous with respect to the subgraph relation. Does it become ubiquitous if we delete its 3-star on the left?

47. Imitate the proof of Theorem 8.2.6 to find a function $f: \mathbb{N} \to \mathbb{N}$ such that whenever an end ω of a graph G contains $f(k)$ disjoint rays, G contains a subdivision of the $k \times \mathbb{N}$ hexagonal grid whose rays all belong to ω.

48. Show that there is no universal locally finite connected graph for the subgraph relation.

49. Construct a universal locally finite connected graph for the minor relation. Is there one for the topological minor relation?

50.[−] Show that each of the following operations performed on the Rado graph R leaves a graph isomorphic to R:

   (i) taking the complement, i.e. changing all edges into non-edges and vice versa;

   (ii) deleting finitely many vertices;

   (iii) changing finitely many edges into non-edges or vice versa;

   (iv) changing all the edges between a finite vertex set $X \subseteq V(R)$ and its complement $V(R) \smallsetminus X$ into non-edges, and vice versa.

51.[−] Prove that the Rado graph is homogeneous.

52. Show that a homogeneous countable graph is determined uniquely, up to isomorphism, by the class of (the isomorphism types of) its finite subgraphs.

53. Recall that subgraphs $H_1, H_2, \ldots$ of a graph G are said to *partition* G if their edge sets form a partition of $E(G)$. Show that the Rado graph can be partitioned into any given countable set of countable locally finite graphs, as long as each of them contains at least one edge.

54.⁻ A linear order is called *dense* if between any two elements there lies a third.

    (i) Find, or construct, a countable dense linear order that has neither a maximal nor a minimal element.

    (ii) Show that this order is unique, i.e. that every two such orders are order-isomorphic. (Definition?)

    (iii) Show that this ordering is universal among the countable linear orders. Is it homogeneous? (Supply appropriate definitions.)

55.   Given a bijection f between $\mathbb{N}$ and $[\mathbb{N}]^{<\omega}$, let G_f be the graph on $\mathbb{N}$ in which $u, v \in \mathbb{N}$ are adjacent if $u \in f(v)$ or vice versa. Prove that all such graphs G_f are isomorphic.

56.   (for set theorists) Show that, given any countable model of set theory, the graph whose vertices are the sets and in which two sets are adjacent if and only if one contains the other as an element, is the Rado graph.

57.⁻ Given sets A, B of vertices in a graph G, show that either G contains infinitely many edge-disjoint A–B paths or there is a finite set of edges separating A from B in G.

58.   Let G be a locally finite graph. Let us say that a finite set S of vertices *separates* two ends ω and ω' if $C(S, \omega) \neq C(S, \omega')$. Use Proposition 8.4.1 to show that if ω can be separated from ω' by $k \in \mathbb{N}$ but no fewer vertices, then G contains k disjoint double rays each with one tail in ω and one in ω'. Is the same true for all graphs that are not locally finite?

59.⁺ Prove the following more structural version of Exercise 43 (i). Let ω be an end of a graph G. Show that either G contains a $TK^{\aleph_0}$ with all its rays in ω, or there are disjoint finite sets $S_0, S_1, \ldots$ such that, if C_i is the component of $G - (S_0 \cup S_i)$ that contains a tail of every ray in ω, we have for all $i < j$ that $C_i \supsetneq C_j$ and $G[S_i \cup C_i]$ contains $|S_i|$ disjoint S_i–S_{i+1} paths for all $i \geqslant 1$.

60.⁺ Is there a planar $\aleph_0$-regular graph all whose ends have infinite vertex-degree?

61.⁻ Let A, B be two vertex sets in a locally finite connected graph G. Can there be an infinite sequence $\mathcal{P}_1, \mathcal{P}_2, \ldots$ of disjoint A–B paths such that each $\mathcal{P}_{n+1}$ arises from $\mathcal{P}_n$ by applying an alternating walk, and such that some edge $e \in G$ lies in $E[\mathcal{P}_n]$ for infinitely many n but not in $E[\mathcal{P}_n]$ for infinitely many other n?

62.   Construct an example of a small limit of large waves. Can you find a locally finite one?

63.⁺ Prove Theorem 8.4.2 for trees.

64.⁺ Prove Pym's theorem (8.4.7).

65. (i)$^-$ Prove the naive extension of Dilworth's theorem to arbitrary infinite posets P: if P has no antichain of order $k \in \mathbb{N}$, then P can be partitioned into fewer than k chains. (A proof for countable P will do.)

(ii)$^-$ Find a poset that has no infinite antichain and no partition into finitely many chains.

(iii) For posets without infinite chains, deduce from Theorem 8.4.8 the following Erdős-Menger-type extension of Dilworth's theorem: every such poset has a partition $\mathcal{C}$ into chains such that some antichain meets all the chains in $\mathcal{C}$.

66. Let G be a countable graph in which for every partial matching there is an augmenting path.

(i) Find an example of G and a sequence $M_0, M_1, \ldots$ of partial matchings, each obtained from the previous as its symmetric difference with the edge set of an augmenting path, so that for every edge e of G we have $e \in M_{n+1} \setminus M_n$ for infinitely many n.

(ii) Show that for every partial matching M there exists a sequence as in (i) such that $\bigcup_m \bigcap_{n>m} M_n$ is the edge set of a 1-factor.

67. Find an uncountable graph in which every partial matching admits an augmenting path (finite or infinite) but which has no 1-factor.

68.$^-$ Let G be a countable graph whose finite subgraphs are all perfect. Show that G is weakly perfect but not necessarily perfect.

69.$^+$ Let G be the incomparability graph of the binary tree. (Thus, $V(G) = V(T_2)$, and two vertices are adjacent if and only if they are incomparable in the tree-order of T_2.) Show that G is perfect but not strongly perfect.

70.$^+$ (i) Show that the vertices of any infinite connected locally finite graph can be enumerated in such a way that every vertex is adjacent to some later vertex.

(ii) Characterize the class of all these graphs, countable but not necessarily locally finite, by their separation properties.

71. Show that a tree has a rank as defined in the second paragraph after the proof of Proposition 8.5.1 if and only if it is recursively prunable, and that it has rank α if and only if α is the maximum of its pruning labels.

72. Let G be a rayless graph, of rank α say, and let U be a finite set of vertices witnessing this, of minimal order. Show that U is unique.

73. (i) Construct a countable tree that has rank ω in the ranking of rayless graphs. Can you find one such tree that contains all the others?

(ii)$^+$ Is there a tree of rank ω that is a subtree of every such tree?

74. A graph $G = (V, E)$ is called *bounded* if for every vertex labelling $\ell: V \to \mathbb{N}$ there exists a function $f: \mathbb{N} \to \mathbb{N}$ that exceeds the labelling along any ray in G eventually. (Formally: for every ray $v_1 v_2 \ldots$ in G there exists an n_0 such that $f(n) > \ell(v_n)$ for every $n > n_0$.) Prove the following assertions:

(i) The ray is bounded.

(ii) Every locally finite connected graph is bounded.

(iii) A countable tree is bounded if and only if it contains no subdivision of the $\aleph_0$-regular tree $T_{\aleph_0}$.

75.[+] Let T be a tree with root r, and let $\leqslant$ denote the tree-order on $V(T)$ associated with T and r. Show that T contains no subdivision of the $\aleph_1$-regular tree $T_{\aleph_1}$ if and only if T has an ordinal labelling $t \mapsto o(t)$ such that $o(t) \geqslant o(t')$ whenever $t < t'$ and no more than countably many vertices of T have the same label.

76. Let G be a countable connected graph with vertices $v_0, v_1, \ldots$. For every $n \in \mathbb{N}$ write $S_n := \{v_0, \ldots, v_{n-1}\}$. Prove the following statements:

(i) For every end ω of G there is a unique sequence $C_0 \supseteq C_1 \supseteq \ldots$ of components C_n of $G - S_n$ such that $C_n = C(S_n, \omega)$ for all n.

(ii) For every infinite sequence $C_0 \supseteq C_1 \supseteq \ldots$ of components C_n of $G - S_n$ there exists a unique end ω such that $C_n = C(S_n, \omega)$ for all n.

77. Let G be a graph, $U \subseteq V(G)$, and $R \in \omega \in \Omega(G)$. Show that G contains a comb with spine R and teeth in U if and only if $\omega \in \overline{U}$.

78. Given graphs $H \subseteq G$, let $\eta \colon \Omega(H) \to \Omega(G)$ assign to every end of H the unique end of G containing it as a subset (of rays). For the following questions, assume that H is connected and $V(H) = V(G)$.

(i) Show that η need not be injective. Must it be surjective?

(ii) Investigate how η relates the subspace $\Omega(H)$ of $|H|$ to its image in $|G|$. Is η always continuous? Is it open? Do the answers to these questions change if η is known to be injective?

(iii) A spanning tree is called *end-faithful* if η is bijective, and *topologically end-faithful* if η is a homeomorphism. Show that every connected countable graph has a topologically end-faithful spanning tree.

The *end space* of a graph G is the subspace $\Omega(G)$ of $|G|$.

79. Consider the end space Ω of the binary tree T_2 shown in Figure 8.1.4, in which its vertices are the finite 0–1 sequences.

(i) Show that Ω is homeomorphic to $2^{\mathbb{N}}$, where $2 = \{0, 1\}$ carries the discrete topology and $2^{\mathbb{N}}$ the product topology.

(ii) Identify in Ω every two ends whose infinite binary sequences encode the same rational. Show that the resulting quotient space of Ω is homeomorphic to the real interval $[0, 1]$.

80. Above every horizontal edge of the plane graph shown in Figure 8.6.1 add infinitely many horizontal edges in the plane, so as to turn every pair of rays whose associated 0–1 sequences define the same rational number into a ladder. Prove or disprove that the end space of the resulting graph is homeomorphic to $[0, 1]$.

81. A compact metric space is a *Cantor set* if the singletons are its only connected subsets and every point is an accumulation point.

   (i) Characterize the trees whose end space is a Cantor set.

   (ii) Show that the end space of a connected locally finite graph is a subset of a Cantor set.

82. (i) Show that if H is a contraction minor of G with finite branch sets, then the end spaces of H and G are homeomorphic.

   (ii) Let T_n denote the *n-ary tree*, the rooted tree in which every vertex has exactly n successors. Show that all these trees have homeomorphic end spaces.

83. Give an independent proof of Proposition 8.6.1 using sequential compactness and the infinity lemma.

84.$^+$ (for topologists) In a locally compact, connected, and locally connected Hausdorff space X, consider sequences $U_1 \supseteq U_2 \supseteq \ldots$ of open, non-empty, connected subsets with compact frontiers such that $\bigcap_{i \in \mathbb{N}} \overline{U}_i = \emptyset$. Call such a sequence *equivalent* to another such sequence if every set of one sequence contains some set of the other sequence and vice versa. Note that this is indeed an equivalence relation, and call its classes the *Freudenthal ends* of X. Now add these to the space X, and define a natural topology on the extended space $\hat{X}$ that makes it homeomorphic to $|X|$ if X is a graph, by a homeomorphism that is the identity on X.

85.$^+$ Let G be a connected countable graph that is not locally finite. Show that $|G|$ is not compact, but that $\Omega(G)$ is compact if and only if for every finite set $S \subseteq V(G)$ only finitely many components of $G - S$ contain a ray.

86.$^+$ Let G be a connected graph. Assuming that G has a normal spanning tree, define a metric on $|G|$ that induces its usual topology. Conversely, use Exercise 41 to show that if $V \cup \Omega \subseteq |G|$ is metrizable then G has a normal spanning tree.

87. Find a graph G for which $|G|$ is not metrizable.

   (Hint. Rather than thinking of metrics directly, recall some properties of metric spaces, and construct a graph G without such a property.)

A topological space X is *locally connected* if for every $x \in X$ and every neighbourhood U of x there is an open connected neighbourhood $U' \subseteq U$ of x. A *continuum* is a compact, connected Hausdorff space. By a theorem of general topology, every locally connected metric continuum is arc-connected.

88.$^+$ Show that, for G connected and locally finite, every connected standard subspace of $|G|$ is locally connected. Using the theorem cited above, deduce Lemma 8.6.5.

89.$^+$ Prove Lemma 8.6.6 directly, without relying on Lemma 8.6.5.

90. Let G be a locally finite graph, and X a standard subspace of $|G|$ spanned by a set of at least two edges. Show that X is a circle if and only if, for every two distinct edges $e, e' \in E(X)$, the subspace $X \setminus \mathring{e}$ is connected but $X \setminus (\mathring{e} \cup \mathring{e}')$ is disconnected.

91. Does every infinite locally finite 2-connected graph contain an infinite circuit? Does it contain an infinite bond?

92. Consider a locally finite graph.

    (i) Show that every infinite circuit meets some infinite bond in exactly one edge.

    (ii) Show that every infinite bond meets some infinite circuit in exactly one edge.

93. Show that the union of all the edges contained in an arc or circle C in $|G|$ is dense in C.

94.[+] Prove that the circle shown in Figure 8.6.1 is really a circle, by exhibiting a homeomorphism with S^1.

95. Every arc induces on its points a linear ordering inherited from $[0, 1]$. Call an arc in $|G|$ *wild* if it induces on some subset of its vertices the ordering of the rationals: between every two there lies another. Show that every arc containing uncountably many ends is wild.

96. Find a locally finite graph G with a connected standard subspace of $|G|$ that is the closure of a union of disjoint circles.

97. Show that, for G locally finite, a closed standard subspace C of $|G|$ is a circle in $|G|$ if and only if C is connected, every vertex in C is incident with exactly two edges in C, and every end in C has topological degree 2.

98. Let T be a locally finite tree. Construct a continuous map $\sigma\colon [0, 1] \to |T|$ that maps 0 and 1 to the root and traverses every edge exactly twice, once in each direction. (Formally: define σ so that every inner point of an edge is the image of exactly two points in $[0, 1]$.)

    (Hint. Define σ as a limit of similar maps σ_n for finite subtrees T_n.)

99. Let G be a connected locally finite graph. Show that the following assertions are equivalent for a spanning subgraph T of G:

    (i) $\overline{T}$ is a topological spanning tree of $|G|$;

    (ii) T is edge-maximal with $\overline{T}$ containing no circle;

    (iii) T is edge-minimal with $\overline{T}$ a connected subspace of $|G|$.

100.[−] (i) Observe that a topological spanning tree need not be homeomorphic to a tree. Is it homeomorphic to the space $|T|$ for a suitable tree T?

    (ii) Find a graph G with an ordinary spanning tree T whose closure in $|G|$ is not arc-connected.

101. Let T be an end-faithful spanning tree of a locally finite graph G. (Such trees are defined in Exercise 78.) Is $\overline{T}$ a topological spanning tree of $|G|$?

102. Let G be locally finite and connected, with vertices $v_0, v_1, \ldots$ say. Let G_n be the minor of G obtained by contracting every component of $G - \{v_0, \ldots, v_n\}$ to a vertex. Construct spanning trees T_n of G_n so that $\bigcup_n E(T_n)$ is the edge set of a topological spanning tree of G.

103. Let F be a set of edges in a locally finite connected graph $G = (V, E)$.

    (i) Show that F is a circuit if and only if F is not contained in the edge set of any topological spanning tree of G and is minimal with this property.

    (ii) Show that F is a finite bond if and only if F meets the edge set of every topological spanning tree of G and is minimal with this property.

104. Extend Exercise 43 of Chapter 1 to characterizations of the bonds, and of the finite bonds, in a locally finite connected graph.

105.$^+$ Prove a topological tree-packing theorem for standard subspaces X of locally finite graphs. Here, a *topological spanning tree of* X is a connected closed subspace of X that contains all its vertices but no circle.

106. To show that Theorem 3.2.6 does not generalize to infinite graphs with its finitary cycle space, construct a 3-connected locally finite planar graph with a separating cycle that is not a finite sum of non-separating induced cycles. Can you find an example where even infinite thin sums of finite non-separating induced cycles do not generate all separating cycles?

107.$^-$ As a converse to Theorem 8.7.1 (iii), show that the fundamental circuits of an ordinary spanning tree T of a locally finite graph G do not generate $\mathcal{C}(G)$ unless $\overline{T}$ is a topological spanning tree.

108.$^-$ Prove that the edge set of a countable graph G can be partitioned into finite circuits if G has no odd cut. Where does your argument break down if G is uncountable?

109.$^-$ Explain why Theorem 8.7.3 is needed in the proof of Corollary 8.7.4: can't we just combine the constituent sums of circuits for the D_i (from our assumption that $D_i \in \mathcal{C}(G)$) into one big family?

110. Deduce Corollary 8.7.4 from Theorem 8.7.1, not using Theorem 8.7.3.

111. If a finite set D of edges meets every finite cut of G evenly, must it also meet every infinite cut evenly?

112. Prove Theorem 8.7.3 by the method used to prove Theorem 8.6.9.

113. Prove Lemma 8.6.5 by the method used to prove Theorem 8.7.3.

For the next ten exercises, let G be a locally finite connected graph. Let $\mathcal{C} = \mathcal{C}(G)$, and define the *cut space* $\mathcal{B} = \mathcal{B}(G)$ of G as in Chapter 1.9. Note that cuts may now be infinite. Define 'generate' for cuts as for circuits, allowing infinite thin sums. Given a set $\mathcal{F} \subseteq \mathcal{E}(G)$, write $\mathcal{F}_{\mathrm{fin}} := \{ F \in \mathcal{F} : |F| < \infty \}$, $\mathcal{F}^{\perp} := \{ D \in \mathcal{E}(G) : |D \cap F| \in 2\mathbb{N} \ \forall F \in \mathcal{F} \}$ and $(\mathcal{F}_{\mathrm{fin}})^{\perp} =: \mathcal{F}_{\mathrm{fin}}^{\perp}$.

114. Show that $\mathcal{C}$ and $\mathcal{B}$ are closed in the edge space $\mathcal{E} = \{0,1\}^E$ of G if $\{0,1\}$ carries the discrete topology and $\{0,1\}^E$ the product topology.

115. Show that the definition of $\mathcal{C}_{\mathrm{fin}}$ given above conincides with that given in the text: that the finite elements of $\mathcal{C}$ are finite sums of finite circuits.

116. (i) Show that the fundamental circuits of any ordinary spanning tree of G generate $\mathcal{C}_{\mathrm{fin}}$ by finite sums, but that they need not generate $\mathcal{C}$.

(ii) Show that the fundamental cuts of any topological spanning tree of G generate $\mathcal{B}_{\mathrm{fin}}$ by finite sums, but that they need not generate $\mathcal{B}$.

117. (i)$^-$ Show that $\mathcal{B}$ is a subspace of $\mathcal{E}(G)$ generated by finite cuts.

(ii) Show that every cut is a disjoint union of bonds.

(iii)$^+$ Show that the fundamental cuts of any ordinary spanning tree of G generate $\mathcal{B}$.

(iv)$^+$ Show that $\mathcal{B}$ is closed under infinite thin sums.

118. (i)$^-$ Find in this book a proof, or sketch of a proof, for each of the following two statements: $\mathcal{C} = \mathcal{B}_{\mathrm{fin}}^{\perp}$ and $\mathcal{B} = \mathcal{C}_{\mathrm{fin}}^{\perp}$.

(ii)$^+$ Show that $\mathcal{B}^{\perp} = \mathcal{C}_{\mathrm{fin}}$ and, if G is 2-edge-connected, $\mathcal{C}^{\perp} = \mathcal{B}_{\mathrm{fin}}$.

119.$^+$ Write $\hat{\mathcal{C}}$ for the set of circuits in G, and $\hat{\mathcal{B}}$ for the set of bonds.

(i) Show that $\hat{\mathcal{C}}_{\mathrm{fin}}^{\perp} = \mathcal{C}_{\mathrm{fin}}^{\perp}$ and $\hat{\mathcal{B}}_{\mathrm{fin}}^{\perp} = \mathcal{B}_{\mathrm{fin}}^{\perp}$.

(ii) Show that every element of $\hat{\mathcal{C}}^{\perp}$ is a disjoint union of finite bonds, and that every element of $\hat{\mathcal{B}}^{\perp}$ is a disjoint union of finite circuits.

(iii) Construct 2-connected graphs with $\mathcal{C}^{\perp} \subsetneq \hat{\mathcal{C}}^{\perp}$ or $\mathcal{B}^{\perp} \subsetneq \hat{\mathcal{B}}^{\perp}$.

120. Extending Gallai's partition theorem of Exercise 46 (ii), Chapter 1, show that $E(G)$ can be partitioned into a set $C \in \mathcal{C}$ and a set $D \in \mathcal{B}$. (This strengthens Exercise 23.)

121. Let $F \subseteq E(G)$ be a set of edges.

(i)$^-$ Show that F extends to a cut if it contains no odd circuit.

(ii)$^+$ Show that F extends to some $D \in \mathcal{C}$ if it contains no odd bond.

122.$^+$ Let H be an abelian group. The group $\mathcal{C}_H$ of H-*circulations* on $|G|$ consists of the maps $\psi \colon \vec{E} \to H$ that satisfy (F1) and $\psi(X, Y) = 0$ for any finite cut $E(X, Y)$ of G. (See Chapter 6.1 for notation.) Extend Exercise 8 of Chapter 6 to $\mathcal{C}_H$, with $\mathcal{E}_H$ and $\mathcal{D}_H$ as defined there.

123. Let X be a connected standard subspace of $|G|$. Call a continuous map $\sigma \colon S^1 \to X$ a *topological Euler tour* of X if it traverses every edge in $E(X)$ exactly once. (Formally: every inner point of an edge in $E(X)$ must be the image of exactly one point in S^1.) Show that X admits a topological Euler tour if and only if $E(X) \in \mathcal{C}(G)$.

124.$^+$ An *open Euler tour* in an infinite connected graph G is a 2-way infinite walk $\ldots e_{-1} v_0 e_0 \ldots$ that contains every edge of G exactly once. Show that G contains an open Euler tour if and only if G is countable, every vertex has even or infinite degree, and any finite cut $F = E(V_1, V_2)$ with both V_1 and V_2 infinite is odd.

125. By Exercise 23 of Chapter 4, every finite 2-connected graph without a K^4 or $K_{2,3}$ minor contains a *Hamilton* cycle, one that contains all its vertices. Show that every locally finite such graph has a *Hamilton circle*, a circle in $|G|$ containing all the vertices (and ends) of G.

126.+ Extend Theorem 2.4.4 to packings and coverings of locally finite graphs with topological spanning trees in appropriate spaces obtained from $|G|$.

127. Where in the proof of Theorem 8.8.2 do we use that G is connected?

128. Use the techniques from Section 8.8 to prove that the wild circle is indeed a circle. Your proof may be informal in its handling of the wild circle graph G—use a picture rather than its formal definition.

129.+ Use the methods from Section 8.8 to prove that connected standard subspaces of $|G|$, for G locally finite, have topological spanning trees: closed connected subspaces containing all their vertices.

130.+ Consider the space $\|G\|$ for the edgeless countably infinite graph G. Have you met this space before in another guise?

Notes

There is no comprehensive monograph on infinite graph theory, but over time several surveys have been published. A relatively wide-ranging collection of survey articles can be found in R. Diestel (ed.), *Directions in Infinite Graph Theory and Combinatorics*, North-Holland 1992. (This has been reprinted as Volume 95 of the journal *Discrete Mathematics*.) Some of the articles there address purely graph-theoretic aspects of infinite graphs, while others point to connections with other fields in mathematics such as differential geometry, topological groups, or logic. A similar collection, edited by R. Diestel, B. Mohar and G. Hahn, appeared in 2011 as a special issue (vol. 311) of *Discrete Mathematics*. It includes a survey of everything to do with $|G|$, the topological space consisting of an infinite graph G and its ends: R. Diestel, Locally finite graphs with ends: a topological approach, arXiv:0912.4213.

A survey of infinite graph theory as a whole was given by C. Thomassen, Infinite graphs, in (L.W. Beineke & R.J. Wilson, eds.) *Selected Topics in Graph Theory 2*, Academic Press 1983. This also treats a number of aspects of infinite graph theory not considered in our chapter here, including problems of Erdős concerning infinite chromatic number, infinite Ramsey theory (also known as *partition calculus*), and reconstruction. The first two of these topics receive much attention also in A. Hajnal's chapter of the *Handbook of Combinatorics* (R.L. Graham, M. Grötschel & L. Lovász, eds.), North-Holland 1995, which has a strong set-theoretical flavour. Péter Komjáth is currently preparing a monograph about this kind of infinite graph theory. A specific survey on reconstruction by Nash-Williams can be found in the *Directions* volume cited above. R. Halin, Miscellaneous problems on infinite graphs, *J. Graph Theory* **35** (2000), 128–151, contains his legacy of unsolved infinite graph problems.

Halin's book, *Graphentheorie* (2nd ed.), Wissenschaftliche Buchgesellschaft 1989, is also good general reference for infinite graphs. A more specific monograph about *simplicial decompositions* of infinite graphs (see Chapter 12) is R. Diestel, *Graph Decompositions*, Oxford University Press 1990. Our Chapter 12.6 closes with two theorems about forbidden minors in infinite graphs.

When sets get bigger than countable, combinatorial set theory offers some interesting ways other than cardinality to distinguish 'small' from 'large' sets.

Among these are the use of *clubs* and *stationary sets*, of *ultrafilters*, and of *measure and category*. See P. Erdős, A. Hajnal, A. Máté & R. Rado, *Combinatorial Set Theory: partition relations for cardinals*, North-Holland 1984; W.W. Comfort & S. Negropontis, *The Theory of Ultrafilters*, Springer 1974; J.C. Oxtoby, *Measure and Category: a survey of the analogies between topological and measure spaces* (2nd ed.), Springer 1980.

Infinite matroids, whose study long lay dormant for want of a clear idea of what exactly they should be, were finally axiomatized in H. Bruhn, R. Diestel, M. Kriesell & P. Wollan, Axioms for infinite matroids, *Adv. Math.* **239** (2013), 18–46, arXiv:1003.3919. Since that paper, the theory of infinite matroids has flourished. It draws much inspiration from topological infinite graph theory as presented in Sections 8.6–8.7, but also has its own problems and agenda. Regular updates can be found in Nathan Bowler's blog at http://matroidunion.org/?author=11.

In addition to its main guiding themes, as followed in this chapter and in the sources just mentioned, infinite graph theory has a number of interesting individual results which, as yet, stand essentially by themselves. One such is a theorem of A. Huck, F. Niedermeyer and S. Shelah, Large κ-preserving sets in infinite graphs, *J. Graph Theory* **18** (1994), 413–426, which says that every infinitely connected graph G has a set S of $|G|$ vertices such that $\kappa(G - S') = \kappa(G)$ for every $S' \subseteq S$. Another is Halin's *bounded graph conjecture* and related problems. (See Exercise 74 for the definition of 'bounded' and the tree case of the conjecture.) A proof can be found in R. Diestel & I.B. Leader, A proof of the bounded graph conjecture, *Invent. math.* **108** (1992), 131–162.

König's infinity lemma, sometimes referred to as *König's lemma*, is as old as the first-ever book on graph theory, which includes it: D. König, *Theorie der endlichen und unendlichen Graphen*, Akademische Verlagsgesellschaft, Leipzig 1936. Appendix A gives a generalization of the infinity lemma to structures of any cardinality, which is still very intuitive and graph-like: the compactness theorem for inverse limits of finite sets. Compactness proofs can also come in the guise of Rado's *selection lemma*, or of Gödel's *compactness theorem* from first-order logic. These two, as well as the generalized infinity lemma, are equivalent to the compactness principle as stated in Appendix A, but stronger than the ordinary infinity lemma. They follow from Tychonoff's theorem (which is one of the many statments equivalent to the axiom of choice) but do not imply it. They do however imply the weakening of Tychonoff's theorem that we typically use in compactness proofs, namely, that spaces of the form S^X with S finite are compact in the product topology.

Theorem 8.1.3 is due to N. G. de Bruijn and P. Erdős, A colour problem for infinite graphs and a problem in the theory of relations, *Indag. Math.* **13** (1951), 371–373. The infinite analogue of the weakening of Hadwiger's conjecture that every graph of chromatic number $\alpha \geqslant \aleph_0$ contains a TK_β for every $\beta < \alpha$, whose proof for $\alpha = \aleph_0$ is asked in Exercise 14, is due to R. Halin, Unterteilungen vollständiger Graphen in Graphen mit unendlicher chromatischer Zahl, *Abh. Math. Sem. Univ. Hamburg* **31** (1967), 156–165.

Unlike for the chromatic number, a bound on the colouring number of all finite subgraphs does not extend to the whole graph by compactness. P. Erdős & A. Hajnal, On the chromatic number of graphs and set systems, *Acta Math. Acad. Sci. Hung.* **17** (1966), 61–99, proved that if every finite subgraph of G has colouring number at most k then G has colouring number at most $2k - 2$, and showed that this is best possible. However, as for finite graphs, the colouring number appears to interact better with other graph invariants than the chromatic number does; compare Theorem 8.6.2. For any cardinal κ, the graphs with colouring number at most κ were characterized by forbidden subgraphs by N. Bowler, J. Carmesin and C. Reiher, The colouring number of infinite graphs, arXiv:1512.02911.

The unfriendly partition conjecture is one of the best-known open problems in infinite graph theory, but there are few results. E.C. Milner and S. Shelah, Graphs with no unfriendly partitions, in (A. Baker, B. Bollobás & A. Hajnal, eds.), *A tribute to Paul Erdős*, Cambridge University Press 1990, construct an uncountable counterexample, but show that every graph has an unfriendly partition into three classes. (The original conjecture, which they attribute to R. Cowan and W. Emerson (unpublished), appears to have asserted for every graph the existence of a vertex partition into any given finite number of classes such that every vertex has at least as many neighbours in other classes as in its own.) Some positive results for bipartitions were obtained by R. Aharoni, E.C. Milner and K. Prikry, Unfriendly partitions of graphs, *J. Comb. Theory B* **50** (1990), 1–10. H. Bruhn, R. Diestel, A. Georgakopoulos and Ph. Sprüssel, Every rayless graph has an unfriendly partition, *Combinatorica* **30** (2010), 521–532, arXiv:0901.4858, used rankings such as defined in Section 8.5 to prove that all rayless graphs have unfriendly partitions. In 2010, Eli Berger announced a proof that every graph not containing a subdivision of an infinite complete graph has an unfriendly partition.

Theorem 8.2.4 is a special case of the result stated in Exercise 41 (i), which is due to H.A. Jung, Wurzelbäume und unendliche Wege in Graphen, *Math. Nachr.* **41** (1969), 1–22. The graphs that admit a normal spanning tree can be characterized by forbidden minors: as shown in R. Diestel & I. Leader, Normal spanning trees, Aronszajn trees and excluded minors, *J. Lond. Math. Soc.* **63** (2001), 16–32, there are two types of graphs that are easily seen not to have normal spanning trees, and one of these must occur as a minor in every graph without a normal spanning tree. Note that such a characterization is possible only because the class of graphs admitting a normal spanning tree is closed under taking connected minors—a consequence of Jung's theorem (see Exercise 41 (ii)) for which, oddly, no direct proof is known. One corollary of this characterization is the equivalence of (ii) and (iii) in Theorem 8.6.2, that a connected graph has a normal spanning tree if and only if all its minors have countable colouring number. A useful sufficient condition for the existence of a normal spanning tree is that the graph contains no subdivision of a *fat* $K^{\aleph_0}$, one in which every edge has been replaced by uncountably many parallel edges. See R. Diestel, A simple existence criterion for normal spanning trees, *Electronic. J. Comb.* **23** (2016), #P2.33, arXiv:1202.4399.

Theorems 8.2.5 and 8.2.6 are from R. Halin, Über die Maximalzahl fremder unendlicher Wege, *Math. Nachr.* **30** (1965), 63–85. Our proof of Theorem 8.2.5 (i) is due to Andreae (unpublished); the proof of Theorem 8.2.5 (ii) given in the hint for Exercise 42 is new. The analogue of Theorem 8.2.5 (ii) for double rays was proved only recently by N. Bowler, J. Carmesin and J. Pott, Edge-disjoint double rays in infinite graphs: a Halin type result, *J. Comb. Theory B* **111** (2015), 1–16.

Our proof of Theorem 8.2.6 is new. Halin's paper also includes a structure theorem for graphs that do not contain infinitely many disjoint rays. Except for a finite set of vertices, such a graph can be written as an infinite chain of rayless subgraphs each overlapping the previous in exactly m vertices, where m is the maximum number of disjoint rays (which exists by Theorem 8.2.5). These overlap sets are disjoint, and there are m disjoint rays containing exactly one vertex from each of them.

A good reference on ubiquity, including the ubiquity conjecture, is Th. Andreae, On disjoint configurations in infinite graphs, *J. Graph Theory* **39** (2002), 222–229.

Universal graphs have been studied mostly with respect to the induced subgraph relation, with numerous but mostly negative results. See G. Cherlin, S. Shelah & N. Shi, Universal graphs with forbidden subgraphs and algebraic closure, *Adv. Appl. Math.* **22** (1999), 454–491, for an overview and a model-theoretic framework for the proof techniques typically applied.

The Rado graph is probably the best-studied single graph in the literature (with the Petersen graph a close runner-up). The most comprehensive source for anything related to it (and far beyond) is R. Fraïssé, *Theory of Relations* (2nd edn.), Elsevier 2000. More accessible introductions are given by N. Sauer in his appendix to Fraïssé's book, and by P.J. Cameron, The random graph, in (R.L. Graham & J. Nešetřil, eds.): *The Mathematics of Paul Erdős*, Springer 1997, and its references.

Theorem 8.3.1 is due to P. Erdős and A. Rényi, Asymmetric graphs, *Acta Math. Acad. Sci. Hung.* **14** (1963), 295–315. The existence part of their proof is probabilistic and will be given in Theorem 11.3.5. Rado's explicit definition of the graph R was given in R. Rado, Universal graphs and universal functions, *Acta Arithm.* **9** (1964), 393–407. However, its universality and that of R^r are already included in more general results of B. Jónsson, Universal relational systems, *Math. Scand.* **4** (1956), 193–208.

Theorem 8.3.3 is due to A.H. Lachlan and R.E. Woodrow, Countable ultrahomogeneous undirected graphs, *Trans. Amer. Math. Soc.* **262** (1980), 51–94. The classification of the countable homogeneous *directed* graphs is much more difficult still. It was achieved by G. Cherlin, The classification of countable homogeneous directed graphs and countable homogeneous n-tournaments, *Mem. Am. Math. Soc.* **621** (1998), which also includes a shorter proof of Theorem 8.3.3. M. Hamann obtained, in his 2014 Habilitation at Hamburg University, the even more difficult classification of the *Connected-homogeneous digraphs*. The thesis is available on the internet and still appearing, spread over a number of papers.

Proposition 8.3.2, too, has a less trivial directed analogue: the countable directed graphs that are isomorphic to at least one of the two sides induced by any bipartition of their vertex set are precisely the edgeless graph, the random tournament, the transitive tournaments of order type ω^α, and two specific orientations of the Rado graph (R. Diestel, I. Leader, A. Scott & S. Thomassé, Partitions and orientations of the Rado graph, *Trans. Amer. Math. Soc.* **359** (2007), 2395–2405.

Theorem 8.3.4 is proved in R. Diestel & D. Kühn, A universal planar graph under the minor relation, *J. Graph Theory* **32** (1999), 191–206. It is not known whether or not there is a universal planar graph for the topological minor relation. However it can be shown that there is no minor-universal graph for embeddability in any closed surface other than the sphere; see the above paper.

When Erdős conjectured his extension of Menger's theorem is not known; C.St.J.A. Nash-Williams, Infinite graphs – a survey, *J. Comb. Theory B* **3** (1967), 286–301, cites the proceedings of a 1963 conference as its source. Its proof as Theorem 8.4.2 by R. Aharoni and E. Berger, Menger's theorem for infinite graphs, *Invent. math.* **176** (2009), 1–62, arXiv:math/0509397, came as the culmination of a long effort over many years, for the most part also due to Aharoni. Our proof of its countable case is adapted from R. Aharoni, Menger's theorem for countable graphs, *J. Comb. Theory B* **43** (1987), 303–313. The theorem has been extended to graphs with ends by H. Bruhn, R. Diestel & M. Stein, Menger's theorem for infinite graphs with ends, *J. Graph Theory* **50** (2005), 199–211.

Theorem 8.4.7 is due to J.S. Pym, A proof of the linkage theorem, *J. Math. Anal. Appl.* **27** (1969), 636–638. The short proof outlined in Exercise 64 can be found in R. Diestel & C. Thomassen, A Cantor-Bernstein theorem for paths in graphs, *Amer. Math. Monthly* **113** (2006), 161–166.

The matching theorems of Chapter 2—König's duality theorem, Hall's marriage theorem, Tutte's 1-factor theorem, and the Gallai-Edmonds matching theorem—extend essentially unchanged to locally finite graphs by compactness; see e.g. Exercises 24–27. For non-locally-finite graphs, matching theory is considerably deeper. A good survey and open problems can be found in R. Aharoni, Infinite matching theory, in the *Directions* volume cited earlier.

Most of the results and techniques for infinite matching were developed first for countable graphs, by Podewski and Steffens in the 1970s. In the 1980s, Aharoni extended them to arbitrary graphs, where things are more difficult still and additional methods are required. Theorem 8.4.8 is due to R. Aharoni, König's duality theorem for infinite bipartite graphs, *J. Lond. Math. Soc.* **29** (1984), 1–12. The proof builds on R. Aharoni, C.St.J.A. Nash-Williams & S. Shelah, A general criterion for the existence of transversals, *Proc. Lond. Math. Soc.* **47** (1983), 43–68. It is described in detail also in M. Holz, K.P. Podewski & K. Steffens, *Injective choice functions*, Lecture Notes in Mathematics **1238** Springer-Verlag 1987. Theorem 8.4.10 is due to E.C. Milner & S. Shelah, Sufficiency conditions for the existence of transversals, *Can. J. Math.* **26** (1974), 948–961; a short proof was given by H. Tverberg, On the Milner-Shelah condition for transversals, *J. Lond. Math. Soc.* **13** (1976), 520–524. Theorem 8.4.11 can be derived from the material in K. Steffens, Matchings in countable graphs, *Can. J. Math.* **29** (1977), 165–168. Theorem 8.4.12 is due to R. Aharoni, Matchings in infinite graphs, *J. Comb. Theory B* **44** (1988), 87–125; a shorter proof was given by F. Niedermeyer and K.P. Podewski, Matchable infinite graphs, *J. Comb. Theory B* **62** (1994), 213–227.

The recursive ranking of rayless graphs defined in Section 8.5 was introduced by R. Schmidt, Ein Ordnungsbegriff für Graphen ohne unendliche Wege mit einer Anwendung auf n-fach zusammenhängende Graphen, *Arch. Math.* **40** (1983), 283–288. The paper offers an interesting structure theory for rayless graphs including applications, such as to reconstruction.

The topology on G introduced in Section 8.6 coincides, when G is locally finite, with the usual topology of a 1-dimensional CW-complex. The space $|G|$ is its *Freudenthal compactification*, as suggested by H. Freudenthal, Über die Enden topologischer Räume und Gruppen, *Math. Zeit.* **33** (1931), 692–713; see Exercise 84. Although $|G|$ has been known for so long and is a familiar object in geometric group theory, its fundamental group was characterized only recently by R. Diestel & Ph. Sprüssel, The fundamental group of a locally finite graph with ends, *Adv. Math.* **226** (2011), 2643–2675, arXiv:0910.5647.

For graphs that are not locally finite, the graph-theoretical notion of an end is more general than the topological one; see R. Diestel & D. Kühn, Graph-theoretical versus topological ends of graphs, *J. Comb. Theory B* **87** (2003), 197–206. For such graphs it can be natural to consider a coarser topology on $|G|$, obtained by taking as basic open sets $\hat{C}_\epsilon(S, \omega)$ only those with $\epsilon = 1$. Under this topology, $|G|$ is no longer Hausdorff, because every vertex dominating an end ω will lie in the closure of every $\hat{C}(S, \omega)$. But $|G|$ can now be compact, and it can have a natural quotient space—in which ends are identified with vertices dominating them and rays converge to vertices—that is both Hausdorff and compact. For details see R. Diestel, On end spaces and spanning trees, *J. Comb. Theory B* **96** (2006), 846–854, where also Theorem 8.6.2 is proved. The fact that $|G|$, and therefore also its closed subspace Ω, is normal also for non-locally-finite G was proved by Ph. Sprüssel, End spaces are normal, *J. Comb. Theory B* **98** (2008), 798–804. Further topological aspects of the subspaces Ω and $V \cup \Omega$ were studied by Polat; see e.g. N. Polat, Ends and multi-endings I & II, *J. Comb. Theory B* **67** (1996), 56–110.

Arbitrary graphs can be compactified by adding their $\aleph_0$-tangles. Tangles are defined in Chapter 12; the $\aleph_0$-tangles of an infinite graph include its ends. For G locally finite, its tangle-compactification is precisely $|G|$; see R. Diestel, Ends and tangles, Special volume in memory of Rudolf Halin, to appear in *Abh. Math. Sem. Univ. Hamburg*.

Lemma 8.6.4, that the image of any x–y path in a Hausdorff space contains an x–y arc, is from D.W. Hall and G.L. Spencer, Elementary Topology. A locally finite graph G with a connected subset of $|G|$ that is not arc-connected has been constructed by A. Georgakopoulos, Connected but not path-connected subspaces of infinite graphs, *Combinatorica* **27** (2007), 683–698.

The construction of highly connected graphs that are left disconnected by the deletion of any finite circuit (Exercise 19) is due to R. Aharoni and C. Thomassen, Infinite highly connected digraphs with no two arc-disjoint spanning trees, *J. Graph Theory* **13** (1989), 71–74. Its consequence that the infinite analogue of the tree packing theorem fails with ordinary spanning trees had already been established by J.G. Oxley, On a packing problem for infinite graphs and independence spaces, *J. Comb. Theory B* **26** (1979), 123–130.

Nash-Williams had conjectured in his original paper that the finite tree packing theorem should extend to infinite graphs verbatim. What Tutte thought about an infinite version is not recorded. In his original paper he does consider the infinite case, but 'backwards': rather than speculating on which infinite graphs might admit edge-decompositions into k spanning trees, he proves that the locally finite graphs satisfying the cross-edges condition decompose into 'semiconnected subgraphs', defined—just for this purpose—as those subgraphs that contain an edge from every finite cut. These subgraphs, of course, are precisely the spanning subgraphs whose closures are connected

(Lemma 8.6.7), so Tutte in fact proved Lemma 8.6.10 without having the topological language to express it.

The companion to the finite tree packing theorem, the tree-covering theorem, extends to locally finite graphs verbatim (Exercise 17). The packing-covering theorem, Theorem 2.4.4, extends to infinite graphs separately in two ways: with ordinary spanning trees, and with topological ones. See Exercises 18 and 126, and their hints.

The example of the wild circle C illustrates why end-degrees in subspaces are defined in terms of arcs rather than rays. Every end on C contains only one ray that also lies in C, but it is the endpoint of two otherwise disjoint arcs in C. If we wish to be able to characterize the circles as those subspaces in which every vertex and every end has degree 2 (Exercise 97), we thus need the topological definition of end-degrees in subspaces.

The supremum in the definition of topological end degrees is in fact a maximum. This is a special case of Menger's ω-Beinsatz (1927), which he proved in the same paper as his now famous Theorem 3.3.1.

The (combinatorial) vertex-degree of an end used to be called its *multiplicity*. The term 'degree' was introduced by H. Bruhn and M. Stein, On end degrees and infinite circuits in locally finite graphs, *Combinatorica* **27** (2007), 269–291. Their main result was that the (entire) edge set of a locally finite graph lies in its topological cycle space if and only if every vertex has even degree and every end has even edge-degree—with a newly found division of the ends of infinite degree into 'even' and 'odd'. This was later generalized to arbitrary edge sets by E. Berger & H. Bruhn, Eulerian edge sets in locally finite graphs, *Combinatorica* **31** (2011), 21–38.

An interesting new aspect of end degrees is that they can make it possible to study extremal-type problems for infinite graphs that would otherwise make sense only for finite graphs. For example, while finite graphs of large enough minimum degree contain any desired topological minor or minor (see Chapter 7), an infinite graph of large minimum degree can be a tree. The ends of a tree, however, have degree 1. An assumption that the degrees of both vertices and ends of an infinite graph are large can still not force a non-planar minor (because such graphs can be planar), but it does force arbitrarily highly connected subgraphs. See R. Diestel, Forcing finite minors in sparse infinite graphs by large-degree assumptions, *Electronic. J. Comb.* **22** (2015), #P1.43, arXiv:1209.5318, as well as M. Stein, Extremal infinite graph theory, *Discrete Math.* **311** (2011), 1472–1496, arXiv:1102.0697, for this and other results in this vein. Another approach to 'extremal' infinite graph theory, which seeks to force infinite substructures by assuming a lower bound for $\|G[v_1, \ldots v_n]\|$ when $V(G) = \{v_1, v_2, \ldots\}$, is taken by J. Czipszer, P. Erdős and A. Hajnal, Some extremal problems on infinite graphs, *Publ. Math. Inst. Hung. Acad. Sci., Ser. A* **7** (1962), 441–457.

Our topological notion of the cycle space $\mathcal{C}(G)$ may appear natural in an infinite setting, but historically it is very young. It was developed in order to extend the classical applications of the cycle space of finite graphs, such as in planarity and duality, to locally finite graphs. As in the case of the tree packing theorem, those extensions fail when only finite circuits and sums are permitted, but they do hold for the topological cycle space. Examples include Tutte's theorem (3.2.6) that the non-separating induced cycles generate the whole cycle space; MacLane's (4.5.1), Kelmans's (4.5.2) and Whitney's (4.6.3) characterizations of planarity; and Gallai's partition theorem of Exercise 46, Chapter 1. There are a couple of papers by Diestel and Sprüssel that extend the notion of the topological cycle space to a general homology theory for locally compact spaces: The homology of locally finite graphs with ends, *Combinatorica* **30** (2010), 681–714, and On the homology of locally compact spaces with ends, *Topology and its Applications* **158** (2011), 1626–1639, arXiv:0910.5650.

For graphs that are not locally finite, there seems to be no one notion of topological cycle space that caters for all needs. A promising approach that takes account of this diversity was suggested by A. Georgakopoulos, Graph topologies induced by edge lengths, *Discrete Math.* **311** (2011), 1523–42, arXiv:0903.1744. For locally finite G this approach builds our familiar $|G|$ as the completion, rather than compactification, of G with a suitable metric— a fact to which the title of Section 8.6 alludes. By varying the metric, however, this approach can also yield other kinds of boundaries, such as the hyperbolic boundary of a hyperbolic graph.

Theorems 8.7.1 and 8.7.3 are from R. Diestel & D. Kühn, On infinite cycles I–II, *Combinatorica* **24** (2004), 69–116. What they say about edge sets $D \in \mathcal{C}(G)$ can be strengthened when $D = E(G)$: if a graph has no odd cut, its edge set decomposes into *finite* circuits. For arbitrary graphs, this is a deep theorem of C.St.J.A. Nash-Williams, Decomposition of graphs into closed and endless chains, *Proc. Lond. Math. Soc.* **10** (1960), 221–238. For countable graphs this is easy (Exercise 108).

Lacking the concept of an infinite circuit as we defined it here, Nash-Williams also sought to generalize the above and other theorems about finite cycles by replacing 'cycle' with '2-regular connected graph' (which may be finite or infinite). The resulting statements are not always as smooth as the finite theorems they generalize, but some substantial work has been done in this direction. C.St.J.A. Nash-Williams, Decompositions of graphs into two-way infinite paths, *Can. J. Math.* **15** (1963), 479–485, characterizes the graphs admitting edge-decompositions into double rays. F. Laviolette, Decompositions of infinite graphs I–II, *J. Comb. Theory B* **94** (2005), 259–333, characterizes the graphs admitting edge-decompositions into cycles and double rays. Results on the existence of spanning rays or double rays are referenced in the notes for Chapter 10.

While Theorems 8.7.1 and 8.7.3 extend the familiar properties of the cycle space of a finite graph to locally finite infinite graphs, the same can be done for the cut space (Exercise 117). The finitary cycle space $\mathcal{C}_{\mathrm{fin}}$, which is clearly a subspace of the topological cycle space $\mathcal{C}$, is in fact the set of *all* its finite elements (Exercise 115), just as the finitary cut space is the set $\mathcal{B}_{\mathrm{fin}}$ of all the finite elements of the entire cut space $\mathcal{B}$. For 3-connected graphs (but not otherwise, see Exercise 119 (iii)), edge sets orthogonal to all the circuits are in fact in $\mathcal{C}^{\perp}$, and sets orthogonal to all the bonds are in fact in $\mathcal{B}^{\perp}$; see R. Diestel & J. Pott, Orthogonality and minimality in the homology of locally finite graphs, *Electronic. J. Comb.* **21** (2014), #P3.36, arXiv:1307.0728.

The orthogonalities between $\mathcal{C}$, $\mathcal{B}$, $\mathcal{C}_{\mathrm{fin}}$ and $\mathcal{B}_{\mathrm{fin}}$ described in Exercise 118 is best captured in terms of matroids; see H. Bruhn & R. Diestel, Infinite matroids in graphs, *Discrete Math.* **311** (2011), 1461–1471, arXiv:1011.4749. Duality for planar infinite graphs is treated in H. Bruhn & R. Diestel, Duality in infinite graphs, *Comb. Probab. Comput.* **15** (2006), 75–90.

Theorem 8.8.2 is folklore—it was probably already known to Freudenthal. As mentioned in the text, one can obtain any standard subspace X' of $X = |G|$ as an inverse limit of the finite subgraphs H_p of G/p induced by the edges in X'. In order to find a topological spanning tree in a standard subspace $X \subsetneq |G|$, however, or even just an arc between two given points (so as to prove Lemma 8.6.5), we will need to construct these H_p by hand: in order to be able to expand a spanning tree or path in H_p to one in H_q for $p < q$, we need to ensure that every dummy vertex of H_p induces a connected subspace in X.

One way to do this is to mimic the G_n inside X: to enumerate $V(X)$, and take as p_n the partition of $V(X)$ consisting of the singleton classes of its first n vertices of X and, as further classes, the vertex sets of the arc-components obtained from X by deleting these n vertices and their incident edges. If G is locally finite and X is connected, then these p_n will be finite partitions, and by the jumping arc lemma they will have only finitely many cross-edges even in G. (One needs to show here that arc-components of standard subspaces are closed; but this is easy, even without Lemma 8.6.5).

One can then show directly that X is the inverse limit of the compact spaces X_n obtained from X by collapsing each of the partition classes in p_n. If one wishes to apply, rather than re-prove, Theorem 8.8.2 to obtain X as such an inverse limit, one has to expand the p_n to partitions $\bar{p}_n$ of $V(G)$, making sure that these $\bar{p}_n$ are cofinal in $P(G)$, and note that the desired X_n arise as the images of X under the given projections f_p. This can be done too, but it needs some care; see the hint for Exercise 129.

For connected graphs G that are not locally finite one can still define $\|G\|$ as in the text. The compact space one obtains can be described as a Hausdorff quotient of the space obtained from G by adding its 'edge ends', the equivalence classes of rays with respect to finite edge separators, minus loops.

9 Ramsey Theory for Graphs

In this chapter we set out from a type of problem which, on the face of it, appears to be similar to the theme of Chapter 7: what kind of substructures are necessarily present in every large enough graph?

The regularity lemma in Chapter 7.4 provides one possible answer to this question: every (large) graph G contains large random-like subgraphs. If we are looking for a concrete given subgraph H, on the other hand, our problem becomes more like Turán's theorem (7.1.1), Wagner's theorem (7.3.4), or Hadwiger's conjecture: we cannot expect an arbitrary graph G to contain a copy of H, but if it does not then this might have some interesting structural implications for G.

The kind of structural implication that will be typical for this chapter is simply that of containing some other (induced) subgraph. For example: given an integer r, does every large enough graph contain either a K^r or an induced $\overline{K^r}$? Does every large enough connected graph contain either a K^r or else a large induced path or star?

Despite its superficial similarity to extremal problems, the above type of question leads to a kind of mathematics with a distinctive flavour of its own. Indeed, the theorems and proofs in this chapter have more in common with similar results in algebra or geometry, say, than with most other areas of graph theory. The study of their underlying methods, therefore, is generally regarded as a combinatorial subject in its own right: the discipline of *Ramsey theory*.

In line with the subject of this book, we shall focus on results that are naturally expressed in terms of graphs. Even from the viewpoint of general Ramsey theory, however, this is not as much of a limitation as it might seem: graphs are a natural setting for Ramsey problems, and the material in this chapter brings out a sufficient variety of ideas and methods to convey some of the fascination of the theory as a whole.

© Reinhard Diestel 2017
R. Diestel, *Graph Theory*, Graduate Texts in Mathematics 173,
DOI 10.1007/978-3-662-53622-3_9

9.1 Ramsey's original theorems

In its simplest version, Ramsey's theorem says that, given an integer $r \geqslant 0$, every large enough graph G contains either K^r or $\overline{K^r}$ as an induced subgraph. At first glance, this may seem surprising: after all, we need a proportion of about $(r-2)/(r-1)$ of all possible edges to force a K^r subgraph in G (Corollary 7.1.3), but neither G nor $\overline{G}$ can be expected to have more than half of all possible edges. However, as the Turán graphs illustrate well, squeezing many edges into G without creating a K^r imposes additional structure on G, which may help us find an induced $\overline{K^r}$.

So how could we go about proving Ramsey's theorem? Let us try to build a K^r or $\overline{K^r}$ in G inductively, starting with an arbitrary vertex $v_1 \in V_1 := V(G)$. If $|G|$ is large, there will be a large set $V_2 \subseteq V_1 \smallsetminus \{v_1\}$ of vertices that are either all adjacent to v_1 or all non-adjacent to v_1. Accordingly, we may think of v_1 as the first vertex of a K^r or $\overline{K^r}$ whose other vertices all lie in V_2. Let us then choose another vertex $v_2 \in V_2$ for our K^r or $\overline{K^r}$. Since V_2 is large, it will have a subset V_3, still fairly large, of vertices that are all 'of the same type' with respect to v_2 as well: either all adjacent or all non-adjacent to it. We then continue our search for vertices inside V_3, and so on (Fig. 9.1.1).

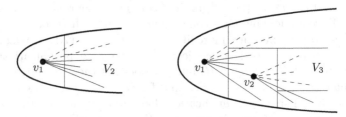

Fig. 9.1.1. Choosing the sequence $v_1, v_2, \ldots$

How long can we go on in this way? This depends on the size of our initial set V_1: each set V_i has at least half the size of its predecessor V_{i-1}, so we shall be able to complete s construction steps if G has order about 2^s. As the following proof shows, the choice of $s = 2r - 3$ vertices v_i suffices to find among them the vertices of a K^r or $\overline{K^r}$.

[9.2.2]
Theorem 9.1.1. (Ramsey 1930)
For every $r \in \mathbb{N}$ there exists an $n \in \mathbb{N}$ such that every graph of order at least n contains either K^r or $\overline{K^r}$ as an induced subgraph.

Proof. The assertion is trivial for $r \leqslant 1$; we assume that $r \geqslant 2$. Let $n := 2^{2r-3}$, and let G be a graph of order at least n. We shall define a sequence $V_1, \ldots, V_{2r-2}$ of sets and choose vertices $v_i \in V_i$ with the following properties:

    (i) $|V_i| = 2^{2r-2-i}$ $(i = 1, \ldots, 2r-2)$;

(ii) $V_i \subseteq V_{i-1} \setminus \{v_{i-1}\}$ $(i = 2, \ldots, 2r - 2)$;

(iii) v_{i-1} is adjacent either to all vertices in V_i or to no vertex in V_i $(i = 2, \ldots, 2r - 2)$.

Let $V_1 \subseteq V(G)$ be any set of 2^{2r-3} vertices, and pick $v_1 \in V_1$ arbitrarily. Then (i) holds for $i = 1$, while (ii) and (iii) hold trivially. Suppose now that V_{i-1} and $v_{i-1} \in V_{i-1}$ have been chosen so as to satisfy (i)–(iii) for $i - 1$, where $1 < i \leqslant 2r - 2$. Since

$$|V_{i-1} \setminus \{v_{i-1}\}| = 2^{2r-1-i} - 1$$

is odd, V_{i-1} has a subset V_i satisfying (i)–(iii); we pick $v_i \in V_i$ arbitrarily.

Among the $2r - 3$ vertices $v_1, \ldots, v_{2r-3}$, there are $r - 1$ vertices that show the same behaviour when viewed as v_{i-1} in (iii), being adjacent either to all the vertices in V_i or to none. Accordingly, these $r - 1$ vertices and v_{2r-2} induce either a K^r or a $\overline{K^r}$ in G, because $v_i, \ldots, v_{2r-2} \in V_i$ for all i. $\square$

The least integer n associated with r as in Theorem 9.1.1 is the *Ramsey number* $R(r)$ of r; our proof shows that $R(r) \leqslant 2^{2r-3}$. In Chapter 11 we shall use a simple probabilistic argument to show that $R(r)$ is bounded below by $2^{r/2}$ (Theorem 11.1.3).

Ramsey number $R(r)$

In other words, the largest clique or independent set of vertices that a graph of order n must contain is, asymptotically, logarithmically small in n. As soon as we forbid some fixed induced subgraph, however, it may be much bigger, of size linear in n: The *Erdős-Hajnal conjecture* says that for every graph H there exists a constant $\delta_H > 0$ such that every graph G not containing an induced copy of H has a set of at least $|G|^{\delta_H}$ vertices that are either independent or span a complete subgraph in G.

Erdős-Hajnal conjecture

It is customary in Ramsey theory to think of partitions as colourings: a *colouring* of (the elements of) a set X *with c colours*, or *c-colouring* for short, is simply a partition of X into c classes (indexed by the 'colours'). In particular, these colourings need not satisfy any non-adjacency requirements as in Chapter 5. Given a c-colouring of $[X]^k$, the set of all k-subsets of X, we call a set $Y \subseteq X$ *monochromatic* if all the elements of $[Y]^k$ have the same colour,[1] i.e. belong to the same of the c partition classes of $[X]^k$. Similarly, if $G = (V, E)$ is a graph and all the edges of $H \subseteq G$ have the same colour in some colouring of E, we call H a *monochromatic subgraph* of G, speak of a red (green, etc.) H in G, and so on.

c-colouring

$[X]^k$

mono-chromatic

In the above terminology, Ramsey's theorem can be expressed as follows: for every r there exists an n such that, given any n-set X,

---

[1] Note that Y is called monochromatic, but it is the elements of $[Y]^k$, not of Y, that are (equally) coloured.

every 2-colouring of $[X]^2$ yields a monochromatic r-set $Y \subseteq X$. Interestingly, this assertion remains true for c-colourings of $[X]^k$ with arbitrary c and k—with almost exactly the same proof!

We first prove the infinite version, which is easier, and then deduce the finite version.

[12.1.1]
Theorem 9.1.2. *Let k, c be positive integers, and X an infinite set. If $[X]^k$ is coloured with c colours, then X has an infinite monochromatic subset.*

Proof. We prove the theorem by induction on k, with c fixed. For $k = 1$ the assertion holds, so let $k > 1$ and assume the assertion for smaller values of k.

Let $[X]^k$ be coloured with c colours. We shall construct an infinite sequence $X_0, X_1, \ldots$ of infinite subsets of X and choose elements $x_i \in X_i$ with the following properties (for all i):

> (i) $X_{i+1} \subseteq X_i \smallsetminus \{x_i\}$;
> (ii) all k-sets $\{x_i\} \cup Z$ with $Z \in [X_{i+1}]^{k-1}$ have the same colour, which we *associate* with x_i.

We start with $X_0 := X$ and pick $x_0 \in X_0$ arbitrarily. By assumption, X_0 is infinite. Having chosen an infinite set X_i and $x_i \in X_i$ for some i, we c-colour $[X_i \smallsetminus \{x_i\}]^{k-1}$ by giving each set Z the colour of $\{x_i\} \cup Z$ from our c-colouring of $[X]^k$. By the induction hypothesis, $X_i \smallsetminus \{x_i\}$ has an infinite monochromatic subset, which we choose as X_{i+1}. Clearly, this choice satisfies (i) and (ii). Finally, we pick $x_{i+1} \in X_{i+1}$ arbitrarily.

Since c is finite, one of the c colours is associated with infinitely many x_i. These x_i form an infinite monochromatic subset of X. $\square$

If desired, the finite version of Theorem 9.1.2 could be proved just like the infinite version above. However to ensure that the relevant sets are large enough at all stages of the induction, we have to keep track of their sizes, which involves a good deal of boring calculation. As long as we are not interested in bounds, the more elegant route is to deduce the finite version from the infinite 'by compactness', that is, using König's infinity lemma (8.1.2).

[9.3.3]
Theorem 9.1.3. *For all $k, c, r \geqslant 1$ there exists an $n \geqslant k$ such that every n-set X has a monochromatic r-subset with respect to any c-colouring of $[X]^k$.*

(8.1.2)
k, c, r
Proof. As is customary in set theory, we denote by $n \in \mathbb{N}$ (also) the set $\{0, \ldots, n-1\}$. Suppose the assertion fails for some k, c, r. Then for every $n \geqslant k$ there exist an n-set, without loss of generality the set n, and a c-colouring $[n]^k \to c$ such that n contains no monochromatic r-set. Let
bad colouring
us call such colourings *bad*; we are thus assuming that for every $n \geqslant k$

there exists a bad colouring of $[n]^k$. Our aim is to combine these into a bad colouring of $[\mathbb{N}]^k$, which will contradict Theorem 9.1.2.

For every $n \geqslant k$ let $V_n \neq \emptyset$ be the set of bad colourings of $[n]^k$. For $n > k$, the restriction $f(g)$ of any $g \in V_n$ to $[n-1]^k$ is still bad, and hence lies in V_{n-1}. By the infinity lemma (8.1.2), there is an infinite sequence $g_k, g_{k+1}, \ldots$ of bad colourings $g_n \in V_n$ such that $f(g_n) = g_{n-1}$ for all $n > k$. For every $m \geqslant k$, all colourings g_n with $n \geqslant m$ agree on $[m]^k$, so for each $Y \in [\mathbb{N}]^k$ the value of $g_n(Y)$ coincides for all $n > \max Y$. Let us define $g(Y)$ as this common value $g_n(Y)$. Then g is a bad colouring of $[\mathbb{N}]^k$: every r-set $S \subseteq \mathbb{N}$ is contained in some sufficiently large n, so S cannot be monochromatic since g coincides on $[n]^k$ with the bad colouring g_n. $\square$

The least integer n associated with k, c, r as in Theorem 9.1.3 is the *Ramsey number* for these parameters; we denote it by $R(k, c, r)$.

Ramsey number $R(k, c, r)$

9.2 Ramsey numbers

Ramsey's theorem may be rephrased as follows: if $H = K^r$ and G is a graph with sufficiently many vertices, then either G itself or its complement $\overline{G}$ contains a copy of H as a subgraph. Clearly, the same is true for any graph H, simply because $H \subseteq K^h$ for $h := |H|$.

However, if we ask for the *least* n such that every graph G of order n has the above property—this is the *Ramsey number* $R(H)$ of H—then the above question makes sense: if H has only few edges, it should embed more easily in G or $\overline{G}$, and we would expect $R(H)$ to be smaller than the Ramsey number $R(h) = R(K^h)$.

Ramsey number $R(H)$

A little more generally, let $R(H_1, H_2)$ denote the least $n \in \mathbb{N}$ such that $H_1 \subseteq G$ or $H_2 \subseteq \overline{G}$ for every graph G of order n. For most graphs H_1, H_2, only very rough estimates are known for $R(H_1, H_2)$. Interestingly, lower bounds given by random graphs (as in Theorem 11.1.3) are often sharper than even the best bounds provided by explicit constructions.

$R(H_1, H_2)$

The following proposition describes one of the few cases where exact Ramsey numbers are known for a relatively large class of graphs:

Proposition 9.2.1. *Let s, t be positive integers, and let T be a tree of order t. Then $R(T, K^s) = (s-1)(t-1) + 1$.*

Proof. The disjoint union of $s - 1$ graphs K^{t-1} contains no copy of T, while the complement of this graph, the complete $(s - 1)$-partite graph K_{t-1}^{s-1}, does not contain K^s. This proves $R(T, K^s) \geqslant (s-1)(t-1) + 1$.

(1.5.4)
(5.2.3)

Conversely, let G be any graph of order $n = (s-1)(t-1) + 1$ whose complement contains no K^s. Then $s > 1$, and in any vertex colouring

of G (in the sense of Chapter 5) at most $s - 1$ vertices can have the
same colour. Hence, $\chi(G) \geqslant \lceil n/(s-1) \rceil = t$. By Lemma 5.2.3, G has
a subgraph H with $\delta(H) \geqslant t - 1$, which by Corollary 1.5.4 contains a
copy of T.   □

As the main result of this section, we shall now prove one of those
rare general theorems providing a relatively good upper bound for the
Ramsey numbers of a large class of graphs, a class defined in terms
of a standard graph invariant. The theorem deals with the Ramsey
numbers of sparse graphs: it says that the Ramsey number of graphs H
with bounded maximum degree grows only linearly in $|H|$—an enormous
improvement on the exponential bound from the proof of Theorem 9.1.1.

Theorem 9.2.2. (Chvátal, Rödl, Szemerédi & Trotter 1983)
For every positive integer Δ there is a constant c such that

$$R(H) \leqslant c\,|H|$$

(7.1.1)
(7.4.1)
(7.5.2)
(9.1.1)

for all graphs H with $\Delta(H) \leqslant \Delta$.

Proof. The basic idea of the proof is as follows. We wish to show that
$H \subseteq G$ or $H \subseteq \overline{G}$ if $|G|$ is large enough (though not too large). Consider
an ϵ-regular partition of G, as provided by the regularity lemma. If
enough of the ϵ-regular pairs in this partition have high density, we may
hope to find a copy of H in G. If most pairs have low density, we try
to find H in $\overline{G}$. Let R, R' and R'' be the regularity graphs of G whose
edges correspond to the pairs of density $\geqslant 0$; $\geqslant 1/2$; $< 1/2$ respectively.[2]
Then R is the edge-disjoint union of R' and R''.

Now to obtain $H \subseteq G$ or $H \subseteq \overline{G}$, it suffices by Lemma 7.5.2 to
ensure that H is contained in a suitable 'inflated regularity graph' R'_s
or R''_s. Since $\chi(H) \leqslant \Delta(H) + 1 \leqslant \Delta + 1$, this will be the case if $s \geqslant \alpha(H)$
and we can find a $K^{\Delta+1}$ in R' or in R''. But that is easy to ensure: we
just need that $K^r \subseteq R$, where r is the Ramsey number of $\Delta + 1$, which
will follow from Turán's theorem because R is dense.

Δ, d
ϵ_0, m
ϵ

For the formal proof let now $\Delta \geqslant 1$ be given. On input $d := 1/2$
and Δ, Lemma 7.5.2 returns an ϵ_0. Let $m := R(\Delta + 1)$ be the Ramsey
number of $\Delta + 1$. Let $\epsilon \leqslant \epsilon_0$ be positive but small enough that for $k = m$
(and hence for all $k \geqslant m$)

$$2\epsilon < \frac{1}{m-1} - \frac{1}{k}; \tag{1}$$

M

then in particular $\epsilon < 1$. Finally, let M be the integer returned by the
regularity lemma (Theorem 7.4.1) on input ϵ and m.

---

[2] In our formal proof later we shall define R'' a little differently, so that it complies
properly with our definition of a regularity graph.

All the quantities defined so far depend only on Δ. We shall prove the theorem with

$$c := \frac{2^{\Delta+1}M}{1-\epsilon}.$$

c

Let H with $\Delta(H) \leqslant \Delta$ be given, and let $s := |H|$. Let G be an arbitrary graph of order $n \geqslant c|H|$; we show that $H \subseteq G$ or $H \subseteq \overline{G}$.

s
G, n

By Theorem 7.4.1, G has an ϵ-regular partition $\{V_0, V_1, \ldots, V_k\}$ with exceptional set V_0 and $|V_1| = \ldots = |V_k| =: \ell$, where $m \leqslant k \leqslant M$. Then

k
ℓ

$$\ell = \frac{n-|V_0|}{k} \geqslant n\,\frac{1-\epsilon}{M} \geqslant cs\,\frac{1-\epsilon}{M} \geqslant 2^{\Delta+1}s = 2s/d^\Delta. \qquad (2)$$

Let R be the regularity graph with parameters $\epsilon, \ell, 0$ corresponding to this partition. By definition, R has k vertices and

R

$$\begin{aligned}
\|R\| &\geqslant \binom{k}{2} - \epsilon k^2 \\
&= \tfrac{1}{2}k^2\left(1 - \frac{1}{k} - 2\epsilon\right) \\
&\underset{(1)}{>} \tfrac{1}{2}k^2\left(1 - \frac{1}{k} - \frac{1}{m-1} + \frac{1}{k}\right) \\
&= \tfrac{1}{2}k^2\,\frac{m-2}{m-1} \\
&\geqslant t_{m-1}(k)
\end{aligned}$$

edges. By Theorem 7.1.1, therefore, R has a subgraph $K = K^m$.

K

We now colour the edges of R with two colours: red if the edge corresponds to a pair (V_i, V_j) of density at least $1/2$, and green otherwise. Let R' be the spanning subgraph of R formed by the red edges, and R'' the spanning subgraph of R formed by the green edges and those whose corresponding pair has density exactly $1/2$. Then R' is a regularity graph of G with parameters ϵ, ℓ and $1/2$. And R'' is a regularity graph of $\overline{G}$, with the same parameters: as one easily checks, every pair (V_i, V_j) that is ϵ-regular for G is also ϵ-regular for $\overline{G}$.

By definition of m, our graph K contains a red or a green K^r, for $r := \chi(H) \leqslant \Delta + 1$. Correspondingly, $H \subseteq R'_s$ or $H \subseteq R''_s$. Since $\epsilon \leqslant \epsilon_0$ and $\ell \geqslant 2s/d^\Delta$ by (2), both R' and R'' satisfy the requirements of Lemma 7.5.2, so $H \subseteq G$ or $H \subseteq \overline{G}$ as desired. $\qquad\square$

So far in this section, we have been asking what is the least order of a complete graph G such that every 2-colouring of its edges yields a monochromatic copy of some given graph H. Rather than keeping G complete and focusing on its order, let us now consider its structure too, i.e., minimize G with respect to the subgraph relation. Given a graph H,

Ramsey-
minimal

let us call a graph G *Ramsey-minimal* for H if G is minimal with the property that every 2-colouring of its edges yields a monochromatic copy of H.

What do such Ramsey-minimal graphs look like? Are they unique? The following result, which we include for its pretty proof, answers the second question for some H:

Proposition 9.2.3. *If T is a tree but not a star, then infinitely many graphs are Ramsey-minimal for T.*

(1.5.4)
(5.2.3)
(5.2.5)
(=11.2.2)

Proof. Let $|T| =: r$. We show that for every $n \in \mathbb{N}$ there is a graph of order at least n that is Ramsey-minimal for T.

By Theorem 5.2.5, there exists a graph G with chromatic number $\chi(G) > r^2$ and girth $g(G) > n$. If we colour the edges of G red and green, then the red and the green subgraph cannot both have an r-(vertex-)colouring in the sense of Chapter 5: otherwise we could colour the vertices of G with the pairs of colours from those colourings and obtain a contradiction to $\chi(G) > r^2$. So let $G' \subseteq G$ be monochromatic with $\chi(G') > r$. By Lemma 5.2.3, G' has a subgraph of minimum degree at least r, which contains a copy of T by Corollary 1.5.4.

Let $G^* \subseteq G$ be Ramsey-minimal for T. Clearly, G^* is not a forest: the edges of any forest can be 2-coloured (partitioned) so that no monochromatic subforest contains a path of length 3, let alone a copy of T. (Here we use that T is not a star, and hence contains a P^3.) So G^* contains a cycle, which has length $g(G) > n$ since $G^* \subseteq G$. In particular, $|G^*| > n$ as desired. $\square$

9.3 Induced Ramsey theorems

Ramsey's theorem can be rephrased as follows. For every graph $H = K^r$ there *exists* a graph G such that every 2-colouring of the edges of G yields a monochromatic $H \subseteq G$; as it turns out, this is witnessed by any large enough complete graph as G. Let us now change the problem slightly and ask for a graph G in which every 2-edge-colouring yields a monochromatic *induced* $H \subseteq G$, where H is now an arbitrary given graph.

This slight modification changes the character of the problem dramatically. What is needed now is no longer a simple proof that G is 'big enough' (as for Theorem 9.1.1), but a careful construction: the construction of a graph that, however we bipartition its edges, contains an induced copy of H with all edges in one partition class. We shall call

Ramsey
graph

such a graph a *Ramsey graph* for H.

The fact that such a Ramsey graph exists for every choice of H is one of the fundamental results of graph Ramsey theory. It was proved around 1973, independently by Deuber, by Erdős, Hajnal & Pósa, and by Rödl.

Theorem 9.3.1. *Every graph has a Ramsey graph. In other words, for every graph H there exists a graph G that, for every partition $\{E_1, E_2\}$ of $E(G)$, has an induced subgraph H with $E(H) \subseteq E_1$ or $E(H) \subseteq E_2$.*

We give two proofs. Each of these is highly individual, yet each offers a glimpse of true Ramsey theory: the graphs involved are used as hardly more than bricks in the construction, but the edifice is impressive.

First proof. In our construction of the desired Ramsey graph we shall repeatedly replace vertices of a graph $G = (V, E)$ already constructed by copies of another graph H. For a vertex set $U \subseteq V$ let $G[U \to H]$ $G[U \to H]$ denote the graph obtained from G by replacing the vertices $u \in U$ with copies $H(u)$ of H and joining each $H(u)$ completely to all $H(u')$ with $H(u)$ $uu' \in E$ and to all vertices $v \in V \smallsetminus U$ with $uv \in E$ (Fig. 9.3.1). Formally,

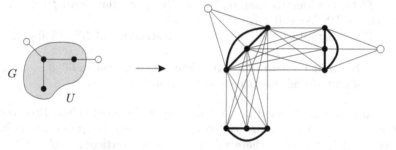

Fig. 9.3.1. A graph $G[U \to H]$ with $H = K^3$

$G[U \to H]$ is the graph on

$$(U \times V(H)) \cup ((V \smallsetminus U) \times \{\emptyset\})$$

in which two vertices (v, w) and (v', w') are adjacent if and only if either $vv' \in E$, or else $v = v' \in U$ and $ww' \in E(H)$.[3]

We prove the following formal strengthening of Theorem 9.3.1:

> For any two graphs H_1, H_2 there exists a graph $G = G(H_1, H_2)$ such that every edge colouring of G with the $G(H_1, H_2)$ colours 1 and 2 yields either an induced $H_1 \subseteq G$ with all   (∗) its edges coloured 1 or an induced $H_2 \subseteq G$ with all its edges coloured 2.

---

[3] The replacement of $V \smallsetminus U$ by $(V \smallsetminus U) \times \{\emptyset\}$ is just a formal device to ensure that all vertices of $G[U \to H]$ have the same form (v, w), and that $G[U \to H]$ is formally disjoint from G.

This formal strengthening makes it possible to apply induction on $|H_1| + |H_2|$, as follows.

If either H_1 or H_2 has no edges (in particular, if $|H_1| + |H_2| \leqslant 1$), then $(*)$ holds with $G = \overline{K^n}$ for large enough n. For the induction step, we now assume that both H_1 and H_2 have at least one edge, and that $(*)$ holds for all pairs (H_1', H_2') with smaller $|H_1'| + |H_2'|$.

x_i
H_i', H_i''

For each $i = 1, 2$, pick a vertex $x_i \in H_i$ that is incident with an edge. Let $H_i' := H_i - x_i$, and let H_i'' be the subgraph of H_i' induced by the neighbours of x_i.

We shall construct a sequence $G^0, \ldots, G^n$ of disjoint graphs; G^n will be the desired Ramsey graph $G(H_1, H_2)$. Along with the graphs G_i, we shall define subsets $V^i \subseteq V(G^i)$ and a map

$$f \colon V^1 \cup \ldots \cup V^n \to V^0 \cup \ldots \cup V^{n-1}$$

such that

$$f(V^i) = V^{i-1} \qquad (1)$$

f^i

for all $i \geqslant 1$. Writing $f^i := f \circ \ldots \circ f$ for the i-fold composition of f, and f^0 for the identity map on $V^0 = V(G^0)$, we thus have $f^i(v) \in V^0$ for all $v \in V^i$. We call $f^i(v)$ the *origin* of v.

origin

The subgraphs $G^i[V^i]$ will reflect the structure of G^0 as follows:

Vertices in V^i with different origins are adjacent in G^i if
and only if their origins are adjacent in G^0. $\qquad (2)$

Assertion (2) will not be used formally in the proof below. However, it can help us to visualize the graphs G^i: every G^i (more precisely, every $G^i[V^i]$; for $i \geqslant 1$ there will also be some vertices $x \in G^i - V^i$) is essentially an inflated copy of G^0 in which every vertex $w \in G^0$ has been replaced by the set of all vertices in V^i with origin w, and the map f links vertices with the same origin across the various G^i.

By the induction hypothesis, there are Ramsey graphs

G_1, G_2

$$G_1 := G(H_1, H_2') \quad \text{and} \quad G_2 := G(H_1', H_2) \,.$$

G^0, V^0
W_i'
n
W_i''

Let G^0 be a copy of G_1, and set $V^0 := V(G^0)$. Let $W_0', \ldots, W_{n-1}'$ be the subsets of V^0 spanning an H_2' in G^0. Thus, n is defined as the number of induced copies of H_2' in G^0, and we shall construct a graph G^i for every set W_{i-1}', $i = 1, \ldots, n$. For $i = 0, \ldots, n-1$, let W_i'' be the image of $V(H_2'')$ under some isomorphism $H_2' \to G^0[W_i']$.

U^{i-1}

Assume now that $G^0, \ldots, G^{i-1}$ and $V^0, \ldots, V^{i-1}$ have been defined for some $i \geqslant 1$, and that f has been defined on $V^1 \cup \ldots \cup V^{i-1}$ and satisfies (1) for all $j \leqslant i$. We construct G^i from G^{i-1} in two steps. For the first step, consider the set U^{i-1} of all the vertices $v \in V^{i-1}$ whose origin $f^{i-1}(v)$ lies in W_{i-1}''. (For $i = 1$, this gives $U^0 = W_0''$.) Expand

G^{i-1} to a new graph $\tilde{G}^{i-1}$ (disjoint from G^{i-1}) by replacing every vertex
$u \in U^{i-1}$ with a copy $G_2(u)$ of G_2, i.e. let $G_2(u)$

$$\tilde{G}^{i-1} := G^{i-1}[U^{i-1} \rightarrow G_2]   \qquad \tilde{G}^{i-1}$$

(see Figures 9.3.2 and 9.3.3). Set $f(u') := u$ for all $u \in U^{i-1}$ and

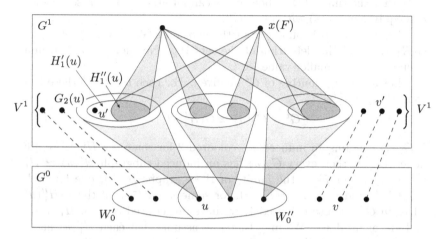

Fig. 9.3.2. The construction of G^1

$u' \in G_2(u)$, and $f(v') := v$ for all $v' = (v, \emptyset)$ with $v \in V^{i-1} \smallsetminus U^{i-1}$.
(Recall that $(v, \emptyset)$ is simply the unexpanded copy of a vertex $v \in G^{i-1}$
in $\tilde{G}^{i-1}$.) Let V^i be the set of those vertices v' or u' of $\tilde{G}^{i-1}$ for which V^i
f has thus been defined, i.e. the vertices that either correspond directly
to a vertex v in V^{i-1} or else belong to an expansion $G_2(u)$ of such a
vertex u. Then (1) holds for i. Also, if we assume (2) inductively for
$i - 1$, then (2) holds again for i (in $\tilde{G}^{i-1}$). The graph $\tilde{G}^{i-1}$ is already
the essential part of G^i: the part that looks like an inflated copy of G^0.

In the second step we now extend $\tilde{G}^{i-1}$ to the desired graph G^i by
adding some further vertices $x \notin V^i$. Let $\mathcal{F}$ denote the set of all families $\mathcal{F}$
F of the form

$$F = \left(H_1'(u) \mid u \in U^{i-1} \right),$$

where each $H_1'(u)$ is an induced subgraph of $G_2(u)$ isomorphic to H_1'. $H_1'(u)$
(Less formally: $\mathcal{F}$ is the collection of ways to select simultaneously from
each $G_2(u)$ exactly one induced copy of H_1'.) For each $F \in \mathcal{F}$, add a
vertex $x(F)$ to $\tilde{G}^{i-1}$ and join it, for every $u \in U^{i-1}$, to all the vertices in $x(F)$
the image $H_1''(u) \subseteq H_1'(u)$ of H_1'' under some isomorphism from H_1' to $H_1''(u)$
the $H_1'(u) \subseteq G_2(u)$ selected by F (Fig. 9.3.2). Denote the resulting graph
by G^i. This completes the inductive definition of the graphs $G^0, \ldots, G^n$. G^i

Let us now show that $G := G^n$ satisfies $(*)$. To this end, we prove
the following assertion $(**)$ about G^i for $i = 0, \ldots, n$:

> For every edge colouring with the colours 1 and 2, G^i con-
> tains either an induced H_1 coloured 1, or an induced H_2
> coloured 2, or an induced subgraph H coloured 2 such that $\quad(**)$
> $V(H) \subseteq V^i$ and the restriction of f^i to $V(H)$ is an isomor-
> phism between H and $G^0[W_k']$ for some $k \in \{i, \ldots, n-1\}$.

Note that the third of the above cases cannot arise for $i = n$, so $(**)$ for n is equivalent to $(*)$ with $G := G^n$.

For $i = 0$, $(**)$ follows from the choice of G^0 as a copy of $G_1 = G(H_1, H_2')$ and the definition of the sets W_k'. Now let $1 \leqslant i \leqslant n$, and assume $(**)$ for smaller values of i.

Let an edge colouring of G^i be given. For each $u \in U^{i-1}$ there is a copy of G_2 in G^i:

$$G^i \supseteq G_2(u) \simeq G(H_1', H_2).$$

If $G_2(u)$ contains an induced H_2 coloured 2 for some $u \in U^{i-1}$, we are done. If not, then every $G_2(u)$ has an induced subgraph $H_1'(u) \simeq H_1'$ coloured 1. Let F be the family of these graphs $H_1'(u)$, one for each $u \in U^{i-1}$, and let $x := x(F)$. If, for some $u \in U^{i-1}$, all the x–$H_1''(u)$ edges in G^i are also coloured 1, we have an induced copy of H_1 in G^i and are again done. We may therefore assume that each $H_1''(u)$ has a vertex y_u for which the edge xy_u is coloured 2. The restriction $y_u \mapsto u$ of f to

$$\hat{U}^{i-1} := \{ y_u \mid u \in U^{i-1} \} \subseteq V^i$$

extends by $(v, \emptyset) \mapsto v$ to an isomorphism from

$$\hat{G}^{i-1} := G^i \left[\hat{U}^{i-1} \cup \{ (v, \emptyset) \mid v \in V(G^{i-1}) \setminus U^{i-1} \} \right]$$

to G^{i-1}, and so our edge colouring of G^i induces an edge colouring of G^{i-1}. If this colouring yields an induced $H_1 \subseteq G^{i-1}$ coloured 1 or an induced $H_2 \subseteq G^{i-1}$ coloured 2, we have these also in $\hat{G}^{i-1} \subseteq G^i$ and are again home.

By $(**)$ for $i-1$ we may therefore assume that G^{i-1} has an induced subgraph H' coloured 2, with $V(H') \subseteq V^{i-1}$, and such that the restriction of f^{i-1} to $V(H')$ is an isomorphism from H' to $G^0[W_k'] \simeq H_2'$ for some $k \in \{i-1, \ldots, n-1\}$. Let $\hat{H}'$ be the corresponding induced subgraph of $\hat{G}^{i-1} \subseteq G^i$ (also coloured 2); then $V(\hat{H}') \subseteq V^i$,

$$f^i(V(\hat{H}')) = f^{i-1}(V(H')) = W_k',$$

and $f^i \colon \hat{H}' \to G^0[W_k']$ is an isomorphism.

If $k \geqslant i$, this completes the proof of $(**)$ with $H := \hat{H}'$; we therefore assume that $k < i$, and hence $k = i - 1$ (Fig. 9.3.3). By definition of U^{i-1} and $\hat{G}^{i-1}$, the inverse image of W_{i-1}'' under the isomorphism

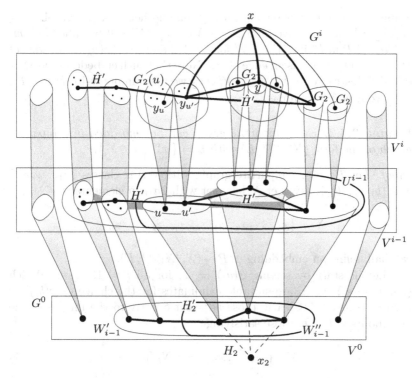

Fig. 9.3.3. A monochromatic copy of H_2 in G^i

$f^i\colon \hat{H}' \to G^0[W'_{i-1}]$ is a subset of $\hat{U}^{i-1}$. Since x is joined to precisely those vertices of $\hat{H}'$ that lie in $\hat{U}^{i-1}$, and all these edges xy_u have colour 2, the graph $\hat{H}'$ and x together induce in G^i a copy of H_2 coloured 2, and the proof of $(**)$ is complete. □

Let us return once more to the reformulation of Ramsey's theorem considered at the beginning of this section: for every graph H there exists a graph G such that every 2-colouring of the edges of G yields a monochromatic $H \subseteq G$. The graph G for which this follows at once from Ramsey's theorem is a sufficiently large complete graph. If we ask, however, that G shall not contain any complete subgraphs larger than those in H, i.e. that $\omega(G) = \omega(H)$, the problem again becomes difficult—even if we do not require H to be induced in G.

Our second proof of Theorem 9.3.1 solves both problems at once: given H, we shall construct a Ramsey graph for H with the same clique number as H.

For this proof, i.e. for the remainder of this section, let us view bipartite graphs P as triples (V_1, V_2, E), where V_1 and V_2 are the two vertex classes and $E \subseteq V_1 \times V_2$ is the set of edges. The reason for this more explicit notation is that we want embeddings between bipartite

bipartite

graphs to respect their bipartitions: given another bipartite graph $P' = (V_1', V_2', E')$, an injective map $\varphi\colon V_1 \cup V_2 \to V_1' \cup V_2'$ will be called an *embedding* of P in P' if $\varphi(V_i) \subseteq V_i'$ for $i = 1, 2$ and $\varphi(v_1)\varphi(v_2)$ is an edge of P' if and only if $v_1 v_2$ is an edge of P. (Note that such embeddings are 'induced'.) Instead of $\varphi\colon V_1 \cup V_2 \to V_1' \cup V_2'$ we may simply write $\varphi\colon P \to P'$.

embedding
$P \to P'$

We need two lemmas.

Lemma 9.3.2. *Every bipartite graph can be embedded in a bipartite graph of the form $(X, [X]^k, E)$ with $E = \{\, xY \mid x \in Y \,\}$.*

E

Proof. Let P be any bipartite graph, with vertex classes $\{a_1, \ldots, a_n\}$ and $\{b_1, \ldots, b_m\}$, say. Let X be a set with $2n + m$ elements, say

$$X = \{x_1, \ldots, x_n,\ y_1, \ldots, y_n,\ z_1, \ldots, z_m\}\,;$$

we shall define an embedding $\varphi\colon P \to (X, [X]^{n+1}, E)$.

Let us start by setting $\varphi(a_i) := x_i$ for all $i = 1, \ldots, n$. Which $(n+1)$-sets $Y \subseteq X$ are suitable candidates for the choice of $\varphi(b_i)$ for a given vertex b_i? Clearly those adjacent exactly to the images of the neighbours of b_i, i.e. those satisfying

$$Y \cap \{x_1, \ldots, x_n\} = \varphi(N_P(b_i))\,. \tag{1}$$

Since $d(b_i) \leqslant n$, the requirement of (1) leaves at least one of the $n + 1$ elements of Y unspecified. In addition to $\varphi(N_P(b_i))$, we may therefore include in each $Y = \varphi(b_i)$ the vertex z_i as an 'index'; this ensures that $\varphi(b_i) \neq \varphi(b_j)$ for $i \neq j$, even when b_i and b_j have the same neighbours in P. To specify the sets $Y = \varphi(b_i)$ completely, we finally fill them up with 'dummy' elements y_j until $|Y| = n + 1$. $\qquad\square$

Our second lemma already covers the bipartite case of the theorem: it says that every bipartite graph has a Ramsey graph—even a bipartite one.

Lemma 9.3.3. *For every bipartite graph P there exists a bipartite graph P' such that for every 2-colouring of the edges of P' there is an embedding $\varphi\colon P \to P'$ for which all the edges of $\varphi(P)$ have the same colour.*

(9.1.3)

P, X, k, E
P', X', k'

Proof. We may assume by Lemma 9.3.2 that P has the form $(X, [X]^k, E)$ with $E = \{\, xY \mid x \in Y \,\}$. We show the assertion for the graph $P' := (X', [X']^{k'}, E')$, where $k' := 2k - 1$, X' is any set of cardinality

$$|X'| = R\left(k',\ 2\binom{k'}{k},\ k\,|X| + k - 1\right),$$

(this is the Ramsey number defined after Theorem 9.1.3), and

$$E' := \{\, x'Y' \mid x' \in Y' \,\}. \qquad\qquad E'$$

Let us then colour the edges of P' with two colours α and β. Of the $\qquad \alpha, \beta$
$|Y'| = 2k - 1$ edges incident with a vertex $Y' \in [X']^{k'}$, at least k must
have the same colour. For each Y' we may therefore choose a fixed k-set
$Z' \subseteq Y'$ such that all the edges $x'Y'$ with $x' \in Z'$ have the same colour; $\qquad Z'$
we shall call this colour *associated* with Y'. $\qquad\qquad$ *associated*

The sets Z' can lie within their supersets Y' in $\binom{k'}{k}$ ways, as follows.
Let X' be linearly ordered. Then for every $Y' \in [X']^k$ there is a unique
order-preserving bijection $\sigma_{Y'}\colon Y' \to \{1, \dots, k'\}$, which maps Z' to one $\qquad \sigma_{Y'}$
of $\binom{k'}{k}$ possible images.

We now colour $[X']^{k'}$ with the $2\binom{k'}{k}$ elements of the set

$$[\{1, \dots, k'\}]^k \times \{\alpha, \beta\}$$

as colours, giving each $Y' \in [X']^{k'}$ as its colour the pair $(\sigma_{Y'}(Z'), \gamma)$,
where γ is the colour α or β associated with Y'. Since $|X'|$ was chosen
as the Ramsey number with parameters k', $2\binom{k'}{k}$ and $k\,|X| + k - 1$, we

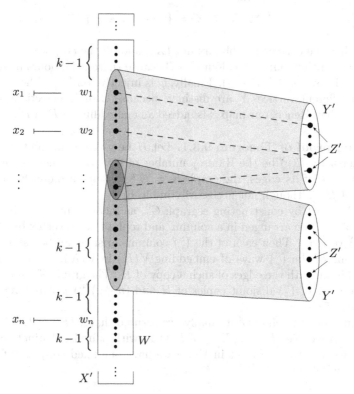

Fig. 9.3.4. The graph of Lemma 9.3.3

W

know that X' has a monochromatic subset W of cardinality $k|X|+k-1$. All Z' with $Y' \subseteq W$ thus lie within their Y' in the same way, i.e. there exists an $S \in [\{1,\dots,k'\}]^k$ such that $\sigma_{Y'}(Z') = S$ for all $Y' \in [W]^{k'}$, and

α

all $Y' \in [W]^{k'}$ are associated with the same colour, say with α.

$\varphi|_X$

We now construct the desired embedding φ of P in P'. We first

x_i, w_i, n

define φ on $X =: \{x_1,\dots,x_n\}$, choosing images $\varphi(x_i) =: w_i \in W$ so that $w_i < w_j$ in our ordering of X' whenever $i < j$. Moreover, we choose the w_i so that exactly $k-1$ elements of W are smaller than w_1, exactly $k-1$ lie between w_i and w_{i+1} for $i = 1,\dots,n-1$, and exactly $k-1$ are bigger than w_n. Since $|W| = kn+k-1$, this can indeed be done (Fig. 9.3.4).

$\varphi|_{[X]^k}$

We now define φ on $[X]^k$. Given $Y \in [X]^k$, we wish to choose $\varphi(Y) =: Y' \in [X']^{k'}$ so that the neighbours of Y' among the vertices in $\varphi(X)$ are precisely the images of the neighbours of Y in P, i.e. the k vertices $\varphi(x)$ with $x \in Y$, and so that all these edges at Y' are coloured α. To find such a set Y', we first fix its subset Z' as $\{\varphi(x) \mid x \in Y\}$ (these are k vertices of type w_i) and then extend Z' by $k'-k$ further vertices $u \in W \setminus \varphi(X)$ to a set $Y' \in [W]^{k'}$, in such a way that Z' lies correctly within Y', i.e. so that $\sigma_{Y'}(Z') = S$. This can be done, because $k-1 = k'-k$ other vertices of W lie between any two w_i. Then

$$Y' \cap \varphi(X) = Z' = \{\varphi(x) \mid x \in Y\},$$

so Y' has the correct neighbours in $\varphi(X)$, and all the edges between Y' and these neighbours are coloured α (because those neighbours lie in Z' and Y' is associated with α). Finally, φ is injective on $[X]^k$: the images Y' of different vertices Y are distinct, because their intersections with $\varphi(X)$ differ. Hence, our map φ is indeed an embedding of P in P'. $\square$

Second proof of Theorem 9.3.1. Let H be given as in the theorem,

r, n

and let $n := R(r)$ be the Ramsey number of $r := |H|$. Then, for every

K

2-colouring of its edges, the graph $K = K^n$ contains a monochromatic copy of H—although not necessarily induced.

We start by constructing a graph G^0, as follows. Imagine the vertices of K to be arranged in a column, and replace every vertex by a row of $\binom{n}{r}$ vertices. Then each of the $\binom{n}{r}$ columns arising can be associated with one of the $\binom{n}{r}$ ways of embedding $V(H)$ in $V(K)$; let us furnish this column with the edges of such a copy of H. The graph G^0 thus arising consists of $\binom{n}{r}$ disjoint copies of H and $(n-r)\binom{n}{r}$ isolated vertices (Fig. 9.3.5).

In order to define G^0 formally, we assume that $V(K) = \{1,\dots,n\}$ and choose copies $H_1,\dots,H_{\binom{n}{r}}$ of H in K with pairwise distinct vertex sets. (Thus, on each r-set in $V(K)$ we have one fixed copy H_j of H.)

G^0

We then define

$$V(G^0) := \{(i,j) \mid i = 1,\dots,n; \ j = 1,\dots,\tbinom{n}{r}\}$$

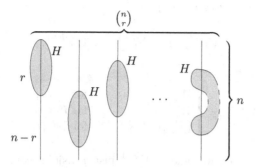

Fig. 9.3.5. The graph G^0

and

$$E(G^0) := \bigcup_{j=1}^{\binom{n}{r}} \left\{ (i,j)(i',j) \mid ii' \in E(H_j) \right\}.$$

The idea of the proof now is as follows. Our aim is to reduce the general case of the theorem to the bipartite case dealt with in Lemma 9.3.3. Applying the lemma iteratively to all the pairs of rows of G^0, we construct a very large graph G such that for every edge colouring of G there is an induced copy of G^0 in G that is monochromatic on all the bipartite subgraphs induced by its pairs of rows, i.e. in which edges between the same two rows always have the same colour. The projection of this $G^0 \subseteq G$ to $\{1,\dots,n\}$ (by contracting its rows) then defines an edge colouring of K. (If the contraction does not yield all the edges of K, colour the missing edges arbitrarily.) By the choice of $|K|$, some $K^r \subseteq K$ will be monochromatic. The H_j inside this K^r then occurs with the same colouring in the jth column of our G^0, where it is an induced subgraph of G^0, and hence of G.

Formally, we shall define a sequence $G^0,\dots,G^m$ of n-partite graphs G^k, with n-partition $\{V_1^k,\dots,V_n^k\}$ say, and then let $G := G^m$. The graph G^0 has been defined above; let $V_1^0,\dots,V_n^0$ be its rows:

$$V_i^0 := \left\{ (i,j) \mid j = 1,\dots,\binom{n}{r} \right\}. \qquad V_i^0$$

Now let $e_1,\dots,e_m$ be an enumeration of the edges of K. For $k = 0,\dots,m-1$, construct G^{k+1} from G^k as follows. If $e_{k+1} = i_1 i_2$, say, let $P = (V_{i_1}^k, V_{i_2}^k, E)$ be the bipartite subgraph of G^k induced by its i_1th and i_2th row. By Lemma 9.3.3, P has a bipartite Ramsey graph $P' = (W_1, W_2, E')$. We wish to define $G^{k+1} \supseteq P'$ in such a way that every (monochromatic) embedding $P \to P'$ can be extended to an embedding $G^k \to G^{k+1}$ respecting their n-partitions. Let $\{\varphi_1,\dots,\varphi_q\}$ be the set of

e_k, m

i_1, i_2

P

P'

W_1, W_2

φ_p, q

all embeddings of P in P', and let

$$V(G^{k+1}) := V_1^{k+1} \cup \ldots \cup V_n^{k+1},$$

where

$$V_i^{k+1} := \begin{cases} W_1 & \text{for } i = i_1 \\ W_2 & \text{for } i = i_2 \\ \bigcup_{p=1}^{q}(V_i^k \times \{p\}) & \text{for } i \notin \{i_1, i_2\}. \end{cases}$$

(Thus for $i \neq i_1, i_2$, we take as V_i^{k+1} just q disjoint copies of V_i^k.) We now define the edge set of G^{k+1} so that the obvious extensions of φ_p to all of $V(G^k)$ become embeddings of G^k in G^{k+1}: for $p = 1, \ldots, q$, let $\psi_p : V(G^k) \to V(G^{k+1})$ be defined by

$$\psi_p(v) := \begin{cases} \varphi_p(v) & \text{for } v \in P \\ (v, p) & \text{for } v \notin P \end{cases}$$

and let

$$E(G^{k+1}) := \bigcup_{p=1}^{q} \{ \psi_p(v)\psi_p(v') \mid vv' \in E(G^k) \}.$$

Now for every 2-colouring of its edges, G^{k+1} contains an induced copy $\psi_p(G^k)$ of G^k whose edges in P, i.e. those between its i_1th and i_2th row, have the same colour: just choose p so that $\varphi_p(P)$ is the monochromatic induced copy of P in P' that exists by Lemma 9.3.3.

We claim that $G := G^m$ satisfies the assertion of the theorem. So let a 2-colouring of the edges of G be given. By the construction of G^m from G^{m-1}, we can find in G^m an induced copy of G^{m-1} such that for $e_m = ii'$ all edges between the ith and the i'th row have the same colour. In the same way, we find inside this copy of G^{m-1} an induced copy of G^{m-2} whose edges between the ith and the i'th row have the same colour also for $ii' = e_{m-1}$. Continuing in this way, we finally arrive at an induced copy of G^0 in G such that, for each pair (i, i'), all the edges between V_i^0 and $V_{i'}^0$ have the same colour. As shown earlier, this G^0 contains a monochromatic induced copy H_j of H. □

9.4 Ramsey properties and connectivity

According to Ramsey's theorem, every large enough graph G has a very dense or a very sparse induced subgraph of given order, a K^r or $\overline{K^r}$. If we assume that G is connected, we can say a little more:

Proposition 9.4.1. *For every $r \in \mathbb{N}$ there is an $n \in \mathbb{N}$ such that every connected graph of order at least n contains K^r, $K_{1,r}$ or P^r as an induced subgraph.*

Proof. Let $d + 1$ be the Ramsey number of r, let $n \geqslant \frac{d}{d-2}(d-1)^r$, and let G be a graph of order at least n. If G has a vertex v of degree at least $d + 1$ then, by Theorem 9.1.1 and the choice of d, either $N(v)$ induces a K^r in G or $\{v\} \cup N(v)$ induces a $K_{1,r}$. On the other hand, if $\Delta(G) \leqslant d$, then by Proposition 1.3.3 G has radius $> r$, and hence contains two vertices at a distance $\geqslant r$. Any shortest path in G between these two vertices contains a P^r. $\square$

(1.3.3)

In principle, we could now look for a similar set of 'unavoidable' k-connected subgraphs for any given connectivity k. To keep these 'unavoidable sets' small, it helps to relax the containment relation from 'induced subgraph' for $k = 1$ (as above) to 'topological minor' for $k = 2$, and on to 'minor' for $k = 3$ and $k = 4$. For larger k, no similar results are known.

Proposition 9.4.2. *For every $r \in \mathbb{N}$ there is an $n \in \mathbb{N}$ such that every 2-connected graph of order at least n contains C^r or $K_{2,r}$ as a topological minor.*

Proof. Let d be the n associated with r in Proposition 9.4.1, and let G be a 2-connected graph with at least $\frac{d}{d-2}(d-1)^r$ vertices. By Proposition 1.3.3, either G has a vertex of degree $> d$ or $\mathrm{diam}(G) \geqslant \mathrm{rad}(G) > r$.

(1.3.3)
(3.3.6)

In the latter case let $a, b \in G$ be two vertices at distance $> r$. By Menger's theorem (3.3.6), G contains two independent a–b paths. These form a cycle of length $> r$.

Assume now that G has a vertex v of degree $> d$. Since G is 2-connected, $G - v$ is connected and thus has a spanning tree; let T be a minimal tree in $G - v$ that contains all the neighbours of v. Then every leaf of T is a neighbour of v. By the choice of d, either T has a vertex of degree $\geqslant r$ or T contains a path of length $\geqslant r$, without loss of generality linking two leaves. Together with v, such a path forms a cycle of length $\geqslant r$. A vertex u of degree $\geqslant r$ in T can be joined to v by r independent paths through T, to form a $TK_{2,r}$. $\square$

Theorem 9.4.3. (Oporowski, Oxley & Thomas 1993)
For every $r \in \mathbb{N}$ there is an $n \in \mathbb{N}$ such that every 3-connected graph of order at least n contains a wheel of order r or a $K_{3,r}$ as a minor.

Let us call a graph of the form $C^n * \overline{K^2}$ ($n \geqslant 4$) a *double wheel*, the 1-skeleton of a triangulation of the cylinder as in Fig. 9.4.1 a *crown*, and the 1-skeleton of a triangulation of the Möbius strip a *Möbius crown*.

Theorem 9.4.4. (Oporowski, Oxley & Thomas 1993)
For every $r \in \mathbb{N}$ there is an $n \in \mathbb{N}$ such that every 4-connected graph with at least n vertices has a minor of order $\geqslant r$ that is a double wheel, a crown, a Möbius crown, or a $K_{4,r}$.

Fig. 9.4.1. A crown and a Möbius crown

Note that the graphs listed in Theorems 9.4.3 and 9.4.4 are themselves 3-connected resp. 4-connected, as required.

At first glance, the 'unavoidable' substructures presented in the four theorems above may seem to be chosen somewhat arbitrarily. In fact, the contrary is true: these sets are smallest possible, and as such unique.

non-trivial property

To make this precise, call a graph property *non-trivial* if it contains graphs of infinitely many isomorphism types. Given two such properties $\mathcal{P}, \mathcal{P}'$ and an order relation $\leqslant$ between graphs (such as the subgraph relation $\subseteq$, or the minor relation $\preccurlyeq$), write $\mathcal{P} \leqslant \mathcal{P}'$ if for every $G \in \mathcal{P}$ there is a $G' \in \mathcal{P}'$ such that $G \leqslant G'$. If $\mathcal{P} \leqslant \mathcal{P}'$ as well as $\mathcal{P} \geqslant \mathcal{P}'$, call $\mathcal{P}$ and $\mathcal{P}'$ *equivalent* and write $\mathcal{P} \sim \mathcal{P}'$. For example, if $\leqslant$ is the subgraph relation, $\mathcal{P}$ is the class of all paths, $\mathcal{P}'$ is the class of paths of even length, and $\mathcal{S}$ is the class of all subdivisions of stars, then $\mathcal{P} \sim \mathcal{P}' \leqslant \mathcal{S} \nleqslant \mathcal{P}$.

Kuratowski set

Given a non-trivial graph property $\mathcal{G}$, call a finite set $\{\mathcal{P}_1, \ldots, \mathcal{P}_k\}$ of non-trivial graph properties $\mathcal{P}_i \subseteq \mathcal{G}$ a *Kuratowski set* for $\mathcal{G}$ and $\leqslant$ if the $\mathcal{P}_i$ are incomparable (i.e., $\mathcal{P}_i \nleqslant \mathcal{P}_j$ whenever $i \neq j$) and for every non-trivial graph property $\mathcal{P} \subseteq \mathcal{G}$ there is an i such that $\mathcal{P}_i \leqslant \mathcal{P}$. Such a Kuratowski set $\{\mathcal{P}_1, \ldots, \mathcal{P}_k\}$ is unique up to equivalence: if $\{\mathcal{Q}_1, \ldots, \mathcal{Q}_\ell\}$ is another Kuratowski set for $\mathcal{G}$ then $\ell = k$ and, with suitable enumeration, $\mathcal{Q}_i \sim \mathcal{P}_i$ for $i = 1, \ldots, k$. (Why?)

The essence of our last four theorems can now be stated more comprehensively, as follows. Let us say *k-connectedness* for the class of all k-connected finite graphs, and *connectedness* for 1-connectedness.

Theorem 9.4.5.

(i) *The stars and the paths form the (2-element) Kuratowski set for connectedness and the subgraph relation.*

(ii) *The cycles and the graphs $K_{2,r}$ ($r \in \mathbb{N}$) form the (2-element) Kuratowski set for 2-connectedness and the topological minor relation.*

(iii) *The wheels and the graphs $K_{3,r}$ ($r \in \mathbb{N}$) form the (2-element) Kuratowski set for 3-connectedness and the minor relation.*

(iv) *The double wheels, the crowns, the Möbius crowns, and the graphs $K_{4,r}$ ($r \in \mathbb{N}$) form the (4-element) Kuratowski set for 4-connectedness and the minor relation.* $\square$

Exercises

1.⁻ Determine the Ramsey number $R(3)$.

2.⁻ Deduce the case $k = 2$ (but c arbitrary) of Theorem 9.1.3 directly from Theorem 9.1.1.

3. An *arithmetic progression* is an increasing sequence of numbers of the form $a, a + d, a + 2d, a + 3d \dots$. Van der Waerden's theorem says that no matter how we partition the natural numbers into two classes, one of these classes will contain arbitrarily long arithmetic progressions. Must there even be an infinite arithmetic progression in one of the classes?

4. Can you improve the exponential upper bound on the Ramsey number $R(n)$ for perfect graphs?

5.⁺ Construct a graph on $\mathbb{R}$ that has neither a complete nor an edgeless induced subgraph on $|\mathbb{R}| = 2^{\aleph_0}$ vertices. (So Ramsey's theorem does not extend to uncountable sets.)

6.⁺ Prove the edge version of the Erdős-Pósa theorem (2.3.2): there exists a function $g: \mathbb{N} \to \mathbb{R}$ such that, given $k \in \mathbb{N}$, every graph contains either k edge-disjoint cycles or a set of at most $g(k)$ edges meeting all its cycles.

   (Hint. Consider in each component a normal spanning tree T. If T has many chords xy, use any regular pattern of how the paths xTy intersect to find many edge-disjoint cycles.)

7.⁺ Use Ramsey's theorem to show that for any $k, \ell \in \mathbb{N}$ there is an $n \in \mathbb{N}$ such that every sequence of n distinct integers has an increasing subsequence of length $k + 1$ or a decreasing subsequence of length $\ell + 1$. Prove that $n = k\ell + 1$ has this property, but that $n = k\ell$ does not.

8. Show that for every $k \in \mathbb{N}$ there is an $n \in \mathbb{N}$ such that among any n points in the plane, no three of them collinear, there are k points spanning a convex k-gon, i.e. such that none of them lies in the convex hull of the others.

9. Show that for every $k \in \mathbb{N}$ there is an $n \in \mathbb{N}$ such that, for every partition of $\{1, \dots, n\}$ into k sets, at least one of the subsets contains numbers x, y, z such that $x + y = z$.

10. Let $(X, \leqslant)$ be a totally ordered set, and let $G = (V, E)$ be the graph on $V := [X]^2$ with $E := \{(x, y)(x', y') \mid x < y = x' < y'\}$.

    (i) Show that G contains no triangle.

    (ii) Show that $\chi(G)$ will get arbitrarily large if $|X|$ is chosen large enough.

11. A family of sets is called a Δ-*system* if every two of the sets have the same intersection. Show that every infinite family of sets of the same finite cardinality contains an infinite Δ-system.

12. Prove that for every $r \in \mathbb{N}$ and every tree T there exists a $k \in \mathbb{N}$ such that every graph G with $\chi(G) \geqslant k$ and $\omega(G) < r$ contains a subdivision of T in which no two branch vertices are adjacent in G (unless they are adjacent in T).

13. Let $m, n \in \mathbb{N}$, and assume that $m - 1$ divides $n - 1$. Show that every tree T of order m satisfies $R(T, K_{1,n}) = m + n - 1$.

14. Prove that $2^c < R(2, c, 3) \leqslant 3c!$ for every $c \in \mathbb{N}$.
    (Hint. Induction on c.)

15. Explain why, in the proof of Theorem 9.2.2, choosing ϵ small enough can ensure that the regularity graph R contains a copy of K^ℓ, although some of the pairs (V_i, V_j) in G may not be ϵ-regular. Your explanation may use that $t_{\ell-1}(k) \approx \frac{\ell-2}{\ell-1}\binom{k}{2}$, but should contain no calculations.

16.⁻ Derive the statement $(*)$ in the first proof of Theorem 9.3.1 from the theorem itself, i.e. show that $(*)$ is only formally stronger than the theorem.

17.⁻ How is n defined in the first proof of Theorem 9.3.1? Could it be zero, and if so how does the proof work then?

18. Show that, given any two graphs H_1 and H_2, there exists a graph $G = G(H_1, H_2)$ such that, for every vertex-colouring of G with colours 1 and 2, there is either an induced copy of H_1 coloured 1 or an induced copy of H_2 coloured 2 in G.

19. Show that the Ramsey graph G for H constructed in the second proof of Theorem 9.3.1 does indeed satisfy $\omega(G) = \omega(H)$.

20. In the second proof of Theorem 9.3.1, is it really necessary to equip G^{k+1} for $i \notin \{i_1, i_2\}$ with separate disjoint copies of V_k^i, one for every p, or could we define G^{k+1} from G^k by just replacing P with P' and joining it to the other V_i^k in the right way?

21.⁻ Show that any Kuratowski set for a non-trivial graph property is unique up to equivalence.

22. Deduce Theorem 9.4.5 (iii) from Theorem 9.4.3, and vice versa.

Notes

Due to increased interaction with research on random and pseudo-random[4] structures (the latter being provided, for example, by the regularity lemma), the Ramsey theory of graphs has recently seen a period of major activity and advance. Theorem 9.2.2 is an early example of this development.

For the more classical approach, the introductory text by R.L. Graham, B.L. Rothschild & J.H. Spencer, *Ramsey Theory* (2nd edn.), Wiley 1990, makes stimulating reading. This book includes a chapter on graph Ramsey theory, but is not confined to it. Surveys of finite and infinite Ramsey theory are given by J. Nešetřil and A. Hajnal in their chapters in the *Handbook of Combinatorics* (R.L. Graham, M. Grötschel & L. Lovász, eds.), North-Holland

---

[4] Concrete graphs whose structure resembles the structure expected of a random graph are called *pseudo-random*. For example, the bipartite graphs spanned by an ϵ-regular pair of vertex sets in a graph are pseudo-random.

1995. The Ramsey theory of infinite sets forms a substantial part of combinatorial set theory, and is treated in depth in P. Erdős, A. Hajnal, A. Máté & R. Rado, *Combinatorial Set Theory*, North-Holland 1984. An attractive collection of highlights from various branches of Ramsey theory, including applications in algebra, geometry and point-set topology, is offered in B. Bollobás, *Graph Theory*, Springer GTM 63, 1979.

Ramsey's original theorem, Theorem 9.1.1, is from F.P. Ramsey, On a problem of formal logic, *Proc. Lond. Math. Soc.* **2** (1930), 264–286. The Erdős-Hajnal conjecture is taken from P. Erdős & A. Hajnal, Ramsey-type theorems, *Discrete Appl. Math.* **25** (1989), 37–52. A survey on the state of the art a couple of years ago was given by M. Chudnovsky, The Erdős-Hajnal conjecture—a survey, *J. Graph Theory* **75** (2014), 178–190, arXiv:1606.08827.

Theorem 9.2.2 is due to V. Chvátal, V. Rödl, E. Szemerédi & W.T. Trotter, The Ramsey number of a graph with bounded maximum degree, *J. Comb. Theory B* **34** (1983), 239–243. Our proof follows the sketch in J. Komlós & M. Simonovits, Szemerédi's Regularity Lemma and its applications in graph theory, in (D. Miklós, V.T. Sós & T. Szőnyi, eds.) *Paul Erdős is 80*, Vol. 2, Proc. Colloq. Math. Soc. János Bolyai (1996). The theorem marks a breakthrough towards a conjecture of Burr and Erdős (1975), which asserts that the Ramsey numbers of graphs with bounded *average* degree in every subgraph are linear: for every $d \in \mathbb{N}$, the conjecture says, there exists a constant c such that $R(H) \leqslant c|H|$ for all graphs H with $d(H') \leqslant d$ for all $H' \subseteq H$. This conjecture has been verified approximately by A. Kostochka and B. Sudakov, On Ramsey numbers of sparse graphs, *Comb. Probab. Comput.* **12** (2003), 627–641, who proved that $R(H) \leqslant |H|^{1+o(1)}$.

Our first proof of Theorem 9.3.1 is based on W. Deuber, A generalization of Ramsey's theorem, in (A. Hajnal, R. Rado & V.T. Sós, eds.) *Infinite and finite sets*, North-Holland 1975. The same volume contains the alternative proof of this theorem by Erdős, Hajnal and Pósa. Rödl proved the same result in his MSc thesis at Charles University, Prague, in 1973. Our second proof of Theorem 9.3.1, which preserves the clique number of H for G, is due to J. Nešetřil & V. Rödl, Simple proof of the existence of restricted Ramsey graphs by means of a partite construction, *Combinatorica* **1** (1981), 199–202. These authors later refined their methods to obtain an even stronger version of Theorem 9.3.1, with a proof that doubles as a construction of graphs of large chromatic number and girth (Theorem 11.2.2); see J. Nešetřil & V. Rödl, Sparse Ramsey graphs, *Combinatorica* **4** (1984), 71–78.

The two theorems in Section 9.4 are based on B. Oporowski, J. Oxley & R. Thomas, Typical subgraphs of 3- and 4-connected graphs, *J. Comb. Theory B* **57** (1993), 239–257. They have been generalized to arbitrary k, but for a weaker 'global' notion of connectivity as often used in graph minor theory, by Benson Joeris, Connectivity, tree-decompositions and unavoidable minors, PhD thesis, University of Waterloo (2015).

10   Hamilton Cycles

In Chapter 1.8 we briefly discussed the problem of when a graph contains
an Euler tour, a closed walk traversing every edge exactly once. The
simple Theorem 1.8.1 solved that problem quite satisfactorily. Let us
now ask the analogous question for vertices: when does a graph G contain
a closed walk that contains every vertex of G exactly once? If $|G| \geqslant 3$,
then any such walk is a cycle: a *Hamilton cycle* of G. If G has a Hamilton
cycle, it is called *hamiltonian*. Similarly, a path in G containing every
vertex of G is a *Hamilton path*.

    *Hamilton
cycle*

    *Hamilton
path*

    To determine whether or not a given graph has a Hamilton cycle is
much harder than deciding whether it is Eulerian, and no good characterization is known[1] of the graphs that do. In the first two sections of
this chapter we present the standard sufficient conditions for the existence of a Hamilton cycle, as well as a more recent non-standard one. The
third section is devoted to the proof of another classic: Fleischner's
theorem that the 'square' of every 2-connected graph has a Hamilton
cycle. We shall present this theorem with an ingenious short proof due
to Georgakopoulos.

10.1 Sufficient conditions

What kind of condition might be sufficient for the existence of a Hamilton
cycle in a graph G? Purely global assumptions, like high edge density,
will not be enough: we cannot do without the local property that every
vertex has at least two neighbours. But neither is any large (but constant) minimum degree sufficient: it is easy to find graphs without a Hamilton cycle whose minimum degree exceeds any given constant bound.

    The following classic result derives its significance from this back-
ground:

---

[1] ... or indeed expected to exist; see the notes for details.

© Reinhard Diestel 2017
R. Diestel, *Graph Theory*, Graduate Texts in Mathematics 173,
DOI 10.1007/978-3-662-53622-3_10

Theorem 10.1.1. (Dirac 1952)
Every graph with $n \geqslant 3$ vertices and minimum degree at least $n/2$ has a Hamilton cycle.

Proof. Let $G = (V, E)$ be a graph with $|G| = n \geqslant 3$ and $\delta(G) \geqslant n/2$. Then G is connected: otherwise, the degree of any vertex in the smallest component C of G would be less than $|C| \leqslant n/2$.

Let $P = v_0 \ldots v_k$ be a longest path in G. Let us call v_i the *left* end of the edge $v_i v_{i+1}$, and v_{i+1} its *right* end. By the maximality of P, each of the $d(v_0) \geqslant n/2$ neighbours of v_0 is the right end of an edge of P, and these $d(v_0)$ edges are distinct. Similarly, at least $n/2$ edges of P are such that their left end is adjacent to v_k. Since P has fewer than n edges, it has an edge $v_i v_{i+1}$ with both properties (Fig. 10.1.1).

Fig. 10.1.1. Finding a Hamilton cycle in the proof Theorem 10.1.1

We claim that the cycle $C := v_0 v_{i+1} P v_k v_i P v_0$ is a Hamilton cycle of G. Indeed, since G is connected, C would otherwise have a neighbour in $G - C$, which could be combined with a spanning path of C into a path longer than P. $\qquad\qquad\square$

Theorem 10.1.1 is best possible in that we cannot replace the bound of $n/2$ with $\lfloor n/2 \rfloor$: if n is odd and G is the union of two copies of $K^{\lceil n/2 \rceil}$ meeting in one vertex, then $\delta(G) = \lfloor n/2 \rfloor$ but $\kappa(G) = 1$, so G cannot have a Hamilton cycle. In other words, the high level of the bound of $\delta \geqslant n/2$ is needed to ensure, if nothing else, that G is 2-connected: a condition just as trivially necessary for hamiltonicity as a minimum degree of at least 2. It would seem, therefore, that prescribing some high (constant) value for κ rather than for δ stands a better chance of implying hamiltonicity. However, this is not so: although every large enough k-connected graph contains a cycle of length at least $2k$ (Ex. 21, Ch. 3), the graphs $K_{k,n}$ show that this is already best possible.

Slightly more generally, a graph G with a separating set S of k vertices such that $G - S$ has more than k components is clearly not hamiltonian. Could it be true that all non-hamiltonian graphs have such a separating set, one that leaves many components compared with its size? We shall return to this question at the end of this section.

For now, just note that such graphs as above also have relatively large independent sets: pick one vertex from each component of $G - S$ to obtain one of order at least $k + 1$. Might we be able to force a Hamilton cycle by forbidding large independent sets?

By itself, the assumption of $\alpha(G) \leqslant k$ already guarantees a cycle of length at least $|G|/k$ (Ex. 13, Ch. 5). And combined with the assumption of k-connectedness, it does indeed imply hamiltonicity:

Proposition 10.1.2. *Every graph G with $|G| \geqslant 3$ and $\alpha(G) \leqslant \kappa(G)$ has a Hamilton cycle.*

Proof. Put $\kappa(G) =: k$, and let C be a longest cycle in G. Enumerate the vertices of C cyclically, say as $V(C) = \{ v_i \mid i \in \mathbb{Z}_n \}$ with $v_i v_{i+1} \in E(C)$ for all $i \in \mathbb{Z}_n$. If C is not a Hamilton cycle, pick a vertex $u \in G - C$ and a u–C fan $\mathcal{F} = \{ P_i \mid i \in I \}$ in G, where $I \subseteq \mathbb{Z}_n$ and each P_i ends in v_i. Let $\mathcal{F}$ be chosen with maximum cardinality; then $uv_j \notin E(G)$ for any $j \notin I$, and

$$|\mathcal{F}| \geqslant \min\{k, |C|\} \tag{1}$$

by Menger's theorem (3.3.4).

(3.3.4)
k

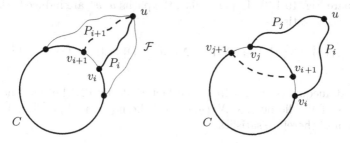

Fig. 10.1.2. Two cycles longer than C

For every $i \in I$, we have $i + 1 \notin I$: otherwise, $(C \cup P_i u P_{i+1}) - v_i v_{i+1}$ would be a cycle longer than C (Fig. 10.1.2, left). Thus $|\mathcal{F}| < |C|$, and hence $|I| = |\mathcal{F}| \geqslant k$ by (1). Furthermore, $v_{i+1} v_{j+1} \notin E(G)$ for all $i, j \in I$, as otherwise $(C \cup P_i u P_j) + v_{i+1} v_{j+1} - v_i v_{i+1} - v_j v_{j+1}$ would be a cycle longer than C (Fig. 10.1.2, right). Hence $\{ v_{i+1} \mid i \in I \} \cup \{u\}$ is a set of $k + 1$ or more independent vertices in G, contradicting $\alpha(G) \leqslant k$. □

Our next result uses the ideas from the proof of Proposition 10.1.2 to establish a *local* degree condition for hamiltonicity, considerably strengthening Dirac's theorem and several similar results proved later in its wake:

Theorem 10.1.3. (Asratian & Khachatrian 1990)
A connected graph G of order at least 3 is hamiltonian if

$$d(u) + d(w) \geqslant |N(u) \cup N(v) \cup N(w)|$$

for every induced path uvw.

Proof. Consider any induced path uvw in G. Since $d(u) + d(w) = |N(u) \cup N(w)| + |N(u) \cap N(w)|$, our degree assumption implies that

$$|N(u) \cap N(w)| \geqslant |N(u) \cup N(v) \cup N(w)| - |N(u) \cup N(w)|$$
$$= |N(v) \smallsetminus N(\{u, w\})| \geqslant |\{u, w\}| \geqslant 2. \qquad (1)$$

In particular, G contains a cycle.

C   Let C be a longest cycle in G. Assuming that G is not hamiltonian,
u   pick a vertex $u \notin C$ that has a neighbour on C and put $V := N(u) \cap V(C)$.
V   For vertices $v \in V$ let v^+ denote the successor of v on C along some fixed
V⁺ orientation of C, and put $V^+ := \{v^+ \mid v \in V\}$.

Since C is a longest cycle, we have $V \cap V^+ = \emptyset$, and

> *no two vertices of* $V^+ \cup \{u\}$ *are adjacent or have a common*
> *neighbour outside* C (2)

(compare Fig. 10.1.2). In particular, the paths uvv^+ are induced. Hence
every $v \in V$ satisfies

$$|N(u) \cap N(v^+)| \underset{(1)}{\geqslant} |N(v) \smallsetminus N(\{u, v^+\})| \geqslant |N(v) \cap V^+| + 1 ;$$

the last inequality comes from the fact that, by (2), both u and the
vertices of V^+ lie outside $N(\{u, v^+\})$. The number $\|V, V^+\|$ of V–V^+
edges of G therefore satisfies

$$\|V, V^+\| = \sum_{v \in V} |N(v) \cap V^+| \leqslant \sum_{v \in V} \big(|N(u) \cap N(v^+)| - 1\big) \underset{(2)}{=} \|V, V^+\| - |V|$$

(a contradiction); for the last equality note that, by (2), v^+ has all its
common neighbours with u in V. $\qquad\qquad\qquad\qquad\qquad\qquad\qquad\qquad$ □

Let us return to the question of whether an assumption that no
small separator leaves many components can guarantee a Hamilton cycle.
t-tough   A graph G is called t-*tough*, where $t > 0$ is any real number, if for every
separator S the graph $G - S$ has at most $|S|/t$ components. Clearly,
hamiltonian graphs must be 1-tough—so what about the converse?

Unfortunately, it is not difficult to find even small graphs that are
1-tough but have no Hamilton cycle (Exercise 6), so toughness does not
provide a characterization of hamiltonian graphs in the spirit of Menger's
theorem or Tutte's 1-factor theorem. However, a famous conjecture asserts that t-toughness for some t will force hamiltonicity:

Toughness Conjecture. (Chvátal 1973)
There exists an integer t *such that every* t-*tough graph has a Hamilton
cycle.*

The toughness conjecture was long expected to hold even with $t = 2$. This was disproved after many years, but the general conjecture remains open. See the exercises for how the conjecture ties in with the results given in the remainder of this chapter.

It may come as a surprise to learn that hamiltonicity is also related to the four colour problem. As we noted in Chapter 6.6, the four colour theorem is equivalent to the non-existence of a planar snark, i.e. to the assertion that every bridgeless planar cubic graph has a 4-flow. It is easily checked that 'bridgeless' can be replaced with '3-connected' in this assertion, and that every hamiltonian graph has a 4-flow (Ex. 16, Ch. 6). For a proof of the four colour theorem, therefore, it would suffice to show that every 3-connected planar cubic graph has a Hamilton cycle!

Unfortunately, this is not the case: the first counterexample was found by Tutte in 1946. Ten years later, Tutte proved the following deep theorem as a best possible weakening:

Theorem 10.1.4. (Tutte 1956)
Every 4-connected planar graph has a Hamilton cycle.

Although, at first glance, it appears that the study of Hamilton cycles is a part of graph theory that cannot possibly extend to infinite graphs, there is a fascinating conjecture that does just that. Recall that a *circle* in an infinite graph G is a homeomorphic copy of the unit circle S^1 in the topological space $|G|$ formed by G and its ends (see Chapter 8.6). A *Hamilton circle* of G is a circle that contains every vertex of G.

Hamilton circle

Conjecture. (Bruhn 2003)
Every locally finite 4-connected planar graph has a Hamilton circle.

10.2 Hamilton cycles and degree sequences

Historically, Dirac's theorem formed the point of departure for the discovery of a series of weaker and weaker degree conditions, all sufficient for hamiltonicity. The development culminated in a single theorem that encompasses all the earlier results: the theorem we shall prove in this section.

If G is a graph with n vertices and degrees $d_1 \leqslant \ldots \leqslant d_n$, then the n-tuple $(d_1, \ldots, d_n)$ is called the *degree sequence* of G. Note that this sequence is unique, even though G has several vertex enumerations giving rise to its degree sequence. Let us call an arbitrary integer sequence $(a_1, \ldots, a_n)$ *hamiltonian* if every graph with n vertices and a degree sequence pointwise greater than $(a_1, \ldots, a_n)$ is hamiltonian. (A sequence $(d_1, \ldots, d_n)$ is *pointwise greater* than $(a_1, \ldots, a_n)$ if $d_i \geqslant a_i$ for all i.)

degree sequence

hamiltonian sequence

pointwise greater

The following theorem characterizes all hamiltonian sequences:

Theorem 10.2.1. (Chvátal 1972)
An integer sequence $(a_1, \ldots, a_n)$ such that $0 \leqslant a_1 \leqslant \ldots \leqslant a_n < n$ and $n \geqslant 3$ is hamiltonian if and only if the following holds for every $i < n/2$:

$$a_i \leqslant i \Rightarrow a_{n-i} \geqslant n-i \,.$$

$(a_1, \ldots, a_n)$ *Proof.* Let $(a_1, \ldots, a_n)$ be an arbitrary integer sequence such that $0 \leqslant a_1 \leqslant \ldots \leqslant a_n < n$ and $n \geqslant 3$. We first assume that this sequence satisfies the condition of the theorem and prove that it is hamiltonian.

Suppose not. Then there exists a graph whose degree sequence
$(d_1, \ldots, d_n)$ $(d_1, \ldots, d_n)$ satisfies

$$d_i \geqslant a_i \qquad \text{for all } i \tag{1}$$

$G = (V, E)$ but which has no Hamilton cycle. Let $G = (V, E)$ be such a graph, chosen with the maximum number of edges.

By (1), our assumptions for $(a_1, \ldots, a_n)$ transfer to the degree sequence $(d_1, \ldots, d_n)$ of G; thus,

$$d_i \leqslant i \Rightarrow d_{n-i} \geqslant n-i \qquad \text{for all } i < n/2. \tag{2}$$

x, y Let x, y be distinct and non-adjacent vertices in G, with $d(x) \leqslant d(y)$ and $d(x) + d(y)$ as large as possible. One easily checks that the degree sequence of $G + xy$ is pointwise greater than $(d_1, \ldots, d_n)$, and hence than $(a_1, \ldots, a_n)$. Hence, by the maximality of G, the new edge xy lies on a
$x_1, \ldots, x_n$ Hamilton cycle H of $G + xy$. Then $H - xy$ is a Hamilton path $x_1, \ldots, x_n$ in G, with $x_1 = x$ and $x_n = y$ say.

As in the proof of Dirac's theorem, we now consider the index sets

$$I := \{ i \mid xx_{i+1} \in E \} \quad \text{and} \quad J := \{ j \mid x_j y \in E \} \,.$$

Then $I \cup J \subseteq \{1, \ldots, n-1\}$, and $I \cap J = \emptyset$ because G has no Hamilton cycle. Hence

$$d(x) + d(y) = |I| + |J| < n \,, \tag{3}$$

h so $h := d(x) < n/2$ by the choice of x.

Since $x_i y \notin E$ for all $i \in I$, all these x_i were candidates for the choice of x (together with y). Our choice of $\{x, y\}$ with $d(x) + d(y)$ maximum thus implies that $d(x_i) \leqslant d(x)$ for all $i \in I$. Hence G has at least $|I| = h$ vertices of degree at most h, so $d_h \leqslant h$. By (2), this implies that $d_{n-h} \geqslant n - h$, i.e. the $h + 1$ vertices with the degrees $d_{n-h}, \ldots, d_n$ all have degree at least $n - h$. Since $d(x) = h$, one of these vertices,
z z say, is not adjacent to x. Since

$$d(x) + d(z) \geqslant h + (n - h) = n \,,$$

this contradicts the choice of x and y by (3).

Let us now show that, conversely, for every sequence $(a_1, \ldots, a_n)$ as in the theorem, but with

$$a_h \leqslant h \quad \text{and} \quad a_{n-h} \leqslant n - h - 1$$

for some $h < n/2$, there exists a graph that has a pointwise greater degree sequence than $(a_1, \ldots, a_n)$ but no Hamilton cycle. As the sequence

$$\big(\underbrace{h, \ldots, h}_{h \text{ times}}, \underbrace{n-h-1, \ldots, n-h-1}_{n-2h \text{ times}}, \underbrace{n-1, \ldots, n-1}_{h \text{ times}}\big)$$

is pointwise greater than $(a_1, \ldots, a_n)$, it suffices to find a graph with this degree sequence that has no Hamilton cycle.

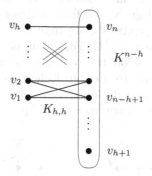

Fig. 10.2.1. Any cycle containing $v_1, \ldots, v_h$ misses v_{h+1}

Figure 10.2.1 shows such a graph, with vertices $v_1, \ldots, v_n$ and the edge set

$$\{v_i v_j \mid i, j > h\} \cup \{v_i v_j \mid i \leqslant h;\ j > n - h\};$$

it is the union of a K^{n-h} on the vertices $v_{h+1}, \ldots, v_n$ and a $K_{h,h}$ with partition sets $\{v_1, \ldots, v_h\}$ and $\{v_{n-h+1}, \ldots, v_n\}$. $\qquad \square$

By applying Theorem 10.2.1 to graphs of the form $G * K^1$, one can easily prove the following adaptation of the theorem to Hamilton paths. Let an integer sequence be called *path-hamiltonian* if every graph with a pointwise greater degree sequence has a Hamilton path.

Corollary 10.2.2. *An integer sequence $(a_1, \ldots, a_n)$ such that $n \geqslant 2$ and $0 \leqslant a_1 \leqslant \ldots \leqslant a_n < n$ is path-hamiltonian if and only if every $i \leqslant n/2$ is such that $a_i < i \Rightarrow a_{n+1-i} \geqslant n - i$.* $\qquad \square$

10.3 Hamilton cycles in the square of a graph

G^d

Given a graph G and a positive integer d, we denote by G^d the graph on $V(G)$ in which two vertices are adjacent if and only if they have distance at most d in G. Clearly, $G = G^1 \subseteq G^2 \subseteq \ldots$ Our goal in this section is to prove the following fundamental result:

Theorem 10.3.1. (Fleischner 1974)
If G is a 2-connected graph, then G^2 has a Hamilton cycle.

The proof of Theorem 10.3.1 will go roughly as follows. We start by finding a cycle C in G. Using induction, we shall cover the remaining vertices by C-paths in G^2. The first and last edges of those paths will be edges of G, like those of C. By deleting some of these edges and doubling others, we turn the union of C and all the C-paths into a multigraph with even degrees, and find an Euler tour in it. This Euler tour W will pass some vertices more than once, but all edges in such multiple passes will be edges of G. For all but one of the passes through a given vertex we can therefore try to replace its two G-edges by an edge of G^2 (Fig. 10.3.1), hoping to turn our Euler tour into a Hamilton cycle of G^2. The main difficulty will be to ensure that these lifts of passes are compatible, i.e., that we do not attempt to lift an edge at both its ends.

Fig. 10.3.1. Reducing the degree of v in W by lifting a pass

Lemma 10.3.2. *For every 2-connected graph G and $x \in V(G)$, there is a cycle $C \subseteq G$ that contains x as well as a vertex $y \neq x$ with $N_G(y) \subseteq V(C)$.*

Proof. If G has a Hamilton cycle, there is nothing more to show. If not, let $C' \subseteq G$ be any cycle containing x; such a cycle exists, since G is 2-connected. Let D be a component of $G - C'$. Assume that C' and D are chosen so that $|D|$ is minimum. Since G is 2-connected, D has at least two neighbours on C'. Then C' contains a path P between two such neighbours u and v, whose interior $\mathring{P}$ does not contain x and has no neighbour in D (Fig. 10.3.2).

Replacing P in C' by a u–v path through D, we obtain a cycle C that contains x and a vertex $y \in D$. If y had a neighbour z in $G - C$, then z would lie in a component $D' \subsetneq D$ of $G - C$, contradicting the choice of C' and D. Hence all the neighbours of y lie on C, and C satisfies the assertion of the lemma. $\square$

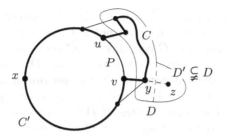

Fig. 10.3.2. The proof of Lemma 10.3.2

For our proof of Theorem 10.3.1 we need some more definitions. Let G be a multigraph, and W a walk in G. A *pass* of W *through* a vertex x is a subwalk of the form $uexfv$, where e and f are edges. (We also count $uexfv$ as a pass of W if $W = xfv \ldots uex$.) By *lifting* this pass we mean replacing it in W by a new u–v edge if $u \neq v$, or by the single vertex u if $u = v$. A *multipath* is a multigraph obtained from a path by replacing some of its edges by double edges. Given $C \subseteq G$, we define a *C-trail* to be either a C-path or a cycle meeting C in exactly one vertex.

pass

lift

multipath
C-trail

Proof of Theorem 10.3.1. Let $G = (V, E)$ be a 2-connected graph. We prove the following stronger assertion by induction on $|G|$:

(1.2.1)
(1.8.1)

$G = (V, E)$

> For every vertex $x \in V$ there is a Hamilton cycle in G^2
> whose edges at x lie in E.

x

If G is hamiltonian, there is nothing more to show. If not, let C and y be as provided by Lemma 10.3.2. For $i = 1, 2$ let $r_i, s_i \in V(C)$ and $g_i, h_i \in E(C)$ be such that

C, y

$$C = xg_1r_1 \ldots s_1h_1yh_2s_2 \ldots r_2g_2x;$$

$r_i, s_i; \; g_i, h_i$

see Figure 10.3.3. (These vertices and edges need not all be distinct.)

Our first aim is to construct for every component D of $G - C$ a set of C-trails in $G^2 + \overline{E}$, where $\overline{E}$ will be a set of additional edges parallel to edges of G. Every vertex of D will lie on exactly one such trail, and every edge of such a trail that is incident with a vertex of C will lie in E or in $\overline{E}$.

If D consists of a single vertex u, we pick any C-trail in G containing u, and let E_D be the set of its two edges. If $|D| > 1$, let $\tilde{D}$ be the (2-connected) graph obtained from G by contracting $G - D$ to a vertex $\tilde{x}$. Applying the induction hypothesis to $\tilde{D}$, we obtain a Hamilton cycle $\tilde{H}$ of $\tilde{D}^2$ whose edges at $\tilde{x}$ lie in $E(\tilde{D})$. Write $\tilde{E}$ for the set of those edges of $\tilde{H}$ that are not edges of G^2; these include its two edges at $\tilde{x}$. Replacing the edges from $\tilde{E}$ by edges of G or new edges $\overline{e} \in \overline{E}$, we shall turn $E(\tilde{H})$ into the edge set of a union of C-trails.

Consider an edge $uv \in \tilde{E}$, with $u \in D$. Then either $v = \tilde{x}$, or u and v have distance at most 2 in $\tilde{D}$ but not in G, and are hence neighbours of $\tilde{x}$ in $\tilde{D}$. In either case, G contains a u–C edge. Let E_D be obtained from $E(\tilde{H}) \setminus \tilde{E}$ by adding at every vertex $u \in D$ as many u–C edges from E as u has incident edges in $\tilde{E}$; if u has two incident edges in $\tilde{E}$ but in G sends only one edge e to C, we add both e and a new edge $\bar{e}$ parallel to e. Then every vertex of D has the same degree (two) in (V, E_D) as in $\tilde{H}$, so E_D is the edge set of a union of C-trails. Let

$\bar{e}$

G_0

$$G_0 := \left(V, E(C) \cup \bigcup_D E_D\right)$$

be the union of C and all these C-trails, for all components D of $G - C$ together.

Our next aim is to turn G_0 into an Eulerian multigraph by doubling some edges of C. Since G_0 is connected, it will suffice to do this in such a way that all degrees become even (Theorem 1.8.1).[2] The vertices of G_0 outside C already have degree 2. To make the degrees even also at the vertices of C we consider these in reverse order, starting with x and ending with r_1. Let u be the vertex currently considered, and let v be

$\bar{e}$

the vertex to be considered next. Add a new edge $\bar{e}$ parallel to $e = uv$ if and only if u has odd degree in the multigraph obtained from G_0 so far. When finally $u = r_1$ is considered, every other vertex has even degree, so r_1 must have even degree too (Lemma 1.2.1), and no edge parallel to g_1 will be added.

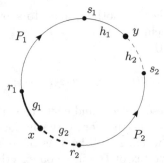

Fig. 10.3.3. Broken edges may not exist in G_1; bold edges are known to be single (if they exist)

From the Eulerian multigraph thus obtained we may now delete one or two double edges, as follows (Fig. 10.3.3). If g_2 has a parallel edge $\bar{g}_2$, we delete both g_2 and $\bar{g}_2$. If h_2 has a parallel edge $\bar{h}_2$, we delete h_2 and $\bar{h}_2$ unless this (together with the deletion of g_2 and $\bar{g}_2$) would

---

[2] To deduce the multigraph version of Theorem 1.8.1, subdivide every edge once to obtain a simple graph.

disconnect our multigraph. We write G_1 for the Eulerian multigraph
thus obtained, $\overline{E} = E(G_1) \smallsetminus E(G^2)$ for the set of all its new parallel
edges, and $C_1 := G_1[V(C)]$.

G_1
$\overline{E}$
C_1

Let us note two properties of G_1, which follow from its construction
and the definition of y:

$$\text{The edges of } G_1 \text{ at vertices of } C \text{ all lie in } E \cup \overline{E}. \tag{1}$$

$$N_{G_1}(y) \subseteq \{s_1, s_2\}; \text{ thus, } y \text{ has degree 2 or 4 in } G_1. \tag{2}$$

Let $P_1 = x_0^1 \ldots x_{\ell_1}^1$ be the (maximal) x–y multipath in C_1 containing g_1, and let $P_2 = x_0^2 \ldots x_{\ell_2}^2$ be the multipath consisting of the
other edges of C_1. Unless P_2 is empty, we think of it as running from
$x_0^2 \in \{x, r_2\}$ to $x_{\ell_2}^2 \in \{y, s_2\}$. We write e_j^i for the x_{j-1}^i–x_j^i edge of P_i
in $E(C)$, and $\overline{e}_j^i$ for its possible parallel edge in $\overline{E}$ ($i = 1, 2$).

$\ell_i, \ x_j^i$
P_i

$e_j^i, \overline{e}_j^i$

Our plan is to find an Euler tour W_1 of G_1 that can be transformed
into a Hamilton cycle of G^2. In order to endow W_1 more easily with
the required properties, we shall not define it directly. Instead, we shall
derive W_1 from an Euler tour W_2 of a related multigraph G_2, which we
define next.

For $i = 1, 2$ and every $j = 1, \ldots, \ell_i - 1$ such that $\overline{e}_{j+1}^i \in G_1$, we delete
e_j^i and $\overline{e}_{j+1}^i$ from G_1 and add a new edge f_j^i joining x_{j-1}^i to x_{j+1}^i; we shall
say that f_j^i represents the path $x_{j-1}^i e_j^i x_j^i \overline{e}_{j+1}^i x_{j+1}^i \subseteq P_i$ (Fig. 10.3.4).
Note that every such replacement leaves the current multigraph con-
nected, and it preserves the parity of all degrees. Hence, the multigraph
G_2 obtained from G_1 by all these replacements is Eulerian.

f_j^i

G_2

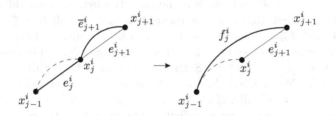

Fig. 10.3.4. Replacing e_j^i and $\overline{e}_{j+1}^i$ by a new edge f_j^i

Pick an Euler tour W_2 of G_2. To transform W_2 into an Euler tour
W_1 of G_1, replace every edge in $E(W_2) \smallsetminus E(G_1)$ by the path it represents.

W_2
W_1

By (2), there are either one or two passes of W_1 through y. If $d(y) = 4$
then G_1 is connected but $G_1 - \{h_2, \overline{h}_2\}$ is not, by definition of G_1. Hence
the proper y–y subwalk W of W_1 starting or ending with the edge h_2
must end or start with the edge $\overline{h}_2$: otherwise it would contain an s_2–y
walk in $G_1 - \{h_2, \overline{h}_2\}$, and deleting h_2 and $\overline{h}_2$ from G_1 could not affect
its connectedness. Therefore W_1 has no pass through y containing both
h_2 and $\overline{h}_2$: this would close the subwalk W and thus imply $W = W_1$,

contrary to the definition of W. Reversing W if necessary, we may thus assume:

> If W_1 contains two passes through y, then one of these contains h_1 and h_2. (3)

Our plan is to transform W_1 into a Hamilton cycle of G^2 by lifting, at every vertex $v \in C$, all but one of the passes of W_1 through v. We begin by marking the one pass at each vertex that we shall not lift. At x we mark an arbitrary pass Q of W_1. At y we mark the pass containing h_1. For every $v \in V(C) \smallsetminus \{x, y\}$ there is a unique pair (i, j) such that $v = x_j^i$. If $j \geqslant 1$, we mark the pass through $v = x_j^i$ that contains e_j^i. If $j = 0$ (which can happen only if $v = x_0^2 = r_2$), we mark the pass through v that contains $\bar{e}_1^2$ if $\bar{e}_1^2 \in G_1$; otherwise we mark an arbitrary pass through v. At every vertex $v \notin C$ we mark the unique pass of W_1 through v. Thus:

> At every vertex $v \in G$ we marked exactly one pass through v. (4)

To avoid conflicts when we later lift the unmarked passes, we need that no edge of W_1 is left unmarked at both its ends:

> For every edge $e = uv$ in W_1 we marked at least one of the two passes of W_1 containing e (one through u, the other through v). If $u = x$, we marked the pass through v. (5)

This is clear for edges not in C_1. For every edge $e \in C_1$, there is a unique pair (i, j) such that $e = e_j^i$ or $e = \bar{e}_j^i$; then $j \geqslant 1$. If $e = e_j^i$, we marked the pass of W_1 through x_j^i that contains e; for $e = h_2$ this follows from (3). If $e = \bar{e}_j^i$, we marked the pass through x_{j-1}^i containing e. Indeed, note first that e is not incident with x: recall that $\bar{g}_1$ never existed, and if $\bar{g}_2$ existed it was deleted in the definition of G_1. Hence unless P_2 starts at r_2 and $e = \bar{e}_1^2$, an edge f_{j-1}^i was defined to represent the path $x_{j-2}^i e_{j-1}^i x_{j-1}^i \bar{e}_j^i x_j^i$. Since W_2 contained f_{j-1}^i, this path is a pass in W_1. We marked this pass, because it is a pass through x_{j-1}^i containing e_{j-1}^i. Finally, if P_2 starts at r_2 and $e = \bar{e}_1^2$, we marked the pass through r_2 containing e explicitly. This completes the proof of (5).

By (1), all unmarked passes lift to edges of G^2. As different unmarked passes never share an edge (5), lifting them all at once turns W_1 into one closed walk $\overline{H}$ in $G^2 + \overline{E}$ (which inherits the cyclic ordering of its edges from W_1). By (4), $\overline{H}$ still contains all the vertices v of G: if uv is an edge of a pass marked at v, then $\overline{H}$ contains either the edge uv or the lift wv of a pass $weufv$. Also by (4), $\overline{H}$ traverses every vertex only once. In particular, $\overline{H}$ cannot contain a pair of parallel edges. We can therefore replace every edge $\bar{e}$ in $\overline{H}$ by its parallel edge $e \in E$, to obtain a Hamilton cycle H of G^2. Since we marked Q, and by (5) no edge of Q was lifted at its other end, $\overline{H}$ contains the edges of Q. By (1), these lie in $E \cup \overline{E}$. Hence the edges of H at x lie in E, as desired. $\square$

Fleischner's theorem has a natural extension to locally finite graphs, which is much harder to prove:

Theorem 10.3.3. (Georgakopoulos 2009)
The square of a 2-connected locally finite graph contains a Hamilton circle.

We close the chapter with a far-reaching conjecture generalizing Dirac's theorem:

Conjecture. (Seymour 1974)
Let G be a graph of order $n \geqslant 3$, and let k be a positive integer. If G has minimum degree

$$\delta(G) \geqslant \frac{k}{k+1} n \,,$$

then G has a Hamilton cycle H such that $H^k \subseteq G$.

For $k = 1$, this is precisely Dirac's theorem. The conjecture was proved for large enough n (depending on k) by Komlós, Sárközy and Szemerédi (1998).

Exercises

1. An oriented complete graph is called a *tournament*. Show that every tournament contains a (directed) Hamilton path.

2. Show that every uniquely 3-edge-colourable cubic graph is hamiltonian. ('Unique' means that all 3-edge-colourings induce the same edge partition.)

3. Given an even positive integer k, construct for every $n \geqslant k$ a k-regular graph of order $2n + 1$.

4. Prove or disprove the following strengthening of Proposition 10.1.2: 'Every k-connected graph G with $|G| \geqslant 3$ and $\chi(G) \geqslant |G|/k$ has a Hamilton cycle.'

5. Let G be a graph, and $H := L(G)$ its line graph.
    (i) Show that H is hamiltonian if G has a spanning Eulerian subgraph.
    (ii)[+] Deduce that H is hamiltonian if G is 4-edge-connected.

6. (i)[-] Show that hamiltonian graphs are 1-tough.
    (ii) Find a graph that is 1-tough but not hamiltonian.

7. Prove the toughness conjecture for planar graphs. Does it hold with $t = 2$, or even with some $t < 2$?

8.⁻ Find a hamiltonian graph whose degree sequence is not hamiltonian.

9.⁻ Let G be a graph with fewer than i vertices of degree at most i, for every $i < |G|/2$. Use Chvátal's theorem to show that G is hamiltonian. (Thus in particular, Chvátal's theorem implies Dirac's theorem.)

10. Prove that the square G^2 of a k-connected graph G is k-tough. Use this to deduce Fleischner's theorem for graphs satisfying the toughness conjecture with $t = 2$.

11. Show that Exercise 6 (i) has the following weak converse: for every non-hamiltonian graph G there exists a graph G' that has a pointwise greater degree-sequence than G but is not 1-tough.

12. (i) Show that, unlike the graphs satisfying Dirac's condition of $\delta \geqslant n/2$, graphs satisfying the degree condition of Theorem 10.1.3 can be sparse: there exists an integer d for which there are arbitrarily large graphs of average degree at most d that satisfy the condition.

   (ii) Show that there is no integer d that bounds the average degrees of arbitrarily large graphs satisfying Chvátal's degree condition.

13. Find a connected graph G whose square G^2 has no Hamilton cycle.

14.⁺ Show by induction on $|G|$ that the third power G^3 of any connected graph G contains a Hamilton cycle.

15.⁻ Deduce from the proof of Fleischner's theorem that the square of a 2-connected graph contains a Hamilton path between any two vertices.

16.⁺ Let G be a graph in which every vertex has odd degree. Show that every edge of G lies on an even number of Hamilton cycles.

   (Hint. Let $xy \in E(G)$ be given. The Hamilton cycles through xy correspond to the Hamilton paths in $G - xy$ from x to y. Consider the set $\mathcal{H}$ of all Hamilton paths in $G - xy$ starting at x, and show that an even number of these end in y. To show this, define a graph on $\mathcal{H}$ so that the desired assertion follows from Proposition 1.2.1.)

Notes

The problem of finding a Hamilton cycle in a graph has the same kind of origin as its Euler tour counterpart and the four colour problem: all three problems come from mathematical puzzles older than graph theory itself. What began as a game invented by W.R. Hamilton in 1857—in which 'Hamilton cycles' had to be found on the graph of the dodecahedron—reemerged over a hundred years later as a combinatorial optimization problem of prime importance: the *travelling salesman problem*. Here, a salesman has to visit a number of customers, and his problem is to arrange these in a suitable circular route. (For reasons not included in the mathematical brief, the route has to be such that after visiting a customer the salesman does not pass through that town again.) Much of the motivation for considering Hamilton cycles comes from variations of this algorithmic problem.

The lack of a good characterization of hamiltonicity also has to do with an algorithmic problem: deciding whether or not a given graph is hamiltonian is NP-hard (indeed, this was one of the early prototypes of an NP-complete decision problem), while the existence of a good characterization would place it in NP ∩ co-NP, which is widely believed to equal P. Thus, unless P = NP, no good characterization of hamiltonicity exists. See the introduction to Chapter 12.7, or the end of the notes for Chapter 12, for more.

The 'proof' of the four colour theorem indicated at the end of Section 10.1, which is based on the (false) premise that every 3-connected cubic planar graph is hamiltonian, is usually attributed to the Scottish mathematician P.G. Tait. Following Kempe's flawed proof of 1879 (see the notes for Chapter 5), it seems that Tait believed to be in possession of at least one 'new proof of Kempe's theorem'. However, when he addressed the Edinburgh Mathematical Society on this subject in 1883, he seems to have been aware that he could not—really— prove the above statement about Hamilton cycles. His account in P.G. Tait, Listing's topologie, *Phil. Mag.* **17** (1884), 30–46, makes some entertaining reading.

A shorter proof of Tutte's theorem that 4-connected planar graphs are hamiltonian has been given by C. Thomassen, A theorem on paths in planar graphs, *J. Graph Theory* **7** (1983), 169–176. Tutte's counterexample to Tait's assumption that even 3-connectedness suffices (at least for cubic graphs) is shown in Bollobás, and in J.A. Bondy & U.S.R. Murty, *Graph Theory with Applications*, Macmillan 1976 (where Tait's attempted proof is discussed in some detail).

Bruhn's conjecture generalizing Tutte's theorem to infinite graphs was first stated in R. Diestel, The cycle space of an infinite graph, *Comb. Probab. Comput.* **14** (2005), 59–79. As the notion of a Hamilton circle is relatively recent, earlier generalizations of Hamilton cycle theorems asked for spanning double rays. Now a ray can pass through a finite separator only finitely often, so a necessary condition for the existence of a spanning ray or double ray is that the graph has at most one or two ends, respectively. Confirming a long-standing conjecture of Nash-Williams, X. Yu, Infinite paths in planar graphs I– V, *J. Graph Theory* (2004–08), proved that a 4-connected planar graph with at most two ends contains a spanning double ray. N. Dean, R. Thomas and X Yu, Spanning paths in infinite planar graphs, *J. Graph Theory* **23** (1996), 163–174, proved Nash-Williams's conjecture that a one-ended 4-connected planar graph has a spanning ray.

Proposition 10.1.2 is due to Chvátal and Erdős (1972). Theorem 10.1.3 was found much later: by A.S. Asratian and N.K. Khachatrian, Some localization theorems on hamiltonian circuits, *J. Comb. Theory B* **49** (1990), 287– 294. Since its hamiltonicity condition is local, the theorem might generalize to Hamilton circles in locally finite graphs, similarly to Fleischner's theorem. This appears to be a hard problem.

The toughness invariant and conjecture were proposed by V. Chvátal, Tough graphs and hamiltonian circuits, *Discrete Math.* **5** (1973), 215–228. If true with $t = 2$, the conjecture would have implied Fleischner's thereom; see Exercise 10. However, it was disproved for $t = 2$ by D. Bauer, H.J. Broersma & H.J. Veldman, Not every 2-tough graph is hamiltonian, *Discrete Appl. Math.* **99** (2000), 317–321. Theorem 10.2.1 is due to V. Chvátal, On Hamilton's ideals, *J. Comb. Theory B* **12** (1972), 163–168.

The extension of Fleischner's theorem to locally finite graphs, Theorem 10.3.3, was proved by A. Georgakopoulos, Infinite Hamilton cycles in squares of locally finite graphs, *Adv. Math.* **220** (2009), 670–705. Our short proof of Fleischner's theorem is a windfall of that proof.

Seymour's conjecture is from P.D. Seymour, Problem 3, in (T.P. McDonough and V.C. Mavron, eds.) *Combinatorics*, Cambridge University Press 1974. Its proof for large n is due to J. Komlós, G.N. Sárközy & E. Szemerédi, Proof of the Seymour conjecture for large graphs, *Ann. Comb.* **2** (1998), 43–60.

Finally, let us mention Thomassen's conjecture (1986) that every 4-connected line graph is hamiltonian. T. Kaiser, Hamilton cycles in 5-connected line graphs, *Eur. J. Comb.* **33** (2012), 924–947, arXiv:1009.3754, proved that 5-connected line graphs of minimum degree at least 6 are hamiltonian.

11 Random Graphs

At various points in this book, we already encountered the following fundamental theorem of Erdős: *for every integer k there is a graph G with $g(G) > k$ and $\chi(G) > k$.* In plain English: there exist graphs combining arbitrarily large girth with arbitrarily high chromatic number.

How could one prove such a theorem? The standard approach would be to construct a graph with those two properties, possibly in steps by induction on k. However, this is anything but straightforward: the global nature of the second property forced by the first, namely, that the graph should have high chromatic number 'overall' but be acyclic (and hence 2-colourable) locally, flies in the face of any attempt to build it up, constructively, from smaller pieces that have the same or similar properties.

In his pioneering paper of 1959, Erdős took a radically different approach: for each n he defined a probability space on the set of graphs with n vertices, and showed that, for some carefully chosen probability measures, the probability that an n-vertex graph has both of the above properties is positive for all large enough n.

This approach, now called the *probabilistic method*, has since unfolded into a sophisticated and versatile proof technique, in graph theory as much as in other branches of discrete mathematics. The theory of *random graphs* is now a subject in its own right. The aim of this chapter is to offer an elementary but rigorous introduction to random graphs: no more than is necessary to understand its basic concepts, ideas and techniques, but enough to give an inkling of the power and elegance hidden behind the calculations.

Erdős's theorem asserts the existence of a graph with certain properties: it is a perfectly ordinary assertion showing no trace of the randomness employed in its proof. There are also results in random graphs that are generically random even in their statement: these are theorems about *almost all* graphs, a notion we shall meet in Section 11.3. In the

© Reinhard Diestel 2017
R. Diestel, *Graph Theory*, Graduate Texts in Mathematics 173,
DOI 10.1007/978-3-662-53622-3_11

last section, we give a detailed proof of a theorem of Erdős and Rényi that illustrates a proof technique frequently used in random graphs, the so-called *second moment method*.

11.1 The notion of a random graph

V

G

Let V be a fixed set of n elements, say $V = \{0, \ldots, n-1\}$. Our aim is to turn the set $\mathcal{G}$ of all graphs on V into a probability space, and then to consider the kind of questions typically asked about random objects: What is the probability that a graph $G \in \mathcal{G}$ has this or that property? What is the expected value of a given invariant on G, say its expected girth or chromatic number?

Intuitively, we should be able to generate G randomly as follows. For each $e \in [V]^2$ we decide by some random experiment whether or not e shall be an edge of G; these experiments are performed independently, and for each the probability of success—i.e. of accepting e as an edge

p

q

for G—is equal to some fixed[1] number $p \in [0, 1]$. Then if G_0 is some fixed graph on V, with m edges say, the elementary event $\{G_0\}$ has a probability of $p^m q^{\binom{n}{2} - m}$ (where $q := 1 - p$): with this probability, our randomly generated graph G is this particular graph G_0. (The probability that G is *isomorphic* to G_0 will usually be greater.) But if the probabilities of all the elementary events are thus determined, then so is the entire probability measure of our desired space $\mathcal{G}$. Hence all that remains to be checked is that such a probability measure on $\mathcal{G}$, one for which all individual edges occur independently with probability p, does indeed exist.[2]

In order to construct such a measure on $\mathcal{G}$ formally, we start by defining for every potential edge $e \in [V]^2$ its own little probability space

Ω_e

$\mathbb{P}_e$

$\mathcal{G}(n, p)$

$\Omega_e := \{0_e, 1_e\}$, choosing $\mathbb{P}_e(\{1_e\}) := p$ and $\mathbb{P}_e(\{0_e\}) := q$ as the probabilities of its two elementary events. As our desired probability space $\mathcal{G} = \mathcal{G}(n, p)$ we then take the product space

Ω

$$\Omega := \prod_{e \in [V]^2} \Omega_e.$$

---

[1] Often, the value of p will depend on the cardinality n of the set V on which our random graphs are generated; thus, p will be the value $p = p(n)$ of some function $n \mapsto p(n)$. Note, however, that V (and hence n) is fixed for the definition of $\mathcal{G}$: for each n separately, we are constructing a probability space of the graphs G on $V = \{0, \ldots, n-1\}$, and within each space the probability that $e \in [V]^2$ is an edge of G has the same value for all e.

[2] Any reader ready to believe this may skip ahead now to the end of Proposition 11.1.1, without missing anything.

Thus, formally, an element of Ω is a map ω assigning to every $e \in [V]^2$ either 0_e or 1_e, and the probability measure $\mathbb{P}$ on Ω is the product measure of all the measures $\mathbb{P}_e$. In practice, of course, we identify ω with the graph G on V whose edge set is

$$E(G) = \{\, e \mid \omega(e) = 1_e \,\},$$

and call G a *random graph* on V with edge probability p.

Following standard probabilistic terminology, we may now call any set of graphs on V an *event* in $\mathcal{G}(n,p)$. In particular, for every $e \in [V]^2$ the set

$$A_e := \{\, \omega \mid \omega(e) = 1_e \,\}$$

of all graphs G on V with $e \in E(G)$ is an event: the event that e is an edge of G. For these events, we can now prove formally what had been our guiding intuition all along:

Proposition 11.1.1. *The events A_e are independent and occur with probability p.*

Proof. By definition,

$$A_e = \{1_e\} \times \prod_{e' \neq e} \Omega_{e'}.$$

Since $\mathbb{P}$ is the product measure of all the measures $\mathbb{P}_e$, this implies

$$\mathbb{P}(A_e) = p \cdot \prod_{e' \neq e} 1 = p.$$

Similarly, if $\{e_1, \ldots, e_k\}$ is any subset of $[V]^2$, then

$$\mathbb{P}(A_{e_1} \cap \ldots \cap A_{e_k}) = \mathbb{P}\Big(\{1_{e_1}\} \times \ldots \times \{1_{e_k}\} \times \prod_{e \notin \{e_1,\ldots,e_k\}} \Omega_e\Big)$$
$$= p^k$$
$$= \mathbb{P}(A_{e_1}) \cdots \mathbb{P}(A_{e_k}).$$
$$\square$$

As noted before, $\mathbb{P}$ is determined uniquely by the value of p and our assumption that the events A_e are independent. In order to calculate probabilities in $\mathcal{G}(n,p)$, it therefore generally suffices to work with these two assumptions: our concrete model for $\mathcal{G}(n,p)$ has served its purpose and will not be needed again.

As a simple example of such a calculation, consider the event that G contains some fixed graph H on a subset of V as a subgraph; let $|H| =: k$ and $\|H\| =: \ell$. The probability of this event $H \subseteq G$ is the product of the probabilities A_e over all the edges $e \in H$, so $\mathbb{P}[H \subseteq G] = p^\ell$. In

contrast, the probability that H is an *induced* subgraph of G is $p^\ell q^{\binom{k}{2}-\ell}$:
now the edges missing from H are required to be missing from G too,
and they do so independently with probability q.

   The probability p_H that G has an induced subgraph *isomorphic*
to H is usually more difficult to compute: since the possible instances
of H on subsets of V overlap, the events that they occur in G are not
independent. However, the sum (over all k-sets $U \subseteq V$) of the probabilities $\mathbb{P}[H \simeq G[U]]$ is always an upper bound for p_H, since p_H is the
measure of the union of all those events. For example, if $H = \overline{K^k}$, we
have the following trivial upper bound on the probability that G contains
an induced copy of H:

[11.2.1]
[11.3.4]
Lemma 11.1.2. *For all integers n, k with $n \geqslant k \geqslant 2$, the probability
that $G \in \mathcal{G}(n,p)$ has a set of k independent vertices is at most*

$$\mathbb{P}[\alpha(G) \geqslant k] \leqslant \binom{n}{k} q^{\binom{k}{2}}.$$

Proof. The probability that a fixed k-set $U \subseteq V$ is independent in
G is $q^{\binom{k}{2}}$. The assertion thus follows from the fact that there are only
$\binom{n}{k}$ such sets U. $\square$

   Analogously, the probability that $G \in \mathcal{G}(n,p)$ contains a K^k is at
most

$$\mathbb{P}[\omega(G) \geqslant k] \leqslant \binom{n}{k} p^{\binom{k}{2}}.$$

Now if k is fixed, and n is small enough that these bounds for the prob-
abilities $\mathbb{P}[\alpha(G) \geqslant k]$ and $\mathbb{P}[\omega(G) \geqslant k]$ sum to less than 1, then $\mathcal{G}$
contains graphs that have neither property: graphs which contain nei-
ther a K^k nor a $\overline{K^k}$ induced. But then any such n is a lower bound for
the Ramsey number of k!

   As the following theorem shows, this lower bound is quite close to
the upper bound of 2^{2k-3} implied by the proof of Theorem 9.1.1:

Theorem 11.1.3. (Erdős 1947)
For every integer $k \geqslant 3$, the Ramsey number of k satisfies

$$R(k) > 2^{k/2}.$$

Proof. For $k = 3$ we trivially have $R(3) \geqslant 3 > 2^{3/2}$, so let $k \geqslant 4$. We show
that, for all $n \leqslant 2^{k/2}$ and $G \in \mathcal{G}(n, \frac{1}{2})$, the probabilities $\mathbb{P}[\alpha(G) \geqslant k]$
and $\mathbb{P}[\omega(G) \geqslant k]$ are both less than $\frac{1}{2}$.

   Since $p = q = \frac{1}{2}$, Lemma 11.1.2 and the analogous assertion for $\omega(G)$
imply the following for all $n \leqslant 2^{k/2}$ (use that $k! > 2^k$ for $k \geqslant 4$):

$$\mathbb{P}\left[\,\alpha(G) \geqslant k\,\right],\ \mathbb{P}\left[\,\omega(G) \geqslant k\,\right] \leqslant \binom{n}{k}\left(\tfrac{1}{2}\right)^{\binom{k}{2}}$$

$$< \left(n^k/2^k\right) 2^{-\frac{1}{2}k(k-1)}$$

$$\leqslant \left(2^{k^2/2}/2^k\right) 2^{-\frac{1}{2}k(k-1)}$$

$$= 2^{-k/2}$$

$$< \tfrac{1}{2}\,.$$

$\qquad\qquad\qquad\qquad\qquad\qquad\qquad\qquad\qquad\qquad\qquad\qquad\square$

In the context of random graphs, each of the familiar graph invariants (like average degree, connectivity, girth, chromatic number, and so on) may be interpreted as a non-negative *random variable* on $\mathcal{G}(n,p)$, a function

random variable

$$X \colon \mathcal{G}(n,p) \to [0,\infty)\,.$$

The *mean* or *expected* value of X is the number

mean

expectation

$$\mathbb{E}(X) := \sum_{G \in \mathcal{G}(n,p)} \mathbb{P}(\{G\}) \cdot X(G)\,.$$

$\mathbb{E}$

If X takes integers as values, we can compute $\mathbb{E}(X)$ alternatively by summing over these values k:

$$\mathbb{E}(X) = \sum_{k \geqslant 1} \mathbb{P}\left[\,X \geqslant k\,\right] = \sum_{k \geqslant 1} k \cdot \mathbb{P}\left[\,X = k\,\right]\,.$$

Note also that the operator $\mathbb{E}$, the *expectation*, is linear: we have $\mathbb{E}(X + Y) = \mathbb{E}(X) + \mathbb{E}(Y)$ and $\mathbb{E}(\lambda X) = \lambda\mathbb{E}(X)$ for any two random variables X, Y on $\mathcal{G}(n,p)$ and $\lambda \geqslant 0$.

Since our probability spaces are finite, the expectation can often be computed by a simple application of *double counting*, a standard combinatorial technique we met before in the proofs of Corollary 4.2.10 and Theorem 5.5.4. For example, if X is a random variable on $\mathcal{G}(n,p)$ that counts the number of subgraphs of G in some fixed set $\mathcal{H}$ of graphs on V, then $\mathbb{E}(X)$, by definition, counts the number of pairs (G, H) such that $H \in \mathcal{H}$ and $H \subseteq G$, each weighted with the probability $\mathbb{P}(\{G\})$. Algorithmically, we compute $\mathbb{E}(X)$ by going through the graphs $G \in \mathcal{G}(n,p)$ in an 'outer loop' and performing, for each G, an 'inner loop' that runs through the graphs $H \in \mathcal{H}$ and counts '$\mathbb{P}(\{G\})$' whenever $H \subseteq G$. Alternatively, we may count the same set of weighted pairs with H in the outer and G in the inner loop. This amounts to adding up, over all $H \in \mathcal{H}$, the probabilities $\mathbb{P}\left[\,H \subseteq G\,\right]$:

$$\mathbb{E}(X) = \sum_{G \in \mathcal{G}(n,p)} \left|\{H \in \mathcal{H} : H \subseteq G\}\right| \cdot \mathbb{P}(\{G\}) = \sum_{H \in \mathcal{H}} \mathbb{P}\left[\,H \subseteq G\,\right].$$

X

To illustrate this once in detail, and to introduce the probabilistic terminology commonly used at this point, let us compute the expected number of cycles of some given length $k \geqslant 3$ in a random graph $G \in \mathcal{G}(n,p)$. (We shall also need this for our proof of Erdős's theorem in Section 11.2.) Let $X: \mathcal{G}(n,p) \to \mathbb{N}$ be the random variable that assigns to a random graph G its number of k-cycles, the number of subgraphs isomorphic to C^k.

$\mathcal{C}_k$

How many potential such cycles are there? In other words, how large is the set $\mathcal{C}_k$ of all k-cycles with vertices in V? Since there are

$(n)_k$

$$(n)_k := n\,(n-1)(n-2)\cdots(n-k+1)$$

ways of choosing a sequence of k distinct vertices in V, and each k-cycle is identified by $2k$ of those sequences, clearly

$$|\mathcal{C}_k| = (n)_k/2k\,. \tag{1}$$

[11.2.2]
[11.4.3]

Lemma 11.1.4. *The expected number of k-cycles in $G \in \mathcal{G}(n,p)$ is*

$$\mathbb{E}(X) = \frac{(n)_k}{2k}\,p^k.$$

Proof. Consider for every fixed $C \in \mathcal{C}_k$ its *indicator random variable* $X_C: \mathcal{G}(n,p) \to \{0,1\}$, defined by

$$X_C\colon\ G \mapsto \begin{cases} 1 & \text{if } C \subseteq G; \\ 0 & \text{otherwise.} \end{cases}$$

Since X_C takes only 1 as a positive value, its expectation $\mathbb{E}(X_C)$ equals the measure $\mathbb{P}[\,X_C = 1\,]$ of the set of all graphs in $\mathcal{G}(n,p)$ that contain C. But this is just the probability that $C \subseteq G$:

$$\mathbb{E}(X_C) = \mathbb{P}[\,C \subseteq G\,] = p^k. \tag{2}$$

Our random variable X assigns to every graph G its number of k-cycles. Hence

$$X(G) = \sum_{C \in \mathcal{C}_k} X_C(G)$$

for every G, or $X = \sum X_C$ for short. Since the expectation is linear, applying this with (1) and (2) yields

$$\mathbb{E}(X) = \sum_{C \in \mathcal{C}_k} \mathbb{E}(X_C) = \sum_{C \in \mathcal{C}_k} \mathbb{P}[\,C \subseteq G\,] = \frac{(n)_k}{2k}\,p^k$$

as claimed.   □

Computing the mean of a random variable X can be a simple and effective way to establish the existence of a graph G such that $X(G) < a$ for some fixed $a > 0$ and, moreover, G has some desired property $\mathcal{P}$. Indeed, if the expected value of X is small, then $X(G)$ cannot be large for more than a few graphs in $\mathcal{G}(n,p)$, because $X(G) \geqslant 0$ for all $G \in \mathcal{G}(n,p)$. Hence X must be small for many graphs in $\mathcal{G}(n,p)$, and it is reasonable to expect that among these we may find one with the desired property $\mathcal{P}$.

This simple idea lies at the heart of countless non-constructive existence proofs using random graphs, including the proof of Erdős's theorem presented in the next section. Quantified, it takes the form of the following lemma, whose proof follows at once from the definition of the expectation and the additivity of $\mathbb{P}$:

Lemma 11.1.5. (Markov's Inequality)
Let $X \geqslant 0$ be a random variable on $\mathcal{G}(n,p)$ and $a > 0$. Then

$$\mathbb{P}[X \geqslant a] \leqslant \mathbb{E}(X)/a.$$

[11.2.2]
[11.4.1]
[11.4.3]

Proof.

$$\mathbb{E}(X) = \sum_{G \in \mathcal{G}(n,p)} \mathbb{P}(\{G\}) \cdot X(G) \;\geqslant\; \sum_{\substack{G \in \mathcal{G}(n,p) \\ X(G) \geqslant a}} \mathbb{P}(\{G\}) \cdot a = \mathbb{P}[X \geqslant a] \cdot a.$$

$\square$

11.2 The probabilistic method

Very roughly, the *probabilistic method* in discrete mathematics has developed from the following idea. In order to prove the existence of an object with some desired property, one defines a probability space on some larger—and certainly non-empty—class of objects, and then shows that an element of this space has the desired property with positive probability. The 'objects' inhabiting this probability space may be of any kind: partitions or orderings of the vertices of some fixed graph arise as naturally as mappings, embeddings and, of course, graphs themselves. In this section, we illustrate the probabilistic method by giving a detailed account of one of its earliest results: of Erdős's classic theorem on large girth and chromatic number (Theorem 5.2.5).

Erdős's theorem says that, given any positive integer k, there is a graph G with girth $g(G) > k$ and chromatic number $\chi(G) > k$. Let us call cycles of length at most k *short*, and sets of $|G|/k$ or more vertices *big*. For a proof of Erdős's theorem, it suffices to find a graph G without short cycles and without big independent sets of vertices: then the colour classes in any vertex colouring of G are *small* (not big), so we need more than k colours to colour G.

short

big/small

How can we find such a graph G? If we choose p small enough, then a random graph in $\mathcal{G}(n,p)$ is unlikely to contain any (short) cycles. If we choose p large enough, then G is unlikely to have big independent vertex sets. So the question is: do these two ranges of p overlap, that is, can we choose p so that, for some n, it is both small enough to give $\mathbb{P}[g \leqslant k] < \frac{1}{2}$ and large enough for $\mathbb{P}[\alpha \geqslant n/k] < \frac{1}{2}$? If so, then $\mathcal{G}(n,p)$ will contain at least one graph without either short cycles or big independent sets.

Unfortunately, such a choice of p is impossible: the two ranges of p do not overlap! As we shall see in Section 11.4, we must keep p below n^{-1} to make the occurrence of short cycles in G unlikely—but for any such p there will most likely be no cycles in G at all (Exercise 18), so G will be bipartite and hence have at least $n/2$ independent vertices.

But all is not lost. In order to make big independent sets unlikely, we shall fix p above n^{-1}, at $n^{\epsilon-1}$ for some $\epsilon > 0$. Fortunately, though, if ϵ is small enough then this will produce only *few* short cycles in G, even compared with n (rather than, more typically, with n^k). If we then delete a vertex in each of those cycles, the graph H obtained will have no short cycles, and its independence number $\alpha(H)$ will be at most that of G. Since H is not much smaller than G, its chromatic number will thus still be large, so we have found a graph with both large girth and large chromatic number.

To prepare for the formal proof of Erdős's theorem, we first show that an edge probability of $p = n^{\epsilon-1}$ is indeed always large enough to ensure that $G \in \mathcal{G}(n,p)$ 'almost surely' has no big independent set of vertices. More precisely, we prove the following stronger assertion:

Lemma 11.2.1. *Let $k > 0$ be an integer, and let $p = p(n)$ be a function of n such that $p(n) \geqslant 16k^2/n$ for n large. Then*

$$\lim_{n\to\infty} \mathbb{P}[\alpha \geqslant \tfrac{1}{2}n/k] = 0.$$

(11.1.2) *Proof.* For all integers $n \geqslant r \geqslant 2$ and $G \in \mathcal{G}(n,p)$, Lemma 11.1.2 implies

$$\mathbb{P}[\alpha \geqslant r] \leqslant \binom{n}{r} q^{\binom{r}{2}} \leqslant 2^n q^{\binom{r}{2}} \leqslant 2^n e^{-p\binom{r}{2}},$$

where the last inequality follows from the fact that $1 - p \leqslant e^{-p}$ for all p. (Compare the functions $x \mapsto e^x$ and $x \mapsto x+1$ for $x = -p$.) Now if $p \geqslant 16k^2/n$ and $r \geqslant \frac{1}{2}n/k \geqslant 2$, then

$$\mathbb{P}[\alpha \geqslant r] \leqslant 2^n e^{-pr^2/4} \leqslant 2^n e^{-pn^2/16k^2} \leqslant 2^n e^{-n} \xrightarrow[n\to\infty]{} 0.$$

As α is an integer and hence $\mathbb{P}[\alpha \geqslant r] = \mathbb{P}[\alpha \geqslant \frac{1}{2}n/k]$ for $r = \lceil \frac{1}{2}n/k \rceil$, this implies the assertion. $\qquad\square$

We are now ready to prove Theorem 5.2.5, which we restate:

Theorem 11.2.2. (Erdős 1959)

For every integer k there exists a graph H with girth $g(H) > k$ and chromatic number $\chi(H) > k$.

Proof. Assume that $k \geqslant 3$, fix ϵ with $0 < \epsilon < 1/k$, and let $p := n^{\epsilon-1}$. Let $X(G)$ denote the number of short cycles in a random graph $G \in \mathcal{G}(n,p)$, i.e. its number of cycles of length at most k.

By Lemma 11.1.4, we have

$$\mathbb{E}(X) = \sum_{i=3}^{k} \frac{(n)_i}{2i} p^i \leqslant \frac{1}{2} \sum_{i=3}^{k} n^i p^i \leqslant \frac{1}{2}(k-2)\, n^k p^k \,;$$

note that $(np)^i \leqslant (np)^k$, because $np = n^\epsilon \geqslant 1$. By Lemma 11.1.5,

$$\begin{aligned}
\mathbb{P}[X \geqslant n/2] &\leqslant \mathbb{E}(X)/(n/2) \\
&\leqslant (k-2)\, n^{k-1} p^k \\
&= (k-2)\, n^{k-1} n^{(\epsilon-1)k} \\
&= (k-2)\, n^{k\epsilon-1} .
\end{aligned}$$

As $k\epsilon - 1 < 0$ by our choice of ϵ, this implies that

$$\lim_{n \to \infty} \mathbb{P}[X \geqslant n/2] = 0 \,.$$

Let n be large enough that $\mathbb{P}[X \geqslant n/2] < \frac{1}{2}$ and $\mathbb{P}[\alpha \geqslant \frac{1}{2}n/k] < \frac{1}{2}$; the latter is possible by our choice of p and Lemma 11.2.1. Then there is a graph $G \in \mathcal{G}(n,p)$ with fewer than $n/2$ short cycles and $\alpha(G) < \frac{1}{2}n/k$. From each of those cycles delete a vertex, and let H be the graph obtained. Then $|H| \geqslant n/2$ and H has no short cycles, so $g(H) > k$. By definition of G,

$$\chi(H) \geqslant \frac{|H|}{\alpha(H)} \geqslant \frac{n/2}{\alpha(G)} > k \,.$$

$\square$

Corollary 11.2.3. *There are graphs with arbitrarily large girth and arbitrarily large values of the invariants κ, ε and δ.*

Proof. Apply Lemma 5.2.3 and Theorem 1.4.3. $\square$

[9.2.3]

(11.1.5)
(11.1.4)

p, ϵ, X

n

(1.4.3)
(5.2.3)

11.3 Properties of almost all graphs

Recall that a *graph property* is a class of graphs that is closed under isomorphism, one that contains with every graph G also the graphs isomorphic to G. If $p = p(n)$ is a fixed function (possibly constant), and $\mathcal{P}$ is a graph property, we may ask how the probability $\mathbb{P}\,[\,G \in \mathcal{P}\,]$ behaves for $G \in \mathcal{G}(n,p)$ as $n \to \infty$. If this probability tends to 1, we say that $G \in \mathcal{P}$ for *almost all* (or *almost every*) $G \in \mathcal{G}(n,p)$, or that $G \in \mathcal{P}$ *almost surely*; if it tends to 0, we say that *almost no* $G \in \mathcal{G}(n,p)$ has the property $\mathcal{P}$. (For example, in Lemma 11.2.1 we proved that, for a certain p, almost no $G \in \mathcal{G}(n,p)$ has a set of more than $\frac{1}{2}n/k$ independent vertices.)

almost all
etc.

To illustrate the new concept let us show that, for constant p, every fixed abstract[3] graph H is an induced subgraph of almost all graphs:

Proposition 11.3.1. *For every constant $p \in (0,1)$ and every graph H, almost every $G \in \mathcal{G}(n,p)$ contains an induced copy of H.*

Proof. Let H be given, and $k := |H|$. If $n \geqslant k$ and $U \subseteq \{0,\dots,n-1\}$ is a fixed set of k vertices of G, then $G[U]$ is isomorphic to H with a certain probability $r > 0$. This probability r depends on p, but not on n (why not?). Now G contains a collection of $\lfloor n/k \rfloor$ disjoint such sets U. The probability that none of the corresponding graphs $G[U]$ is isomorphic to H is $(1-r)^{\lfloor n/k \rfloor}$, since these events are independent by the disjointness of the edges sets $[U]^2$. Thus

$$\mathbb{P}\,[\,H \nsubseteq G \text{ induced}\,] \leqslant (1-r)^{\lfloor n/k \rfloor} \xrightarrow[n\to\infty]{} 0,$$

which implies the assertion. $\square$

The following lemma is a simple device enabling us to deduce that quite a number of natural graph properties (including that of Proposition 11.3.1) are shared by almost all graphs. Given $i, j \in \mathbb{N}$, let $\mathcal{P}_{i,j}$ denote the property that the graph considered contains, for any disjoint vertex sets U, W with $|U| \leqslant i$ and $|W| \leqslant j$, a vertex $v \notin U \cup W$ that is adjacent to all the vertices in U but to none in W.

$\mathcal{P}_{i,j}$

Lemma 11.3.2. *For every constant $p \in (0,1)$ and $i, j \in \mathbb{N}$, almost every graph $G \in \mathcal{G}(n,p)$ has the property $\mathcal{P}_{i,j}$.*

---

[3] The word 'abstract' is used to indicate that only the isomorphism type of H is known or relevant, not its actual vertex and edge sets. In our context, it indicates that the word 'subgraph' is used in the usual sense of 'isomorphic to a subgraph'.

Proof. For fixed U, W and $v \in G - (U \cup W)$, the probability that v is adjacent to all the vertices in U but to none in W, is

$$p^{|U|}q^{|W|} \geqslant p^i q^j.$$

Hence, the probability that no suitable v exists for these U and W, is

$$(1 - p^{|U|}q^{|W|})^{n-|U|-|W|} \leqslant (1 - p^i q^j)^{n-i-j}$$

(for $n \geqslant i + j$), since the corresponding events are independent for different v. As there are no more than n^{i+j} pairs of such sets U, W in $V(G)$ (encode sets U of fewer than i points as non-injective maps $\{0, \ldots, i-1\} \to \{0, \ldots, n-1\}$, etc.), the probability that some such pair has no suitable v is at most

$$n^{i+j}(1 - p^i q^j)^{n-i-j},$$

which tends to zero as $n \to \infty$ since $1 - p^i q^j < 1$. $\square$

Corollary 11.3.3. *For every constant $p \in (0,1)$ and $k \in \mathbb{N}$, almost every graph in $\mathcal{G}(n, p)$ is k-connected.*

Proof. By Lemma 11.3.2, it is enough to show that every graph in $\mathcal{P}_{2,k-1}$ is k-connected. But this is easy: any graph in $\mathcal{P}_{2,k-1}$ has order at least $k + 2$, and if W is a set of fewer than k vertices, then by definition of $\mathcal{P}_{2,k-1}$ any other two vertices x, y have a common neighbour $v \notin W$; in particular, W does not separate x from y. $\square$

In the proof of Corollary 11.3.3, we showed substantially more than was asked for: rather than finding, for any two vertices $x, y \notin W$, some x–y path avoiding W, we showed that x and y have a common neighbour outside W; thus, all the paths needed to establish the desired connectivity could in fact be chosen of length 2. What seemed like a clever trick in this particular proof is in fact indicative of a more fundamental phenomenon for constant edge probabilities: by an easy result in logic, any statement about graphs expressed by quantifying over vertices only (rather than over sets or sequences of vertices)[4] is either almost surely true or almost surely false. All such statements, or their negations, are in fact immediate consequences of an assertion that the graph has property $\mathcal{P}_{i,j}$, for some suitable i, j.

As a last example of an 'almost all' result we now show that almost every graph has a surprisingly high chromatic number:

---

[4] In the terminology of logic: any first order sentence in the language of graph theory

Proposition 11.3.4. *For every constant $p \in (0,1)$ and every $\epsilon > 0$, almost every graph $G \in \mathcal{G}(n,p)$ has chromatic number*

$$\chi(G) > \frac{\log(1/q)}{2+\epsilon} \cdot \frac{n}{\log n} \,.$$

Proof. For any fixed $n \geqslant k \geqslant 2$, Lemma 11.1.2 implies

(11.1.2)

$$
\begin{aligned}
\mathbb{P}\,[\,\alpha \geqslant k\,] &\leqslant \binom{n}{k} q^{\binom{k}{2}} \\
&\leqslant n^k q^{\binom{k}{2}} \\
&= q^{k\frac{\log n}{\log q} + \frac{1}{2}k(k-1)} \\
&= q^{\frac{k}{2}\left(-\frac{2\log n}{\log(1/q)} + k - 1\right)}.
\end{aligned}
$$

For

$$k := (2+\epsilon)\,\frac{\log n}{\log(1/q)}$$

the exponent of this expression tends to infinity with n, so the expression itself tends to zero. Hence, almost every $G \in \mathcal{G}(n,p)$ is such that in any vertex colouring of G no k vertices can have the same colour, so every colouring uses more than

$$\frac{n}{k} = \frac{\log(1/q)}{2+\epsilon} \cdot \frac{n}{\log n}$$

colours.   □

By a result of Bollobás (1988), Proposition 11.3.4 is sharp in the following sense: if we replace ϵ by $-\epsilon$, then the lower bound given for χ turns into an upper bound.

We finish this section with a little gem, the one and only theorem about infinite random graphs. Let $\mathcal{G}(\aleph_0, p)$ be defined exactly like $\mathcal{G}(n,p)$ for $n = \aleph_0$, as the (product) space of random graphs on $\mathbb{N}$ whose edges are chosen independently with probability p.

As we saw in Lemma 11.3.2, the properties $\mathcal{P}_{i,j}$ hold almost surely for finite random graphs with constant edge probability. It will therefore hardly come as a surprise that an infinite random graph almost surely (which now has the usual meaning of 'with probability 1') has all these properties at once. However, in Chapter 8.3 we saw that, up to isomorphism, there is exactly one countable graph, the Rado graph R, that has property $\mathcal{P}_{i,j}$ for all $i,j \in \mathbb{N}$ simultaneously; this joint property was denoted as (∗) there. Combining these facts, we get the following rather bizarre result:

Theorem 11.3.5. (Erdős & Rényi 1963)
With probability 1, a random graph $G \in \mathcal{G}(\aleph_0, p)$ with $0 < p < 1$ is isomorphic to the Rado graph R.

Proof. Given fixed disjoint finite sets $U, W \subseteq \mathbb{N}$, the probability that a (8.3.1) vertex $v \notin U \cup W$ is not joined to $U \cup W$ as expressed in property $(*)$ of Chapter 8.3 (i.e., is not joined to all of U or is joined to some vertex in W) is some number $r < 1$ depending only on U and W. The probability that none of k given vertices v is joined to $U \cup W$ as in $(*)$ is r^k, which tends to 0 as $k \to \infty$. Hence the probability that all the (infinitely many) vertices outside $U \cup W$ fail to witness $(*)$ for these sets U and W is 0.

Now there are only countably many choices for U and W as above. Since the union of countably many sets of measure 0 again has measure 0, the probability that $(*)$ fails for *any* sets U and W is still 0. Therefore G satisfies $(*)$ with probability 1. By Theorem 8.3.1 this means that, almost surely, $G \simeq R$. □

How can we make sense of the paradox that the result of infinitely many independent choices can be so predictable? The answer, of course, lies in the fact that the uniqueness of R holds only up to isomorphism. Now, constructing an automorphism for an infinite graph with property $(*)$ is a much easier task than finding one for a finite random graph, so in this sense the uniqueness is no longer that surprising. Viewed in this way, Theorem 11.3.5 expresses not a lack of variety in infinite random graphs but rather the abundance of symmetry that glosses over this variety when the graphs $G \in \mathcal{G}(\aleph_0, p)$ are viewed only up to isomorphism.

11.4 Threshold functions and second moments

The results we saw in Section 11.3 have an interesting common feature: the values of p played no role as long as they were constant, that is, independent of n. For example, if almost every graph in $\mathcal{G}(n, p)$ with $p = 0.99$ had the property considered, then the same was true for $p = 0.01$. How could this happen?

Such insensitivity of our random model to changes of p was certainly not intended. For most properties, however, the critical order of magnitude of p around which the property will 'just' occur or not occur lies below any constant value of p: it is a function of n tending to zero as $n \to \infty$. In the proof of Erdős's theorem, for example, this critical probability for the two properties we were trying to relate was $p(n) = 1/n$.

Let us call a positive real function $t = t(n)$ a *threshold function* for a graph property $\mathcal{P}$ if the following holds for all $p = p(n)$ and $G \in \mathcal{G}(n, p)$:

$$\lim_{n \to \infty} \mathbb{P}[G \in \mathcal{P}] = \begin{cases} 0 & \text{if } p/t \to 0 \text{ as } n \to \infty \\ 1 & \text{if } p/t \to \infty \text{ as } n \to \infty. \end{cases}$$

If $\mathcal{P}$ has a threshold function t, then clearly any positive multiple ct of t is also a threshold function for $\mathcal{P}$; thus, threshold functions in the above sense are only ever unique up to a multiplicative constant.[5]

Bollobás & Thomason (1987) have shown that, trivial exceptions aside, all *increasing* graph properties have threshold functions, properties that are closed under the addition of edges. Properties of the form $\{\, G \mid G \supseteq H \,\}$, with H fixed, are a common example; we shall compute their threshold functions in this section.

For the purpose of computing its threshold function, it is convenient to cast the graph property $\mathcal{P}$ considered in the form

$$\mathcal{P} = \{\, G \mid X(G) \geqslant 1 \,\},$$

$X \geqslant 0$ where $X \geqslant 0$ is a suitable random variable on $\mathcal{G}(n, p)$. For example, we could take the indicator random variable of $\mathcal{P}$ on $\mathcal{G}(n, p)$. But other choices of X are allowed too; if $\mathcal{P}$ is connectedness, for example, we might have $X(G)$ count the number of spanning trees of G.

How could we prove that some given t is a threshold function of $\mathcal{P}$? Any such proof will consist of two parts: a proof that almost no $G \in \mathcal{G}(n, p)$ has $\mathcal{P}$ when p is small compared with t, and one showing that almost every G has $\mathcal{P}$ when p is large.

Since $X \geqslant 0$, we may use Markov's inequality for the first part of the proof and find an upper bound for $\mathbb{E}(X)$ instead of $\mathbb{P}[X \geqslant 1]$: if $\mathbb{E}(X)$ is much smaller than 1 then $X(G)$ can be at least 1 only for few $G \in \mathcal{G}(n, p)$, and for almost no G if $\mathbb{E}(X) \to 0$ as $n \to \infty$. Besides, the expectation is much easier to calculate than probabilities: without worrying about such things as independence or incompatibility of events, we may compute the expectation of a sum of random variables—for example, of indicator random variables—simply by adding up their individual expected values.

For the second part of the proof, things are more complicated. In order to show that $\mathbb{P}[X \geqslant 1]$ is large, it is not enough to bound $\mathbb{E}(X)$ from below: since X is not bounded above, $\mathbb{E}(X)$ may be large simply because X is very large on just a few graphs G—so X may still be zero for most $G \in \mathcal{G}(n, p)$.[6] In order to prove that $\mathbb{P}[X \geqslant 1] \to 1$, we thus

---

[5] Our notion of threshold reflects only the crudest interesting level of screening: for some properties, such as connectedness, one can define sharper thresholds where the constant factor is crucial.

[6] For some p between n^{-1} and $(\log n)n^{-1}$, for example, almost every $G \in \mathcal{G}(n, p)$ has an isolated vertex (and hence no spanning tree), but its expected number of spanning trees tends to infinity with n. See the Exercise 11 for details.

have to show that this cannot happen, i.e. that X does not deviate a lot from its mean too often.

The following tool from probability theory achieves just that. As is customary, we write

$$\mu := \mathbb{E}(X)$$

and define $\sigma \geqslant 0$ by setting

$$\sigma^2 := \mathbb{E}\big((X - \mu)^2\big).$$

This quantity σ^2 is called the *variance* or *second moment* of X; by definition, it is a measure of how much X deviates from its mean. Since $\mathbb{E}$ is linear, the defining term for σ^2 expands to

$$\sigma^2 = \mathbb{E}(X^2 - 2\mu X + \mu^2) = \mathbb{E}(X^2) - \mu^2.$$

Note that μ and σ^2 always refer to a random variable on some fixed probability space. In our setting, where we consider the spaces $\mathcal{G}(n, p)$, both quantities are functions of n.

The following lemma says exactly what we need: that X cannot deviate a lot from its mean too often.

Lemma 11.4.1. (Chebyshev's Inequality)
For all real $\lambda > 0$,

$$\mathbb{P}\big[\, |X - \mu| \geqslant \lambda \,\big] \;\leqslant\; \sigma^2/\lambda^2.$$

Proof. By Lemma 11.1.5 and definition of σ^2, (11.1.5)

$$\mathbb{P}\big[\, |X - \mu| \geqslant \lambda \,\big] = \mathbb{P}\big[\, (X - \mu)^2 \geqslant \lambda^2 \,\big] \leqslant \sigma^2/\lambda^2. \qquad \square$$

For a proof that $X(G) \geqslant 1$ for almost all $G \in \mathcal{G}(n, p)$, Chebyshev's inequality can be used as follows:

Lemma 11.4.2. *If* $\mu > 0$ *for all large enough* n, *and* $\sigma^2/\mu^2 \to 0$ *as* $n \to \infty$, *then* $X(G) > 0$ *for almost all* $G \in \mathcal{G}(n, p)$.

Proof. Any graph G with $X(G) = 0$ satisfies $|X(G) - \mu| = \mu$. Hence Lemma 11.4.1 implies with $\lambda := \mu$ that

$$\mathbb{P}\big[\, X = 0 \,\big] \;\leqslant\; \mathbb{P}\big[\, |X - \mu| \geqslant \mu \,\big] \;\leqslant\; \sigma^2/\mu^2 \;\xrightarrow[n \to \infty]{}\; 0.$$

Since $X \geqslant 0$, this means that $X > 0$ almost surely, i.e. that $X(G) > 0$ for almost all $G \in \mathcal{G}(n, p)$. $\qquad \square$

As the main result of this section, we now prove a theorem that will

$\mathcal{P}_H$ give us a threshold function for all graph properties of the form $\mathcal{P}_H$, the property of containing a copy of a fixed graph H as a subgraph.

$H; k, \ell$ Let H be given, put $k := |H|$ and $\ell := \|H\|$, and assume that $\ell \geqslant 1$. Write $X(G)$ for the number of subgraphs of a graph G that are
X isomorphic to H.

Given $n \in \mathbb{N}$, let $\mathcal{H}$ denote the set of all copies of H on subsets of $\{0, \ldots, n-1\}$, the vertex set of the graphs in $\mathcal{G}(n,p)$:

$\mathcal{H}$ $$\mathcal{H} := \left\{ H' \mid H' \simeq H, \ V(H') \subseteq \{0, \ldots, n-1\} \right\}.$$

Given $H' \in \mathcal{H}$ and $G \in \mathcal{G}(n,p)$, we shall write $H' \subseteq G$ to express that H' itself—not just an isomorphic copy of H'—is a subgraph of G. As in the proof of Lemma 11.1.4, double counting gives

$$\mathbb{E}(X) = \sum_{H' \in \mathcal{H}} \mathbb{P}\left[H' \subseteq G \right]. \tag{1}$$

And for every fixed $H' \in \mathcal{H}$ we have

$$\mathbb{P}\left[H' \subseteq G \right] = p^\ell, \tag{2}$$

because $\|H\| = \ell$.

h Let h denote the number of isomorphic copies of H on a fixed k-set; clearly, $h \leqslant k!$. As there are $\binom{n}{k}$ possible vertex sets for the graphs in $\mathcal{H}$, we thus have

$$|\mathcal{H}| = \binom{n}{k} h \leqslant \binom{n}{k} k! \leqslant n^k. \tag{3}$$

Given a probability $p = p(n)$ and a candidate $t = t(n)$ for a threshold
γ function, we write $\gamma := p/t$. Our first lemma treats the case of $\gamma \to 0$:

Lemma 11.4.3. *If $t = n^{-1/\varepsilon(H)}$, and p is such that $\gamma \to 0$ as $n \to \infty$, then almost no $G \in \mathcal{G}(n,p)$ lies in $\mathcal{P}_H$.*

(11.1.5)
(11.1.4)
Proof. Our aim is to find an upper bound for $\mathbb{E}(X)$, and to show that this tends to zero as $n \to \infty$. By our choice of t and definition of γ we have

$$p = \gamma t = \gamma n^{-k/\ell},$$

so

$$\mathbb{E}(X) \underset{(1,2)}{=} |\mathcal{H}|\, p^\ell \underset{(3)}{\leqslant} n^k (\gamma n^{-k/\ell})^\ell = \gamma^\ell.$$

Thus if $\gamma \to 0$ as $n \to \infty$, then

$$\mathbb{P}\left[G \in \mathcal{P}_H \right] = \mathbb{P}\left[X \geqslant 1 \right] \leqslant \mathbb{E}(X) \leqslant \gamma^\ell \underset{n\to\infty}{\longrightarrow} 0$$

by Markov's inequality (11.1.5). $\qquad\qquad\qquad\qquad\qquad \square$

Unlike the function t in Lemma 11.4.3, our threshold function for $\mathcal{P}_H$ will not be expressed in terms of $\varepsilon(H)$, but of

$$\varepsilon'(H) := \max\{\,\varepsilon(H') \mid H' \subseteq H\,\}. \qquad \varepsilon'(H)$$

Our second lemma treats the case of $\gamma \to \infty$:

Lemma 11.4.4. *If* $t = n^{-1/\varepsilon'(H)}$, *and* p *is such that* $\gamma \to \infty$ *as* $n \to \infty$, *then almost every* $G \in \mathcal{G}(n,p)$ *lies in* $\mathcal{P}_H$.

Proof. By our new choice of t and definition of γ we now have

$$p = \gamma n^{-1/\varepsilon'}, \qquad (4)$$

where $\varepsilon' := \varepsilon'(H)$. $\qquad\qquad \varepsilon'$

Before we start on the main proof, let us note an inequality for later use. For all $n \geqslant k$,

$$
\begin{aligned}
\binom{n}{k} n^{-k} &= \frac{1}{k!}\left(\frac{n}{n}\cdots\frac{n-k+1}{n}\right)\\
&\geqslant \frac{1}{k!}\left(\frac{n-k+1}{n}\right)^k\\
&\geqslant \frac{1}{k!}\left(1-\frac{k-1}{k}\right)^k. \qquad (5)
\end{aligned}
$$

When n gets large and k is bounded, as in our case, the upshot is that n^k exceeds $\binom{n}{k}$ by no more than a constant factor, one independent of n.

Our goal is to apply Lemma 11.4.2, and hence to bound $\sigma^2/\mu^2 = \big(\mathbb{E}(X^2) - \mu^2\big)/\mu^2$ from above. To help us estimate $\mathbb{E}(X^2)$, we begin by rewriting X^2 in a strange way, as

$$X^2(G) = \big|\{H \in \mathcal{H} : H \subseteq G\}\big|^2 = \big|\{(H',H'') \in \mathcal{H}^2 : H' \subseteq G \text{ and } H'' \subseteq G\}\big|.$$

We can now calculate $\mathbb{E}(X^2)$ by double counting, just as in (1):

$$\mathbb{E}(X^2) = \sum_{(H',H'') \in \mathcal{H}^2} \mathbb{P}\,[\,H' \cup H'' \subseteq G\,]. \qquad (6)$$

Let us then calculate these probabilities $\mathbb{P}\,[\,H' \cup H'' \subseteq G\,]$. Given $H', H'' \in \mathcal{H}$, we have

$$\mathbb{P}\,[\,H' \cup H'' \subseteq G\,] = p^{2\ell - \|H' \cap H''\|}.$$

As $\|H' \cap H''\| \leqslant i\varepsilon'$ for $i := |H' \cap H''|$ by definition of ε', this yields

$$\mathbb{P}[H' \cup H'' \subseteq G] \leqslant p^{2\ell - i\varepsilon'}. \tag{7}$$

We have now estimated the individual summands in (6); what does this imply for the sum as a whole? Since (7) depends on the parameter $i = |H' \cap H''|$, we partition the range $\mathcal{H}^2$ of the sum in (6) into the subsets

$$\mathcal{H}^2_i := \{(H', H'') \in \mathcal{H}^2 : |H' \cap H''| = i\}, \qquad i = 0, \ldots, k,$$

and calculate for each $\mathcal{H}^2_i$ the corresponding sum

$$A_i := \sum_i \mathbb{P}[H' \cup H'' \subseteq G]$$

by itself. (Here, as below, we use $\sum_i$ to denote sums over all pairs $(H', H'') \in \mathcal{H}^2_i$.)

If $i = 0$ then H' and H'' are disjoint, so the events $H' \subseteq G$ and $H'' \subseteq G$ are independent. Hence,

$$
\begin{aligned}
A_0 &= \sum_0 \mathbb{P}[H' \cup H'' \subseteq G] \\
&= \sum_0 \mathbb{P}[H' \subseteq G] \cdot \mathbb{P}[H'' \subseteq G] \\
&\leqslant \sum_{(H', H'') \in \mathcal{H}^2} \mathbb{P}[H' \subseteq G] \cdot \mathbb{P}[H'' \subseteq G] \\
&= \Big(\sum_{H' \in \mathcal{H}} \mathbb{P}[H' \subseteq G]\Big) \cdot \Big(\sum_{H'' \in \mathcal{H}} \mathbb{P}[H'' \subseteq G]\Big) \\
&\underset{(1)}{=} \mu^2. \tag{8}
\end{aligned}
$$

Let us now estimate A_i for $i \geqslant 1$. Note that $\sum_i$ can be written as $\sum_{H' \in \mathcal{H}} \sum_{H'' \in \mathcal{H}: |H' \cap H''| = i}$. For fixed H', the second sum ranges over

$$\binom{k}{i}\binom{n-k}{k-i} h$$

summands: the number of graphs $H'' \in \mathcal{H}$ with $|H'' \cap H'| = i$. Hence, for all $i \geqslant 1$ and suitable constants c_1, c_2 independent of n,

$$
\begin{aligned}
A_i &= \sum_i \mathbb{P}[H' \cup H'' \subseteq G] \\
&\underset{(7)}{\leqslant} \sum_{H' \in \mathcal{H}} \binom{k}{i}\binom{n-k}{k-i} h\, p^{2\ell} p^{-i\varepsilon'} \\
&\underset{(4)}{=} |\mathcal{H}| \binom{k}{i}\binom{n-k}{k-i} h\, p^{2\ell} \big(\gamma\, n^{-1/\varepsilon'}\big)^{-i\varepsilon'}
\end{aligned}
$$

$$\leqslant |\mathcal{H}|\, p^\ell c_1\, n^{k-i}\, h\, p^\ell \gamma^{-i\varepsilon'}\, n^i$$

$$\underset{(1,2)}{=} \mu\, c_1 n^k h\, p^\ell \gamma^{-i\varepsilon'}$$

$$\underset{(5)}{\leqslant} \mu\, c_2 \binom{n}{k} h\, p^\ell \gamma^{-i\varepsilon'}$$

$$\underset{(1,2,3)}{=} \mu^2 c_2 \gamma^{-i\varepsilon'}$$

$$\leqslant \mu^2 c_2 \gamma^{-\varepsilon'}$$

if $\gamma \geqslant 1$. By definition of the A_i, this implies with $c_3 := kc_2$ that

$$\mathbb{E}(X^2)/\mu^2 \underset{(6)}{=} \left(A_0/\mu^2 + \sum_{i=1}^{k} A_i/\mu^2 \right) \underset{(8)}{\leqslant} 1 + c_3 \gamma^{-\varepsilon'}$$

and hence

$$\frac{\sigma^2}{\mu^2} = \frac{\mathbb{E}(X^2) - \mu^2}{\mu^2} \leqslant c_3 \gamma^{-\varepsilon'} \xrightarrow[\gamma \to \infty]{} 0,$$

since $\varepsilon' \geqslant \varepsilon > 0$ by our assumption that $\ell = \|H\| > 0$. By Lemma 11.4.2, therefore, $X > 0$ almost surely, i.e. almost every $G \in \mathcal{G}(n,p)$ has a subgraph isomorphic to H and hence lies in $\mathcal{P}_H$. $\square$

Theorem 11.4.5. (Erdős & Rényi 1960; Bollobás 1981)
Let H be a graph with at least one edge. Then $t = n^{-1/\varepsilon'(H)}$ is a threshold function for $\mathcal{P}_H$.

Proof. We have to show that almost no $G \in \mathcal{G}(n,p)$ lies in $\mathcal{P}_H$ if $\gamma \to 0$ as $n \to \infty$, and that almost all $G \in \mathcal{G}(n,p)$ lie in $\mathcal{P}_H$ if $\gamma \to \infty$ as $n \to \infty$. This latter assertion was proved in Lemma 11.4.4.

To prove that almost no $G \in \mathcal{G}(n,p)$ lies in $\mathcal{P}_H$ if $\gamma \to 0$, we apply Lemma 11.4.3 to a subgraph $H' \subseteq H$ for which the maximum in the definition of $\varepsilon'(H)$ is attained, i.e. which is such that $\varepsilon(H') = \varepsilon'(H)$. The lemma implies that almost no $G \in \mathcal{G}(n,p)$ contains a copy of H'. Since any graph containing H also contains H', this implies that almost no $G \in \mathcal{G}(n,p)$ contains a copy of H, as desired. $\square$

The bound in Theorem 11.4.5 is particularly easy to compute for *balanced* graphs H, those for which $\varepsilon'(H) = \varepsilon(H)$. Cycles and trees are examples of balanced graphs. For cycles, we have the threshold familiar from the proof of Erdős's theorem:

Corollary 11.4.6. If $k \geqslant 3$, then $t(n) = n^{-1}$ is a threshold function for the property of containing a k-cycle. $\square$

Note that t does not depend on k. (See also Exercise 18.)

For trees, there is a similar phenomenon. Here, the threshold function does depend on the order of the tree, but not on its shape:

Corollary 11.4.7. *If T is a tree of order $k \geqslant 2$, then $t(n) = n^{-k/(k-1)}$ is a threshold function for the property of containing a copy of T.* $\square$

The systematic study of threshold functions has led to an overall picture of how the typical properties of a graph $G \in \mathcal{G}(n,p)$ unfold as the growth rate of $p = p(n)$ increases. This picture, dubbed the *evolution of random graphs*, is quite fascinating: as in the evolution of species, changes happen 'in jumps', marked by the times the growth rate of p crosses that of a threshold function.

For a very rough sketch, let us begin with edge probabilities p whose order of magnitude lies below n^{-2}; for such p, a random graph $G \in \mathcal{G}(n,p)$ almost surely has no edges at all. But as p grows, it acquires more and more structure. As p approaches n^{-1}, its components become larger and larger trees (Corollary 11.4.7), until at $p = n^{-1}$ the first cycles are born (Exercise 18). Soon, some of these will have several crossing chords, making the graph non-planar. At the same time, one component outgrows the others, until it devours them around $p = (\log n)n^{-1}$, making the graph connected. Hardly later, at a mere $p = (1+\epsilon)(\log n)n^{-1}$, our random graph already almost surely has a Hamilton cycle...

Exercises

1.⁻ What is the probability that a random graph in $\mathcal{G}(n,p)$ has exactly m edges, for $0 \leqslant m \leqslant \binom{n}{2}$ fixed?

2. What is the expected number of edges in $G \in \mathcal{G}(n,p)$?

3. What is the expected number of K^r-subgraphs in $G \in \mathcal{G}(n,p)$?

4. Characterize the graphs that occur as a subgraph in every graph of sufficiently large average degree.

5. In the usual terminology of measure spaces (and in particular, of probability spaces), the phrase 'almost all' is used to refer to a set of points whose complement has measure zero. Rather than considering a limit of probabilities in $\mathcal{G}(n,p)$ as $n \to \infty$, would it not be more natural to define a probability space on the set of *all* finite graphs (one copy of each) and to investigate properties of 'almost all' graphs in this space, in the sense above?

6. Show that if almost all $G \in \mathcal{G}(n,p)$ have a graph property $\mathcal{P}_1$ and almost all $G \in \mathcal{G}(n,p)$ have a graph property $\mathcal{P}_2$, then almost all $G \in \mathcal{G}(n,p)$ have both properties, i.e. have the property $\mathcal{P}_1 \cap \mathcal{P}_2$.

7. Show that, for constant $p \in (0,1)$, almost every graph in $\mathcal{G}(n,p)$ has diameter 2.

8.   Show that, for constant $p \in (0,1)$, almost no graph in $\mathcal{G}(n,p)$ has a separating complete subgraph.

9.   Derive Proposition 11.3.1 from Lemma 11.3.2.

10. Show that for every graph H there exists a function $p = p(n)$ such that $\lim_{n \to \infty} p(n) = 0$ but almost every $G \in \mathcal{G}(n,p)$ contains an induced copy of H.

11.$^+$ (i) Show that, for every $0 < \epsilon \leqslant 1$ and $p = (1-\epsilon)(\ln n)n^{-1}$, almost every $G \in \mathcal{G}(n,p)$ has an isolated vertex.

   (ii) Find a probability $p = p(n)$ such that almost every $G \in \mathcal{G}(n,p)$ is disconnected but the expected number of spanning trees of G tends to infinity as $n \to \infty$.

   (Hint for (ii): A theorem of Cayley states that K^n has exactly n^{n-2} spanning trees.)

12.$^+$ Given $r \in \mathbb{N}$, find a $c > 0$ such that, for $p = cn^{-1}$, almost every $G \in \mathcal{G}(n,p)$ has a K^r minor. Can c be chosen independently of r?

13. Find an increasing graph property without a threshold function, and a property that is not increasing but has a threshold function.

14.$^-$ Let H be a graph of order k, and let h denote the number of graphs isomorphic to H on some fixed set of k elements. Show that $h \leqslant k!$. For which graphs H does equality hold?

15.$^-$ For every $k \geqslant 1$, find a threshold function for $\{\, G \mid \Delta(G) \geqslant k \,\}$.

16.$^-$ For every $d \in \mathbb{N}$, determine the threshold function for the property of containing the d-dimensional cube (see Exercise 2, Chapter 1), and for the property of containing the complete graph K^d.

17. Does the property of containing any tree of order k (for $k \geqslant 2$ fixed) have a threshold function? If so, which? If not, why not?

18. Show that $t(n) = n^{-1}$ is also a threshold function for the property of containing any cycle.

19. Consider the terms A_0 and A_1 in the proof of Lemma 11.4.4, which are both functions of n. Recall that $\mathbb{P}\,[\,H' \cup H'' \subseteq G\,] = p^{2\ell}$ both for $H' \cap H'' = \emptyset$ and for $|H' \cap H''| = 1$.

   (i) Show that $A_0 \not\to 0$ (while $A_1 \to 0$) as $n \to \infty$.

   (ii) Explain the difference without doing any formal calculations.

20.$^+$ Given a graph H, let $\mathcal{P}$ be the property of containing an induced copy of H. Show that $\mathcal{P}$ has no threshold function unless H is complete.

Notes

There are a number of monographs and texts on the subject of random graphs. The first comprehensive monograph was B. Bollobás, *Random Graphs*, Academic Press 1985. Another advanced monograph is S. Janson, T. Łuczak & A. Ruciński, *Random Graphs*, Wiley 2000; this concentrates on areas developed since *Random Graphs* was published. E.M. Palmer, *Graphical Evolution*, Wiley 1985, covers material similar to parts of *Random Graphs* but is written in a more elementary way. Compact introductions going beyond what is covered in this chapter are given by B. Bollobás, *Modern Graph Theory*, Springer GTM 184, 1998, and by M. Karoński, *Handbook of Combinatorics* (R.L. Graham, M. Grötschel & L. Lovász, eds.), North-Holland 1995.

A stimulating advanced introduction to the use of random techniques in discrete mathematics more generally is given by N. Alon & J.H. Spencer, *The Probabilistic Method*, Wiley 1992. One of the attractions of this book lies in the way it shows probabilistic methods to be relevant in proofs of entirely deterministic theorems, where nobody would suspect it. Other examples for this phenomenon are Alon's proof of Theorem 5.4.1, or the proof of Theorem 1.3.4; see the notes for Chapters 5 and 1, respectively.

The probabilistic method had its first origins in the 1940s, one of its earliest results being Erdős's probabilistic lower bound for Ramsey numbers (Theorem 11.1.3). Lemma 11.3.2 about the properties $\mathcal{P}_{i,j}$ is taken from Bollobás's Springer text cited above. A very readable rendering of the proof that, for constant p, every first order sentence about graphs is either almost surely true or almost surely false, is given by P. Winkler, Random structures and zero-one laws, in (N.W. Sauer et al., eds.) *Finite and Infinite Combinatorics in Sets and Logic* (NATO ASI Series C **411**), Kluwer 1993.

Theorem 11.3.5 is due to P. Erdős and A. Rényi, Asymmetric graphs, *Acta Math. Acad. Sci. Hungar.* **14** (1963), 295–315. For further references about the infinite random graph R see the notes in Chapter 8.

The seminal paper on graph evolution is P. Erdős & A. Rényi, On the evolution of random graphs, *Publ. Math. Inst. Hungar. Acad. Sci.* **5** (1960), 17–61. This paper also includes Theorem 11.4.5 for balanced graphs. The generalization to unbalanced subgraphs was first proved by Bollobás in 1981; see Karoński's *Handbook* chapter. The fact that all 'non-trivial' increasing graph properties have a threshold function was proved by B. Bollobás and A.G. Thomason, Threshold functions, *Combinatorica* **7** (1987), 35–38.

There is another way of defining a random graph G, which is just as natural and common as the model we considered. Rather than choosing the edges of G independently, we choose the entire graph G uniformly at random from among all the graphs on $\{0, \ldots, n-1\}$ that have exactly $M = M(n)$ edges: then each of these graphs occurs with the same probability of $\binom{N}{M}$, where $N := \binom{n}{2}$. Just as we studied the likely properties of the graphs in $\mathcal{G}(n,p)$ for different functions $p = p(n)$, we may investigate how the properties of G in the other model depend on the function $M(n)$. If M is close to pN, the expected number of edges of a graph in $\mathcal{G}(n,p)$, then the two models behave very similarly. It is then largely a matter of convenience which of them to consider; see Bollobás for details.

In order to study threshold phenomena in more detail, one often considers the following *random graph process*: starting with a $\overline{K^n}$ as stage zero, one chooses additional edges one by one (uniformly at random) until the graph is complete. This is a simple example of a Markov chain, whose Mth stage corresponds to the 'uniform' random graph model described above. A survey about threshold phenomena in this setting is given by T. Łuczak, The phase transition in a random graph, in (D. Miklós, V.T. Sós & T. Szőnyi, eds.) *Paul Erdős is 80*, Vol. 2, Proc. Colloq. Math. Soc. János Bolyai (1996).

12

Graph Minors

Our goal in this last chapter is a single theorem, one which dwarfs any other result in graph theory and may doubtless be counted among the deepest theorems that mathematics has to offer: *in every infinite set of graphs there are two such that one is a minor of the other.* This *graph minor theorem*, inconspicuous though it may look at first glance, has made a fundamental impact both outside graph theory and within. Its proof, due to Neil Robertson and Paul Seymour, takes well over 500 pages.

So we have to be modest: of the actual proof of the graph minor theorem this chapter will convey only a very rough impression. However, as with most truly fundamental results, the proof has sparked off the development of methods of quite independent interest and potential. This is true particularly for the use of *tree-decompositions*, a concept that is not only central to graph minor theory but has found algorithmic applications too, and *tangles*, a radically new notion of high connectivity somewhere inside a given graph.

Section 12.1 gives an introduction to *well-quasi-ordering*, a concept central to the graph minor theorem. In Section 12.2 we apply this to prove the graph minor theorem for trees. We study tree-decompositions in Sections 12.3–12.4, and tangles in Section 12.5. In Section 12.6 we look at *forbidden-minor theorems*: results in the spirit of Kuratowski's theorem (4.4.6) or Wagner's theorem (7.3.4), which describe the structure of the graphs not containing some specified graph or graphs as a minor.

In Section 12.7 we give a direct proof of the 'generalized Kuratowski' theorem that embeddability in any fixed surface can be characterized by forbidding finitely many minors. We conclude with an overview of the proof and implications of the graph minor theorem itself.

© Reinhard Diestel 2017
R. Diestel, *Graph Theory*, Graduate Texts in Mathematics 173,
DOI 10.1007/978-3-662-53622-3_12

12.1 Well-quasi-ordering

well-quasi-
ordering

A reflexive and transitive relation is called a *quasi-ordering*. A quasi-ordering $\leqslant$ on X is a *well-quasi-ordering*, and the elements of X are *well-quasi-ordered* by $\leqslant$, if for every infinite sequence $x_0, x_1, \ldots$ in X

good pair

there are indices $i < j$ such that $x_i \leqslant x_j$. Then (x_i, x_j) is a *good pair* of this sequence. A sequence containing a good pair is a *good sequence*;

good/bad
sequence

thus, a quasi-ordering on X is a well-quasi-ordering if and only if every infinite sequence in X is good. An infinite sequence is *bad* if it is not good.

Proposition 12.1.1. *A quasi-ordering $\leqslant$ on X is a well-quasi-ordering if and only if X contains neither an infinite antichain nor an infinite strictly decreasing sequence $x_0 > x_1 > \ldots$.*

(9.1.2)

Proof. The forward implication is trivial. Conversely, let $x_0, x_1, \ldots$ be any infinite sequence in X. Let K be the complete graph on $\mathbb{N} = \{0, 1, \ldots\}$. Colour the edges ij ($i < j$) of K with three colours: green if $x_i \leqslant x_j$, red if $x_i > x_j$, and amber if x_i, x_j are incomparable. By Ramsey's theorem (9.1.2), K has an infinite induced subgraph whose edges all have the same colour. If there is neither an infinite antichain nor an infinite strictly decreasing sequence in X, then this colour must be green, i.e. $x_0, x_1, \ldots$ has an infinite subsequence in which every pair is good. In particular, the sequence $x_0, x_1, \ldots$ is good. □

In the proof of Proposition 12.1.1, we showed more than was needed: rather than finding a single good pair in $x_0, x_1, \ldots$, we found an infinite increasing subsequence. We have thus shown the following:

Corollary 12.1.2. *If X is well-quasi-ordered, then every infinite sequence in X has an infinite increasing subsequence.* □

The following lemma, and the idea of its proof, are fundamental to the theory of well-quasi-ordering. Let $\leqslant$ be a quasi-ordering on a set X.

$\leqslant$

For finite subsets $A, B \subseteq X$, write $A \leqslant B$ if there is an injective mapping $f \colon A \to B$ such that $a \leqslant f(a)$ for all $a \in A$. This naturally extends $\leqslant$ to

$[X]^{<\omega}$

a quasi-ordering on $[X]^{<\omega}$, the set of all finite subsets of X.

[12.2.1]
[12.7.1]

Lemma 12.1.3. *If X is well-quasi-ordered by $\leqslant$, then so is $[X]^{<\omega}$.*

Proof. Suppose that $\leqslant$ is a well-quasi-ordering on X but not on $[X]^{<\omega}$. We start by constructing a bad sequence $(A_n)_{n \in \mathbb{N}}$ in $[X]^{<\omega}$, as follows. Given $n \in \mathbb{N}$, assume inductively that A_i has been defined for every $i < n$, and that there exists a bad sequence in $[X]^{<\omega}$ starting with $A_0, \ldots, A_{n-1}$. (This is clearly true for $n = 0$: by assumption, $[X]^{<\omega}$ contains a bad sequence, and this has the empty sequence as an initial segment.) Choose $A_n \in [X]^{<\omega}$ so that some bad sequence in $[X]^{<\omega}$ starts with $A_0, \ldots, A_n$ and $|A_n|$ is as small as possible.

Clearly, $(A_n)_{n \in \mathbb{N}}$ is a bad sequence in $[X]^{<\omega}$; in particular, $A_n \neq \emptyset$ for all n. For each n pick an element $a_n \in A_n$ and set $B_n := A_n \smallsetminus \{a_n\}$.

By Corollary 12.1.2, the sequence $(a_n)_{n \in \mathbb{N}}$ has an infinite increasing subsequence $(a_{n_i})_{i \in \mathbb{N}}$. By the minimal choice of A_{n_0}, the sequence

$$A_0, \ldots, A_{n_0 - 1}, B_{n_0}, B_{n_1}, B_{n_2}, \ldots$$

is good; consider a good pair. Since $(A_n)_{n \in \mathbb{N}}$ is bad, this pair cannot have the form (A_i, A_j) or (A_i, B_j), as $B_j \leqslant A_j$. So it has the form (B_i, B_j). Extending the injection $B_i \to B_j$ by $a_i \mapsto a_j$, we deduce again that (A_i, A_j) is good, a contradiction. $\square$

12.2 The graph minor theorem for trees

The graph minor theorem can be expressed by saying that the finite graphs are well-quasi-ordered by the minor relation $\preccurlyeq$. Indeed, by Proposition 12.1.1 and the obvious fact that no strictly descending sequence of minors can be infinite, being well-quasi-ordered is equivalent to the non-existence of an infinite antichain, the formulation used earlier.

In this section, we prove a strong version of the graph minor theorem for trees:

Theorem 12.2.1. (Kruskal 1960) [12.7.1]
The finite trees are well-quasi-ordered by the topological minor relation.

We shall base the proof of Theorem 12.2.1 on the following notion of an embedding between rooted trees, which strengthens the usual embedding as a topological minor. Consider two trees T and T', with roots r and r' say. Let us write $T \leqslant T'$ if there exists an isomorphism φ, from $T \leqslant T'$ some subdivision of T to a subtree T'' of T', that preserves the tree-order on $V(T)$ associated with T and r. (Thus if $x < y$ in T then $\varphi(x) < \varphi(y)$ in T'; see Fig. 12.2.1.) As one easily checks, this is a quasi-ordering on the class of all rooted trees.

Proof of Theorem 12.2.1. We show that the rooted trees are well- (12.1.3) quasi-ordered by the relation $\leqslant$ defined above; this clearly implies the theorem.

Suppose not. To derive a contradiction, we proceed as in the proof of Lemma 12.1.3. Given $n \in \mathbb{N}$, assume inductively that we have chosen a sequence $T_0, \ldots, T_{n-1}$ of rooted trees such that some bad sequence of rooted trees starts with this sequence. Choose as T_n a minimum-order T_n

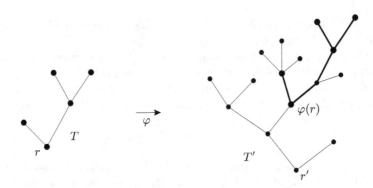

Fig. 12.2.1. An embedding of T in T' showing that $T \leqslant T'$

rooted tree such that some bad sequence starts with $T_0, \ldots, T_n$. For each
r_n $n \in \mathbb{N}$, denote the root of T_n by r_n.
A_n Clearly, $(T_n)_{n \in \mathbb{N}}$ is a bad sequence. For each n, let A_n denote the
set of components of $T_n - r_n$, made into rooted trees by choosing the
neighbours of r_n as their roots. Note that the tree-order of these trees
A is that induced by T_n. Let us prove that the set $A := \bigcup_{n \in \mathbb{N}} A_n$ of all
these trees is well-quasi-ordered.
T^k Let $(T^k)_{k \in \mathbb{N}}$ be any sequence of trees in A. For every $k \in \mathbb{N}$ choose
$n(k)$ an $n = n(k)$ such that $T^k \in A_n$. Pick a k with smallest $n(k)$. Then

$$T_0, \ldots, T_{n(k)-1}, T^k, T^{k+1}, \ldots$$

is a good sequence, by the minimal choice of $T_{n(k)}$ and $T^k \subsetneqq T_{n(k)}$. Let
(T, T') be a good pair of this sequence. Since $(T_n)_{n \in \mathbb{N}}$ is bad, T cannot
be among the first $n(k)$ members $T_0, \ldots, T_{n(k)-1}$ of our sequence: then
T' would be some T^i with $i \geqslant k$, i.e.

$$T \leqslant T' = T^i \leqslant T_{n(i)} \, ;$$

since $n(k) \leqslant n(i)$ by the choice of k, this would make $(T, T_{n(i)})$ a good
pair in the bad sequence $(T_n)_{n \in \mathbb{N}}$. Hence (T, T') is a good pair also in
$(T^k)_{k \in \mathbb{N}}$, completing the proof that A is well-quasi-ordered.
         By Lemma 12.1.3,[1] the sequence $(A_n)_{n \in \mathbb{N}}$ in $[A]^{<\omega}$ has a good pair
i, j (A_i, A_j); let $f \colon A_i \to A_j$ be injective with $T \leqslant f(T)$ for all $T \in A_i$. Now
extend the union of the embeddings $T \to f(T)$ to a map φ from $V(T_i)$
to $V(T_j)$ by letting $\varphi(r_i) := r_j$. This map φ preserves the tree-order
of T_i, and it defines an embedding to show that $T_i \leqslant T_j$, since the edges
$r_i r \in T_i$ map naturally to the paths $r_j T_j \varphi(r)$. Hence (T_i, T_j) is a good
pair in our original bad sequence of rooted trees, a contradiction. $\square$

---

[1] Any readers worried that we might need the lemma for sequences or multisets
rather than just sets here, note that isomorphic elements of A_n are not identified: we
always have $|A_n| = d(r_n)$.

12.3 Tree-decompositions

Trees are graphs with some very distinctive and fundamental properties; consider Theorem 1.5.1 and Corollary 1.5.2, or the more sophisticated example of Kruskal's theorem. It is therefore legitimate to ask to what degree those properties can be transferred to more general graphs, graphs that are not themselves trees but tree-like in some sense.[2] In this section, we study a concept of tree-likeness that permits generalizations of all the tree properties referred to above (including Kruskal's theorem), and which plays a crucial role in the proof of the graph minor theorem.

Let G be a graph, T a tree, and let $\mathcal{V} = (V_t)_{t \in T}$ be a family of vertex sets $V_t \subseteq V(G)$ indexed by the nodes t of T. The pair $(T, \mathcal{V})$ is called a *tree-decomposition* of G if it satisfies the following three conditions: *tree-decomposition*

(T1) $V(G) = \bigcup_{t \in T} V_t$;

(T2) for every edge $e \in G$ there exists a $t \in T$ such that both ends of e lie in V_t;

(T3) $V_{t_1} \cap V_{t_3} \subseteq V_{t_2}$ whenever $t_1, t_2, t_3 \in T$ satisfy $t_2 \in t_1 T t_3$.

Conditions (T1) and (T2) together say that G is the union of the subgraphs $G[V_t]$; we call these subgraphs and the sets V_t themselves the *parts* of $(T, \mathcal{V})$ and say that $(T, \mathcal{V})$ is a tree-decomposition of G *into*   *parts* these parts. Condition (T3) implies that the parts of $(T, \mathcal{V})$ are organized roughly like a tree (Fig. 12.3.1).

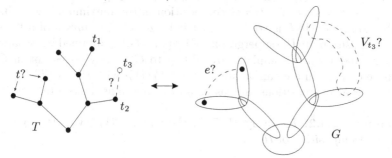

Fig. 12.3.1. Edges and parts ruled out by (T2) and (T3)

Before we discuss the role that tree-decompositions play in the proof of the minor theorem, let us note some of their basic properties. Consider a fixed tree-decomposition $(T, \mathcal{V})$ of G, with $\mathcal{V} = (V_t)_{t \in T}$ as above. $(T, \mathcal{V})$, V_t

Perhaps the most important feature of a tree-decomposition is that it transfers the separation properties of its tree to the graph decomposed:

---

[2] What exactly this 'sense' should be will depend both on the property considered and on its intended application.

[12.4.3]
[12.6.5]

Lemma 12.3.1. *Let t_1t_2 be any edge of T and let T_1, T_2 be the components of $T - t_1t_2$, with $t_1 \in T_1$ and $t_2 \in T_2$. Then $V_{t_1} \cap V_{t_2}$ separates $U_1 := \bigcup_{t \in T_1} V_t$ from $U_2 := \bigcup_{t \in T_2} V_t$ in G (Fig. 12.3.2).*

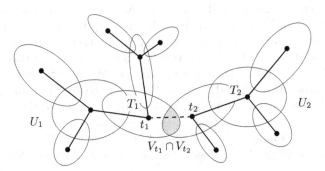

Fig. 12.3.2. $V_{t_1} \cap V_{t_2}$ separates U_1 from U_2 in G

Proof. Both t_1 and t_2 lie on every t–t' path in T with $t \in T_1$ and $t' \in T_2$. Therefore $U_1 \cap U_2 \subseteq V_{t_1} \cap V_{t_2}$ by (T3), so all we have to show is that G has no edge u_1u_2 with $u_1 \in U_1 \setminus U_2$ and $u_2 \in U_2 \setminus U_1$. If u_1u_2 is such an edge, then by (T2) there is a $t \in T$ with $u_1, u_2 \in V_t$. By the choice of u_1 and u_2 we have neither $t \in T_2$ nor $t \in T_1$, a contradiction. $\square$

induced separation

We shall say that the separation $\{U_1, U_2\}$ of G in Lemma 12.3.1 is *induced* by the edge t_1t_2 of T, or more generally by $(T, \mathcal{V})$. Its separator $U_1 \cap U_2 = V_{t_1} \cap V_{t_2}$ is the *adhesion set* of V_{t_1} and V_{t_2}.

adhesion

torsos

The *adhesion* of a tree-decomposition is the maximum size of its adhesion sets. (If T is trivial, we let it be zero.) The *torsos* of a tree-decomposition are the supergraphs of its parts $G[V_t]$ obtained by making their adhesion sets complete: by adding to $G[V_t]$ any edges not in G whose ends lie in a common adhesion set $V_t \cap V_{t'}$ with $tt' \in E(T)$.

Tree-decompositions are passed on to subgraphs, indeed to minors:

[12.4.1]
[12.4.4]
[12.4.3]
[12.6.2]

Lemma 12.3.2. *For every $H \subseteq G$, the pair $\big(T, (V_t \cap V(H))_{t \in T}\big)$ is a tree-decomposition of H.* $\square$

[12.4.1]
[12.4.3]

Lemma 12.3.3. *Suppose that $G = IH$ with branch sets U_h, $h \in V(H)$. Let $f : V(G) \to V(H)$ be the map assigning to each vertex of G the index of the branch set containing it. For each $t \in T$ let $W_t := \{ f(v) \mid v \in V_t \}$, and put $\mathcal{W} := (W_t)_{t \in T}$. Then $(T, \mathcal{W})$ is a tree-decomposition of H.*

Proof. The assertions (T1) and (T2) for $(T, \mathcal{W})$ follow immediately from the corresponding assertions for $(T, \mathcal{V})$. Now let $t_1, t_2, t_3 \in T$ be as in (T3), and consider a vertex $h \in W_{t_1} \cap W_{t_3}$ of H; we show that $h \in W_{t_2}$. By definition of W_{t_1} and W_{t_3}, the set U_h meets both V_{t_1} and V_{t_3}. As U_h is connected in G, this implies by Lemma 12.3.1 that U_h also meets V_{t_2}. Hence $h \in W_{t_2}$, by definition of W_{t_2}. $\square$

Here is another useful consequence of Lemma 12.3.1:

Lemma 12.3.4. *Any set of vertices not contained in a part of $(T, \mathcal{V})$*   [12.4.3]
contains two vertices that are separated by an adhesion set of $(T, \mathcal{V})$.

Proof. Given $W \subseteq V(G)$, orient the edges of T as follows. For each edge
$t_1 t_2 \in T$, define U_1, U_2 as in Lemma 12.3.1; then $V_{t_1} \cap V_{t_2}$ separates U_1
from U_2. Unless $V_{t_1} \cap V_{t_2}$ separates two vertices from W, we can find an
$i \in \{1, 2\}$ such that $W \subseteq U_i$, and orient $t_1 t_2$ towards t_i.

Let t be the last node of a maximal directed path in T; then all the
edges of T at t are oriented towards t. We claim that $W \subseteq V_t$. Given
$w \in W$, let $t' \in T$ be such that $w \in V_{t'}$. If $t' \neq t$, then the edge e at t
that separates t' from t in T is directed towards t, so w also lies in $V_{t''}$
for some t'' in the component of $T - e$ containing t. Therefore $w \in V_t$
by (T3). □

The following special case of Lemma 12.3.4 is used particularly often:

Corollary 12.3.5. *Every complete subgraph of G is contained in some*   [12.6.2]
part of $(T, \mathcal{V})$. □

The tree-decomposition $(T, \mathcal{V})$ of G is called *simplicial* if all the   *simplicial*
separators $V_{t_1} \cap V_{t_2}$ induce complete subgraphs in G. This assumption
can enable us to lift assertions about the parts of the decomposition to
G itself. For example, if all the parts in a simplicial tree-decomposition
of G are k-colourable, then so is G (Exercise 19). The same applies to
the property of not containing a K^r minor for some fixed r.

Conversely, if G can be constructed recursively from a set $\mathcal{H}$ of
graphs by pasting along complete subgraphs, then G has a simplicial
tree-decomposition into elements of $\mathcal{H}$. For example, by Wagner's Theorem 7.3.4, any graph without a K^5 minor has a supergraph with a
simplicial tree-decomposition into plane triangulations and copies of the
Wagner graph W, and similarly for graphs without K^4 minors (see Proposition 12.6.2).

Tree-decompositions may thus lead to intuitive structural characterizations of graph properties. A particularly simple example is the
following characterization of chordal graphs:

Proposition 12.3.6. *G is chordal if and only if G has a tree-decompo-*   [12.4.4]
sition into complete parts.   [12.6.2]

Proof. We apply induction on $|G|$. We first assume that G has a tree-   (5.5.1)
decomposition $(T, \mathcal{V})$ such that $G[V_t]$ is complete for every $t \in T$; let
us choose $(T, \mathcal{V})$ with $|T|$ minimum. If $|T| \leqslant 1$, then G is complete
and hence chordal. So let $t_1 t_2 \in T$ be an edge, and for $i = 1, 2$ define
T_i and $G_i := G[U_i]$ as in Lemma 12.3.1. Then $G = G_1 \cup G_2$ by (T1)

and (T2), and $V(G_1 \cap G_2) = V_{t_1} \cap V_{t_2}$ by the lemma; thus, $G_1 \cap G_2$ is complete. Since $(T_i, (V_t)_{t \in T_i})$ is a tree-decomposition of G_i into complete parts, both G_i are chordal by the induction hypothesis. (By the choice of $(T, \mathcal{V})$, neither G_i is a subgraph of $G[V_{t_1} \cap V_{t_2}] = G_1 \cap G_2$, so both G_i are indeed smaller than G.) Since $G_1 \cap G_2$ is complete, any induced cycle in G lies in G_1 or in G_2 and hence has a chord, so G too is chordal.

Conversely, assume that G is chordal. If G is complete, there is nothing to show. If not then, by Proposition 5.5.1, G is the union of smaller chordal graphs G_1, G_2 with $G_1 \cap G_2$ complete. By the induction hypothesis, G_1 and G_2 have tree-decompositions $(T_1, \mathcal{V}_1)$ and $(T_2, \mathcal{V}_2)$ into complete parts. By Corollary 12.3.5, $G_1 \cap G_2$ lies inside one of those parts in each case, say with indices $t_1 \in T_1$ and $t_2 \in T_2$. As one easily checks, $((T_1 \cup T_2) + t_1 t_2, \mathcal{V}_1 \cup \mathcal{V}_2)$ is a tree-decomposition of G into complete parts. $\square$

Let us wind up this section with an application of tree-decompositions to connectivity that generalizes the idea of the block-cutvertex tree from Lemma 3.1.4. A set $U \subseteq V(G)$ of at least k vertices is $(< k)$-*inseparable* in G if no two vertices from U can be separated in G by fewer than k other vertices (which may or may not lie in U).

k-block   A maximal $(< k)$-inseparable set of vertices is a *k-block*. Thus, the 1-blocks of a graph are its components; its 2-blocks are the non-singleton vertex sets spanning a block as defined in Chapter 3. In general, a k-block need not induce a highly connected subgraph: the many paths between its vertices that are needed to make it $(< k)$-inseparable can all lie outside it. Its 'connectivity' is thus measured in the ambient graph; its vertices themselves may even be independent.

Theorem 12.3.7. *For every integer $k \geqslant 1$, every graph G has a tree-decomposition $(T, \mathcal{V})$ with the following properties:*

(i) *$(T, \mathcal{V})$ has adhesion $< k$.*

(ii) *Distinct k-blocks lie in different parts. Moreover, every two blocks are separated by an adhesion set that is no larger than the smallest set of vertices that separates them in G.*

(iii) *Every automorphism of G acts on the set of parts of $(T, \mathcal{V})$, and the action on $V(T)$ which this induces is an automorphism of T.*

Note that, by (i) and Lemma 12.3.4, every k-block is contained in some part. This is unique by (i) and (T3), so the parts in (ii) are well defined. Assertion (iii) means that, for every automorphism φ of G and every $t \in T$, the set $\varphi(V_t)$ is another part $V_{t'}$ (possibly V_t), and this map $\varphi \colon t \mapsto t'$ is an automorphism of T. More about this in the notes.

12.4 Tree-width

As indicated by Figure 12.3.1, the parts of $(T, \mathcal{V})$ reflect the structure of the tree T, so in this sense the graph G decomposed resembles a tree. However, this is valuable only inasmuch as the structure of G within each part is negligible: the smaller the parts, the closer the resemblance.

This observation motivates the following definition. The *width* of $(T, \mathcal{V})$ is the number

$$\max \big\{ \, |V_t| - 1 : t \in T \big\} \, ,$$

and the *tree-width* tw(G) of G is the least width of any tree-decomposition of G. As one easily checks,[3] trees themselves have tree-width 1.

By Lemmas 12.3.2 and 12.3.3, the tree-width of a graph will never be increased by deletion or contraction:

Lemma 12.4.1. *If $H \preccurlyeq G$ then* tw$(H) \leqslant$ tw(G). $\qquad\qquad$ □

width

tree-width
tw(G)

(12.3.2)
(12.3.3)

[12.6.2]

Graphs of bounded tree-width are sufficiently similar to trees that it becomes possible to adapt the proof of Kruskal's theorem to the class of these graphs; very roughly, one has to iterate the 'minimal bad sequence' argument from the proof of Lemma 12.1.3 tw(G) times. This takes us a step further towards a proof of the graph minor theorem:

Theorem 12.4.2. (Robertson & Seymour 1990)
For every integer $k > 0$, the graphs of tree-width $< k$ are well-quasi-ordered by the minor relation.

[12.7.1]
[12.7.3]

In order to make use of Theorem 12.4.2 for a proof of the full graph minor theorem, we should be able to say something about the graphs which it does not cover, i.e., to deduce some information about a graph from the assumption that its tree-width is large. Our next result, the *tree-width duality theorem*, achieves just that: it identifies a canonical obstruction to small tree-width, a structural phenomenon that occurs in a graph if and only if its tree-width is large.

Let us say that two subsets of $V(G)$ *touch* if they have a vertex in common or G contains an edge between them. A set of mutually touching connected vertex sets in G is a *bramble*. Extending our terminology of Chapter 2, we say that a subset of $V(G)$ *covers* (or is a *cover* of) a bramble $\mathcal{B}$ if it meets every element of $\mathcal{B}$. The least number of vertices covering a bramble is the *order* of that bramble. A *k-bramble* is one of order k.

touch

bramble
cover

order

---

[3] Indeed the '-1' in the definition of width serves no other purpose than to make this statement true.

A typical example of a bramble is the set of crosses in a grid. The
k × k grid is the graph on $\{1, \ldots, k\}^2$ with the edge set

$$\{(i,j)(i',j') : |i - i'| + |j - j'| = 1\}.$$

The *crosses* of this grid are the k^2 sets

$$C_{ij} := \{(i, \ell) \mid \ell = 1, \ldots, k\} \cup \{(\ell, j) \mid \ell = 1, \ldots, k\}.$$

Thus, the cross C_{ij} is the union of the grid's ith row and its jth column.
Clearly, the crosses of the $k \times k$ grid form a k-bramble: they are covered
by any row or column, while any set of fewer than k vertices misses both
a row and a column, and hence a cross.

Theorem 12.4.3. (Seymour & Thomas 1993)
*Let $k \geqslant 0$ be an integer. A graph has tree-width $< k$ if and only if it
contains no bramble of order $> k$.*

Proof. Let $G = (V, E)$ be a graph. For the forward implication, let $\mathcal{B}$ be
any bramble in G. We show that every tree-decomposition $(T, (V_t)_{t \in T})$
of G has a part that covers $\mathcal{B}$.

As in the proof of Lemma 12.3.4 we start by orienting the edges $t_1 t_2$
of T. If $X := V_{t_1} \cap V_{t_2}$ covers $\mathcal{B}$, we are done. If not, then for each $B \in \mathcal{B}$
disjoint from X there is an $i \in \{1, 2\}$ such that $B \subseteq U_i \smallsetminus X$ (defined as
in Lemma 12.3.1); recall that B is connected. This i is the same for all
such B, because they touch. We now orient the edge $t_1 t_2$ towards t_i.

If every edge of T is oriented in this way, it has a node t all whose
incident edges are oriented towards it. Then V_t covers $\mathcal{B}$—just as in the
proof of Lemma 12.3.4.

For a proof of the converse implication, consider a tree-decomposi-
tion $(T, \mathcal{V})$ of G, with $\mathcal{V} = (V_t)_{t \in T}$, say. If x is a leaf of T with $|V_x| > k$
and incident edge $xy \in T$, we call the set $V_x \smallsetminus V_y$ the *petal* of x. The
decomposition $(T, \mathcal{V})$ is *good* if $|V_t| \leqslant k$ for at least one t and $|V_t| \leqslant k$
whenever t is not a leaf. Then the neighbour y of any leaf x with $|V_x| > k$
satisfies $|V_y| \leqslant k$. Hence, petals in good tree-decompositions are non-
empty and have at most k neighbours.

Suppose now that $\mathrm{tw}(G) \geqslant k$, and let us find a bramble of order $> k$.
To get started, consider good tree-decompositions $(T_1, \mathcal{V}_1)$ and $(T_2, \mathcal{V}_2)$,
where $T_1 \cap T_2 = \emptyset$, with petals X of $x \in T_1$ and Y of $y \in T_2$. Assume that
the part of $(T_1, \mathcal{V}_1)$ corresponding to x is exactly $X \cup N(X)$, that the
part of $(T_2, \mathcal{V}_2)$ corresponding to y is exactly $Y \cup N(Y)$, that no petal
in $(T_1, \mathcal{V}_1)$ contains Y, that no petal in $(T_2, \mathcal{V}_2)$ contains X, and that X
and Y do not touch. Then we claim the following:

> There is a good tree-decomposition $(T, \mathcal{V})$ all whose petals
> are contained in petals of $(T_1, \mathcal{V}_1)$ and $(T_2, \mathcal{V}_2)$ and in which $(*)$
> neither X nor Y is a petal.

(margin labels:)
grid

(3.3.1)
(12.3.1)
(12.3.2)
(12.3.3)

petal
good

$(T_i, \mathcal{V}_i)$
$x, X; y, Y$

Indeed, as X and Y do not touch, the set $N(X)$ is disjoint from both X and Y and separates them in G. Hence G has a separation $\{A, B\}$ such that $X \subseteq A \smallsetminus B$ and $Y \subseteq B \smallsetminus A$. As $|N(X)| \leqslant k$ since X is a petal, choosing $\{A, B\}$ of minimum order ensures that $S := A \cap B$ has size at most k. By the minimality of S and Menger's Theorem 3.3.1, there is a family $\{ P_s \mid s \in S \}$ of disjoint S–$N(X)$ paths in $G[A]$ and a family $\{ Q_s \mid s \in S \}$ of disjoint S–$N(Y)$ paths in $G[B]$. $\qquad$ A, B $\quad$ S $\quad$ P_s, Q_s

Let H be the minor of G obtained by deleting $A \smallsetminus \bigcup_{s \in S} V(P_s)$ and contracting each of the paths P_s. Identifying the contracted branch sets $V(P_s)$ with their representatives s, we may think of H as obtained from $G[B]$ by adding some edges on S. Let $(T_1, \mathcal{V}_1')$ be the tree-decomposition which $(T_1, \mathcal{V}_1)$ induces on H as in Lemmas 12.3.2 and 12.3.3, and think of it as a tree-decomposition of $G[B]$. Thus for any $t \in T_1$, with the part $V_t^1 \in \mathcal{V}_1$ say, its part V_t in $\mathcal{V}_1'$ is $\qquad$ H

$$V_t = (V_t^1 \cap B) \cup \{ s \in S \mid V_t^1 \cap V(P_s) \neq \emptyset \} \qquad (1)$$

(Fig. 12.4.1). In particular, $V_x = S$, since $V_x = X \cup N(X) \subseteq A$ and $N(X)$ meets every P_s. Similarly, let J be the minor of G obtained by deleting $B \smallsetminus \bigcup_{s \in S} V(Q_s)$ and contracting the paths Q_s, and let $(T_2, \mathcal{V}_2')$ be the tree-decomposition which $(T_2, \mathcal{V}_2)$ induces on J. As before, think of this as a tree-decomposition of $G[A]$ in which S is the part corresponding to y. $\qquad$ J

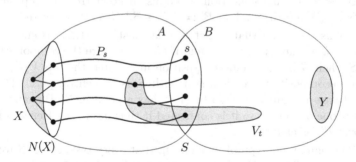

Fig. 12.4.1. To obtain V_t' from $V_t \cap B$, add two vertices from S

Let T be obtained from the (disjoint) trees T_1 and T_2 by identifying x and y into a new node r. As Y and X are non-empty, x is not the only node of T_1 and y is not the only node of T_2, so r is not a leaf of T. Let $V_r := S$. For all $t \in T - r$ let V_t be the part in $\mathcal{V}_1'$ or $\mathcal{V}_2'$ that corresponds to t there, thought of as a subset of B if $t \in T_1$, or of A if $t \in T_2$. We claim that $(T, \mathcal{V})$ with $\mathcal{V} = (V_t)_{t \in T}$ is a good tree-decomposition of G satisfying $(*)$. $\qquad$ T $\quad$ V_t $\quad$ $(T, \mathcal{V})$

Using that $(T_1, \mathcal{V}_1')$ and $(T_2, \mathcal{V}_2')$ are tree-decompositions of $G[B]$ and $G[A]$, it is easy to check that $(T, \mathcal{V})$ is indeed a tree-decomposition of G. The non-leaves of T are precisely those of T_1 and T_2, plus r. We have already seen that $|S| \leqslant k$. For $t \in T_1 - x$, its part V_t in $\mathcal{V}$ is no larger than its part V_t^1 in $\mathcal{V}_1$: by (1), there exists for every $s \in V_t \smallsetminus V_t^1$ a vertex of P_s in $V_t^1 \smallsetminus V_t$. Similarly, the parts V_t with $t \in T_2$ are no larger than their corresponding parts in $\mathcal{V}_2$. Thus, $(T, \mathcal{V})$ is good.

To show that $(T, \mathcal{V})$ satisfies $(*)$, consider a petal Z in $(T, \mathcal{V})$ of a leaf z of T. Then z is a leaf also of T_1 or T_2. Let us assume that $z \in T_1$; then $V_z \subseteq B$. By axiom (T3) for $(T, \mathcal{V})$, any vertex of $V_z \cap V_r$ lies in the adhesion set of V_z, so $Z \cap S = \emptyset$. Hence $Z \subseteq B \smallsetminus A$. But this implies that Z lies inside the petal Z_1 of z in $(T_1, \mathcal{V}_1)$: recall that the part V_z^1 corresponding to z in $\mathcal{V}_1$ has size at least $|V_z| > k$ (so it can have a petal) and differs from V_z only by vertices in A, and likewise for the neighbour of z in T_1 and T. Finally, $Z \neq X$ because $X \subseteq A \smallsetminus B$, and $Z \neq Y$ since the petal $Z_1 \supseteq Z$ of z in $(T_1, \mathcal{V}_1)$ does not contain Y by assumption. This completes the proof of $(*)$.

Let us now use $(*)$ to find a bramble of order $> k$. Since $\mathrm{tw}(G) \geqslant k$, every good tree-decomposition of G has a petal. In particular, the set $\mathcal{B}$ of all petals of good tree-decompositions satisfies

(i) $\mathcal{B}$ contains a petal of every good tree-decomposition;

(ii) $\mathcal{B}$ is closed under taking supersets among petals: if $X \subseteq X'$ are both petals of good tree-decompositions and $X \in \mathcal{B}$, then $X' \in \mathcal{B}$.

Let $\mathcal{B}$ be a minimal collection of petals of good tree-decompositions satisfying (i) and (ii), and let $\mathcal{B}' := \{ X \in \mathcal{B} \mid G[X] \text{ is connected} \}$.

Let us show first that no set S of at most k vertices covers $\mathcal{B}'$. As $\mathrm{tw}(G) \geqslant k$ we have $V \smallsetminus S \neq \emptyset$, and if $C_1, \ldots, C_n$ are the components of $G - S$, then G has a tree-decomposition into S and the sets $V(C_i) \cup S$, whose decomposition tree is a star with S as its central part. This is a good tree-decomposition, so by (i) it has a petal in $\mathcal{B}$. But any such petal is one of the C_i, so it is connected it thus lies in $\mathcal{B}'$. Hence S fails to cover $\mathcal{B}'$, as claimed.

We complete our proof by showing that every two sets in $\mathcal{B}$ touch. If not, we can find $X, Y \in \mathcal{B}$ that do not touch and are $\subseteq$-minimal in $\mathcal{B}$. Since $\mathcal{B} \smallsetminus \{X\}$ and $\mathcal{B} \smallsetminus \{Y\}$ still satisfy (ii), the minimality of $\mathcal{B}$ implies that they violate (i). So there exists a good tree-decomposition $(T_1, \mathcal{V}_1)$ whose only petal in $\mathcal{B}$ is X, and a good tree-decomposition $(T_2, \mathcal{V}_2)$ whose only petal in $\mathcal{B}$ is Y. Deleting any vertices outside $X \cup N(X)$ and $Y \cup N(Y)$ from their corresponding parts, we may assume that these parts are exactly $X \cup N(X)$ and $Y \cup N(Y)$. (The trimmed parts still have size $> k$, since otherwise the modified tree-decomposition would have no petal in $\mathcal{B}$.) Since any petal in $(T_1, \mathcal{V}_1)$ containing Y would lie in $\mathcal{B} \smallsetminus \{X\}$, by (ii), there is no such petal; similarly, no petal in $(T_2, \mathcal{V}_2)$ contains X. Thus, $(T_1, \mathcal{V}_1)$ and $(T_2, \mathcal{V}_2)$ satisfy the premise of $(*)$.

By $(*)$, $(T_1, \mathcal{V}_1)$ and $(T_2, \mathcal{V}_2)$ give rise to a good tree-decomposition $(T, \mathcal{V})$ of G that has no petal in $\mathcal{B}$. This contradicts (i). $\qquad \square$

Often, Theorem 12.4.3 is stated in terms of the *bramble number* of a graph, the largest order of any bramble in it. The theorem then says that the tree-width of a graph is exactly one less than its bramble number.

How useful even the easy forward implication of Theorem 12.4.3 can be is exemplified once more by our example of the crosses bramble in the $k \times k$ grid: this bramble has order k, so by the theorem the $k \times k$ grid has tree-width at least $k - 1$. (Try to show this without the theorem!)

In fact, the $k \times k$ grid has tree-width k (Exercise 34). But more important than its precise value is the fact that the tree-width of grids tends to infinity with their size. For as we shall see, large grid minors pose another canonical obstruction to small tree-width: not only do large grids (and hence all graphs containing large grids as minors; cf. Lemma 12.4.1) have large tree-width, but conversely every graph of large tree-width has a large grid minor (Theorem 12.6.3).

In Section 12.5 we shall place these within the wider framework of *tangles*, another central notion in graph minor theory. Using tangles one can formulate a more general duality theory between highly connected substructures and trees, of which Theorem 12.4.3 is but a special case.

Tree-width can also be expressed as follows:

Proposition 12.4.4. $\mathrm{tw}(G) = \min\big\{ \omega(H) - 1 \mid G \subseteq H;\ H \text{ chordal} \big\}.$

Proof. By Corollary 12.3.5 and Proposition 12.3.6, each of the graphs H considered for the minimum has a tree-decomposition of width $\omega(H) - 1$. Every such tree-decomposition induced one of G by Lemma 12.3.2, so $\mathrm{tw}(G) \leqslant \omega(H) - 1$ for every H.

Conversely, let us construct an H as above with $\omega(H) - 1 \leqslant \mathrm{tw}(G)$. Let $(T, \mathcal{V})$ be a tree-decomposition of G of width $\mathrm{tw}(G)$. For every $t \in T$ let K_t denote the complete graph on V_t, and put $H := \bigcup_{t \in T} K_t$. Clearly, $(T, \mathcal{V})$ is also a tree-decomposition of H. By Proposition 12.3.6, H is chordal, and by Corollary 12.3.5, $\omega(H) - 1$ is at most the width of $(T, \mathcal{V})$, i.e. at most $\mathrm{tw}(G)$. $\qquad\square$

(12.3.2)
(12.3.5)
(12.3.6)

A tree-decomposition $(T, \mathcal{V})$ of G with $\mathcal{V} = (V_t)_{t \in T}$ is *linked*, or *lean*, if it satisfies the following condition: *linked/lean*

(T4) Given $t_1, t_2 \in T$ and vertex sets $Z_1 \subseteq V_{t_1}$ and $Z_2 \subseteq V_{t_2}$ such that $|Z_1| = |Z_2| =: k$, either G contains k disjoint Z_1–Z_2 paths or there exists an edge $tt' \in t_1 T t_2$ with $|V_t \cap V_{t'}| < k$.

The 'branches' in a lean tree-decomposition are thus stripped of any bulk not necessary to maintain their connecting qualities. Indeed, if a branch is thick (i.e. the adhesion sets $V_t \cap V_{t'}$ along a path in T are all large), then G is highly connected along this branch, and the parts themselves are no larger than their 'external connectivity' in G requires: for $t_1 = t_2$,

(T4) says that two k-sets of vertices will only lie in a common part if they cannot be separated (in G) by fewer than k vertices.

In our quest for tree-decompositions into 'small' parts, we now have two criteria to choose between: the global 'worst case' criterion of width, and the more subtle local criterion of leanness. Surprisingly, it is always possible to find a tree-decomposition that is optimal with respect to both criteria at once:

Theorem 12.4.5. (Thomas 1990)
Every graph G has a lean tree-decomposition of width $\mathrm{tw}(G)$.

Another natural feature one might ask of a tree-decomposition is that its parts, as induced subgraphs of G, be connected. Let us call such a tree-decomposition *connected*, and let the *connected tree-width* $\mathrm{ctw}(G)$ of G be the least width of a connected tree-decomposition of G.

The connected tree-width of most graphs is greater than their ordinary tree-width. For example, every cycle has tree-width 2, but $\mathrm{ctw}(C^n) = \lceil n/2 \rceil$ (Exercise 33). And unlike ordinary tree-width, the connected tree-width of a subgraph of G can be greater than that of G. (For example, let G be obtained from a long cycle by adding a chord between opposite vertices.) However, if $C \subseteq G$ is a *geodesic* cycle, i.e., if $d_C(u,v) = d_G(u,v)$ for all vertices $u, v \in C$, then $\mathrm{ctw}(C) \leqslant \mathrm{ctw}(G)$.

The presence of long geodesic cycles in a graph thus is an obstruction to small connected tree-width—as is, trivially, large ordinary tree-width. By the following theorem, however, these are the only obstructions:

Theorem 12.4.6. *There is a function $f \colon \mathbb{N}^2 \to \mathbb{N}$ such that every graph of tree-width $\leqslant w \in \mathbb{N}$ that has no geodesic cycle of length $> k \in \mathbb{N}$ has connected tree-width at most $f(w, k)$.*

12.5 Tangles

We have already in this chapter met a few types of substructures of possibly sparse graphs that are highly connected in some sense, but which are not just k-connected subgraphs or minors for some large k. Minors of large grids are an example, the k-blocks defined at the end of Section 12.3 are another, as are brambles of high order. All these substructures have one feature in common: for every low-order separation of the graph they lie essentially, though not necessarily entirely, on one of its two sides.

$G = (V, E)$ For example, given a bramble $\mathcal{B}$ of order k in a graph $G = (V, E)$, if $\{A, B\}$ is a separation of order $< k$ then $A \smallsetminus B$ or $B \smallsetminus A$, but not both, contains a set from $\mathcal{B}$. This helped us prove the easy implication of the tree-width duality theorem: $\mathcal{B}$ orients every edge of the decomposition tree of any tree-decomposition $(T, \mathcal{V})$ of adhesion $< k$ 'towards' the side of its induced separation that contains one of the sets in $\mathcal{B}$, and the edges thus oriented point to a central node t of T for which V_t covers $\mathcal{B}$.

It has turned out that, sometimes, the only feature of a highly connected substructure that we really care about is the information of how it 'orients' the low-order separations of G in this way. Collecting just this information together leads to a more abstract notion of a highly connected substructure, called a *tangle*. The purpose of this section is to make this precise, to prove a duality theorem for tangles in the spirit of Theorem 12.4.3, and to point out how this setting can be used to express the duality between tree-structure and highly connected substructures more generally.

The *orientations* of a separation $\{A, B\}$ of G are the two *oriented separations* (A, B) and (B, A). We say that (A, B) is *oriented*, or *point-ing, towards* B and its subsets. Given a set S of separations, we write $\vec{S} := \{(A, B) \mid \{A, B\} \in S\}$ for the set of all their orientations. An *orientation of S* is a subset O of $\vec{S}$ that contains for every element of S exactly one of its two orientations. We say that O *avoids* a collection $\mathcal{F}$ of sets of oriented separations if no subset of O lies in $\mathcal{F}$.

oriented separation

$\vec{S}$

avoids

Given oriented separations (A, B) and (C, D) of G, let us write $(A, B) \leqslant (C, D)$ if $A \subseteq C$ and $B \supseteq D$. A set σ of oriented separations of G is *consistent* if it does *not* contain (B, A) whenever $(A, B) \leqslant (C, D)$ with $(C, D) \in \sigma$.[4] And σ is a *star* of oriented separations if $(A, B) \leqslant (D, C)$ for all distinct $(A, B), (C, D) \in \sigma$ (Fig. 12.5.1).

$\leqslant$

consistent
star

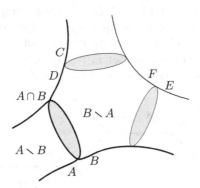

Fig. 12.5.1. The separations $(A, B), (C, D), (E, F)$ form a star

For example, if $(T, \mathcal{V})$ is a tree-decomposition of G with $\mathcal{V} = (V_t)_{t \in T}$, then orienting the separations induced by the edges of T towards some fixed V_t orients them consistently. And the separations corresponding to edges at t become, with this orientation, a star of oriented separations. A

---

[4] Intuitively, σ is consistent if no two of its elements point away from each other. In particular, it will not contain both orientations of any given separation.

S_k

k-block X of G even defines a consistent orientation of the entire set S_k of separations of order $< k$ in G: the orientation $\{(A, B) \in \vec{S_k} \mid X \subseteq B\}$.

A bramble $\mathcal{B}$ of order $n \geqslant k$ also defines an orientation of S_k: the set $O = \{(A, B) \in \vec{S_k} \mid \exists X \in \mathcal{B}: X \subseteq B \smallsetminus A\}$. Unlike in the case of k-blocks, there need not be any fixed bramble set that lies in every B with $(A, B) \in O$; indeed the intersection of all these B may be empty. (Example?) But O shows the large order of $\mathcal{B}$ in another way, in that we cannot cover $\mathcal{B}$ by few sets A with $(A, B) \in O$: since any bramble set meeting A also meets $A \cap B$ (because it touches the bramble set in $B \smallsetminus A$), and $|A \cap B| \leqslant k - 1$, we need at least $n/(k-1)$ sets A to cover $\mathcal{B}$. The idea that this is enough not only to reflect but to constitute a kind of highly connected substructure in G has led to the following concept.

tangle

A *tangle of order k*, or *k-tangle*, is an orientation of S_k that avoids

$$\mathcal{T} := \big\{\{(A_1, B_1), (A_2, B_2), (A_3, B_3)\} \mid G[A_1] \cup G[A_2] \cup G[A_3] = G\big\}.$$

in G

Tangles of unspecified order will be referred to as tangles *in G*.

Note that tangles are consistent, since $(A, B) \leqslant (C, D)$ implies that

T^*

$G[B] \cup G[C] \supseteq G[D] \cup G[C] = G$. Let $T^* := \{\sigma \in \mathcal{T} \mid \sigma$ is a star$\}$.

The set $\vec{E}(T)$ of the oriented edges of a tree T is partially ordered

$\vec{e} \leqslant \vec{f}$

by letting $\vec{e} \leqslant \vec{f}$ whenever $\vec{e} = (e, x, y)$ and $\vec{f} = (f, u, v)$ are such that the unique $\{x, y\}$–$\{u, v\}$ path in T starts in y and ends in u.

S-tree

Given a set S of separations of G, an *S-tree* is a pair (T, α) such that T is a tree and $\alpha: \vec{E}(T) \to \vec{S}$ respects the orderings on these sets and commutes with inversion: that $\alpha(\vec{e}) \leqslant \alpha(\vec{f})$ if $\vec{e} \leqslant \vec{f}$ (Fig. 12.5.2), and $\alpha(\overleftarrow{e}) = (B, A)$ whenever $\alpha(\vec{e}) = (A, B)$. A tree-decomposition $(T, \mathcal{V})$, for example, makes T into an S-tree for the set S of separations it induces.

over

We say that (T, α) is an S-tree *over* a set $\mathcal{F}$ of stars in $\vec{S}$ if for every

$\vec{F_t}$

node t of T the map α sends the set $\vec{F_t} := \{(e, s, t) \in \vec{E}(T) \mid e = st \in T\}$ of its incoming incident edges to an element of $\mathcal{F}$.

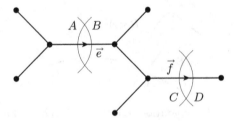

Fig. 12.5.2. An S-tree with $\alpha(\vec{e}) = (A, B) \leqslant (C, D) = \alpha(\vec{f})$

tangle
duality
theorem

Theorem 12.5.1. (Robertson & Seymour 1991)

The following assertions are equivalent for all graphs G and integers $k > 0$:

(i) *G has a tangle of order k;*

(ii) *S_k has a consistent orientation that avoids T^*;*

(iii) *G has no S_k-tree over T^*.*

For our proof of Theorem 12.5.1 we need an observation about the order of 'crossing' separations (Fig. 12.5.3). Given two separations $\{A, B\}$ and $\{C, D\}$ of G, it is easy to check that also $\{A \cup C, B \cap D\}$ and $\{A \cap C, B \cup D\}$ are separations of G. The orders of these *corner separations* sum to

corner separations

$$|(A \cup C) \cap (B \cap D)| + |(A \cap C) \cap (B \cup D)| = |A \cap B| + |C \cap D|. \quad (\dagger)$$

The important part of ($\dagger$) is the inequality '$\leqslant$', sometimes referred to as *submodularity*. It implies, for example, that of any two opposite corner separations of two separations in S_k one must again be in S_k. We shall need this twice in the proof of Theorem 12.5.1.

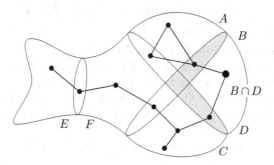

Fig. 12.5.3. The separation $\{A \cup C, B \cap D\}$ is one of the four corner separations of $\{A, B\}$ and $\{C, D\}$

Lemma 12.5.2. *Every consistent orientation O of S_k that has a subset in $\mathcal{T}$ also has a subset in $\mathcal{T}^*$.*

Proof. Consider any set $\sigma = \{(A_i, B_i) \mid i = 1, 2, 3\} \subseteq O$ in $\mathcal{T}$. We show that unless σ is a star we can replace one of its separations by a strictly smaller separation from O while keeping it in $\mathcal{T}$. In finitely many steps this will turn σ into a star—a subset of O in $\mathcal{T}^*$.

If σ is not a star we may assume that $(A_1, B_1) \not\leqslant (B_2, A_2)$. By ($\dagger$) and $\sigma \subseteq \vec{S}_k$ we may further assume that $(C, D) := (A_1 \cap B_2, B_1 \cup A_2) \in \vec{S}_k$; the other case, that $(B_1 \cap A_2, A_1 \cup B_2) \in \vec{S}_k$, is analogous. Since O is consistent and $(C, D) \leqslant (A_1, B_1) \in O$, we cannot have $(D, C) \in O$. Thus, $(C, D) \in O$. But $(C, D) < (A_1, B_1)$, since either $A_1 \not\subseteq B_2$ or $B_1 \not\supseteq A_2$ by assumption. Replacing (A_1, B_1) in σ with (C, D), to obtain σ' say, gives the desired reduction: since any vertex or edge of $G[A_1]$ that does not lie in $G[C]$ lies in $G[A_2]$, and $(A_2, B_2) \in \sigma'$, we have $\sigma' \in \mathcal{T}$. $\square$

Proof of Theorem 12.5.1. Since tangles are consistent, (i) and (ii) are equivalent by Lemma 12.5.2.

(ii)→(iii) Suppose G has an S_k-tree (T, α) over $\mathcal{T}^*$. Then any orientation O of S_k defines via α an orientation of the edges of T. Let $t \in T$

be the last node of any maximal path in T whose edges are all oriented forward. Then all the edges at t are oriented towards t, and α maps these oriented edges to a star in T^*. Hence O does not avoid T^*.

O^- (iii)$\rightarrow$(ii) Consider a set $O^- \subseteq \vec{S_k}$ that is closed down in the ordering of $\vec{S_k}$, i.e., which contains every $(A,B) \in \vec{S_k}$ for which there exists a $(C,D) \in O^-$ with $(A,B) \leqslant (C,D)$, and which contains all the oriented separations (A,V) with $|A| < k$: those whose inverses (V,A) form singleton stars in T^*. Let $T^+ := \big\{\{(B,A)\} : (A,B) \in O^- \big\} \smallsetminus T^*$. Thus,

$$O^- = \big\{(A,B) : \{(B,A)\} \in T^* \cup T^+\big\}. \tag{$*$}$$

We show that one of the following two statements holds:

(1) S_k has a consistent orientation that avoids $T^* \cup T^+$;

(2) G has an S_k-tree over $T^* \cup T^+$.

Note that this yields (iii)$\rightarrow$(ii) when $T^+ = \emptyset$.

If O^- contains a separation (X,Y) together with its inverse (Y,X), then (T,α) with $T = K^2$ and $\alpha\colon \vec{E}(T) \to \{(X,Y),(Y,X)\}$ satisfies (2), by $(*)$. We now assume that O^- contains no inverse pair of separations.

S^- Then O^- is an orientation of some set $S^- \subseteq S_k$, and $(*)$ implies

$$O^- \text{ contains no } (B,A) \text{ such that } \{(B,A)\} \in T^* \cup T^+. \tag{$**$}$$

Let us apply induction on $|S_k \smallsetminus S^-|$ to show that (1) or (2) holds. At the induction start, O^- is an orientation of all of S_k. Since it is closed down in $\vec{S_k}$, it is consistent. Hence if (1) fails then O^- has a subset $\sigma \in T^* \cup T^+$, with $|\sigma| \geqslant 2$ by $(**)$. Let T be the star $K_{1,n}$ with $n = |\sigma|$ leaves, and let α map its oriented edges (e,s,t) with s a leaf bijectively to the elements of σ. Then (T,α) satisfies (2), by definition of σ and $(*)$.

In the induction step we have $S_k \smallsetminus S^- \neq \emptyset$. Choose $\{U_1,W_1\}$ and U_i, W_i $\{U_2,W_2\}$ in $S_k \smallsetminus S^-$ so that both (U_i,W_i) are minimal in $\vec{S_k} \smallsetminus S^-$ and $(U_1,W_1) \leqslant (W_2,U_2)$.[5] Then the (U_i,W_i) are minimal even in $\vec{S_k} \smallsetminus O^-$: for any $(U,W) < (U_i,W_i)$ in $\vec{S_k} \smallsetminus O^-$ we would have $(W,U) \in O^-$ by the minimality of (U_i,W_i) in $\vec{S_k} \smallsetminus S^-$, so $(W_i,U_i) < (W,U)$ would be in O^- (this being closed down), contradicting the fact that $(U_i,W_i) \notin \vec{S^-}$.

O_i^- Thus, the sets $O_i^- = O^- \cup \{(U_i,W_i)\}$ are again closed down in $\vec{S_k}$, and are orientations of sets $S_i^- \subseteq S_k$ that contain S^- properly. We may therefore apply the induction hypothesis to these O_i^-, to obtain (1) or (2) for $T_i^+ = \big\{\{(B,A)\} : (A,B) \in O_i^- \big\} \smallsetminus T^*$.

Since (1) holds with T^+ as soon as it holds with T_1^+ or T_2^+, we may (T_i, α_i) assume that both T_i^+ satisfy (2). Let (T_i,α_i) be the corresponding S_k-trees over $T^* \cup T_i^+$. If one of these is in fact over $T^* \cup T^+$ we are done,

---

[5] It is easy to see that such separations exist, just choose them in turn.

so we assume not. Then each T_i has a node u_i such that $\alpha_i(e_i, u_i, w_i) = $ e_i, u_i, w_i
(U_i, W_i) for every edge $e_i = u_i w_i$ of T_i at u_i.

Every such u_i must be a leaf. For otherwise $(W_i, U_i) \leqslant (U_i, W_i)$, and
therefore $W_i \subseteq U_i$. But then $U_i = V$ and hence $(W_i, U_i) \in O^-$, contrary
to the choice of (U_i, W_i). Similarly, these leaves u_i are unique. Indeed if
u_i' is another leaf, with incident edge $e_i' = u_i' w_i'$ say, then $(e_i', u_i', w_i') \leqslant$
(e_i, w_i, u_i) and therefore $\alpha_i(e_i', u_i', w_i') \leqslant \alpha_i(e_i, w_i, u_i) = (W_i, U_i)$. Hence
if $\alpha_i(e_i', u_i', w_i') = (U_i, W_i)$ then $(U_i, W_i) \leqslant (W_i, U_i)$, with a contradiction
as earlier.

Thus, our S_k-trees (T_i, α_i) are nearly over $T^* \cup T^+$: they are, except
at their leaf u_i.

Choose $\{X_1, X_2\} \in S_k \setminus S^-$ of minimum order with X_1, X_2

$$(U_1, W_1) \leqslant (X_1, X_2) \leqslant (W_2, U_2);$$

such a separation exists, because (U_1, W_1) is a candidate. We shall
modify the maps α_i to maps α_i' defining S_k-trees (T_i, α_i'). These will
again be over $T^* \cup T^+$ except at u_i, where we shall have $\alpha_i'(e_i, w_i, u_i) =$
(X_{3-i}, X_i). Our plan will then be to join the newly labelled trees $T_i - u_i$
by adding the edge $w_i w_{3-i}$ and mapping it to their common separation
(X_i, X_{3-i}), to obtain our desired S_k-tree over $T^* \cup T^+$.

To define α_i', consider an edge e of T_i. Name its ends t, t' so that
$(e_i, u_i, w_i) \leqslant (e, t, t')$. Then if $\alpha_i(e, t, t') = (A, B)$, say, let

$$\alpha_i'(e, t, t') = (A', B') := (A \cup X_i, B \cap X_{3-i})$$ α_i'

and $\alpha_i'(e, t', t) = (B', A')$. (Fig. 12.5.4).

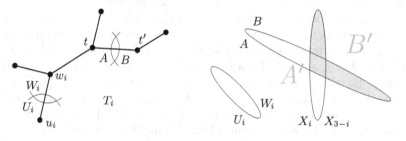

Fig. 12.5.4. Shifting (A, B) to (A', B')

Let us show that α_i', like α_i, respects the orderings of $\vec{E}(T_i)$ and $\vec{S_k}$.
Consider any $\vec{e} \leqslant \vec{f} \in \vec{E}(T_i)$, with $(A, B) = \alpha_i(\vec{e}) \leqslant \alpha_i(\vec{f}) = (C, D)$
say. Then $A \subseteq C$ and $B \supseteq D$. If $(e_i, u_i, w_i) \leqslant \vec{e}$, then $\alpha_i'(\vec{e}) =:$
$(A', B') \leqslant (C', D') := \alpha_i'(\vec{f})$, because $A' = A \cup X_i \subseteq C \cup X_i = C'$ and
$B' = B \cap X_{3-i} \supseteq D \cap X_{3-i} = D'$. On the other hand if $(e_i, u_i, w_i) \leqslant \overleftarrow{e}$
but $(e_i, u_i, w_i) \leqslant \vec{f}$, then $A' = A \cap X_{3-i} \subseteq A \subseteq C \subseteq C \cup X_i = C'$ while
$B' = B \cup X_i \supseteq B \supseteq D \supseteq D \cap X_{3-i} = D'$, so again $(A', B') \leqslant (C', D')$.

Since α_i', by definition, commutes with inversions of orientations, this covers all the cases to be considered.

Next, let us show that α_i' maps $\vec{E}(T_i)$ to $\vec{S_k}$. Given an edge $e \in E(T_i)$, let $\vec{e}$ be such that $(e_i, u_i, w_i) \leqslant \vec{e}$, as in the definition of α_i'. Then for $\alpha_i(\vec{e}) = (A, B)$ we have $\alpha_i'(\vec{e}) = (A', B') = (A \cup X_i, B \cap X_{3-i})$. By (†) we have $|A' \cap B'| \leqslant |A \cap B| \leqslant k$, as desired, if the order of

$$(Y_i, Y_{3-i}) := (A \cap X_i, B \cup X_{3-i})$$

is no less than the order of $\{X_1, X_2\}$. And it cannot be less, as that would contradict our choice of $\{X_1, X_2\}$ since $(U_1, W_1) \leqslant (Y_1, Y_2) \leqslant (W_2, U_2)$. Indeed, recall that $U_i \subseteq X_i$ as well as $U_i \subseteq A$, and $W_i \supseteq X_{3-i}$ as well as $W_i \supseteq B$, because $(U_i, W_i) \leqslant (X_i, X_{3-i})$ as well as $(U_i, W_i) \leqslant (A, B)$. Hence $(U_i, W_i) \leqslant (Y_i, Y_{3-i})$, and $(Y_i, Y_{3-i}) \leqslant (X_i, X_{3-i}) \leqslant (W_{3-i}, U_{3-i})$.

So (T_i, α_i') is indeed an S_k-tree. Let us show that it is an S_k-tree over $\mathcal{T}^* \cup \mathcal{T}^+$ except at u_i, where

$$\alpha_i'(e_i, w_i, u_i) = (W_i \cap X_{3-i}, U_i \cup X_i) = (X_{3-i}, X_i).$$

Given a leaf $t \neq u_i$, with incident edge $e = st$ and $\alpha_i(e, s, t) = (A, B)$ say, we have $(e_i, u_i, w_i) \leqslant (e, s, t)$ and hence

$$\alpha_i'(e, s, t) = (A \cup X_i, B \cap X_{3-i}) \geqslant (A, B) = \alpha_i(e, s, t).$$

Since (T_i, α_i) is over $\mathcal{T}^* \cup \mathcal{T}^+$ except at u_i, we have $\alpha_i(e, t, s) \in O^-$ by (∗). As O^- is closed down in $\vec{S_k}$, this means that $\alpha_i'(e, t, s) \leqslant \alpha_i(e, t, s)$ is also in O^-. Thus, $\{\alpha_i'(e, s, t)\} \in \mathcal{T}^* \cup \mathcal{T}^+$, as desired.

Now consider a non-leaf node t, with $\vec{F_t} = \{(f_0, s_0, t), \dots, (f_n, s_n, t)\}$ its set of incoming edges (and $n \leqslant 2$). Since α_i' respects the orderings of $\vec{E}(T_i)$ and $\vec{S_k}$, we already know that $\alpha_i'(\vec{F_t})$ is a star. To show $\alpha_i'(\vec{F_t}) \in \mathcal{T}$, assume that $s_0 \in u_i T_i t$. Then $(e_i, u_i, w_i) \leqslant (f_0, s_0, t) \leqslant (f_j, t, s_j)$ for all $j \geqslant 1$. So with $\alpha_i(f_j, s_j, t) =: (A_j, B_j)$ and $\alpha_i'(f_j, s_j, t) =: (A_j', B_j')$ we have $A_0' = A_0 \cup X_i$, while $A_j' = A_j \cap X_{3-i}$ for $j \geqslant 1$. Hence any vertex or edge of any $G[A_j]$ that is not in $G[A_j']$ lies in $G[X_i] \subseteq G[A_0']$. As α_i maps $\vec{F_t}$ to $\mathcal{T}^*$ because t is not a leaf, this implies that so does α_i'.

Let T be the tree obtained from the disjoint union of $T_1 - u_1$ and $T_2 - u_2$ by joining w_1 to w_2 by a new edge e. Let $\alpha: \vec{E}(T) \to \vec{S_k}$ map (e, w_{3-i}, w_i) to $\alpha_i'(e_i, u_i, w_i) = (X_i, X_{3-i})$ for $i = 1, 2$, and otherwise extend the α_i'. Then α commutes with the inversion of $\vec{e}$ and of (X_1, X_2), and $\alpha(\vec{F_t}) = \alpha_i'(\vec{F_t}) \in \mathcal{T}^* \cup \mathcal{T}^+$ for all $t \in T$, in particular for $t = w_i$. Hence, (T, α) satisfies (2). $\square$

Theorem 12.5.1, as stated above, is a special case of a more general result in which $\mathcal{T}^*$ can be replaced by other collections $\mathcal{F}$ of 'forbidden'

stars of oriented separations. Given a set S of separations, an $\mathcal{F}$-*tangle* $\qquad$ $\mathcal{F}$-*tangle*
of S is a consistent orientation of S that has no subset $\sigma \in \mathcal{F}$. For

$$\mathcal{F}_k := \left\{ \sigma \subseteq \vec{S} : \sigma \text{ is a star and } \left| \bigcap \{ B : (A, B) \in \sigma \} \right| < k \right\}$$

we obtain the following more tangle-like duality theorem for tree-width:

Theorem 12.5.3. *The following assertions are equivalent for all graphs G and integers $k > 0$:*

(i) *G has an S_k-tree over $\mathcal{F}_k$;*

(ii) *G has no $\mathcal{F}_k$-tangle of S_k.*

It is not hard to show that (i) is equivalent to G having tree-width less than $k - 1$. More generally, while every tree-decomposition gives rise to an S-tree for the set S of separations it induces, it is also easy to reobtain this tree-decomposition from that S-tree (Exercise 45).

Like the k-blocks of a graph, its tangles can be 'separated' by a tree-decomposition. Since all that a tangle in a graph refers to are its separations, this is most conveniently expressed not by constructing the tree-decomposition as such but just the set of separations it induces: from these, the full tree-decomposition can easily be constructed if desired. Let us make this more precise.

Two separations in G are *nested* if they have comparable orienta- $\qquad$ *nested*
tions; otherwise the *cross*. The separation $\{E, F\}$ in Figure 12.5.3, for $\qquad$ *cross*
example, is nested with the two crossing separations $\{A, B\}$ and $\{C, D\}$.
The separations induced by a tree-decomposition are clearly nested, and every nested set of separations of a graph is induced by some tree-decomposition (Exercises 16 and 42).

Let us say that a separation *distinguishes* two tangles, not necessar- $\qquad$ *distinguish*
ily of the same order, if they orient it differently. It distinguishes them *efficiently* if they are not distinguished by any separation of smaller order. $\qquad$ *efficient*
A set S of separations *distinguishes* some set of tangles in a graph G
efficiently if every two tangles in this set that are distinguished by some separation in G are distinguished efficiently by a separation in S.

Theorem 12.5.4. (Robertson & Seymour 1991)
Every graph G has a nested set of separations that distinguishes all the $\qquad$ tangle-tree
tangles in G efficiently. $\qquad$ theorem

For the proof we need an easy observation, known as the *fish lemma*, about nested and crossing separations as in Figure 12.5.3:

Lemma 12.5.5. *Let $\{A, B\}$ and $\{C, D\}$ be two crossing separations. Every separation $\{E, F\}$ that is nested with both $\{A, B\}$ and $\{C, D\}$ is also nested with their four corner separations.*

Proof. Since $\{E,F\}$ is nested with both $\{A,B\}$ and $\{C,D\}$, it has an orientation that is $\leqslant$ some orientation of $\{A,B\}$, and another that is $\leqslant$ some orientation of $\{C,D\}$. If these orientations of $\{E,F\}$ cannot be chosen the same, then there exist orientations $(A,B) \leqslant (E,F) \leqslant (C,D)$ showing that $\{A,B\}$ and $\{C,D\}$ are nested, contrary to our assumption.

Hence $\{E,F\}$ has an orientation (E,F) such that $(E,F) \leqslant (A,B)$ as well as $(E,F) \leqslant (C,D)$, for suitable orientations (A,B) of $\{A,B\}$ and (C,D) of $\{C,D\}$. Then $(E,F) \leqslant (A\cap C, B\cup D)$ as well as, more trivially, $(E,F) \leqslant (A\cup C, B\cap D)$ and $(E,F) \leqslant (A\cup D, B\cap C)$ and $(E,F) \leqslant (C\cup B, A\cap D)$. $\qquad\square$

S **Proof of Theorem 12.5.4.** Choose a maximal sequence S of nested separations in G with the following properties, writing $|s|$ for the order of a separation s:

(i) every $s \in S$ distinguishes some two tangles in G efficiently;

(ii) for every $s \in S$, the separations preceding s in S distinguish all the tangles in G of order at most $|s|$ efficiently.

We claim that S distinguishes all the tangles in G efficiently.

Suppose there exist tangles τ, τ' in G that are distinguished by some separation t in G but not distinguished efficiently by any separation in S.

$k,\,\tau,\tau'$ Choose τ, τ' and t so that t has minimum order, k say. Then τ and τ' have order $> k$ and t distinguishes them efficiently; in particular $t \notin S$.

t Among the separations of order k that distinguish τ and τ', choose t nested with as many separations in S as possible.

All the separations in S have order at most k. Indeed, if $s \in S$ had order $|s| > k$, then by (ii) the tangles $\tau \cap \vec{S}_{k+1}$ and $\tau' \cap \vec{S}_{k+1}$ of order $k+1 \leqslant |s|$ would be distinguished by some separation in S. This would then also distinguish τ from τ', a contradiction.

$\ell,\,s$ Since we cannot append t to S, there exists some $s \in S$, of order ℓ say, which t crosses. (Indeed, by the minimality of k, appending t to S cannot invalidate (ii) even if all the separations in S have order $< k$.) By (i),

σ,σ' the separation s distinguishes some two tangles σ and σ' efficiently.

Let $t = \{A,B\}$, with $(A,B) \in \tau$ and $(B,A) \in \tau'$ say, and $s = \{C,D\}$ with $(C,D) \in \sigma$ and $(D,C) \in \sigma'$. Since s does not distinguish τ from τ',

r,r' we may assume that they both contain (C,D). Let us show that the corner separations $r = \{A\cup C, B\cap D\}$ and $r' = \{B\cup C, A\cap D\}$ have order $> k$.

Suppose $|r| \leqslant k$. By Lemma 12.5.5, r is nested with every separation in S that t is nested with, because every such separation $\{E,F\}$ is nested not only with t but also with s, since they both lie in S. In addition, r is nested with s, which t crosses. So r is nested with more separations in S than t is. This contradicts the choice of t, since r also distinguishes τ from τ': as $(A,B) \in \tau$ and $(C,D) \in \tau$ we have $(A\cup C, B\cap D) \in \tau$, since

τ is a tangle, whereas $(B \cap D, A \cup C) \leqslant (B, A) \in \tau'$ puts $(B \cap D, A \cup C)$ in τ' by the consistency of τ'.

So r and r' have order $> k$. By submodularity (†), therefore, the other two corner separations of the crossing pair t and s have order less than $\ell = |s|$. Hence neither of these distinguishes σ from σ', because s does so efficiently. Their orientations $(A \cap C, B \cup D) \leqslant (C, D) \in \sigma$ and $(B \cap C, A \cup D) \leqslant (C, D) \in \sigma$, which lie in σ by its consistency, therefore also lie in σ'. As $(D, C) \in \sigma'$ but $G[A \cap C] \cup G[B \cap C] \cup G[D] = G$, this contradicts the fact that σ' is a tangle.   □

Theorem 12.5.4 can be strengthened to a 'canonical' version in the spirit of Theorem 12.3.7. This finds a nested set of separations which, in addition, is invariant under the automorphisms of G. See the notes.

12.6 Tree-decompositions and forbidden minors

If $\mathcal{H}$ is any set or class of graphs, then the class

$$\mathrm{Forb}_{\preccurlyeq}(\mathcal{H}) := \{\, G \mid G \not\succcurlyeq H \text{ for all } H \in \mathcal{H} \,\} \qquad\qquad \mathrm{Forb}_{\preccurlyeq}(\mathcal{H})$$

of all graphs without a minor in $\mathcal{H}$ is a graph property, i.e. is closed under isomorphism.[6] When it is written as above, we say that this property is expressed by specifying the graphs $H \in \mathcal{H}$ as *forbidden* (or *excluded*) *minors*.

By Proposition 1.7.1, $\mathrm{Forb}_{\preccurlyeq}(\mathcal{H})$ is closed under taking minors, or *minor-closed*: if $G' \preccurlyeq G \in \mathrm{Forb}_{\preccurlyeq}(\mathcal{H})$ then $G' \in \mathrm{Forb}_{\preccurlyeq}(\mathcal{H})$. Every minor-closed property can in turn be expressed by forbidden minors:

forbidden minors

(1.7.1)

Lemma 12.6.1. *A graph property $\mathcal{P}$ can be expressed by forbidden minors if and only if it is closed under taking minors.*

[5.2]

Proof. For the 'if' part, note that $\mathcal{P} = \mathrm{Forb}_{\preccurlyeq}(\overline{\mathcal{P}})$, where $\overline{\mathcal{P}}$ is the complement of $\mathcal{P}$.   □

$\overline{\mathcal{P}}$

In Section 12.7, we shall return to the general question of how a given minor-closed property is best represented by forbidden minors. In this section, we begin by looking at a particular example of such a property: bounded tree-width.

Consider the property of having tree-width less than some given integer k. By Lemmas 12.4.1 and 12.6.1, this property can be expressed by forbidden minors. Choosing their set $\mathcal{H}$ as small as possible, we find that $\mathcal{H} = \{K^3\}$ for $k = 2$: the graphs of tree-width < 2 are precisely the forests. For $k = 3$, we have $\mathcal{H} = \{K^4\}$:

---

[6] As usual, we abbreviate $\mathrm{Forb}_{\preccurlyeq}(\{H\})$ to $\mathrm{Forb}_{\preccurlyeq}(H)$.

Proposition 12.6.2. *A graph has tree-width < 3 if and only if it has no K^4 minor.*

(7.3.1)
(12.3.2)
(12.3.5)
(12.3.6)
(12.4.1)

Proof. By Corollary 12.3.5, we have $\mathrm{tw}(K^4) \geqslant 3$. By Lemma 12.4.1, therefore, a graph of tree-width < 3 cannot contain K^4 as a minor.

Conversely, let G be a graph without a K^4 minor; we assume that $|G| \geqslant 3$. Add edges to G until the graph G' obtained is edge-maximal without a K^4 minor. By Proposition 7.3.1, G' can be constructed recursively from triangles by pasting along K^2s. By induction on the number of recursion steps and Corollary 12.3.5, every graph constructible in this way has a tree-decomposition into triangles (as in the proof of Proposition 12.3.6). Such a tree-decomposition of G' has width 2, and by Lemma 12.3.2 it is also a tree-decomposition of G.   □

As k grows, the list of forbidden minors characterizing the graphs of tree-width $< k$ seems to grow fast. They are known explicitly only up to $k = 4$; see the notes.

A question converse to the above is to ask for which H (other than K^3 and K^4) the tree-width of the graphs in $\mathrm{Forb}_{\preccurlyeq}(H)$ is bounded. This is the case, for example, when H is grid:

[12.7.1]
[12.7.3]

Theorem 12.6.3. (Robertson & Seymour 1986)
For every integer r there is an integer k such that every graph of tree-width at least k has an $r \times r$ grid minor.

This *grid theorem* may, at first glance, look like just a specific and technical result. But it has a sweeping consequence:

Corollary 12.6.4. *Given a graph H, the graphs without an H minor have bounded tree-width if and only if H is planar.*

(4.4.6)

Proof. Since all grids and their minors are planar, every class $\mathrm{Forb}_{\preccurlyeq}(H)$ with a non-planar H contains all grids, which have unbounded tree-width (see after Theorem 12.4.3).

Conversely, every planar graph H is a minor of some grid: take a drawing of the graph, fatten its vertices and edges, and superimpose a sufficiently fine plane grid. Hence, by Theorem 12.6.3, the graphs in $\mathrm{Forb}_{\preccurlyeq}(H)$ have bounded tree-width as soon as H is planar.   □

Theorem 12.6.3 has another interesting application. Recall that a class $\mathcal{H}$ of graphs has the *Erdős-Pósa property* if the number of vertices in a graph needed to cover all its subgraphs in $\mathcal{H}$ is bounded by a function of its maximum number of disjoint subgraphs in $\mathcal{H}$. Now let H be a fixed connected graph, and consider the class $\mathcal{H} = IH$ of graphs that contract to a copy of H. (Thus, G has a subgraph in $\mathcal{H}$ if and only if $H \preccurlyeq G$.)

H
$\mathcal{H}$

Theorem 12.6.5. (Robertson & Seymour 1986)
$\mathcal{H}$ has the Erdős-Pósa property if H is planar.

Proof. We have to find a function $f \colon \mathbb{N} \to \mathbb{N}$ such that, given $k \in \mathbb{N}$ and (12.3.1)
a graph G, either G contains k disjoint models of H or there is a set U
of at most $f(k)$ vertices in G such that $H \npreceq G - U$.

By Corollary 12.6.4, there exists for every $k \geq 1$ an integer w_k such
that every graph of tree-width at least w_k contains the disjoint union of
k copies of H (which is again planar) as a minor. Define

$$f(k) := 2f(k-1) + w_k$$

inductively, starting with $f(0) = f(1) = 0$.

To verify that f does what it should, we apply induction on k. For
$k \leq 1$ there is nothing to show. Now let k and G be given for the
induction step. If $\mathrm{tw}(G) \geq w_k$, we are home by definition of w_k. So
assume that $\mathrm{tw}(G) < w_k$, and let $(T, (V_t)_{t \in T})$ be a tree-decomposition
of G of width $< w_k$. Let us direct the edges $t_1 t_2$ of the tree T as
follows. Let T_1, T_2 be the components of $T - t_1 t_2$ containing t_1 and t_2,
respectively, and put

$$G_1 := G\Big[\bigcup_{t \in T_1} (V_t \smallsetminus V_{t_2})\Big] \quad \text{and} \quad G_2 := G\Big[\bigcup_{t \in T_2} (V_t \smallsetminus V_{t_1})\Big].$$

We direct the edge $t_1 t_2$ towards G_i if $H \preceq G_i$, thereby giving $t_1 t_2$ either
one or both or neither direction.

If every edge of T receives at most one direction, we follow these to
a node $t \in T$ such that no edge at t in T is directed away from t. As H is
connected, this implies by Lemma 12.3.1 that V_t meets every $IH \subseteq G$.
This completes the proof with $U = V_t$, since $|V_t| \leq w_k \leq f(k)$ by the
choice of our tree-decomposition.

Suppose now that T has an edge $t_1 t_2$ that received both directions.
For each $i = 1, 2$ let us ask if we can cover all the models of H in G_i by at
most $f(k-1)$ vertices. If we can, for both i, then by Lemma 12.3.1 the
two covers combine with $V_{t_1} \cap V_{t_2}$ to the desired cover U for G. Suppose
now that G_1 has no such cover. Then, by the induction hypothesis,
G_1 contains $k - 1$ disjoint models of H. Since $t_1 t_2$ was also directed
towards t_2, there is another such model in G_2. This gives the desired
total of k disjoint models of H in G. $\square$

Theorem 12.6.5 contains the Erdős-Pósa theorem 2.3.2 as the special
case that $H = K^3$. It is best possible in that if H is non-planar, then
$\mathcal{H} = IH$ does not have the Erdős-Pósa property (Exercise 48).

We conclude this section with statements of the structure theorems for the graphs not containing a given complete graph as a minor. These are far more difficult to prove than any of the results we have seen so far, and they are not even that easy to state. But it's worth an effort: the statement of the excluded-K^n theorem is interesting, it is central to the proof of the graph minor theorem, and it can be applied elsewhere.

linear decomposition

A *linear decomposition* of G is a family $(V_i)_{i \in I}$ of vertex sets indexed by some linear order I such that $\bigcup_{i \in I} V_i = V(G)$, every edge of G has both its ends in some V_i, and $V_i \cap V_k \subseteq V_j$ whenever $i < j < k$. When G is finite, this is just a tree-decomposition whose decomposition tree is a path, and usually called a *path-decomposition*. If each V_i contains at most k vertices and k is minimal with this property, then $(V_i)_{i \in I}$ has *width* $k - 1$.

$C_1, \ldots, C_k$
$S - k$

Let S' be a subspace of a surface[7] S obtained by removing the interiors of finitely many disjoint closed discs, with boundary circles $C_1, \ldots, C_k$ say. This space is determined up to homeomorphism by S and the number k, and we denote it by $S - k$. Each C_i is the image of a continuous map $f_i : [0, 1] \to S'$ that is injective except for $f_i(0) = f_i(1)$.

cuffs

We call $C_1, \ldots, C_k$ the *cuffs* of S' and the points $f_1(0), \ldots, f_k(0)$ their *roots*. The other points of each C_i are linearly ordered by f_i as images of $(0, 1)$; when we use cuffs as index sets for linear decompositions below, we shall be referring to these linear orders. An *embedding* of a graph in S (or in $S - k$) is defined analogously to embeddings in the plane.

k-near embedding

Let H be a graph, S a surface, and $k \in \mathbb{N}$. We say that H is *k-nearly embeddable* in S if H has a set X of at most k vertices such that $H - X$ can be written as $H_0 \cup H_1 \cup \ldots \cup H_k$ so that

(N1) there exists an embedding $\sigma \colon H_0 \hookrightarrow S - k$ that maps only vertices to cuffs and no vertex to the root of a cuff;

(N2) the graphs $H_1, \ldots, H_k$ are pairwise disjoint (and may be empty), and $H_0 \cap H_i = \sigma^{-1}(C_i)$ for each i;

(N3) every H_i with $i \geqslant 1$ has a linear decomposition $(V_z^i)_{z \in C_i \cap \sigma(H_0)}$ of width $< k$ such that $\sigma^{-1}(z) \in V_z^i$ for all z.

Here, then, is the structure theorem for the graphs without a K^n minor.[8] For $n = 5$, the original result of Wagner, Theorem 7.3.4, remains more precise. The case of $n = 4$ is covered by Proposition 7.3.1.

Theorem 12.6.6. (Robertson & Seymour 2003)
For every $n \geqslant 5$ there exists a $k \in \mathbb{N}$ such that every graph not containing K^n as a minor has a tree-decomposition whose torsos are k-nearly embeddable in a surface in which K^n is not embeddable.

---

[7] A compact connected 2-manifold without boundary; see Appendix B.

[8] Robertson and Seymour proved several versions of this theorem, of which Theorem 12.6.6 is the simplest. See the notes for references to yet more powerful versions.

Note that there are only finitely many surfaces in which K^n is not embeddable. The set of those surfaces in the statement of Theorem 12.6.6 could therefore be replaced by just two surfaces: the orientable and the non-orientable surface of maximum genus in this set.

Theorem 12.6.6 also has a converse, though only a qualitative one. A decomposition as described does not by itself prevent the presence of a K^n minor. But for every n there is an r such that no graph with such a decomposition has a K^r minor. This is because the adhesion sets of the tree-decomposition have bounded size, e.g. by $2k + n$, since they induce complete subgraphs in the torsos which are k-nearly embeddable in a surface that does not accommodate K^n. We remark that Theorem 12.6.6, as stated above, is true also for infinite graphs.

For graphs without a given topological minor, there is a related structure theorem:

Theorem 12.6.7. (Grohe & Marx 2012)
For every $n \geqslant 5$ there exists a $k \in \mathbb{N}$ such that every graph not containing K^n as a topological minor has a tree-decomposition whose torsos are either k-nearly embeddable in a surface of Euler genus $\leqslant k$ or have at most k vertices of degree $> k$.

(See Appendix B for the definition of the Euler genus of a surface.)

There are also structure theorems for excluding infinite minors, and we now state two of these.

First, the structure theorem for excluding $K^{\aleph_0}$. Call a graph H *nearly planar* if H has a finite set X of vertices such that $H - X$ can be written as $H_0 \cup H_1$ so that (N1–2) hold with $S = S^2$ (the sphere) and $k = 1$, while (N3) holds with $k = |X|$. (In other words, deleting a bounded number of vertices makes H planar except for a subgraph of bounded linear width sewn on to the unique cuff of $S^2 - 1$.) A tree-decomposition $(T, (V_t)_{t \in T})$ of a graph G has *finite adhesion* if all its adhesion sets are finite and for every infinite path $t_1 t_2 \ldots$ in T the value of $\liminf_{i \to \infty} |V_{t_i} \cap V_{t_{i+1}}|$ is finite.

nearly planar

finite adhesion

Unlike its counterpart for K^n, the excluded-$K^{\aleph_0}$ structure theorem has a direct converse. It thus characterizes the graphs without a $K^{\aleph_0}$ minor, as follows:

Theorem 12.6.8. *A graph G has no $K^{\aleph_0}$ minor if and only if G has a tree-decomposition of finite adhesion whose torsos are nearly planar.*

Finally, a structure theorem for excluding $K^{\aleph_0}$ as a topological minor. Let us say that G has *finite tree-width* if G admits a tree-decomposition $(T, (V_t)_{t \in T})$ into finite parts such that for every infinite path $t_1 t_2 \ldots$ in T the set $\bigcup_{j \geqslant 1} \bigcap_{i \geqslant j} V_{t_i}$ is finite.

finite tree-width

Theorem 12.6.9. *The following assertions are equivalent for connected graphs G:*

(i) *G does not contain $K^{\aleph_0}$ as a topological minor;*

(ii) *G has finite tree-width;*

(iii) *G has a normal spanning tree T such that for every ray R in T there are only finitely many vertices v such that G contains an infinite v–$(R-v)$ fan.*

12.7 The graph minor theorem

Graph properties that are closed under taking minors occur frequently in graph theory. Among the most natural examples are the properties of being embeddable in some fixed surface, such as planarity.

By Kuratowski's theorem, planarity can be expressed by forbidding the minors K^5 and $K_{3,3}$. This is a *good characterization* of planarity in the following sense. Suppose we wish to persuade someone that a certain graph is planar: this is easy (at least intuitively) if we can produce a drawing of the graph. But how do we persuade someone that a graph is non-planar? By Kuratowski's theorem, there is also an easy way to do that: we just have to exhibit an IK^5 or $IK_{3,3}$ in our graph, as an easily checked 'certificate' for non-planarity. Our simple Proposition 12.6.2 is another example of a good characterization: if a graph has tree width < 3, we can prove this by exhibiting a suitable tree-decomposition; if not, we can produce an IK^4 as evidence.

Theorems that characterize a property $\mathcal{P}$ by a set of forbidden minors are doubtless among the most attractive results in graph theory. As we saw in Lemma 12.6.1, such a characterization exists whenever $\mathcal{P}$ is minor-closed: then $\mathcal{P} = \mathrm{Forb}_{\preccurlyeq}(\overline{\mathcal{P}})$, where $\overline{\mathcal{P}}$ is the complement of $\mathcal{P}$. However, one naturally seeks to make the set of forbidden minors as small as possible. And there is indeed a unique smallest such set: the set

(12.6.1)

Kuratowski set $\mathcal{K}_{\mathcal{P}}$

$$\mathcal{K}_{\mathcal{P}} := \{\, H \mid H \text{ is } \preccurlyeq\text{-minimal in } \overline{\mathcal{P}} \,\}$$

satisfies $\mathcal{P} = \mathrm{Forb}_{\preccurlyeq}(\mathcal{K}_{\mathcal{P}})$ and is contained in every other set $\mathcal{H}$ such that $\mathcal{P} = \mathrm{Forb}_{\preccurlyeq}(\mathcal{H})$. We call $\mathcal{K}_{\mathcal{P}}$ the *Kuratowski set* for $\mathcal{P}$.

Clearly, the elements of $\mathcal{K}_{\mathcal{P}}$ are incomparable under the minor relation $\preccurlyeq$. Now the *graph minor theorem* of Robertson & Seymour says that any set of $\preccurlyeq$-incomparable graphs must be finite:

graph minor theorem

Theorem 12.7.1. (Robertson & Seymour 1986–2004)
The finite graphs are well-quasi-ordered by the minor relation $\preccurlyeq$.

We shall give a sketch of the proof of the graph minor theorem at the end of this section.

Corollary 12.7.2. *The Kuratowski set for any minor-closed graph property is finite.* □

As a special case of Corollary 12.7.2 we have, at least in principle, a Kuratowski-type theorem for every surface S: the property $\mathcal{P}(S)$ of embeddability in S is characterized by the finite set $\mathcal{K}_{\mathcal{P}(S)}$ of forbidden minors. $\qquad \mathcal{P}(S)$

Corollary 12.7.3. *For every surface S there exists a finite set of graphs $H_1, \ldots, H_n$ such that a graph is embeddable in S if and only if it contains none of $H_1, \ldots, H_n$ as a minor.* □

While Corollary 12.7.3 is immediate from the graph minor theorem, it can also be proved more directly. It is our next goal to do this. The main step is to prove that the graphs in $\mathcal{K}_{\mathcal{P}(S)}$ do not contain arbitrarily large grids as minors (Lemma 12.7.4). Then their tree-width is bounded (Theorem 12.6.3), so $\mathcal{K}_{\mathcal{P}(S)}$ is well-quasi-ordered (Theorem 12.4.2) and therefore finite.

The proof of Lemma 12.7.4 gives a good impression of the interplay between graph minors and surface topology, which—by way of Theorem 12.6.6, which we could not prove here—is also one of the key ingredients of the proof of the graph minor theorem. Appendix B summarizes the necessary background on surfaces, including a lemma used in the proof. For convenience (cf. Proposition 1.7.3 (ii)), we shall work with hexagonal rather than square grids.

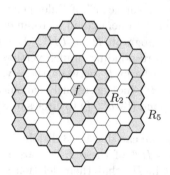

Fig. 12.7.1. The hexagonal grid H^6 with central face f and rings R_2 and R_5

Denote by H^r the plane hexagonal grid whose dual has radius r $\qquad H^r$ (Figure 12.7.1). The face corresponding to the central vertex of its dual is its *central face*. (Generally, when we speak of the *faces* of H^r, we $\qquad$ *faces*

mean its hexagonal faces, not its outer face.) A subgrid H^k of H^r is
canonical if their central faces coincide. We write S_k for the perimeter
cycle of the canonical subgrid H^k in H^r; for example, S_1 is the hexagon
bounding the central face of H^r. The *ring* R_k is the subgraph of H^r
formed by S_k and S_{k+1} and the edges between them.

Lemma 12.7.4. *For every surface S there exists an integer r such that
no graph that is minimal with the property of not being embeddable
in S contains H^r as a topological minor.*

Proof. Let G be a graph that cannot be embedded in S and is minimal
with this property. Our proof will run roughly as follows. Since G
is minimally not embeddable in S, we can embed it in an only slightly
larger surface S'. If G contains a very large H^r grid, then by Lemma B.6
some large H^m subgrid will be flat in S', that is, the union of its faces
in S' will be a disc D'. We then pick an edge e from the middle of this
H^m grid and embed $G - e$ in S. Again by Lemma B.6, one of the rings
of our H^m will be flat in S. In this ring we can embed the (planar)
subgraph of G which our first embedding had placed in D'; note that
this subgraph contains the edge e. The rest of G can then be embedded
in S outside this ring much as before, yielding an embedding of all of G
in S (a contradiction).

More formally, let $\varepsilon := \varepsilon(S)$ denote the Euler genus of S. Let r
be large enough that H^r contains $\varepsilon + 3$ disjoint copies of H^{m+1}, where
$m := 3\varepsilon + 4$. We show that G has no TH^r subgraph.

Let $e' = u'v'$ be any edge of G, and choose an embedding σ' of
$G - e'$ in S. Choose a face with u' on its boundary, and another with v'
on its boundary. Cut a disc out of each face and add a handle between
the two holes, to obtain a surface S' of Euler genus $\varepsilon + 2$ (Lemma B.3).
Embedding e' along this handle, extend σ' to an embedding of G in S'.
Suppose G has a subgraph $H = TH^r$. Let $f : H^r \to H$ map the
vertices of H^r to the corresponding branch vertices of H, and its edges
to the corresponding paths in H between those vertices. Let us show
that H^r has a subgrid H^m (not necessarily canonical) whose hexagonal
face boundaries correspond (by $\sigma' \circ f$) to circles in S' that bound disjoint
open discs there.

By the choice of r, we can find $\varepsilon + 3$ disjoint copies of H^{m+1} in H^r.
The canonical subgrids H^m of these H^{m+1} are not only disjoint, but
sufficiently spaced out in H^r that their deletion leaves a tree $T \subseteq H^r$
that sends an edge to each of them (Figure 12.7.2). Hence whenever
we pick one hexagon from each of these H^m and delete the images C
of those hexagons in S', the component D_0 of the remainder of S' that
contains $(\sigma' \circ f)(T)$ meets all those C in its boundary. By Lemma B.6
and $\varepsilon(S') = \varepsilon + 2$, therefore, it cannot be true that none of our circles C
bounds a disc in S' that is disjoint from $(\sigma' \circ f)(T)$.

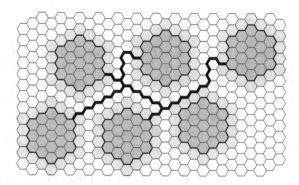

Fig. 12.7.2. Disjoint copies of H^m ($m = 3$) linked up by a tree in the rest of H^r

Hence for one of our copies of H^m in H^r, the image of every hexagon in S' bounds an open disc that is disjoint from $(\sigma' \circ f)(T)$. Let us show that these discs are disjoint. If not, then one of them, D say, contains a point x from the boundary of another such disc. But then D also contains $(\sigma' \circ f)(T)$, contrary to assumption, because we can walk from x to $(\sigma' \circ f)(T)$ in $(\sigma' \circ f)(H^r) \subseteq S'$ avoiding the boundary of D.

From now on, we shall work with this fixed H^m and will no longer consider its supergraph H^r. We write $C_i := f(S_i)$ for the images in G of the concentric cycles S_i of this H^m ($i = 1, \ldots, m$).

Pick an edge $e = uv$ of C_1, and choose an embedding σ of $G - e$ in S. As before, Lemma B.6 implies that one of the $\varepsilon + 1$ disjoint rings R_{3i+2} in our H^m ($i = 0, \ldots, \varepsilon$), R_k say, has the property that its hexagons correspond (by $\sigma \circ f$) to circles in S that bound disjoint open discs there (Figure 12.7.3). Let $R \supseteq (\sigma \circ f)(R_k)$ be the closure in S of the union of those discs, which is a cylinder in S. One of its two boundary circles

C_i

e

$\sigma : G - e \hookrightarrow S$

k

R

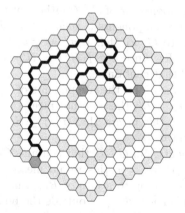

Fig. 12.7.3. A tree linking up hexagons selected from the rings $R_2, R_5, R_8 \ldots$

C

H'

is the image under σ of the cycle $C := C_{k+1}$ in G to which f maps the perimeter cycle S_{k+1} of our special ring $R_k \subseteq H^m$.

Let $H' := f(H^{k+1}) \subseteq G$, where H^{k+1} is canonical in our H^m. Recall that $\sigma' \circ f$ maps the hexagons of H^{k+1} to circles in S' bounding disjoint open discs there. The closure in S' of the union of these discs is a disc

D'
R'

D' in S', bounded by $\sigma'(C)$. Deleting a small open disc inside D' that does not meet $\sigma'(G)$, we obtain a cylinder $R' \subseteq S'$ that contains $\sigma'(H')$.

We shall now combine the embeddings $\sigma \colon G - e \hookrightarrow S$ and $\sigma' \colon G \hookrightarrow S'$

σ''

to an embedding $\sigma'' \colon G \hookrightarrow S$, which will contradict the choice of G. Let $\varphi \colon \sigma'(C) \to \sigma(C)$ be a homeomorphism between the images of C in S' and in S that commutes with these embeddings, i.e., is such that

φ

$\sigma|_C = (\varphi \circ \sigma')|_C$. Then extend this to a homeomorphism $\varphi \colon R' \to R$. The idea now is to define σ'' as $\varphi \circ \sigma'$ on the part of G which σ' maps to D' (which includes the edge e on which σ is undefined), and as σ on the rest of G (Fig. 12.7.4).

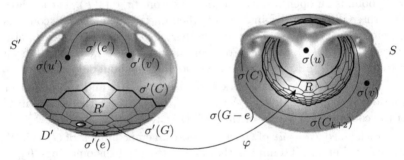

Fig. 12.7.4. Combining $\sigma' \colon G \hookrightarrow S'$ and $\sigma \colon G - e \hookrightarrow S$ to $\sigma'' \colon G \hookrightarrow S$

To make these two partial maps compatible, we start by defining σ'' on C as $\sigma|_C = (\varphi \circ \sigma')|_C$. Next, we define σ'' separately on the components of $G - C$. Since $\sigma'(C)$ bounds the disc D' in S', we know that σ' maps each component J of $G - C$ either entirely to D' or entirely to $S' \smallsetminus D'$. On all the components J such that $\sigma'(J) \subseteq D'$, and on all the edges they send to G, we define σ'' as $\varphi \circ \sigma'$. Thus, σ'' embeds these components in R. Since $e \in f(H^k) = H' - C$, this includes the component of $G - C$ that contains e.

It remains to define σ'' on the components of $G - C$ which σ' maps to $S' \smallsetminus D'$. As $\sigma'(C_k) \subseteq D'$, these do not meet C_k. Since $\sigma(C \cup C_k)$ is the frontier of R in S, this means that $\sigma(J) \subseteq S \smallsetminus R$ or $\sigma(J) \subseteq R$ for every such component J.

J_0

For the component J_0 of $G - C$ that contains C_{k+2} we cannot have $\sigma(J_0) \subseteq R$: as $S_{k+2} \cap R_k = \emptyset$, this would mean that $\sigma(C_{k+2})$ lies in a disc $D \subseteq R$ corresponding to a face of R_k, which is impossible since S_{k+2} sends edges to vertices of S_{k+1} outside the boundary of that face. We thus have $\sigma(J_0) \subseteq S \smallsetminus R$, and define σ'' as σ on J_0 and on all the J_0–C edges of G.

Next, consider any remaining component J of $G - C$ that sends no edge to C. If $\sigma(J) \subseteq S \smallsetminus R$, we define σ'' on J as σ. If $\sigma(J) \subseteq R$, then J is planar. Since J sends no edge to C, we can have σ'' map J to any open disc in R that has not so far been used by σ''.

It remains to define σ'' on the components $J \neq J_0$ of $G - C$ which σ' maps to $S' \smallsetminus D'$ and for which G contains a J–C edge. Let $\mathcal{J}$ be the set of all those components J. We shall group them by the way they attach to C, and define σ'' for these groups in turn.

Since $m \geqslant k + 2$, the disc D' lies inside a larger disc in S', which is the union of D' and closed discs D'' bounded by the images under $\sigma' \circ f$ of the hexagons in R_{k+1}. By definition of $\mathcal{J}$, the embedding σ' maps every $J \in \mathcal{J}$ to such a disc D'' (Fig. 12.7.5). On the path P in C such that $\sigma'(P) = \sigma'(C) \cap D''$ (which is the image under f of one or two consecutive edges on S_{k+1}), let $v_1, \ldots, v_n$ be the vertices with a neighbour in J_0, in their natural order along P, and write P_i for the segment of P from v_i to v_{i+1}. For any v_i with $1 < i < n$, pick a v_i–J_0 edge and extend it through J_0 to a path Q from v_i to C_{k+2} (which exists by definition of J_0); let w be its first vertex that σ' maps to the boundary circle of D''. By Lemma 4.1.2 applied to $\sigma'(v_i Q w)$ and the two arcs joining $\sigma'(v_i)$ to $\sigma'(w)$ along the boundary circle of D'', there is no arc through D'' that links $\sigma'(P_{i-1})$ to $\sigma'(P_i)$ but avoids $\sigma'(v_i Q w)$. Hence, every $J \in \mathcal{J}$ with $\sigma'(J) \subseteq D''$ has all its neighbours on C in the same P_i, and σ' maps J to the face f_i of the plane graph $\sigma'(G[J_0 \cup C]) \cap D''$ whose boundary contains P_i. We shall define σ'' jointly on all those $J \in \mathcal{J}$ which σ' maps to this f_i, for $i = 1, \ldots, n - 1$ in turn.

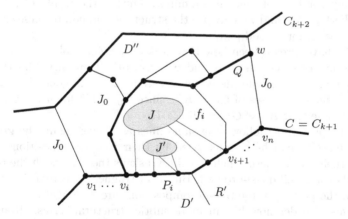

Fig. 12.7.5. Define σ'' jointly for the components $J, J' \in \mathcal{J}$ that attach to the same $P_i \subseteq C$

To do so, we choose an open disc D_i in $S \smallsetminus R$ that has a boundary circle containing $\sigma(P_i)$ and avoids the image of σ'' as defined until now. Such D_i exists in a strip neighbourhood of $\sigma(C)$ in S, because components $J' \in \mathcal{J}$ attaching to a segment $P_j \neq P_i$ of C send no edge to $\mathring{P}_i$. Choose a homeomorphism φ_i from the boundary circle of f_i to that of D_i so that $\sigma|_{P_i} = (\varphi_i \circ \sigma')|_{P_i}$, and extend this to a homeomorphism φ_i from the closure of f_i in S' to the closure of D_i in S. For every $J \in \mathcal{J}$ with $\sigma'(J) \subseteq f_i$, and for all J–C edges of G, define σ'' as $\varphi_i \circ \sigma'$. $\qquad\square$

(1.7.3)
(12.4.2)
(12.6.3)

Proof of Corollary 12.7.3. By their minimality, the graphs in $\mathcal{K}_{\mathcal{P}(S)}$ are incomparable under the minor-relation. If their tree-width is bounded, then $\mathcal{K}_{\mathcal{P}(S)}$ is well-quasi-ordered by the minor relation (Theorem 12.4.2), and hence must be finite. So assume their tree-width is unbounded, and let r be as in Lemma 12.7.4. By Theorem 12.6.3, some $H \in \mathcal{K}_{\mathcal{P}(S)}$ has a grid minor large enough to contain H^r. By Proposition 1.7.3, H^r is a topological minor of H, contrary to the choice of r. $\qquad\square$

We finally come to the proof of the graph minor theorem itself. The complete proof would still fill a book or two, but we are well equipped now to get a good understanding of its main ideas and overall structure. For background on surfaces, we once more refer to Appendix B.

(12.1.3)
(12.2.1)
(12.4.2)
(12.6.3)

Proof of the graph minor theorem (sketch). We have to show that every infinite sequence

$$G_0, G_1, G_2, \ldots$$

of finite graphs contains a good pair: two graphs $G_i \preccurlyeq G_j$ with $i < j$. We may assume that $G_0 \not\preccurlyeq G_i$ for all $i \geqslant 1$, since G_0 forms a good pair with any graph G_i of which it is a minor. Thus all the graphs $G_1, G_2, \ldots$ lie in $\mathrm{Forb}_{\preccurlyeq}(G_0)$, and we may use the structure common to these graphs in our search for a good pair.

We have already seen how this works when G_0 is planar: then the graphs in $\mathrm{Forb}_{\preccurlyeq}(G_0)$ have bounded tree-width (Corollary 12.6.4) and are therefore well-quasi-ordered by Theorem 12.4.2. In general, we need only consider the cases of $G_0 = K^n$: since $G_0 \preccurlyeq K^n$ for $n := |G_0|$, we may assume that $K^n \not\preccurlyeq G_i$ for all $i \geqslant 1$.

The proof now follows the same lines as above: again the graphs in $\mathrm{Forb}_{\preccurlyeq}(K^n)$ can be characterized by their tree-decompositions, and again their tree structure helps, as in Kruskal's theorem, with the proof that they are well-quasi-ordered. But as in Wagner's theorem (7.3.4) for $n = 5$, the parts in these tree-decompositions are no longer constrained in terms of order now but in more subtle structural terms. Roughly speaking, for every n there exists a finite set $\mathcal{S}$ of surfaces such that every graph without a K^n minor has a tree-decomposition into parts each 'nearly' embeddable in one of the surfaces $S \in \mathcal{S}$; see Theorem 12.6.6. By a generalization of Theorem 12.4.2—and hence of Kruskal's theorem—it now suffices, essentially, to prove that the set of all the parts in these tree-decompositions is well-quasi-ordered: then the graphs decomposing into these parts are well-quasi-ordered, too. Since $\mathcal{S}$ is finite, every infinite sequence of such parts has an infinite subsequence whose members are all (nearly) embeddable in the same surface $S \in \mathcal{S}$. Thus all we have to show is that, given any surface S, all the graphs embeddable in S are well-quasi-ordered by the minor relation.

This is shown by induction on the Euler genus of S, using the same approach as before: if $H_0, H_1, H_2, \ldots$ is an infinite sequence of graphs embeddable in S, we may assume that none of the graphs $H_1, H_2, \ldots$ contains H_0 as a minor. If $S = S^2$ we are back in the case that H_0 is planar, so the induction starts. For the induction step we now assume that $S \neq S^2$. Again, the exclusion of H_0 as a minor constrains the structure of the graphs $H_1, H_2, \ldots$, this time topologically: each H_i with $i \geqslant 1$ has an embedding in S which meets some circle $C_i \subseteq S$ that does not bound a disc in S in no more than a bounded number of vertices (and no edges), say in $X_i \subseteq V(H_i)$. (The bound on $|X_i|$ depends on H_0, but not on H_i.) Cutting along C_i and capping the hole(s), we obtain one or two new surfaces of smaller Euler genus. If the cut produces only one new surface S_i, then our embedding of $H_i - X_i$ still counts as a near-embedding of H_i in S_i (since X_i is small). If this happens for infinitely many i, then infinitely many of the surfaces S_i are also the same, and the induction hypothesis gives us a good pair among the corresponding graphs H_i. On the other hand, if we get two surfaces S_i' and S_i'' for infinitely many i (without loss of generality the same two surfaces), then H_i decomposes accordingly into subgraphs H_i' and H_i'' embedded in these surfaces, with $V(H_i' \cap H_i'') = X_i$. The set of all these subgraphs taken together is again well-quasi-ordered by the induction hypothesis, and hence so are the pairs (H_i', H_i'') by Lemma 12.1.3. Using a sharpening of the lemma that takes into account not only the graphs H_i' and H_i'' themselves but also how X_i lies inside them, we finally obtain indices i, j not only with $H_i' \preccurlyeq H_j'$ and $H_i'' \preccurlyeq H_j''$, but also such that these minor embeddings extend to the desired minor embedding of H_i in H_j—completing the proof of the graph minor theorem.

The graph minor theorem does not extend to graphs of arbitrary cardinality, but it might extend to countable graphs. Whether or not it does appears to be a difficult problem. It may be related to the following intriguing conjecture, which easily implies the graph minor theorem for finite graphs (Exercise 54). Call a graph H a *proper minor* of G if there is a contraction from a subgraph of G onto H that is not an isomorphism from G to H.

Self-minor conjecture. (Seymour 1980s)
Every countably infinite graph is a proper minor of itself.

In addition to its impact on 'pure' graph theory, the graph minor theorem has had far-reaching algorithmic consequences. Using their structure theorem for the graphs in $\text{Forb}_{\preccurlyeq}(K^n)$, Theorem 12.6.6, Robertson and Seymour have shown that testing for any fixed minor is 'fast': for every graph H there is a polynomial-time algorithm[9] that decides whether or not the input graph contains H as a minor. By the minor theorem, then, every minor-closed graph property $\mathcal{P}$ can be decided in polynomial (even cubic) time: if $\mathcal{K}_\mathcal{P} = \{H_1, \dots, H_k\}$ is the corresponding set of forbidden minors, then testing a graph G for membership in $\mathcal{P}$ reduces to testing the k assertions $H_i \preccurlyeq G$.

The following example gives an indication of how deeply this algorithmic corollary affects the complexity theory of graph algorithms. Let us call a graph *knotless* if it can be embedded in $\mathbb{R}^3$ so that none of its cycles forms a non-trivial knot. Before the graph minor theorem, it was an open problem whether knotlessness is decidable, that is, whether *any* algorithm exists (no matter how slow) that decides for any given graph whether or not that graph is knotless. To this day, no such algorithm is known. The property of knotlessness, however, is easily 'seen' to be closed under taking minors: contracting an edge of a graph embedded in 3-space will not create a knot where none had been before. Hence, by the minor theorem, there *exists* an algorithm that decides knotlessness— even in polynomial (cubic) time!

However spectacular such unexpected solutions to long-standing problems may be, viewing the graph minor theorem merely in terms of its corollaries will not do it justice. At least as important are the techniques developed for its proof, the various ways in which minors are handled or constructed. Most of these have not even been touched upon here, yet they seem set to influence the development of graph theory for many years to come.

Exercises

1.⁻ Let $\leqslant$ be a quasi-ordering on a set X. Call two elements $x, y \in X$ *equivalent* if both $x \leqslant y$ and $y \leqslant x$. Show that this is indeed an equivalence relation on X, and that $\leqslant$ induces a partial ordering on the set of equivalence classes.

2. Let $(A, \leqslant)$ be a quasi-ordering. For subsets $X \subseteq A$ write

$$\text{Forb}_{\leqslant}(X) := \{\, a \in A \mid a \not\geqslant x \text{ for all } x \in X \,\}.$$

Show that $\leqslant$ is a well-quasi-ordering on A if and only if every subset $B \subseteq A$ that is closed under $\leqslant$ (i.e. such that $x \leqslant y \in B \Rightarrow x \in B$) can be written as $B = \text{Forb}_{\leqslant}(X)$ with finite X.

3. Prove Proposition 12.1.1 and Corollary 12.1.2 directly, without using Ramsey's theorem.

---

[9] indeed a cubic one—although with an enormous constant depending on H

4.⁻ Show that the relation $\leqslant$ between rooted trees defined in the text is indeed a quasi-ordering.

5. Show that the finite trees are not well-quasi-ordered by the subgraph relation.

6. The last step of the proof of Kruskal's theorem considers a 'topological' embedding of T_m in T_n that maps the root of T_m to the root of T_n. Suppose we assume inductively that the trees of A_m are embedded in the trees of A_n in the same way, with roots mapped to roots. We thus seem to obtain a proof that the finite rooted trees are well-quasi-ordered by the subgraph relation, even with roots mapped to roots. Where is the error?

7. Extend Kruskal's theorem to trees whose vertices are labelled from a well-quasi-ordered set. The tree embedding is defined as before but in addition respects the ordering of the labels.

8. Are the connected finite graphs well-quasi-ordered by contraction alone (i.e. by taking minors without deleting edges or vertices)?

9.⁺ Relax the minor relation by not insisting that branch sets be connected. Show that the finite graphs are well-quasi-ordered by this relation.

10.⁺ Show that the finite graphs are not well-quasi-ordered by the topological minor relation.

11.⁺ Given $k \in \mathbb{N}$, is the class $\{ G \mid G \not\supseteq P^k \}$ well-quasi-ordered by the subgraph relation?

12.⁻ Let G be a graph, T a tree, and $\mathcal{V} = (V_t)_{t \in T}$ a family of subsets of $V(G)$. Show that $(T, \mathcal{V})$ is a tree-decomposition of G if and only if

   (i) for every $v \in V(G)$ the set $T_v := \{ t \mid v \in V_t \}$ is connected in T;

   (ii) $T_u \cap T_v \neq \emptyset$ for every edge uv of G.

13.⁻ Consider a tree-decomposition of a graph G in which some parts contain other parts. Modify it into a tree-decomposition whose parts are the $\subseteq$-maximal parts of the first decomposition. How does the new tree arise from the old?

14. Let G be a graph, T a set, and $(V_t)_{t \in T}$ a family of subsets of $V(G)$ satisfying (T1) and (T2) from the definition of a tree-decomposition. Show that there exists a tree on T that makes (T3) true if and only if there exists an enumeration $t_1, \ldots, t_n$ of T such that for every $k = 2, \ldots, n$ there is a $j < k$ satisfying $V_{t_k} \cap \bigcup_{i<k} V_{t_i} \subseteq V_{t_j}$.

   (The new condition tends to be more convenient to check than (T3). It can help, for example, with the construction of a tree-decomposition into a given set of parts.)

15. Prove the following converse of Lemma 12.3.1: if $(T, \mathcal{V})$ satisfies condition (T1) and the statement of the lemma, then $(T, \mathcal{V})$ is a tree-decomposition of G.

16. Recall that two separations $\{U_1, U_2\}$ and $\{W_1, W_2\}$ of G are *nested* if we can choose $i, j \in \{1, 2\}$ so that $U_i \subseteq W_j$ and $U_{3-i} \supseteq W_{3-j}$.

    (i) Show that the separations $S_e := \{U_1, U_2\}$ in Lemma 12.3.1 are pairwise nested (for different choices of the edge $e = t_1 t_2 \in T$).

    (ii)$^+$ Conversely, show that given a set $\mathcal{S}$ of nested separations of G there is a tree-decomposition $(\mathcal{V}, T)$ of G such that $\mathcal{S} = \{\, S_e \mid e \in E(T) \,\}$.

17.$^+$ Prove Theorem 12.3.7 for $k = 3$. Specifically, prove Tutte's theorem that every 2-connected graph has a tree-decomposition of adhesion 2 whose torsos are each either 3-connected or a cycle. Conversely, show that every graph with such a tree-decomposition is 2-connected.

    (Hint. Try the tree-decomposition defined, as in Exercise 16 (ii), by the set of all separations of order 2 that are nested with all other such separations.)

18. Describe the tree-decomposition of a contraction minor H of G which a given tree-decomposition of G induces as in Lemma 12.3.3, in terms subtrees of T (as in Exercise 12).

19.$^-$ Show that any graph with a simplicial tree-decomposition into k-colourable parts is itself k-colourable.

20. Let $\mathcal{H}$ be a set of graphs, and let G be constructed recursively from elements of $\mathcal{H}$ by pasting along complete subgraphs. Show that G has a simplicial tree-decomposition into elements of $\mathcal{H}$.

21. Use the previous exercise to show that G has no K^5 minor if and only if G has a tree-decomposition in which every torso is either planar or a copy of the Wagner graph W (Figure 7.3.1).

22.$^+$ Call a graph *irreducible* if it is not separated by any complete subgraph. Every finite graph G can be decomposed into irreducible induced subgraphs, as follows. If G has a separating complete subgraph S, then decompose G into proper induced subgraphs G' and G'' with $G = G' \cup G''$ and $G' \cap G'' = S$. Then decompose G' and G'' in the same way, and so on, until all the graphs obtained are irreducible. By Exercise 20, G has a simplicial tree-decomposition into these irreducible subgraphs. Show that they are uniquely determined if the complete separators were all chosen minimal.

23. If $\mathcal{F}$ is a family of sets, then the graph G on $\mathcal{F}$ with $XY \in E(G) \Leftrightarrow X \cap Y \neq \emptyset$ is called the *intersection graph* of $\mathcal{F}$. Show that a graph is chordal if and only if it is isomorphic to the intersection graph of a family of (vertex sets of) subtrees of a tree.

24. Show that for $n \geqslant 3$ the graphs K^n, C^n, an arbitrary tree of order n, and the $n \times n$ grid have tree-decompositions of widths $n - 1$, 2, 1, and n, respectively. For K^n and C^n show that these values are best possible.

25. Can the tree-width of a subdivision of a graph G be smaller than tw(G)? Can it be larger?

26. Show that the tree-width of a finite graph is at least its minimum degree. Is this still true for infinite graphs?

27.[+] Show that if a graph has circumference $k \neq 0$, then its tree-width is at most $k - 1$.

28.[+] A graph is called *outerplanar* if it has a drawing in which every vertex lies on the boundary of the outer face. Show that outerplanar graphs can have arbitrarily large tree-width, or find the best upper bound.

A tree-decomposition whose tree is a path is a *path-decomposition*. The *path-width* $\text{pw}(G)$ of G is the least width of a path-decomposition of G.

29. Show that a graph has a path-decomposition into complete graphs if and only if it is isomorphic to an interval graph. (Interval graphs are defined in Exercise 42, Chapter 5.)

30. (continued)

Prove the following analogue of Proposition 12.4.4 for path-width: every graph G satisfies $\text{pw}(G) = \min \omega(H) - 1$, where the minimum is taken over all interval graphs H containing G.

31.[+] Do trees have unbounded path-width?

A *transaction* of a sequence $(v_1, \ldots, v_n)$ of vertices is a set of disjoint paths from an initial segment $\{v_1, \ldots, v_i\}$ to the rest, $\{v_{i+1}, \ldots, v_n\}$.

32.[+] Given $k \in \mathbb{N}$ and a sequence $v_1, \ldots, v_n$ of vertices in a graph G, show that G has a path-decomposition $(V_1, \ldots, V_n)$ of adhesion $\leqslant k$, with $v_i \in V_i$ for all i, if and only if G contains no transaction $\mathcal{P}$ of $(v_1, \ldots, v_n)$ of order $|\mathcal{P}| > k$.

33. Show that the cycle C^n has connected tree-width $\lceil n/2 \rceil$.

34. Show that the $n \times n$ grid has tree-width n.

35.[−] Let $\mathcal{B}$ be a maximum-order bramble in a graph G. Show that every minimum-width tree-decomposition of G has a unique part covering $\mathcal{B}$.

36.[−] Let $\mathcal{P}$ be a minor-closed graph property. Show that strengthening the notion of a minor (for example, to that of topological minor) increases the set of forbidden minors required to characterize $\mathcal{P}$.

37. Deduce from the graph minor theorem that every minor-closed property can be expressed by forbidding finitely many topological minors. Is the same true for every property that is closed under taking topological minors?

Call a set $X \subseteq V(G)$ of vertices *k-connected in G* if $|X| \geqslant k$ and for all subsets $Y, Z \subseteq X$ with $|Y| = |Z| \leqslant k$ there are $|Y|$ disjoint Y–Z paths in G.

38.[+] Show that the tree-width of a graph G is large if and only if it contains a large set of vertices that is k-connected in G for some large k. For example, show that graphs of tree-width $< k$ contain no $(k+1)$-connected set of $3k$ vertices, and that graphs containing no $(k+1)$-connected set of $3k$ vertices have tree-width $< 4k$.

39. (continued)

   $(i)^+$ Find an $\mathbb{N} \to \mathbb{N}^2$ function $k \mapsto (h, \ell)$ such that every graph with an ℓ-connected set of h vertices contains a bramble of order $> k$.

   $(ii)^-$ Using the last exercise, deduce the following weakening of the difficult implication of Theorem 12.4.3: given k, every graph of large enough tree-width $f(k)$ contains a bramble of order $> k$.

40. Does every set of separations in a graph admit a consistent orientation?

41. Show that orienting the edges of a tree $T = (V, E)$ towards some fixed node t is consistent for the partial ordering on $\vec{E}$ defined in Section 12.5. Is this a bijection between V and the consistent orientations of E?

42. (continued)

   Prove the assertion of Exercise 16 (ii) by taking as the nodes of T the consistent orientations of S. How can you define the edges of T?

When θ is a tangle and $(A, B) \in \theta$, we call A the *small side* of $\{A, B\}$ in θ.

43. Let G be a graph with a tangle θ of order k.

   (i) Justify the notion of a 'small side' by showing that if $(A, B) \in \theta$ and $\{A', B'\}$ is a separation of order $< k$ with $A' \subseteq A$ or $B' \supseteq B$, then $(A', B') \in \theta$.

   (ii) Show that when $(A, B), (A', B') \in \theta$ also $(A \cup A', B \cap B') \in \theta$, as long as $\{A \cup A', B \cap B'\}$ is a separation of order $< k$.

   (iii) Deduce that, for every set X of fewer than k vertices, exactly one of the components C of $G - X$ is 'big', in the sense that $(V(G - C), X \cup V(C)) \in \theta$.

   $(iv)^+$ Is the same true in infinite graphs (with k finite)?

44. Show the following implications for a graph G:

   (i) G contains a k-block $\Rightarrow$ G has a bramble of order k.

   (ii) G has a tangle of order k $\Rightarrow$ G has a bramble of order k.

   (iii) G has a bramble of order $3k$ $\Rightarrow$ G has a tangle of order k.

   Is there a function $f : \mathbb{N} \to \mathbb{N}$ such that, for every $k \in \mathbb{N}$, if G has a tangle of order at least $f(k)$ then it contains a k-block?

45. Show the equivalence of Theorems 12.4.3 and 12.5.3 by proving the following assertions for all graphs G:

   (i) G has tree-width $< k - 1$ if and only if it has an S_k-tree over $\mathcal{F}_k$.

   (ii) G has a k-bramble if and only if it admits an $\mathcal{F}_k$-tangle of S_k.

46.$^+$ Modify the proof of Theorem 12.5.1 to obtain a proof of Theorem 12.5.3.

47. Extend Theorem 12.6.5 as follows. Let H be a connected planar graph, let $\mathcal{X}$ be any set of connected graphs including H, and let $\mathcal{H} := \{ IX \mid X \in \mathcal{X} \}$. Show that $\mathcal{H}$ has the Erdős-Pósa property, witnessed by the same function f as defined in the proof of Theorem 12.6.5. Explain how it is possible that f depends on H but not on any of the other graphs in $\mathcal{X}$.

48.$^{+}$ Show that, for every non-planar graph H, the class IH fails to have the Erdős-Pósa property.

(Hint. Embed H in a surface S, and consider only graphs embedded in S.)

49.$^{+}$ Extend Theorem 12.6.5 to disconnected graphs H, or find a counter-example.

50.$^{+}$ Show that the four ingredients to the structure of the graphs in $\text{Forb}_{\preccurlyeq}(K^n)$ as described in Theorem 12.6.6—tree-decomposition, an *apex set* X, arbitrary surfaces $S \not\hookrightarrow K^n$, and *vortices* $H_1, \ldots, H_k$—are all needed to capture all the graphs in $\text{Forb}_{\preccurlyeq}(K^n)$. More precisely, find examples of graphs in $\text{Forb}_{\preccurlyeq}(K^n)$ showing that Theorem 12.6.6 becomes false if we require in addition that the tree-decomposition has only one part, or that X is always empty, or that S is always the sphere, or that $H_1, \ldots, H_k$ are always empty. No exact proofs are required.

51.$^{+}$ (continued)

Show that, unlike in Theorem 12.6.6, the surfaces used in Theorem 12.6.7 cannot be limited to those in which K^n cannot be drawn. (As before, no exact proofs are required.)

52. Without using the graph minor theorem, show that the chromatic number of the graphs in any $\preccurlyeq$-antichain is bounded.

53. Let S_g denote the orientable surface obtained from the sphere by adding g handles. Find a lower bound for $|\mathcal{K}_{\mathcal{P}(S)}|$ in terms of g.

(Hint. The smallest g such that a given graph can be embedded in S_g is its *orientable genus*. Use the theorem that the orientable genus of a graph is equal to the sum of the genera of its blocks.)

54. Deduce the graph minor theorem from the self-minor conjecture.

55. Prove Theorem 12.6.9, assuming that G has a normal spanning tree.

56. Let G be a locally finite graph obtained from the $\mathbb{Z} \times \mathbb{Z}$ grid H by adding an infinite set of edges xy with $d_H(x, y)$ unbounded. Show that $G \succcurlyeq K^{\aleph_0}$. Can you do the same if the distances $d_H(x, y)$ are bounded (but at least 3)?

57. Is the infinite $\mathbb{Z} \times \mathbb{Z}$ grid a minor of the $\mathbb{Z} \times \mathbb{N}$ grid? Is the latter a minor of the $\mathbb{N} \times \mathbb{N}$ grid?

58.$^{+}$ Extend Proposition 12.3.6 to infinite graphs not containing an infinite complete subgraph.

59. Using the previous exercise, prove that if every finite subgraph of G has tree-width less than $k \in \mathbb{N}$ then so does G.

60. Show that no assumption of large finite connectivity can ensure that a countable graph has a K^r minor when $r \geqslant 5$. However, using the previous exercise show that sufficiently large finite connectivity forces an infinite graph to contain any given planar minor.

Notes

Robertson & Seymour have traditionally referred to the graph minor theorem as *Wagner's conjecture*. Wagner did indeed discuss this problem in the 1960s with his then students, Halin and Mader, and it seems that Mader conjectured a positive solution. Wagner himself always insisted that he did not—even after the graph minor theorem had been proved.

Robertson & Seymour's proof of the graph minor theorem is given in the numbers IV–VII, IX–XII and XIV–XXII of their series of over 20 papers under the common title of *Graph Minors*, most of which appeared in the *Journal of Combinatorial Theory, Series B*, between 1983 and 2012. Of their theorems cited in this chapter, Theorem 12.4.2 is from Graph Minors IV, Theorems 12.5.1 and 12.5.4 from Graph Minors X, Theorems 12.6.3 and 12.6.5 from Graph Minors V, and Theorem 12.6.6 from Graph Minors XVI.

Kruskal's theorem on the well-quasi-ordering of finite trees was first published in J.B. Kruskal, Well-quasi ordering, the tree theorem, and Vászonyi's conjecture, *Trans. Amer. Math. Soc.* **95** (1960), 210–225. Our proof is due to Nash-Williams, who introduced the versatile proof technique of choosing a 'minimal bad sequence'. This technique was also used in our proof of Higman's Lemma 12.1.3.

Nash-Williams generalized Kruskal's theorem to infinite graphs. This extension is much more difficult than the finite case. Its proof introduces as a tool the notion of *better-quasi-ordering*, a concept that has profoundly influenced well-quasi-ordering theory. The graph minor theorem is false for uncountable graphs; this was shown by R. Thomas, A counterexample to 'Wagner's conjecture' for infinite graphs, *Math. Proc. Camb. Phil. Soc.* **103** (1988), 55–57. Whether or not the countable graphs are well-quasi-ordered as minors, and whether the finite (or the countable) graphs are better-quasi-ordered as minors, are related questions that remain wide open. Both are related also to the self-minor conjecture. This, too, was originally intended to include graphs of arbitrary cardinality, but was disproved for uncountable graphs by B. Oporowski, A counterexample to Seymour's self-minor conjecture, *J. Graph Theory* **14** (1990), 521–524.

The notions of tree-decomposition and tree-width were first introduced (under different names) by R. Halin, S-functions for graphs, *J. Geometry* **8** (1976), 171–186. Among other things, Halin showed that grids can have arbitrarily large tree-width. Robertson & Seymour reintroduced the two concepts, apparently unaware of Halin's paper, with direct reference to K. Wagner, Über eine Eigenschaft der ebenen Komplexe, *Math. Ann.* **114** (1937), 570–590. (This is the seminal paper that introduced simplicial tree-decompositions to prove Theorem 7.3.4; cf. Exercise 21.) Simplicial tree-decompositions are treated in depth in R. Diestel, *Graph Decompositions*, Oxford University Press 1990.

An instructive introductory survey on tree-width, brambles and tangles is given by B.A. Reed in (R.A. Bailey, ed) *Surveys in Combinatorics 1997*, Cambridge University Press 1997, 87–162. Reed introduced the term 'bramble'; in Seymour & Thomas's original paper they are called 'screens'.

Theorem 12.3.7 is extracted from J. Carmesin, R. Diestel, F. Hundertmark & M. Stein, Connectivity and tree structure in finite graphs, *Combina-*

torica **34** (2014), 1–35, arXiv:1105.1611. The Aut(G)-invariance of T in the theorem is an important feature. It is easy to show that every two k-blocks can be separated by some separation of order $< k$. Theorem 12.3.7 says that we can find a nested subset of these separations that will still separate every two k-blocks: the separations induced by the tree-decomposition. When we try to construct this nested subset, we often have to choose between several 'crossing' (non-nested) separations. The point now is that we can make these choices canonically: neither arbitrarily, nor by appealing to an artificial tie-breaker such as a fixed vertex enumeration, but with reference to the structure of the graph only.

Under mild additional assumptions one can show that the tree-decompositions constructed for the proof of Theorem 12.3.7 refine each other as k grows: the decomposition for $k + 1$ induces tree-decompositions of the torsos of the decomposition for k and is therefore compatible with that decomposition. Just as in Theorem 12.5.4, one thus obtains one overall tree-decomposition whose induced separations separate every two blocks that can be separated at all, i.e., that are not just some k-block contained in a larger ℓ-block (for $\ell < k$).

The tree-width duality theorem, Theorem 12.4.3, is due to P.D. Seymour and R. Thomas, Graph searching and a min-max theorem for tree-width, *J. Comb. Theory B* **58** (1993), 22–33. A short version of the this proof was included in earlier editions of this book and can be found in P. Bellenbaum & R. Diestel, Two short proofs concerning tree-decompositions, *Comb. Probab. Comput.* **11** (2002), 541–547 (which also offers a short proof of Theorem 12.4.5). The proof presented in the text follows an idea of F. Mazoit, personal communication 2013. The simplest proof, perhaps—and the only one not using Menger's theorem—is via Theorem 12.5.3; see Exercises 45–46 and their hints.

Historically, tree-width duality evolved with a few quirks. As Robertson and Seymour developed the theory of tree-decompositions, they simultaneously looked for witnesses to large tree-width, as a way to proceed with the proof of the graph minor theorem when the graphs in question have unbounded tree-width. The result of this search was the notion of a tangle – with hindsight, perhaps the deepest single innovation for graph theory stemming from this proof. Numerically, however, the duality did not exactly fit: while large tree-width implies the existence of a large-order tangle and vice versa, one loses a small constant factor in the conversion. Instead of adjusting the notion of a tangle to repair this, however (e.g., as in Exercise 46), Robertson and Seymour simply changed the notion of a tree-decomposition to a new concept called *branch-decompositions*, which are exactly dual to tangles (except for very small k). To tie up the loose end, Seymour and Thomas later introduced brambles and Theorem 12.4.3 to provide exact duality for tree-width too; but brambles, though interesting, never assumed the significance of tangles.

Theorem 12.4.5 is from R. Thomas, A Menger-like property of tree-width; the finite case, *J. Comb. Theory B* **48** (1990), 67–76. Theorem 12.4.6 is from R. Diestel & M. Müller, Connected tree-width, *Combinatorica* (2017⁺), arXiv:1211.7353. This paper also includes a proof that ctw(C) $\leqslant$ ctw(G) if C is a geodesic cycle in G.

Our proof of Theorem 12.5.1 is adapted from R. Diestel & S. Oum, Unifying duality theorems for width parameters, arXiv:1406.3797. In this paper, a duality theory is developed for tangles in abstract separation systems, not necessarily of graphs. Its main result contains Theorems 12.5.1 and 12.5.3 as special cases.

Such abstract separation systems, their duality theorem for tangles, and a canonical tangle-tree theorem one can prove for them (see below), can be applied to problems in cluster analysis; see R. Diestel & G. Whittle, Tangles and the Mona Lisa, arXiv:1603.06652.

The tangle-tree theorem, Theorem 12.5.4, is one of the cornerstones of the proof of the graph minor theorem. Our short proof is from J. Carmesin, A short proof that every finite graph has a tree-decomposition displaying its tangles, Eur. J. Comb. **58** (2016), 61–65, arXiv:1511.02734. The canonical strengthening mentioned after the proof follows from a more general theorem about tangles in abstract separation systems proved in R. Diestel, F. Hundertmark & S. Lemanczyk, Profiles of separations in graphs and matroids, *Combinatorica* (2017+), arXiv:1110.6207.

The Kuratowski set for the graphs of tree-width < 4 have been determined by S. Arnborg, D.G. Corneil and A. Proskurowski, Forbidden minors characterization of partial 3-trees, *Discrete Math.* **80** (1990), 1–19. They are: K^5, the octahedron $K_{2,2,2}$, the 5-prism $C^5 \times K^2$, and the Wagner graph W. The Kuratowski set $\mathcal{K}_{\mathcal{P}(S)}$ for a given surface S has been determined explicitly for only one surface other than the sphere, the projective plane. It consists of 35 forbidden minors; see D. Archdeacon, A Kuratowski theorem for the projective plane, *J. Graph Theory* **5** (1981), 243–246. It is not difficult to show that $|\mathcal{K}_{\mathcal{P}(S)}|$ grows rapidly with the genus of S (Exercise 53).

A survey of finite forbidden minor theorems is given in Chapter 6.1 of R. Diestel, *Graph Decompositions*, Oxford University Press 1990. More recent developments are surveyed in R. Thomas, Recent excluded minor theorems, in (J.D. Lamb & D.A. Preece, eds) *Surveys in Combinatorics 1999*, Cambridge University Press 1999, 201–222. A survey of infinite forbidden minor theorems was given by N. Robertson, P.D. Seymour & R. Thomas, Excluding infinite minors, *Discrete Math.* **95** (1991), 303–319.

The first short proof of the grid theorem, Theorem 12.6.3, was given by R. Diestel, K.Yu. Gorbunov, T.R. Jensen & C. Thomassen, Highly connected sets and the excluded grid theorem, *J. Comb. Theory B* **75** (1999), 61–73. This proof was included in editions 2–4 of this book. It was further simplified by A. Leaf and P.D. Seymour, Treewidth and planar minors, *J. Comb. Theory B* **111** (2015) 38–53. The first proof with polynomial bound was obtained by C. Chekuri and J. Chuzhoy, Polynomial bounds for the grid-minor theorem, see arXiv:1602.02629.

As a forerunner to the grid theorem, Robertson & Seymour proved its following analogue for path-width (Graph Minors I): excluding a graph H as a minor bounds the path-width of a graph if and only if H is a forest. A short proof of this result, with optimal bounds, can be found in the first edition of this book, or in R. Diestel, Graph Minors I: a short proof of the path width theorem, *Comb. Probab. Comput.* **4** (1995), 27–30. It also follows from the abstract tangle duality theorem of Diestel and Oum cited earlier.

Theorem 12.6.6 is the earliest version of Robertson and Seymour's structure theorem for the graphs without a K^n minor. It has become known as the 'red herring' version—a phrase coined by Robertson and Seymour themselves, referring to its role in their proof of the graph minor theorem. It nonetheless remains the most-often applied version of the structure theorem, especially in algorithmic contexts. The strongest version so far, designed with future applications in mind, is given in R. Diestel, K. Kawarabayashi, Th. Müller & P. Wollan, On the excluded minor structure theorem for graphs of large tree-width, *J. Comb. Theory B* **102** (2012), 1189–1210, arXiv:0910.0946. Its proof is based on Theorem 12.6.6. The structure Theorem 12.6.7 for excluding topological minors is due to M. Grohe and D. Marx, Structure theorem and isomorphism test for graphs with excluded topological subgraphs, *Proc. 44th ann. ACM symp. theory of computing* (STOC 2012), 173–192, arXiv:1111.1109.

The existence of normal spanning trees for graphs with no topological $K^{\aleph_0}$ minor was proved by R. Halin, Simplicial decompositions of infinite graphs, in: (B. Bollobás, ed.) *Advances in Graph Theory, Annals of Discrete Mathematics* **3**, North-Holland 1978. Its strengthening, part (iii) of Theorem 12.6.9, was observed in R. Diestel, The depth-first search tree structure of $TK_{\aleph_0}$-free graphs, *J. Comb. Theory B* **61** (1994), 260–262. Part (iii) easily implies part (ii), which had been proved independently by N. Robertson, P.D. Seymour & R. Thomas, Excluding infinite clique subdivisions, *Trans. Amer. Math. Soc.* **332** (1992), 211–223. Theorem 12.6.8 and the infinite case of Theorem 12.6.6 were proved in R. Diestel & R. Thomas, Excluding a countable clique, *J. Comb. Theory B* **76** (1999), 41–67. The proof of Theorem 12.6.8 builds on the main result of N. Robertson, P.D. Seymour & R. Thomas, Excluding infinite clique minors, *Mem. Amer. Math. Soc.* **118** (1995).

Our proof of the 'generalized Kuratowski theorem', Corollary 12.7.3, was inspired by J. Geelen, B. Richter & G. Salazar, Embedding grids in surfaces, *Eur. J. Comb.* **25** (2004), 785–792. An alternative proof, which bypasses Theorem 12.4.2 by proving directly that the graphs in $\mathcal{K}_{\mathcal{P}(S)}$ have bounded order, is given by B. Mohar & C. Thomassen, *Graphs on Surfaces*, Johns Hopkins University Press 2001. Mohar (see there) also developed a set of algorithms, one for each surface, that decide embeddability in that surface in linear time. As a corollary, he obtains an independent and constructive proof of Corollary 12.7.3.

For every graph X, Graph Minors XIII gives an explicit algorithm that decides in cubic time for every input graph G whether $X \preccurlyeq G$. The constants in the cubic polynomials bounding the running time of these algorithms depend on X but are constructively bounded from above.

The concept of a 'good characterization' of a graph property was first suggested by J. Edmonds, Minimum partition of a matroid into independent subsets, *J. Research of the National Bureau of Standards (B)* **69** (1965) 67–72. In the language of complexity theory, a characterization is *good* if it specifies two assertions about a graph such that, given any graph G, the first assertion holds for G if and only if the second fails, and such that each assertion, if true for G, provides a certificate for its truth that can be checked in polynomial time. Thus every good characterization has the corollary that the decision problem corresponding to the property it characterizes lies in NP ∩ co-NP.

A   Infinite sets

This appendix gives a minimum-fuss summary of the set-theoretic notions and facts, such as Zorn's lemma and transfinite induction, that are used in Chapter 8.

Let A, B be sets. If there exists a bijective map between A and B, we write $|A| = |B|$ and say that A and B have *the same cardinality*. This is clearly an equivalence relation between sets, and we may think of the *cardinality* $|A|$ of A as the equivalence class containing A. We write *cardinality* $|A| \leqslant |B|$ if there exists an injective map $A \to B$. This is clearly well defined, and it is a partial ordering: if there are injective maps $A \to B$ and $B \to A$, there is also a bijection $A \to B$.[1] For every set there exists another that is bigger; for example, $|A| < |B|$ when B is the power set of A, the set of all its subsets.

The natural numbers are defined inductively as $n := \{0, \ldots, n-1\}$, $\mathbb{N}$ starting with $0 := \emptyset$. The usual expression of $|A| = n$ can then be read more formally as an abbreviation for $|A| = |n|$.

A set A is *finite* if there is a natural number n such that $|A| = n$; otherwise it is *infinite*. A is *countable* if $|A| \leqslant |\mathbb{N}|$, and *countably infinite* if $|A| = |\mathbb{N}|$. A bijection $\mathbb{N} \to A$ is an *enumeration* of A. If A is infinite then $|\mathbb{N}| \leqslant |A|$. Thus, $|\mathbb{N}|$ is the smallest infinite cardinality; it is denoted by $\aleph_0$. There is also a smallest uncountable cardinality, denoted by $\aleph_1$. If $|A| = |\mathbb{R}|$ then A is uncountable, and we say that A has *continuum many* elements. For example, there are continuum many infinite 0–1 sequences. (Whether $|\mathbb{R}|$ is equal to $\aleph_1$ or greater depends on the axioms of set theory assumed; in our context, this question does not arise.) We remark that if A is infinite and its elements are countable sets, then the union of all these sets is no bigger than A itself: $|\bigcup A| \leqslant |A|$.

---

[1] This is the Cantor-Bernstein theorem; a simple graph-theoretic proof is given in Proposition 8.4.6.

© Reinhard Diestel 2017   393
R. Diestel, *Graph Theory*, Graduate Texts in Mathematics 173,
DOI 10.1007/978-3-662-53622-3

An element x of a partially ordered set X is *minimal* in X if there is no $y \in X$ with $y < x$, and *maximal* if there is no $z \in X$ with $x < z$. A partially ordered set may have one or many elements that are maximal or minimal, or none at all. An *upper bound* in X of a subset $Y \subseteq X$ is any $x \in X$ such that $y \leqslant x$ for all $y \in Y$.

A *chain* is a partially ordered set in which every two elements are comparable. If $(C, \leqslant)$ is a chain, and if $x, y \in C$ satisfy $x < y$ but no element z of C is such that $x < z < y$, then x is called the *predecessor* *successor*   of y in C, and y the *successor* of x. A set of the form $\{x \in C \mid x < z\}$, for a given $z \in C$, is a proper *initial segment* of C.

A partially ordered set $(X, \leqslant)$ is *well-founded* if every non-empty subset of X has a minimal element, and a well-founded chain is said *well-*   to be *well-ordered*. For example, $\mathbb{N}$, $\mathbb{Z}$ and $\mathbb{R}$ are all chains (with their *ordering*   usual orderings), but only $\mathbb{N}$ is well-ordered. Note that every element x of a well-ordered set X has a successor (unless x is maximal in X): the unique minimal element of $\{y \in X \mid x < y\} \subset X$. However, an element of a well-ordered set need not have a predecessor, even if it is not minimal. *limit*   An element that has no predecessor is called a *limit*; for example, the number 1 is a limit in the well-ordered set

$$A = \{1 - \tfrac{1}{n+1} \mid n \in \mathbb{N}\} \cup \{2 - \tfrac{1}{n+1} \mid n \in \mathbb{N}\}$$

of rationals.

One of the many statements equivalent to the axiom of choice (which we assume throughout) is that for every set X there exists a relation by which X is well-ordered:

Well-ordering theorem. *Every set can be well-ordered.*

Two well-ordered sets are said to have *the same order type* if there is a bijection between them which preserves their orders. Thus $\mathbb{N}$ and the set of even natural numbers have the same order type, but this differs from the order type of the set A defined above. Having the same order type is clearly an equivalence relation, which justifies the term if we think of those order types themselves as equivalence classes.

When one considers properties shared by all well-ordered sets of the same order type, it is convenient to represent each order type by a *ordinals*   specially chosen set of that type, its *ordinal*. The ordinal representing *ω*   the order type of $\mathbb{N}$, for instance, is by custom denoted as ω; our example above thus says that the set of even natural numbers has (the) order type (of) ω. Finite chains of the same cardinality always have the same order type; we choose n as the ordinal representing the chains of order n.

There is an *addition* of ordinals, defined by taking as the sum $\alpha + \beta$ the ordinal representing the order type of the concatenation of α with β (in this order); note that this is again a well-ordered set. For example, $\alpha + 1$ is the successor of α. Note that no inverse operation '$-$' is defined.

If an ordinal β has the same order type as a proper initial segment of another ordinal α, we write $\beta < \alpha$. For example, we have $0 \leqslant n < \omega$ for every natural number n. It can be shown that $<$ defines an ordering, even a well-ordering, on every set of ordinals. On $\mathbb{N}$, this ordering coincides with the usual one, so our notation is unambiguous.

Since a set S of ordinals is itself well-ordered, it has an order type—just like any other well-ordered set. If the ordinal α is a strict upper bound for S, then the order type of S is at most α; it is equal to α if S consists of all the ordinals up to (but excluding) α. In fact, just like the natural numbers, infinite ordinals are usually defined in such a way that α and $\{\beta \mid \beta < \alpha\}$ are actually identical; then our ordering $<$ for ordinals coincides with the relation $\in$.

This makes it natural to write a well-ordered set S, of order type α say, as a family $S = \{s_\beta \mid \beta < \alpha\}$ with $s_\gamma < s_\beta$ for all $\gamma < \beta < \alpha$. This is common practice when one proves statements about the elements of S by *transfinite induction*, which works as follows.

transfinite induction

Suppose we want to show that every $s \in S$ satisfies some proposition P; let us write $P(s)$ to express that it does. Just as in ordinary induction we prove, for every $\beta < \alpha$, that *if P holds for every s_γ with $\gamma < \beta$ then P also holds for s_β*. In practice, we usually have to distinguish the two cases of β being a limit ordinal or a successor. Checking $P(s_0)$ from first principles, as in ordinary induction, is part of the first case, because 0 counts as a limit and the premise of P_γ for all $\gamma < 0$ is void. The conclusion then is that $P(s_\beta)$ for every $\beta < \alpha$, that is, every $s \in S$ satisfies P.

This is certainly simple—but is it correct? Well, any proper justification of transfinite induction requires a formal treatment of set theory, but so does ordinary induction. Informally, what we have shown is that the set

$$\{\beta < \alpha \mid P(s_\beta) \text{ fails}\}$$

has no least element. Since it is well-ordered, it must therefore be empty, so $P(s_\beta)$ holds for all $\beta < \alpha$.

Similarly, we may define things inductively. Such a *recursive definition* specifies for each ordinal α some object x_α, in a way that may refer to the objects x_β with $\beta < \alpha$ (which we think of as 'having been defined earlier'). Our definition of the natural numbers at the start of this appendix is a simple example.

recursive definition

In practice, the definition of x_α often makes sense only for ordinals α less than some fixed ordinal α^*, although the smallest such α^* may not be known in advance. For example, if the x_α are to be distinct vertices picked recursively from a graph G according to some given rules, it is clear that we shall not be able to find such x_α for all $\alpha < \alpha^*$ when $|\alpha^*| > |G|$, because $\alpha \mapsto x_\alpha$ would be an injective map from α^* to $V(G)$ showing that $|\alpha^*| \leqslant |G|$. Since there exist ordinals larger than $|G|$, such as any ordinal equivalent to a well-ordering of the power set of $V(G)$, this means that our recursion cannot go on indefinitely, i.e. we shall not be able to define x_α for all ordinals α. We may not know in advance when our recursion will get stuck, i.e., which is the smallest ordinal α for which x_α cannot be found in compliance with our rules. But this does not matter: we simply *define* α^* as the first ordinal α for which x_α cannot be found, content ourselves with having defined x_α for all $\alpha < \alpha^*$, and say that our recursion *terminates* at step α^*. (In fact, we usually *want* a recursive definition to terminate. In our example, we might wish to consider the set of all vertices $x \in G$ that got picked by our definition, and this will be the set $\{x_\alpha \mid \alpha < \alpha^*\}$.)

Note that our recursive definition for x_α may involve choices. In our example, x_α might be required to be a neighbour of some x_β with $\beta < \alpha$, but there may be several such x_β, each with several neighbours that have not yet been picked. This does not cause our recursion to get stuck at step α: we just pick one eligible vertex as x_α, and proceed. In other words, we accept $\{x_\alpha \mid \alpha < \alpha^*\}$ as a properly defined set even though we may not 'know' its elements x_α constructively.

Back to proving things, here is a formal statement of Zorn's lemma:

Zorn's Lemma. *Let $(X, \leqslant)$ be a partially ordered set such that every chain in X has an upper bound in X. Then X contains at least one maximal element.*

Note that, in applications of Zorn's lemma, the relation $\leqslant$ need not correspond to an intuitive notion of 'smaller than'. Applied to sets or to graphs, for example, it can stand for '$\supseteq$' just as much as for '$\subseteq$'. Then the 'upper bound' of a chain $\mathcal{C}$ is typically its overall intersection $\bigcap \mathcal{C}$.

Finally, compactness. The infinity lemma discussed in Chapter 8 generalizes as follows.[2] As before, we consider a collection $\{X_p \mid p \in P\}$ of finite sets, but rather than indexing these by natural numbers we have one such set X_p for every element p of some partially ordered set $(P, \leqslant)$. All we assume about P is that every two elements have a common upper bound: for all p, q there exists an r such that $p \leqslant r$ and $q \leqslant r$. Furthermore, we have maps $f_{qp} \colon X_q \to X_p$ for all $q > p$, which are compatible in that $f_{qp} \circ f_{rq} = f_{rp}$ whenever $r > q > p$.

Generalized Infinity Lemma. *For every such family $\{X_p \mid p \in P\}$ of finite sets there exists a family $\{x_p \mid p \in P\}$ of representatives $x_p \in X_p$ such that $f_{qp}(x_q) = x_p$ whenever $q > p$.*

The infinity lemma is clearly a special case of this, with $P = \mathbb{N}$ and the f_{qp} defined by iterating the lemma's predecessor function f.

---

[2] In category theory, our 'generalized infinity lemma' is known as the fact that the inverse limit of any directed inverse system of finite sets is non-empty. Instead of finite sets one can take any other compact spaces; see Chapter 8.8 for an application.

The following more combinatorial encoding for compactness arguments brings out particularly well the choices involved, and their interdependence. Let X be any set, S a finite set, and $\mathcal{F}$ a set of finite subsets of X. Assume that every $Y \in \mathcal{F}$ comes with a fixed set $\mathcal{A}(Y)$ of $Y \to S$ functions, its *admissible functions*. Call $\mathcal{Y} \subseteq \mathcal{F}$ *compatible* if there exists a function $f: X \to S$ all whose restrictions to the sets in $\mathcal{Y}$ are admissible, i.e. which satisfies $f|_Y \in \mathcal{A}(Y)$ for all $Y \in \mathcal{Y}$.

Compactness Principle. $\mathcal{F}$ *is compatible if every finite* $\mathcal{Y} \subseteq \mathcal{F}$ *is compatible.*

The proofs of the generalized infinity lemma and of the compactness principle are each just a few lines based on Tychonoff's theorem that every product of compact spaces is compact. We illustrate this for the compactness principle.

Think of the set of all $X \to S$ functions as a product of $|X|$ copies of S, so they form a compact space. For every finite $Y \subseteq X$, the set of all functions $f: X \to S$ with $f|_Y \in \mathcal{A}(Y)$ is closed (as well as open) in this product space. The result now follows from the 'finite intersection property' of compact spaces, that a family of closed sets has a non-empty intersection as soon as all its finite subfamilies do.

B   Surfaces

This appendix offers a summary of background information about surfaces, as needed for an understanding of their role in the proof of the
graph minor theorem or the proof of the 'general Kuratowski theorem'
for arbitrary surfaces given in Chapter 12.7. In order to be read at a
rigorous level it requires familiarity with some basic definitions of general
topology (such as of the product and the identification topology), but
no more.

A *surface*, for the purpose of this book, is a compact connected[1]   *surface*
Hausdorff topological space S in which every point has a neighbourhood
homeomorphic to the Euclidean plane $\mathbb{R}^2$. An *arc*, a *circle*, and a *disc*   *arc*
in S are subsets that are homeomorphic in the subspace topology to the   *circle* S^1
real interval $[0,1]$, to the unit circle $S^1 = \{x \in \mathbb{R}^2 : \|x\| = 1\}$, and to   *disc*
the unit disc $\{x \in \mathbb{R}^2 : \|x\| \leqslant 1\}$ or $\{x \in \mathbb{R}^2 : \|x\| < 1\}$, respectively.

The *components* of a subset X of S are the equivalence classes of   *component*
points in X where two points are *equivalent* if they can be joined by an
arc in X. The surface S itself, being connected, has only one component.

The *frontier* of X is the set of all points y in S such that every   *frontier*
neighbourhood of y meets both X and $S \smallsetminus X$. The frontier F of X
separates $S \smallsetminus X$ from X: since $X \cup F$ is closed, every arc from $S \smallsetminus X$ to
X has a first point in $X \cup F$, which must lie in F. A component of the
frontier of X that is a circle in S is a *boundary circle* of X. A boundary   *boundary*
circle of a disc in S is said to *bound* that disc.   *circle*

There is a fundamental theorem about surfaces, their *classification*.
This says that, up to homeomorphism, every surface can be obtained
from the sphere $S^2 = \{x \in \mathbb{R}^3 : \|x\| = 1\}$ by 'adding finitely many   *sphere* S^2
handles or finitely many crosscaps', and that surfaces obtained by adding
different numbers of handles or crosscaps are distinct. We shall not need
the classification theorem, but to form a picture let us see what the

---

[1] Throughout this appendix, 'connected' means 'arc-connected'.

© Reinhard Diestel 2017   399
R. Diestel, *Graph Theory*, Graduate Texts in Mathematics 173,
DOI 10.1007/978-3-662-53622-3

handle   above operations mean. To *add a handle* to a surface S, we remove two open discs whose closures in S are disjoint, and identify[2] their boundary circles with the circles $S^1 \times \{0\}$ and $S^1 \times \{1\}$ of a copy of $S^1 \times [0,1]$

crosscap   disjoint from S. To *add a crosscap*, we remove one open disc, and then identify opposite points on its boundary circle in pairs.

In order to see that these operations do indeed give new surfaces, we have to check that every identification point ends up with a neighbourhood homeomorphic to $\mathbb{R}^2$. To do this rigorously, let us first look at circles more generally.

cylinder   A *cylinder* is the product space $S^1 \times [0,1]$, or any space homeomorphic to it. Its *middle circle* is the circle $S^1 \times \{\frac{1}{2}\}$. A *Möbius strip* is

Möbius strip   any space homeomorphic to the product space $[0,1] \times [0,1]$ after identification of $(1,y)$ with $(0, 1-y)$ for all $y \in [0,1]$. Its *middle circle* is the set $\{(x, \frac{1}{2}) \mid 0 < x < 1\} \cup \{p\}$, where p is the point resulting from the identification of $(1, \frac{1}{2})$ with $(0, \frac{1}{2})$. It can be shown[3] that every circle C

strip neighbourhood   in a surface S is the middle circle of a suitable cylinder or Möbius strip N in S, which can be chosen small enough to avoid any given compact subset of $S \smallsetminus C$. If this *strip neighbourhood* is a cylinder, then $N \smallsetminus C$

two-sided   has two components and we call C *two-sided*; if it is a Möbius strip, then

one-sided $N \smallsetminus C$ has only one component and we call C *one-sided*.

Using small neighbourhoods inside a strip neighbourhood of the (two-sided) boundary circle of the disc or discs we removed from S in order to attach a crosscap or handle, one can show easily that both operations do produce new surfaces.

separating circle   Since S is connected, $S \smallsetminus C$ cannot have more components than $N \smallsetminus C$. If $S \smallsetminus C$ has two components, we call C a *separating* circle in S; if it has only one, then C is *non-separating*. While one-sided circles are obviously non-separating, two-sided circles can be either separating or non-separating. For example, the middle circle of a cylinder added to S as a 'handle' is a two-sided non-separating circle in the new surface obtained. When S' is obtained from S by adding a crosscap in place of a disc D, then every arc in S that runs half-way round the boundary circle of D becomes a one-sided circle in S'.

The classification theorem thus has the following corollary:

Lemma B.1. *Every surface other than the sphere contains a non-separating circle.*

---

[2] This is made precise by the *identification topology*, whose formal definition can be found in any topology book. Since S^1 has two possible orientations, two copies of S^1 can be identified in two essentially different ways. The corresponding two ways of adding a handle yield different new surfaces. For the classification one only uses one of these, the way that preserves the orientability of the surface (as in Figure B.1).

[3] In principle, the strip neighbourhood N is constructed as in the proof of Lemma 4.2.2, using the compactness of C. However since we are not in a piecewise linear setting now, the construction is considerably more complicated.

We shall see below that, in a sense, our two examples of non-separating circles are all there are: cutting a surface along any non-separating circle (and patching up the holes) will always produce a surface with fewer handles or crosscaps.

An *embedding* $G \hookrightarrow S$ of a graph G in S is a map σ that maps the   *embedding* vertices of G to distinct points in S and its edges xy to $\sigma(x)$–$\sigma(y)$ arcs $\sigma: G \hookrightarrow S$ in S, so that no inner point of such an arc is the image of a vertex or lies on another arc. We then write $\sigma(G)$ for the union of all those points and arcs in S. A *face* of G in S is a component of $S \smallsetminus \sigma(G)$, and the   *face* subgraph of G that σ maps to the frontier of this face is its *boundary*.   *boundary* Note that while faces in the sphere are always discs (if G is connected), in general they need not be.

One can prove that in every surface one can embed a suitable graph so that every face becomes a disc. The following general version of Euler's theorem 4.2.9 therefore applies to all surfaces:

Theorem B.2. *For every surface S there exists an integer $\chi(S)$ such that whenever a graph G with n vertices and m edges is embedded in S so that there are ℓ faces and every face is a disc, we have*

$$n - m + \ell = \chi(S).$$

This invariant χ of S is its *Euler characteristic*. For computational simplicity we usually work instead with the derived invariant

$$\varepsilon(S) := 2 - \chi(S),$$
  $\varepsilon(S)$

the *Euler genus* of S, because χ is negative for most surfaces but ε takes   *Euler genus* its values in $\mathbb{N}$ (see below).

Perhaps the most striking feature of Euler's theorem is that it works with almost any graph embedded in S. This makes it easy to see how the Euler genus is affected by the addition of a handle or crosscap.

Indeed, let D and D' be two open discs in S that we wish to remove in order to attach a handle there. Let G be any graph embedded in S so that every face is a disc. If necessary, shift G on S so that D and D' each lie inside a face, f and f', say. Add cycles C and C' on the boundary circles of D and D', and join them by an edge to the old boundaries of f and f', respectively. Then every face of the resulting graph is again a disc, and D and D' are among these. Now remove D and D', and add a handle with an additional C–C' edge running along it. This operation makes the new handle into one new face, which is a disc. It thus reduces the total number of faces by 1 (since we lost D and D' but gained the new face on the handle) and increases the number of edges by 1, but leaves the number of vertices unchanged. As a result, ε grows by 2.

Similarly, replacing a disc D bounded by a cycle $C \subseteq G$ with a crosscap decreases the number of faces by 1 (since we lose D), but leaves $n - m$ unchanged if we arrange the cycle C in such a way that vertices get identified with vertices when we identify opposite points.

We have thus shown the following:

Lemma B.3.

(i) *Adding a handle to a surface raises its Euler genus by 2.*

(ii) *Adding a crosscap to a surface raises its Euler genus by 1.* $\square$

Since the sphere has Euler genus 0 (Theorem 4.2.9), the classification theorem and Lemma B.3 tell us that ε has all its values in $\mathbb{N}$. We may thus try to prove theorems about surfaces by induction on ε. For the induction step, we could simply undo the addition of a handle or crosscap described earlier, cutting along the new non-separating circle it produced (which runs around the new handle or 'half-way' around the crosscap) and restoring the old surface by putting back the disc or discs we removed. A problem with this is that we do not normally know where on our surface this circle lies, say with respect to a given graph embedded in it.

However, the genus-reducing cut-and-paste operation can be carried out with any non-separating circle: we do not have to use one that we know came from a new handle or crosscap. This is an example of a more general technique known as *surgery*, and works as follows.

Let C be a non-separating circle in a surface $S \neq S^2$. To *cut* S *cutting* *along* C, we form a new space S' from S by replacing every point $x \in C$ with two points x', x'' and defining the topology on the modified set as follows.[4] Let N be any strip neighbourhood of C in S, and put $X' := \{x' \mid x \in C\}$ and $X'' := \{x'' \mid x \in C\}$. If N is a cylinder, then $N \smallsetminus C$ has two components N' and N'', and we choose the neighbourhoods of the new points x' and x'' in S' so that X' and X'' become boundary circles of N' and N'' in S', respectively, and $N' \cup X'$ and $N'' \cup X''$ become disjoint cylinders in S'. If N is a Möbius strip, we choose these neighbourhoods so that X' and X'' each form an arc in S' and $X' \cup X''$ is a boundary circle of $N \smallsetminus C$ in S', with $(N \smallsetminus C) \cup X' \cup X''$ forming one *capping* cylinder in S'. Finally, we turn S' into a surface by *capping its holes*: for each of the (two or one) boundary circles X' and X'' or $X' \cup X''$ of $S \smallsetminus C$ in S' we take a disc disjoint from S' and identify its boundary circle with X', X'' or $X' \cup X''$, respectively, so that the space obtained is again a surface.

---

[4] The description that follows may sound complicated, but it is not: working in our concrete models of the cylinder and the Möbius strip it is easy to write down explicit neighbourhood bases that define a topology with the properties stated. As all we want is to obtain some surface of smaller genus, we do not care about uniqueness (which will follow anyhow from Lemma B.4 and the classification).

Computing how these operations affect the Euler genus of S is again easy, assuming we can embed a graph in S so that every face is a disc and C is the image of a cycle. (This can always be done, but it is not easy to prove.[5]) Indeed, by doubling C we left $n - m$ unchanged, because a cycle has the same number of vertices as edges. So all we changed was ℓ, which increased by 2 in the first case and by 1 in the second.

Lemma B.4. *Let C be any non-separating circle in a surface S, and let S' be obtained from S by cutting along C and capping the hole or holes.*

(i) *If C is one-sided in S, then $\varepsilon(S') = \varepsilon(S) - 1$.*

(ii) *If C is two-sided in S, then $\varepsilon(S') = \varepsilon(S) - 2$.* □

Lemma B.4 gives us a large supply of circles to cut along in an induction on the Euler genus. Still, it is sometimes more convenient to cut along a separating circle, and many of these can be used too:

Lemma B.5. *Let C be a separating circle in a surface S, and let S' and S'' be the two surfaces obtained from S by cutting along C and capping the holes. Then*

$$\varepsilon(S) = \varepsilon(S') + \varepsilon(S'').$$

In particular, if C does not bound a disc in S, both S' and S'' have smaller Euler genus than S.

Proof. As before, embed a graph G in S so that every face is a disc and C is the image of a cycle in G, and let $G' \hookrightarrow S'$ and $G'' \hookrightarrow S''$ be the two graphs obtained in the surgery. Thus, G' and G'' both contain a copy of the cycle on C, which we assume to have k vertices and edges. Then, with the obvious notation, we have

$$\begin{aligned}
\varepsilon(S') + \varepsilon(S'') &= (2 - n' + m' - \ell') + (2 - n'' + m'' - \ell'') \\
&= 4 - (n + k) + (m + k) - (\ell + 2) \\
&= 2 - n + m - \ell \\
&= \varepsilon(S).
\end{aligned}$$

Now if S' (say) is a sphere, then $S' \cap S$ was a disc in S bounded by C. Hence, if C does not bound a disc in S then $\varepsilon(S')$ and $\varepsilon(S'')$ are both non-zero, giving the second statement of the lemma. □

We now apply these techniques to prove a lemma for our direct proof in Chapter 12 of the 'Kuratowski theorem for arbitrary surfaces', Corollary 12.7.3.

---

[5] Perhaps the simplest proof was given by C. Thomassen, The Jordan-Schoenflies theorem and the classification of surfaces, *Amer. Math. Monthly* **99** (1992), 116–130.

[12.7.4]
Lemma B.6. *Let S be a surface, and let $\mathcal{C}$ be a finite set of disjoint circles in S. Assume that $S \smallsetminus \bigcup \mathcal{C}$ has a component D_0 whose closure in S meets every circle in $\mathcal{C}$, and that no circle in $\mathcal{C}$ bounds a disc in S that is disjoint from D_0. Then $\varepsilon(S) \geqslant |\mathcal{C}|$.*

Proof. We begin with the observation that the closure of D_0 not only meets but even contains every circle $C \in \mathcal{C}$. This is because C has a strip neighbourhood N disjoint from all the other circles in $\mathcal{C}$ (since their union is compact), and each of the (one or two) components of $N \smallsetminus C$ has all of C in its closure. Since D_0 meets, and hence contains, at least one component of $N \smallsetminus C$, its closure contains C.

$\mathcal{C}_1, \mathcal{C}_2^1, \mathcal{C}_2^2$ Let us partition $\mathcal{C}$ as $\mathcal{C} = \mathcal{C}_1 \cup \mathcal{C}_2^1 \cup \mathcal{C}_2^2$, where the circles in $\mathcal{C}_1$ are one-sided, those in $\mathcal{C}_2^1$ are two-sided but non-separating, and those in $\mathcal{C}_2^2$ are separating. We shall, in turn, cut along all the circles in $\mathcal{C}_1$, some $|\mathcal{C}_2^2|$ non-separating circles not in $\mathcal{C}$, and at least half the circles in $\mathcal{C}_2^1$. This
$S_0, \dots, S_n$ will give us a sequence $S_0, \dots, S_n$ of surfaces, where $S_0 = S$, and S_{i+1}
C_i is obtained from S_i by cutting along a circle C_i and capping the hole(s). Our task will be to ensure that C_i is non-separating in S_i for every $i = 0, \dots, n-1$. Then Lemma B.4 will imply that $\varepsilon(S_{i+1}) \leqslant \varepsilon(S_i) - 1$ for all i and $\varepsilon(S_{i+1}) \leqslant \varepsilon(S_i) - 2$ whenever $C_i \in \mathcal{C}_2^1$, giving

$$\varepsilon(S) \geqslant \varepsilon(S_n) + |\mathcal{C}_1| + |\mathcal{C}_2^2| + 2\,|\mathcal{C}_2^1|/2 \geqslant |\mathcal{C}|$$

as desired.

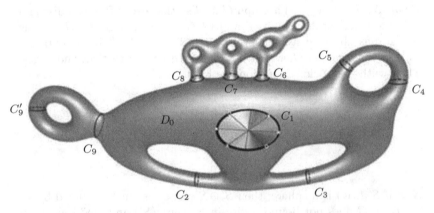

Fig. B.1. Cutting the 1-sided circle C_1 and the 2-sided circles C_2, C_3 and C_5, C_7, C_8 and C_9' does not separate S

Cutting along the circles in $\mathcal{C}_1$ (and capping the holes) is straightforward: since these circles are one-sided, they are always non-separating.

Next, we consider the circles in $\mathcal{C}_2^2$, such as C_9 in Figure B.1. For every $C \in \mathcal{C}_2^2$, denote by $D(C)$ the component of $S \smallsetminus C$ that does not contain D_0. Since every circle in $\mathcal{C}$ lies in the closure of D_0 but no point of $D(C)$ does, these $D(C)$ are also components of $S \smallsetminus \bigcup \mathcal{C}$. In particular, they are disjoint for different C. Thus, each $D(C)$ will also be a component of $S_i \smallsetminus C$, where S_i is the current surface after any surgery performed on the circles in $\mathcal{C}_1$ and inside $D(C')$ for some $C' \neq C$. Given a fixed circle $C \in \mathcal{C}_2^2$, let S' be the surface obtained from $D(C)$ by capping its hole. Since C does not bound a disc in S that is disjoint from D_0, we know that S' is not a sphere and hence contains a non-separating circle C' (Lemma B.1). We choose C' so that it avoids the cap we added to form S', i.e. so that $C' \subseteq S \smallsetminus C$. Then C' is also non-separating in the current surface S_i (since every point of $S_i \smallsetminus C'$ can be joined by an arc in $S_i \smallsetminus C'$ to C, which is connected), and we may select C' as a circle C_i to cut along.

It remains to select at least half of the circles in $\mathcal{C}_2^1$ as circles C_i to cut along. We begin by selecting all those whose entire strip neighbourhoods (i.e., both their 'sides') lie in D_0. (In Figure B.1, these are the circles C_2 and C_3.) These circles C are non-separating also in the surface S_i current before they are cut, because D_0 will lie inside a component of $S_i \smallsetminus C$. Every other $C \in \mathcal{C}_2^1$ lies in the closure also of a component $D(C) \neq D_0$ of $S \smallsetminus \bigcup \mathcal{C}$. (In Figure B.1, these are the circles $C_4, \dots, C_8$.) For every component D of $S \smallsetminus \bigcup \mathcal{C}$ we select all but one of the circles $C \in \mathcal{C}_2^1$ with $D(C) = D$ as a cutting circle C_i. Clearly, each of these C_i will be non-separating also in its current surface S_i, and their total number at least $|\mathcal{C}_2^1|/2$. □

Hints for all the Exercises

At this point in the book there used to be a collection of hints for all the exercises. Their intention was to help those who had already spent some time over an exercise but somehow failed to make progress, by putting them on the right track. In order to not spoil the fun of an exercise, or its usability in class, I tried to design these hints so as to make them somewhat unintelligible to those that had not made such an initial attempt of their own.

However, this is a tricky brief, and in some cases the hints will invariably still give away some key ideas, or narrow a student's mind in pursuit of alternatives. It therefore seemed best to leave the decision of which hints to include to the lecturer setting the problems.

The Hints section still exists, and is regularly updated. As of this fifth edition of the book, however, I have relegated it to the Professional eBook edition. This is available to lecturers and professional mathematicians via

http://diestel-graph-theory.com/

and, for iPad users, from the book's dedicated iOS app, *Graph Theory*. (This gives continued access to earlier editions too, which contain some proofs no longer included in later editions.)

I apologize to those readers that are using the book for self-study and would have liked a peek at a hint every now and then too....

© Reinhard Diestel 2017
R. Diestel, *Graph Theory*, Graduate Texts in Mathematics 173,
DOI 10.1007/978-3-662-53622-3

Index

Page numbers in italics refer to definitions; in the case of author names, they refer to theorems due to that author. The alphabetical order ignores letters that stand as variables; for example, 'k-chromatic' is listed under the letter c.

© Reinhard Diestel 2017

R. Diestel, *Graph Theory*, Graduate Texts in Mathematics 173,
DOI 10.1007/978-3-662-53622-3

Symbol Index

The entries in this index are divided into two groups. Entries involving only mathematical symbols (i.e. no letters except variables) are listed on the first page, grouped loosely by logical function. The entry '[ ]', for example, refers to the definition of induced subgraphs $H[U]$ on page 4 as well as to the definition of face boundaries $G[f]$ on page 94.

Entries involving fixed letters as constituent parts are listed on the second page, in typographical groups ordered alphabetically by those letters. Letters standing as variables are ignored in the ordering.

© Reinhard Diestel 2017
R. Diestel, *Graph Theory*, Graduate Texts in Mathematics 173,
DOI 10.1007/978-3-662-53622-3

Reinhard Diestel received a PhD from the University of Cambridge, following research 1983–86 as a scholar of Trinity College under Béla Bollobás. He was a Fellow of St. John's College, Cambridge, from 1986 to 1990. Research appointments and scholarships have taken him to Bielefeld (Germany), Oxford and the US. He became a professor in Chemnitz in 1994 and has held a chair at Hamburg since 1999.

Reinhard Diestel's main area of research is graph theory, including infinite graph theory. He has published numerous papers and a research monograph, *Graph Decompositions* (Oxford 1990).

Printed in the United States
by Bookmasters

Printed in the United States
By Bookmasters